CONNECTIONS

This is a beautifully curated book on the work and influence of Patsy Healey. The selections from her prolific archive, and the essays from distinguished planning theorists, compellingly make the case for urban planning in a neoliberal and elite world, but planning of a distinctive sort: dialogical, collaborative, and democratic. A marvellous and important book.

Ash Amin, Cambridge, UK
author of *Arts of the Political* (with Nigel Thrift)

This is so much more than a collection of Patsy Healey's work. The editorial essays and careful selection of papers draw out threads of connection across her work, and the commissioned responses show its continuing relevance. Topics such as property development and institutionalism get renewed attention. The result is a major contribution to the field of study that Patsy played such a central role in shaping.

Yvonne Rydin, University College London, UK

Connections

Exploring Contemporary Planning Theory and Practice with Patsy Healey

Edited by

JEAN HILLIER
RMIT University, Australia

JONATHAN METZGER
KTH, Stockholm, Sweden

Routledge
Taylor & Francis Group
LONDON AND NEW YORK

First published 2015 by Ashgate Publishing

2 Park Square, Milton Park, Abingdon, Oxon, OX14 4RN
605 Third Avenue, New York, NY 10017

Routledge is an imprint of the Taylor & Francis Group, an informa business

First issued in paperback 2020

British Library Cataloguing in Publication Data
A catalogue record for this book is available from the British Library

Library of Congress Cataloging-in-Publication Data has been applied for

ISBN 13 : 978-1-4724-3194-3 (hbk)
ISBN 13 : 978-0-367-78736-3 (pbk)

Contents

List of Figures and Tables

Figures

Tables

List of Contributors

Janice Barry
Janice Barry is an Assistant Professor of city planning at the University of Manitoba (Winnipeg, Canada), with strong interests in how more collaborative forms of land use decision-making are influenced by larger institutional structures and discourses. Her research is currently centred on Indigenous peoples' experiences of state-directed land use planning, with a particular interest in the growing shift towards government-to-government approaches. She is currently collaborating on a book manuscript that explores how Indigenous rights and title are being recognised, in both urban and natural resource planning contexts. Janice also explores Indigenous peoples' complex relationships with planning in her postgraduate teaching and is actively involved in a service-based learning partnership with several Manitoba First Nations.

Benno Engels
Benno Engels is senior lecturer in Urban and Regional Planning, in the School of Global, Urban and Social Studies, RMIT University, Melbourne, Australia. His research interests include urban locational disadvantage especially with regard to accessibility, historical developments in Australian and international planning theory and practice, plus Marxian theoretical methods. His areas of teaching specialisation include urban political economy, planning history, planning theory and social planning.

Jean Hillier
Jean Hillier is Emeritus Professor of Sustainability and Urban Planning at RMIT University, Melbourne, Australia. Her research interests include poststructural urban planning theory and methodology for adaptive strategic practice in conditions of uncertainty. Recent books include *Deleuze and Guattari for Planners* (InPlanning e-book, 2013 with AESOP Young Academics); *The Planning Theory Tradition: An international account* (2015, in Mandarin) with Patsy Healey; *Complexity and the Planning of the Built Environment* (2012) edited with Gert de Roo and Joris Van Wezemael, the *Ashgate Research Companion to Planning Theory: Conceptual Challenges for Spatial Planning* (2010) edited with Patsy Healey, *Critical Essays in Planning Theory* (2008) three volumes, edited with Patsy Healey. Jean also guest edited the Special Issue on Poststructuralism and Planning in the Chinese language *Urban Planning International* (2010), and the special issue of *Town Planning Review* on Planning in Uncertainty (2011).

Francesco Lo Piccolo
Francesco Lo Piccolo is Full Professor of Urban Planning and Director of the PhD Programme in the Department of Architecture at the Università degli Studi di Palermo, Italy. His research is mostly directed to social exclusion, ethnicity and diversity, identity and local communities, public and civic participation, with specific interests in ethics and justice. He is the President of AESOP, Association of European Schools of Planning, for the years 2014–2016.

Paul McNamara
Following degrees in geography at the Universities of St Andrews and Edinburgh and several years as researcher and lecturer at Oxford Polytechnic, Paul joined Prudential plc in 1987 becoming Head of Research in 1990 and Board Director in 2005. After twice chairing its Research Committee, Paul was Chairman of the Investment Property Forum 2005/6 and was made a Life Member in 2007. Since retiring in 2012, he is now a part-time consultant. Paul is an Honorary Fellow, past President and past Chair of the Society of Property Researchers; a Visiting Professor at Oxford Brookes University; and a past chair of property working groups for the Institutional Investors Group on Climate Change and United Nations Environment Programme Finance

Initiative. Paul was awarded an OBE for services to the property industry in 2003, a lifetime achievement award from IPE magazine in 2008 and an Honorary Doctorate from Kingston University in 2014.

Jonathan Metzger
Jonathan Metzger is Associate Professor of Urban and Regional Studies at the KTH Royal Institute of Technology in Stockholm, Sweden. Jonathan has a PhD in economic history and a professional background as a practising planner, and much of his research also centres on ethnographic studies of planning practice. In his work he often explores the disciplinary common ground between planning studies, human geography and Science and Technology Studies (STS), while drawing upon the wider social sciences and humanities for inspiration. His most recent publications include the edited volumes *Planning Against the Political*, with Phil Allmendinger and Stijn Oosterlynck (Routledge, 2014), and *Sustainable Stockholm*, with Amy Rader Olsson (Routledge, 2013).

Mee Kam Ng
Mee Kam Ng is Vice-Chair of the Department of Geography and Resource Management, the Director of the Urban Studies Programme, Associate Director of the Institute of Future Cities and the Hong Kong Institute of Asian Pacific Studies at the Chinese University of Hong Kong. She is a member of the Royal Town Planning Institute, a fellow of the Hong Kong Institute of Planners (HKIP) and academic advisor of the Hong Kong Institute of Urban Design. She has published widely on planning, governance and sustainability issues. Her publications have earned her six HKIP Awards. At the international level, she has been consultant to the UN and EU. She is an associate editor for the Current Research on Cities supplement of *Cities* and a member of the editorial board of *Town Planning Review, Planning Theory and Practice, DisP – The Planning Review* and *City, Culture and Society,* and *Business Strategy and the Environment.*

Ananya Roy
Ananya Roy is Professor of City and Regional Planning and Distinguished Chair in Global Poverty and Practice at the University of California, Berkeley. Roy's research is concerned with a transnational and transhistorical analysis of urban poverty and inequality. Her book, *Poverty Capital: Microfinance and the Making of Development* (Routledge, 2010), was the recipient of the 2011 Paul Davidoff Book Award of the Association of Collegiate Schools of Planning, a book award for research that advances social justice. Her forthcoming book, *Territories of Poverty* (co-edited with Emma Shaw Crane, University of Georgia Press), presents a global history of welfare and development and foregrounds the relationship between poor people's movements and bureaucracies of poverty. Roy's public scholarship is concerned with new formats of pedagogy and public debate. One example is The #GlobalPOV Project, which combines critical social theory with improvised art to provoke discussion about poverty and inequality: http://blumcenter.berkeley.edu/globalpov/

Leonie Sandercock
Leonie Sandercock joined the School of Community and Regional Planning at UBC in 2001 after teaching at Macquarie University in Sydney, UCLA, and the University of Melbourne. Her focus now is in working with First Nations, through collaborative community planning, using the medium of film as a catalyst for dialogue, on the possibilities of healing, reconciliation, and partnership for social and economic development. Other research interests include the possibilities of a more therapeutic model of planning; the importance of stories and storytelling in planning theory and practice; and the role of multimedia in planning. Leonie received the Davidoff Award (2005) from the American Collegiate Schools of Planning, for the best book in the field of urban, regional, and community planning, for *Cosmopolis 2: Mongrel Cities of the 21st Century*; the BMW Group Award for Intercultural Learning for her writing on Cosmopolitan Urbanism and the film *Where Strangers Become Neighbours* (2007). In 2012 she was awarded an Honorary Doctorate for lifetime contribution to planning scholarship by Roskilde University in Denmark.

André Sorensen

André Sorensen is Professor and Chair of the Department of Human Geography, University of Toronto Scarborough. He has published extensively on Japanese urbanisation, urban planning, and planning history. His monograph *The Making of Urban Japan: Cities and Planning from Edo to the 21st Century* (Routledge, 2002) won the book prize of the International Planning History Association in 2004. In 2007 he was elected a Fellow of the University of Tokyo School of Engineering in recognition of his research on Japanese urbanism and urban planning. His current research examines institutions, urban space, and temporal processes in urbanisation and urban governance from a historical institutionalist perspective, with a focus on urban land and property development, infrastructure management, and the creation of increasingly differentiated property rights in urban settings.

Wendy Steele

Wendy Steele is Associate Professor and Australian Research Council (DECRA) Fellow located in the School of Global, Urban and Social Studies and Centre for Urban Research at RMIT University, Melbourne, Australia. Her research focuses on the planning, governance and justice dimensions of global cities with an emphasis on critical urban infrastructure and cross-border planning. Since 2010 she has been an international editorial board member of the journal *Urban Policy and Research.*

Zeinab Noureddine Tag-Eldeen

Zeinab Noureddine Tag-Eldeen is a post-doctoral researcher at the School of Architecture and the Built Environment at the Royal Institute of Technology in Stockholm. In her PhD work, she focuses on cross-cultural practices of collaborative planning knowledge in the Egyptian context and examines the potential of a new context for joint development of that knowledge. Her dual role as researcher/practitioner has developed her interest in practice-oriented research. She argues that a creative practice approachable to the history and cultural values of a specific context can offer new ways of argumentation in field-specific research and develop new epistemological stances in planning knowledge. She is currently involved in research on the relevance of planning knowledge and the role of planners in post-crisis recovery. Her practice-oriented research spans Egypt, Jordan, Lebanon, Indonesia, Thailand and the Philippines.

Geoff Vigar

Geoff Vigar is Professor of Urban Planning at Newcastle University in the UK. Geoff's research focuses on the following overlapping elements: spatial strategy-making; professionalism and ethics in planning; the politics of mobility, transport policy and infrastructure supply. He is the author of three books in these areas: *The New Spatial Planning*, with G. Haughton, P., Allmendinger and D. Counsell (London: Routledge 2010); *The Politics of Mobility: Transport, the environment and public policy* (London: Spon Press 2002); and, *Planning, Governance and Spatial Strategy in Britain: An institutionalist analysis*, with P. Healey, A.D. Hull and S. Davoudi (London: Macmillan 2000).

Acknowledgements

Copyright Providers

1) Prof. Catedrático Paulo Pinho, Faculdade de Engenharia, Universidade do Porto, for permission to reproduce:
   Healey, P. (2010) 'Re-thinking the relations between planning, state and market in unstable times', in Paulo Pinho and Vitor Oliveira (eds), *Planning in Times of Uncertainty*, Porto: FEUP Edições, pp. 11–22.

2) SAGE Publications for permission to reproduce:
   Healey, P. and Barrett, S.M. (1990) 'Structure and agency in land and property development processes', *Urban Studies*, 27(1): 89–104.
   Healey, P. (2012) 'The universal and the contingent: Some reflections on the transnational flow of planning ideas and practices', *Planning Theory*, 11(2): 188–207.

3) Taylor and Francis for permission to reproduce:
   Healey, P. (1979) 'Networking as a normative principle with particular reference to local government and land use planning', *Local Government Studies*, 5(1).
   Healey, P. and Gilroy, R. (1990) 'Towards a people-sensitive planning', *Planning Practice and Research*, 5(2): 21–9.
   Healey, P. (1992) 'Development plans and markets', *Planning Practice and Research*, 7(2): 13–20.
   Healey, P. (2006) 'Transforming governance: Challenges of institutional adaptation and a new politics of space', *European Planning* Studies, 14(3): 299–319.
   Healey, P. (2012) 'Re-enchanting democracy as a mode of governance', *Critical Policy Studies*, 6(1): 19–39.

4) Elsevier Ltd for permission to reproduce:
   Healey, P. (1974) 'Planning and change', *Progress in Planning*, 2(3), Oxford: Pergamon, pp. 143–237 (selected parts).
   Healey, P. and Underwood, J. (1978) 'Professional ideals and planning practice', *Progress in Planning*, 9(2): 73–127 (selected parts).

5) Liverpool University Press for permission to reproduce:
   Healey, P. (1992) 'Planning through debate: The communicative turn in planning theory', *Town Planning Review*, 63(2).

6) John Wiley and Sons:
   Healey, P. (2004) 'The treatment of space and place in the new strategic spatial planning in Europe', *International Journal of Urban and Regional Studies*, 28(1): 45–67.

PART I
Introduction

Connections: An Introduction

Jean Hillier and Jonathan Metzger

> The central concern … is with the nature and potentials of planning as a basis for action … In other words … planning and change. The emphasis [is] … on examining the activity of planners and planning agencies in relation to these problems and to the wider society of which they are a part … within the context of the social and political structure of society. This leads to a consideration of the terms and purposes of … planning.
>
> (Healey, 1974a: 147).

Cited above are the opening lines of Patsy Healey's first major contribution as a planning scholar, entitled 'Planning and Change', which appeared in *Progress in Planning* in 1974. In hindsight, it also reads as an overarching description of the broad research agenda which she has pursued for more than four decades and through more than 250 scholarly publications, among which are included some of the most influential texts in planning studies – and in this process generating one of the most distinguished careers within this academic field. As evinced by the opening quote, throughout her career, Patsy Healey has always had the practice of planning as her central concern; the nitty-gritty 'doing' of planning work. She has always understood this practice of planning as inevitably situated in an institutional and geographic context, which in turn generates a demand for critical attention to the 'where of things' and the difference this makes. Further, she has approached the study of planning practice with an understanding of it as being a fundamentally normative activity, even if it is not always recognised as such (a problem in itself according to Healey, see e.g. 1974b), which further means that we always have to keep tabs on exactly which norms and values are being enacted and propagated in the micro of planning 'episodes' and the macro of larger 'endeavours'.

As Healey herself notes in the Epilogue to the present volume, these ideas were painfully out of sync with academic fashions in planning research at the point of time when she started out. But, to a large extent the landscape of this field has shifted significantly, and several of the ideas that were seen as contrarian at the time have been regarded in hindsight as avant-garde and have been accepted into the mainstream of contemporary planning thought. As a consequence of this, Patsy Healey is today one of the most widely cited and translated planning scholars of all times. Her work has been central to the formation of 'planning studies' as a distinct academic subject and recognised international field of research. It may therefore seem bizarre to claim that Healey's work has not yet received the full recognition that it deserves. Nevertheless, this is what André Sorensen does, when he argues (in Chapter 25) that at some future point, it is likely that 'historians of social science will recognise Healey's body of work as a major contribution to sociological institutionalist thought'. We agree with this statement and would like to add that there are also other contributions that she has made that yet remain to be fully recognised. Most of the attention to Patsy Healey's thinking has, as of the present, focused on one text, the seminal, but also somewhat controversial, book *Collaborative Planning* (Healey, 1997a), perhaps the single most cited book in the history of planning studies with 3462 citations in Google Scholar as of May 2014.

As Healey herself notes in her contribution to this book, '[s]ome planning ideas flourish because they are "in their time"', but this flourishing may well be a mixed blessing as evinced by the sometimes curious ways in which *Collaborative Planning* was picked up, interpreted and co-opted as support for various and often conflicting projects and ambitions. With *Collaborative Planning*, Patsy Healey suddenly found herself catapulted into the canon of planning studies. As has been noted by Campbell, Healey's 'metamorphosis to being seen as part of the intellectual establishment, and hence fair game for attack by up-coming generations' must have made for 'a strange personal journey' (Campbell, 2009: 148). Her mix of frustration and amusement at this situation can be read between the lines (of, for instance, Healey, 2003), but she also appears to have productively made use of the experience to reconnect to some of her earlier thinking about how ideas and concepts travel across different contexts, and the translations and transformations to which they are subject in these processes (see e.g. Healey, 1974a, 2012a, 2013; Healey and Upton, 2010, Healey and Underwood, 1978).

Seen over the span of her career, *Collaborative Planning* notwithstanding, Healey's metaphorical 'curse' does not appear to have been that of being 'in her time', but rather ahead of it. Nevertheless, read through the present, many of these ideas and insights are still highly relevant for today and the future, and therefore deserve a better fate than to be forgotten. This book, therefore, partly focuses on some of Patsy Healey's key contributions which predate the publication of *Collaborative Planning* and which have, to some degree, become overshadowed by the latter. In addition, this volume highlights the many ground-breaking ideas contained in the texts that Healey has published in more recent years, which demonstrate a critical scholarly mind which continues to pick up and reflect on ideas, reworking them in creative ways, connecting and reconnecting both recurring, as well as novel, themes and issues with remarkable intellectual agility. In line with the title of the book, this is perhaps the one trait that stands out most clearly throughout Patsy Healey's whole career: her penchant for and skill at making connections, as well as studying them. Throughout her work, from her earliest days as a scholar, a fundamental relational appreciation of the world shines through, understanding phenomena as part of a broader fabric of constitutive connections to other places, things, people and ideas. And connections are also what she has so actively generated all along: connections between different people, including academics, practitioners and lay people, between different realms of ideas, different academic disciplines, between theory and practice, between normative and analytical perspectives – between different places, cultures, and generations. All this done with an understanding that the connections we make, and how we make them, make a difference in the world.

Rationale of the Book: Structure and Organisation

In 1974, in one of her first published texts, Healey ominously wrote that 'planners who merely accept the bureaucratic niche and pay little attention to professional debates or public criticism may find their security given a jolt if a party were elected to power which so undervalued the planning in which most of us are involved that they repealed all the legislative and procedural support which presently surrounds us', and that in the face of such a situation, 'practitioners and educators alike' in the field of planning would 'be searching … desperately for legitimation' (Healey, 1974b: 604). With hindsight, we know that this is the situation that played out across much of the western world in the 1980s, putting planning in the doldrums – perhaps particularly in Healey's own Britain, 'the poster-child of 'neo-liberal' ideology' at the time. Healey's persistent labours played a role in searching out new directions for planning thought and identifying potentials for making a reinvented planning practice relevant in the context of the emerging late-capitalist democratic societies; efforts that led to a revitalisation of planning as an academic discipline and societal practice at the end of the twentieth century, even if temporarily in an economic climate dominated by the global financial crisis and individual State recessions. As a consequence, if we trace Patsy's oeuvre in the form of a heterogeneous network encompassing ideas, practices, texts, etc. it becomes obvious that her achievements stand out as a critical node or anchor point in the broader, and partially overlapping network that could be labelled as the emerging academic discipline of 'Planning Studies'. It is largely her 'reticulist' skills as well as her commitment, engagement and enthusiasm for the subject at hand that have generated and stabilised this network.

As editors, we want to present this book as a showcase for this remarkable scholarly achievement, not as any form of closing of accounts, but rather to stretch this trajectory of thinking further into the future – to provide some potential avenues of exploration that can function to display how Patsy Healey's work may be able to fuel fruitful scholarly investigations into planning theory and practice for many years to come. By placing her thinking in new contexts, through reconsideration and reinterpretation, connections are generated to the emerging futures of different social, political and cultural contexts of human sociality at the beginning of the twenty-first century. Based on this rationale, the volume contains a selection of 13 less readily available, but nevertheless, key texts by Patsy Healey, together with 12 original invited contributions which take the ideas in the reprinted papers as points of departure for their own work, tracing out their continuing relevance for contemporary and future directions in planning scholarship. The project has been developed in close cooperation with Patsy Healey herself who has been very generous with suggestions and ideas concerning the selection of papers to be reproduced, invited contributions, development of the key themes, as well as with the overall structure of the volume.

The reprinted material has been selected to represent the trajectory of Patsy Healey's work across the several decades of her research career. In collaboration with Patsy, we have endeavoured to select material which makes an important contribution to the field, but which is less widely available than other, well-known texts. The aim of the original contributions is thus to tease out the themes and interests in Healey's work which are still highly relevant to the planning project. The authors, who have contributed original texts written specifically for the volume, represent a wide range of scholars from different parts of the world, including global leaders in the field as well as emerging young scholars working with a wide range of research approaches. Some have a background as colleagues and co-authors with Patsy, others have applied or developed her work in various ways. All have been inspired by her. The book is organised into five main Parts, according to main themes of her work running through her whole oeuvre. The first Part is this overall introduction.

The bulk of the content of the book is contained in Parts II–IV, which all include thematically selected reproduced and original papers relating to a specific theme or topic. These Parts are also introduced by original essays contributed by the editors, which situate the selected reprinted papers, as well as the original contributions contained in the Parts, in a broader historical and intellectual context. Part II Normative Perspectives covers approaches to ethics (emphasising decisions in terms of duty, rules, moral standards etc.), the planning profession, the role of politics, and democratisation. Part III Places and Practice consists of two subsections. III.1 The Planning/Development Nexus: How Places are Produced and Changed, concentrates on issues of land and property development, relations between planning and land and property markets and the implementation impacts of planning instruments. Part III.2 Doing Planning Work is concerned with theory and action both with regard to local planning authorities' regulatory, local area, development plans and also to broader practices of strategy formation and policy processes. Part IV Transformation Processes deals with issues which have long interested Patsy Healey, including planning culture, governance and institutional capacity. Reflecting both her international research and the international audience for and application of her work, this Part includes examination and reflections on the ways in which ideas travel, ideas pathways and comparative studies. Finally, Part V is an Epilogue in the form of an original paper written by Patsy Healey especially for this volume.

Please note that Patsy Healey's papers reproduced in this volume have been re-typeset. Quotations from these papers in other chapters cite page numbers from the original publications and therefore differ from corresponding page numbers in this volume.

The remainder of this overall introduction consists of a brief comprehensive academic biography for Healey, placing the various themes of her scholarship in relation to each other in a chronological order. It is rounded off with a short reflection on what stands out – across all the diverse topics she has investigated as a scholar – as something of an ever-present driving force behind her energy and commitment: the energy to make connections across time and place, in pursuit of the caring and nurturing of the promise of a more democratic planning to come.

An Extraordinary Career[1]

Early Career: Late 1960s to mid-1980s

Patsy Healey studied Geography at London University (University College) from 1958–1961 where she was inspired by Mary Douglas, a rare senior academic woman social anthropologist, who had undertaken ethnographic research with the Lele people in the Congo, and whose major work (drawing on the pragmatism of William James – a scholar to whom Patsy Healey will return in the early twenty-first century) traced the

1 We should note a much more in-depth *Conversation with Patsy Healey* by Mona Abdelwahab in the AESOP InPlanning series in 2015, plus Patsy Healey's chapter in Haselberger et al.'s (2015) collection, *The Future of Planning: Personal stories in the evolution of planning thought.*

meaning of dirt in different cultural contexts (Douglas, 1966). On graduation, Patsy Healey studied a Diploma of Education at University College, Swansea and became a secondary school geography teacher, first at a girls' grammar school in Swansea and then at a comprehensive school in Eltham, London, at the time of major restructuring of secondary education.

At this time the London Boroughs were expanding their staffing after reorganisation of the English and Welsh planning system in 1965. Frustrated by the difficult teaching conditions generated by the turmoil in the aftermath of a thorough (and much-needed) overhaul of the British education system by the Labour government (Thomas and Healey, 1991: xii), Patsy Healey began work with the London Borough of Lewisham (1965–1968) and then with the Greater London Council (1968–1969). In her own words, 'she found the whole world of local government intriguing'. She was puzzled by the work of the planning authority, having previously assumed that planning was about making the world a better place for people. Yet such considerations seemed to have no connection to the work she became engaged in, namely 'doing surveys for a Borough Plan' (Thomas and Healey, 1991: xii). This quotation appears to demonstrate something of a naiveté about the spirit and purpose of planning and how it actually materialises in practice, probably linked to a distinct lack of research critically analysing planning as practised. Yet it is this simple statement that planning 'was about making the world a better place for people' which has driven so much of Patsy Healey's work for the last 40 years or so. It also links her strong tradition of empirical work with practitioners and cases from planning practice, her interest in planning education and the professionalism of planners.

While she was practising planning, Patsy Healey studied part-time, as many people still do, for her Diploma of Town Planning at what was then Regents Street Polytechnic (now the University of Westminster) in London, before undertaking her PhD at the London School of Economics (1969–1973). As she says, even though she had gained her professional qualification in planning, 'I still felt I knew little about the planning activity I was involved in. I did, however, come from an academic family, so I thought that by doing a PhD I might get a better idea of the nature of the planning endeavour' (Healey, 2014).

In the late 1960s and early 1970s, in what ex-British Prime Minister Harold Wilson had called the 'white heat of the technological revolution', societal conditions were rapidly changing, both in countries of what we now, albeit tentatively, term the global North and the global South (see, for example, Watson, 2014). Then, typically referred to as 'developing countries', those in the Global South were urbanising at a tremendous rate, often linked with the aims of post-colonial governments to stamp their mark of Progress on their capital and major cities.

Taking the opportunity to go to South America, Patsy Healey researched her doctorate on 'Urban Planning Under Conditions of Rapid Urban Growth' in Valencia, Venezuela and Cali, Colombia. It was in this work – founded upon *in-situ* observations of planning practice – that Healey saw, first-hand, planning practice as a set of potentially transformative processes and ideas, often travelling from elsewhere. This notion, first published in *Progress in Planning* (1974a) – sections reproduced in Part IV – and in Healey (1975) has stayed with Patsy Healey ever since (see Healey, 2015a). As she wrote (2012a: 196):

> My hypothesis … is that transnational learning works most productively through rich narratives – in-depth cases – rather than through 'best practice' summaries or attempts at typologies which systematize qualities of context and try to match them with qualities of experiences. This is an important reason why 'thick' case narratives are celebrated in the planning field.

From the very beginning of her scholarship, Patsy Healey recognised the importance of planning practice in relation to its wider societal (political, cultural) and economic contexts. Her work demonstrated that the procedural rational decision making model of planning was not readily applicable in rapidly changing situations where outcomes could not be predicted with much accuracy.

On her return to England, she took up a position in the School of Planning at Kingston Polytechnic (now University) and from 1974 onwards at Oxford Polytechnic (Oxford Brookes University) which was already becoming something of a hub for planning theorists led by Andreas Faludi, Glen McDougall and Michael Thomas.

Patsy Healey had been influenced, as were many other academics, by student politics of the mid-1960s and debates leading up to the now (in)famous events of 1968 and the turn to Marxist-inspired political economy approaches, especially Andre Gunder Frank's work in Latin America (1966, 1967, 1969) and David Harvey's book, *Social Justice and the City* (1973). Having encountered Joe Bailey at Kingston Polytechnic, Patsy Healey then met Brian McLoughlin who was at the Centre for Environmental Studies (CES) in London. The CES funded innovative research, such as Healey's project 'Planners' Use of Theory in Practice' (1974–1976) which overtly linked theory and practice; something which her work has done ever since. As noted by Huxley (2009: 139), 'in the early 1970s, research into the actual implementation of planning, its effects on land markets, or the experiences of planners in local government offices across the country, was far from common, and in studying these issues, Patsy was at the forefront of social science research in planning', although not always recognised as so.

Healey, McDougall et al. (1982: 17) argued that it was possible, in the early 1980s, to discern two 'tendencies' in theoretical debate about the nature, purpose and method of planning. The dominant tendency was that of procedural planning theory and its technocratic, managerial practice relations. The second tendency was socially democratic, focused on resource redistribution, advocacy and social justice. In response to this duopoly of understandings, the authors added the tradition of political economy with its deeper and broader approach to social and political critique. 'Planners' Use of Theory in Practice' was published as a reaction to a situation in which there was 'no academic capacity for critical evaluation' (Healey, McDougall et al., 1982: 5) of planning practice. In the 1970s and early 1980s, political economy approaches were popular amongst academics, due to their explanatory power: 'it is the only position which offers a coherent and relevant account of the way changes in the economy are reflected in the dilemmas facing government, and how these in turn shape the practices in which planners find themselves' (Healey, McDougall et al., 1982: 16). Yet, as the authors lamented, whereas political economy analyses were common in other local government fields, there was at this time very little such analysis relating to planning.

Patsy Healey's own work 'found inspiration … in a mixture of Marxist-inspired urban political economy and phenomenology' (Healey, 2007: x). She (Healey, 1982: 185–6) cites 'powerful concepts' in Manuel Castells' work: 'one cannot analyse a social or political process independently from its structural context and from the levels of structural interests which determine it' (1977: 75). Clearly, political economy ideas were useful to Healey in asking questions about the role of land as an input to production and about the ways in which land transactions may contribute to the realisation of surplus value (1982: 187). She cites Cynthia Cockburn (1977), James O'Connor (1973) and Claus Offe (1974, 1975) as contributing to an understanding of the work of governments with relevance to land use planning. As she reflects more recently, 'I was guided by the Marxian concept that agency has power to make history, but not in circumstances of our own choosing. So, in analysing and considering the potential progressive transformative contribution the planning project could play, I was interested specifically in how circumstances shaped how the planning project was understood and institutionalised and the shaping role it then played' (Healey, 2015a).

Healey's research, especially that with Jacky Underwood (1978), attempted to relate the activities of practising planners to wider societal structures and the unequal relations of economic power. Her research into planning as actually practised on the ground, engaged her in talking with practitioners in local government offices about their experiences of the newly revised English planning system (1978, reproduced in Part II) and, more importantly, as Albrechts (2009: 143) recalls, 'listen[ing] to planning practice and practitioners rather than simply looking at it (or them)'. However, talking extensively with practitioners about their individual activities in this and other research projects (see, for example, Healey, 1983, 1991a; Healey, Davis et al., 1982; Healey et al., 1985), led Patsy Healey to question political economy ideas as being over-dependent on the role of economic structure, to the detriment of considering the role of agents. Her 1982 chapter, reveals the theoretical struggles that she was having in terms of what made sense with regard to what she was seeing in practice: 'a simple class analysis does not account for the significant activities of local authorities in distributing resources independently of these dominant class forces' (1982: 190). Two years before Giddens' (1984) work on structuration, Patsy Healey had already been searching for something which could explain the detailed practices of individual government agencies as they encountered social and economic structural forces (see also Huxley, 2009). As she concluded in her 1982 chapter: 'the discussion of class categories, of social movements, of the relations between the political and the economic seem to fall all too frequently into the traps of mechanistic theorising which can only be connected to actual experience in crude and often naïve ways' (1982: 193). She then referred to herself as a 'Weberian fellow-traveller', but would probably now prefer to think in terms of Dewey's pragmatism (see below). Nevertheless, throughout the rest of her career, we find that Healey continuously acknowledges the crucial insights of political economy and the influence it has had on her thinking (see e.g. Healey, 2010a and Healey, 2015b). As such, she has never 'disavowed' the crucial insights of the political economy perspective on planning.

Sociological Institutionalism: c.1988 onwards

In 1988 Patsy Healey moved from Oxford Polytechnic to the Chair of Town and Country Planning at the University of Newcastle-upon-Tyne. Here she rapidly made her mark on the Department, with her inclusive management style, nurturing women and early career academics, such as Rose Gilroy, Suzanne Speak and Simin Davoudi in particular. She also Chaired a group which produced a seminal paper for the Education for Planning Association and Royal Town Planning Institute (RTPI) in Britain on 'Gender Issues in Planning Education' (Healey et al., 1988).

Patsy Healey's relatively little known work on land markets (see Part III.1, this volume), both sole-authored and with Sue Barrett, Martin Elson and Paul McNamara in particular, demonstrates her grappling with structural and agential theoretical approaches (Healey, 1991b; Healey and Barrett, 1985, 1990; Healey and Gilbert, 1985; Healey et al., 1988). In her struggle to find ways to bridge the troubling structure/agency divide in relation to land development and land use planning, Patsy Healey came to develop an 'institutional account' of planning and practices, focusing on development interests and institutional practices of mediating amongst conflicting interests at various scales (Healey, 2003: 103).

By the late 1980s, Margaret Thatcher and her Conservative regime had been in power for almost a decade and their neo-liberal economic policies of 'rolling back the State' were biting hard in those areas of Britain (such as the North East) already hit by structural unemployment. Patsy Healey researched the impact of Thatcherite urban policy on real estate development and urban regeneration initiatives in Newcastle-upon-Tyne, a research site to which she has returned on several occasions. At the same time, lobbied by industry and housebuilding interests, such as the Confederation of British Industry, the Housebuilders Association and the Property Advisory Group, the Thatcher regime began[2] questioning the value of the English planning system and the competence of local planners. As Healey (2003: 103) notes, 'strategic spatial planning … was dismissed as a time-wasting constraint on innovation and market efficiency'.

Clearly, structural, capital-oriented explanations were relevant, but Patsy Healey was especially interested in how local planning authorities and practitioners were reacting. She argued that 'there is more to the construction of social worlds than just economic forces' (Healey, 2003: 112). Gender, ethnicity, class and other structuring forces can be identified. Healey thus turned to a combination of Anthony Giddens' (1984) concepts of the binding flows linking agency to structuring processes, Maarten Hajer's (1995) notions of discourse structuration and institutionalisation, Steven Lukes' (1974) and Torben Dyrberg's (1997) work on levels of power and Judith Innes'(1985) ideas about institutional capacity-building (see the Introduction to Part IV). In particular Giddens' 1984 book appears to have offered her a 'peg' to which she could attach her own institutional thinking, which was by then already well-developed. Healey (2003, 2006) retrospectively explains these influences in detail, emphasising how Giddens drew on traditions of Marxism, phenomenology and cultural anthropology, as did Patsy Healey herself, to offer a way of understanding how structures are maintained in the flow of institutional practices and how the active work of agents becomes institutionalised through structuration processes. By relating to Giddens' thinking Patsy Healey was able to find a language and conceptual repertoire which helped reinforce her relational perspective on what she recognised as fundamentally interactive policy processes: 'the fine grain of the daily routines, discourses and practices of governance, between structuring driving forces and what people do in specific episodes of governance' (Healey, 2003: 109).

Studying what goes on, how and why, however, has never been sufficient for Patsy Healey. Her aim is to understand these issues in order to challenge and change them in the interests of those disadvantaged by prevailing systems. For Healey, the co-constitution of structure and agency, the big patterns and seemingly 'small' events are mutually formative; which is why she argues that we 'need to give some attention not just

2 As has since been exponentially heightened under Prime Minister, David Cameron, by Eric Pickles, Secretary of State for Communities and Local Government.

to the micro-dynamics, but to the wider structuring forces which create moments of opportunity and limits on inventive possibilities' (2012a: 197). This is thus not a perspective which posits structuration *against* political economy, but rather as a necessary corrective to a fundamentally political economy inspired perspective – albeit with a strong belief in democratic, pluralistic collective governance and visioning.

Patsy Healey has also always been interested in the institutions of planning education and professionalisation (see Part II this volume). Some of her earliest work in the 1970s followed planning students in their professional careers, critically asking them what they were doing in practice, how and why, at a time of 'bedding in' a new planning system. By 1985, as outlined in the Introduction to Part II, she had become quite critical of the institutionalised form of planning professionalism, suggesting that it had inhibited the development of practitioners' capabilities to think for themselves (Healey, 1985: 503). She was elected to the RTPI Council in 1987, on which she served until 1992.

In 1991, Healey (1991a) claimed that the RTPI, although possessing a 'Research Board', had struggled to define an appropriate agenda for academic planning research. Subsequently, she instigated useful cross-fertilisation of information and links between researchers and practitioners when in 1999 she became founding Senior Editor of the RTPI-sponsored journal, *Planning Theory and Practice*, with its focus on integrating theory *and* practice and its path-breaking Interface section with dialogues between academics and practitioners. Since 2009 she has been Associate Editor for the RTPI Library Series of books published by Routledge.

Even if regarded by the planning 'establishment' in the late 1980s as something of a contentious female, Patsy Healey has since been widely recognised for her contribution to professional planning practice. She was awarded the Order of the British Empire (OBE) for Services to Planning in 1999 and presented with the British Royal Town Planning Institute's Gold Medal in 2007. Healey is the first woman to receive this medal, with its citation 'for outstanding achievement in the field of Town and Country Planning'.

Although Britain and various other European countries had flourishing associations of planning schools and educators, it made evident sense to a group of senior academics to form a Europe-wide association that would bring together planning schools from across the Continent. Accordingly, Klaus Kunzmann (from T.U. Dortmund) gathered a small group of academics at the Schloβ Cappenberg, near Dortmund, in January 1987 and the idea of the Association of European Schools of Planning (AESOP) was born. Patsy Healey was one of these founding academics. She went on to hold Vice- and Presidential roles in AESOP between 1992 and 1997, serving as President from 1994–1996. It was during her Presidency that the extremely successful PhD workshops/Summer Schools for Doctoral Students and Young Academics were established. Healey was made an Honorary Fellow of AESOP in 2004, only the second person to receive such an honour.

After Britain joined the European Union in 1973, Patsy Healey, along with many other academics, began collaborative research projects with scholars in various European countries. This work was facilitated and enhanced by the establishment of AESOP which effectively created networking opportunities across Europe for academics. Healey's work with Dick Williams (Healey and Williams, 1993) and with Abdul Khakee, Barrie Needham and Alain Motte (Healey et al., 1997) provided early pan-European surveys of planning systems and strategic planning issues, which she later updated with co-founders of AESOP, Louis Albrechts and Klaus Kunzman (Albrechts et al., 2003).

With several colleagues from Newcastle University, Patsy Healey investigated how planning policies and implementation were evolving in England, using an institutionalist approach (Vigar et al. 2000). Healey's institutional interests have continued to the present. The paper which she presented to the Association of Collegiate Schools of Planning in 2004 – a forerunner to her 2007 book, *Urban Complexity and Spatial Strategies* – draws together the three critical strands of her interest – the sorting, inventing and filtering processes through which strategic concepts and representations of the city region are formed and focused; the persuasive and mobilising institutional capacity of formation processes, ideas and images, once formed; and their capacity to travel to other arenas. The book then developed new understandings from in-depth case studies of strategic spatial planning in Cambridge, Milan and Amsterdam.

Collaborative Planning: c.1992 onwards

At the beginning of the 1990s, the future looked uncertain for British planning both in practice and academia. Having become something of a bête noire for Margaret Thatcher and her governments, as well as an ideologically important symbolic target of vested interest neoliberalism, the planning profession was struggling to deflect and fend off repeated and sustained administrative assaults by central government and 'big business'. Meanwhile, in the academic world, 'postmodern' theorists tore to shreds any form of coherent 'grand narrative' of progress and development, which – in the eyes of many commentators – left planning theory bereft of its previously, far-too-often and somewhat leisurely, taken-for-granted normative and ontological foundations. Accepting the broader ramifications of this academic critique within the contemporary political context in Britain at the time, Healey (1992) asked herself what could be the purpose and relevance of the practice of planning under such societal and intellectual conditions? Could there be anything more, anything else, to planning than mere piecemeal management of urban environments? And if so, what?

As Patsy Healey stated in her Introductory chapter to *Collaborative Planning* (1997: 5), the institutional approach 'emphasises the range of stakes which people have in local environments, and the diversity of ways we have of asserting claims for policy attention'. Two key elements are apparent in this statement: diversity and communication. From the late 1960s onwards, with the rise of social and political movements claiming, and gaining, voice for women, minority ethnic groups, gays, lesbians and transgender people, people with disabilities, nature, and so on, there have been moves towards more inclusive forms of planning theory and practice which understand and valorise the situated knowledges of diverse stakeholders (e.g. Healey, 1995). Material on identity and diversity in relation to spatial planning has become increasingly rich since the 1980s in particular (see Hillier and Healey, 2008 for a brief overview).

Diversity refers not only to the range of stakeholders affecting and affected by planning policies and practices, but also to the roles and activities of planners themselves. From technocratic experts, leading the 'best' way forward, to facilitators, mediators, advocates and so on, planners perform a range of diverse roles in a range of diverse organisations. Planning thus concerns 'collective endeavours to shape place qualities' (Healey, 2011: 2). As Healey continues, planning 'involves the efforts of many people, not just those trained as 'planners'. And it can take place in many different arenas, not just planning offices' (ibid.).

Planning, then, centres on ways in which human and non-human actors can manage their collective concerns, with respect to the sharing of space and time (Healey, 1992: 145). The role of communication between actors is crucial in this: 'where collective 'acting in the world' is our concern, we need to engage in argumentation and debate' (Healey, 1992: 151). Influenced by John Forester's (1985, 1989) work on communicative planning practice, but also – and importantly – by her extensive empirical work talking with planning practitioners on the ground and watching decision making processes in action (Healey, 1991a, 2003), Patsy Healey began to think through what 'planning as a communicative enterprise' (1992) might potentially look like. Her early writing on the topic (1992, 1994) discussed Jürgen Habermas' ideas on communicative action, but also acknowledged the importance of work by Foucault and Bourdieu.

Her ten propositions for a new planning direction (1992: 154–6) stress Habermasian-influenced processes of communicative action, but do not mention consensus-creation. She is more concerned with thinking through what inclusionary, democratic, argumentative, reflective processes might comprise – hence her term, 'planning through debate'. Furthermore, the tenth proposition advocates building in criteria of critique – including opportunities for reflexive challenge to agreements as to 'what to do next' as circumstances change. As she says:

> we may be able to agree on what to do next, on how to 'start out' and 'travel along' for a while. We cannot know where this will take us. … Neither the 'comprehensive plan' nor 'goal-directed' programmes have more than a temporary existence in such a conception of communicative and potentially transformative environmental planning. (Healey, 1992: 156; also 1994)

In this important paper, we can see that Patsy Healey was discussing ideas of 'temporary fixity' and 'provisionality' over a decade before they were brought to attention by scholars such as Jean Hillier (2005, 2007) and John Pløger (2010). Her concluding sub-heading in the 1992 paper, 'The dialectics of a new planning' and use of the term 'struggle' further suggest that Healey's political economy background had not disappeared entirely.

Concerned by what she saw as predominantly adversarial forms of communication in planning practice and the violence such forms inflicted on those actors unused to, and/or unversed in, planning language (see, for example, Healey and Hillier, 1995), Patsy Healey briefly engaged with Habermasian ideas of consensus-formation in *Collaborative Planning* (1997: 263 ff), but soon regretted having done so (see Healey, 2003) as commentators critically picked up on the term 'consensus' rather than all the other material in the book on argumentation and relationality (see also Part III.2 this volume). In a recent comment, Patsy Healey (2014, pers. comm.) clarifies how she understands the term 'consensus' with regard to its confused adoption by planning theorists and practitioners. She understands consensus to have two meanings. At the macro-societal level, it refers to the culture/values and so on, which a society holds/agrees about. Habermas asked himself could there be perhaps some such shared values which a society – rather than people involved in local or regional disputes – could reach towards. At the micro-level, consensus means 'what we can agree on *for now*', in a situation where people with different positions/arguments are in dispute of some kind. Any such consensus is inevitably provisional. Healey's engagement with consensus is, therefore, as something that is inevitably context-dependent and which fundamentally concerns the questions 'what' and 'how'. She constantly reiterates that struggle is crucial and conflict can be important.

The debate around the book came to focus around the ideas of communicative rationality and consensus seeking. However, these were not the issues which Healey regarded as key elements in her argument. As she states: 'located in complex interactions between multiple webs of relations, it was important, it seemed to me, to focus on the communicative dynamics of the relations and interactions, and the ways these lodged or displaced particular modes of thought and practice' (2015a). Her frustration with critics who had misunderstood her work and overlooked its important material on economic, social and environmental relationalities are apparent in her response (Healey, 2003) and in the postscript to the second edition of the book (2006).

Nevertheless, the term collaborative planning has become part of planning practice internationally. The book which gave its name to the term has been translated into several different languages. In the UK and internationally, however, both the term and its underlying concepts have often been misinterpreted as mere 'public participation' and simply grafted onto rational comprehensive based processes in order to afford them legitimacy.

Interpretive Policy Analysis, Pragmatism and Civic Capacity: 2000s onwards

From the very beginning up until today, Patsy Healey's understanding of planning in particular and public policy in general thoroughly emphasises relationality and interpretation throughout: 'planning is an interactive and interpretive process' (1992: 154). Influenced by the development of interpretive policy analysis (IPA) in the 1980s and 1990s (e.g. Geertz, 1983; Fischer and Forester, 1993), linked to the increasing recognition of the diversity of identities, values, meanings and understandings mentioned above, and resonating with her undergraduate study with anthropologist Mary Douglas, Healey has always recognised that 'there are potentially many epistemologies, each with their own ways of "making sense" of observations and experiences, resulting in particular logics and "rationalities" through which arguments and claims for collective action may be structured and implemented' (Healey, 2007: 244).

From the late 1990s onwards, Healey herself began to contribute to this emerging area of policy scholarship by way of appearances at IPA conferences and publications in journals and books. Through such activities, she has also succeeded in opening the awareness of this wider field of scholars to the relevance of planning practice as an interesting object of study with broader ramifications for policy scholarship than is perhaps generally recognised within contemporary academia (see further Healey, 2015b).

This influence perhaps came to the greatest fruition in her 2007 book, *Urban Complexity and Spatial Strategies: towards a relational planning for our times*. Healey personally regards this as her 'best' book (Healey, 2015a), and this key text of a still emergent 'relational planning' paradigm promises to constitute a crucial reference point for future theoretical developments in the planning field. In the book, Healey presents a lucid discussion on the processes for developing spatial strategies through 'framing', and further relates these insights to a relational understanding of space. This latter element was also influenced by the work of Doreen Massey (1984) and the development of the 'relational turn' in human geography, to which Healey herself also contributed together with colleagues such as Stephen Graham (see, e.g. Graham and Healey, 1999, and further the Introduction to Part III.2).

From the early 2000s, the Blair and Brown Labour regimes in Britain created a rhetorical push for more localised decision making. The election of the Conservative-Liberal Democrat Coalition government in 2010 saw this push continued and the Localism Act (DCLG, 2011) has granted local communities powers to develop Neighbourhood Development Plans, which may become statutory planning policy, and to use 'Community Rights' to take control of local assets (such as buildings and services). Rural Community Councils had already existed since 1988 in England, under the auspices of what was then the Rural Development Commission. By the early twenty-first century, there were some thirty-eight charitable Rural Community Councils with a rural development function, covering areas such as community planning, community buildings support, rural transport schemes and rural affordable housing (Moseley, 2003).

One such community is the Glendale Gateway Trust, in Wooler, North East England, originally formed in 1996 to raise funds for provision of a community hub for the area, together with space for small enterprises. Patsy Healey moved to Wooler in 1996 and was drawn to the Trust and its work, becoming a Director of the Trust in 2010 and its Chair in 2012. The Trust is currently actively developing affordable housing and affordable business space for micro-businesses.

Patsy Healey is one of those scholars who, as John Forester (2009: 134) puts it so well, leads 'by example as much as by argument'. The Glendale Trust epitomises the ideas of capacity building through networking and connecting people and ideas, together with civic democracy, all of which Patsy Healey (e.g. 2011) espouses in her work. The process of gleaning and reviewing local people's 'strategic ideas' about the challenges the area faces and what to do about them, has enabled her to put her ideas for place-focused collective action and of 'place governance with a planning orientation' (2010b) into practice. As she comments (2015a): 'I am back to "doing" planning work, drawing on all my planning and managerial experience'.

A key driver for Patsy Healey is the enhancement of civic capacity for attention to place qualities, 're-configuring the relationship between formal government – the worlds of formal law and politics, and of public administration, with the associations and relations of civil society' (2011: 11). This has led her to take an interest in the development of current theories of 'democratic network governance' which are mirrored in, for instance, Healey (2006) and (2011), but – as should be duly noted – are also in focus in her work from 1979, long before its current advent into academic fashion. In her search for ways of thinking and communicating about these issues she has, in more recent years, discovered a keen affinity to 'classical' American pragmatist philosophers such as William James, Charles Pierce and, particularly, John Dewey (Healey, 2009).

This school of thought, with its focus on practice and also on ethical betterment as a practical endeavour, was always present in her thinking to some degree – filtered through, for example, the early influences of Mary Douglas and later, for instance, the thinking of Jürgen Habermas, John Forester, John Friedmann and Richard Bernstein – but it has recently come to exert a more direct inspiration on her work, helping her to further refine her own thinking about fundamental ethical and political entanglements, as well as the mutual inseparability of questions concerning 'what' should be done and 'how' to do it. Healey for instance relates to Dewey's (1927) conception of 'cultivating' democratic practices in her Introduction to the edited collection, *Crossing Borders* (Healey and Upton, 2010). The topic of her text concerned how ideas 'travel' around the globe, in similar fashion to the extensive transnational travelling of Patsy Healey's own ideas, and it also, in a meta-sense, succinctly illustrates the sometimes surprising careers and trajectories of ideas by bringing in and making relevant Dewey's work in an intellectual context very different from that in which it was developed. In an interesting way, this recent work on 'travelling' concepts, practices and ideas (see the Introduction and papers in Part IV, this volume) brings us almost full circle to her doctoral research on Venezuela and Colombia and the international travelling of Western planning ideas, such as those of Le Corbusier's Regulatory Plan for Bogotá of 1951.

Conclusions: A Living Legacy of Practical Thinking for a Democratic Planning to Come

A Critical Friend of Planning 'Without Guarantees'

'Rich in grand narratives, particularly of micro-practices'; 'underpinned by a strong commitment to expose the interaction between political initiatives and people's actual experiences'; 'seeks to reveal how meanings and mentalities get to shape the materiality of the lived experience of particular places and vice versa'; 'feeds into a practical concern with how planning work – place governance – should be performed to achieve particular normative directions, and how formal institutional designs might foster the emergence of particular kinds of practices' (Healey, 2015b). Summarising here the main tenets of interpretive ideas, Patsy Healey could easily be describing her own research endeavour. Similarly, she herself embodies the set of attributes (2012a) or 'qualities to foster' (2011) which she advocates for more progressive place focused governance (see the Introduction to Part II, this volume). She has continually displayed 'curiosity about what planners do, how they do it and why' (Hajer, 2009: 138) combined with a willingness to 'read against the grain ... to keep one ear on practitioners' experiences and observations and another on the theorists' claims' (Forester, 2009: 134). In combination with a fundamental openness of mind and a willingness to discuss and consider an incredibly wide array of ideas, she has acted as catalyst for the development of so many scholars' contributions to the discipline of planning.

With a preparedness not only to reason and explain, but also to reconsider her own thinking, she appears to have never stopped being amazed at the liveliness and variability of this world we live in, guided by a 'humility that there are always new things to be learnt' (Campbell, 2009: 149). Her own trajectory as practitioner, educator, researcher and 'practitioner' once again demonstrates a keen interest in, and awareness of, the complexities of the worlds of planning practice and how they might inform planning theories and vice versa. As stated already in her very early thinking (e.g. 1974b), any conceptual template inevitably simplifies reality in both necessary but also always dangerous ways; an insight which also intimates that our world is always more strange and astonishing than we can ever fully fathom or grasp. Accordingly, Patsy Healey (2015a) refers to herself as an intellectual 'explorer', observing the world in its multiplicity of manifestations as she goes about in it, but recognising that 'an intellectual explorer can never fully arrive, but must continue searching into the always partly unknowable intricacies of living in the world'. It could be speculated that this is the fundamental insight which underpins her unwavering ability for critical engagement with her own, as well as with others' work, an engagement in which she always displays awareness of problems and limitations which beset practitioners and academics alike, including the reductionism of models and templates, which can offer 'useful "thinking tools"' (Healey, 2011: 15), but should not be replicated blindly or applied as copy/pastes of 'best practice', even in the face of decreasing time, personnel and/or financial resources. As highlighted by Forester (2009: 134), this evinces of an ability to pay 'critical attention', while not succumbing to corrosive critique (Metzger, 2014).

In accordance with Patsy Healey's own intellectual key tenets, we should not go searching for grand meta-structural explanations behind her work. Nevertheless, as she herself suggests in her Epilogue, there are signs of some 'contingent universals' (cf. Healey, 2012a) which appear to shine through and become reiterated throughout most of her thinking and writing as a planning scholar. One of the themes which recurs throughout her whole career, in one guise or another, is care. Specifically; care for the 'planning project'. As noted by Campbell (2009: 147), Healey's work does display 'a profound underlying commitment to planning, as a societal endeavour which can better people's lives', and accordingly, from the early days of her academic career, she has never represented herself as a detached outside commentator to the practice of planning, but rather actively identified herself as a committed *planner* (see e.g. Healey and Underwood, 1978).

But this is in no way an unconditional commitment. Rather, her relation to the planning profession or 'actually existing' planning practice could perhaps be best described as a paradigmatic example of the *critical friend* (cf. Forester, 1997, 1999). Her persistence in holding on to the second of these terms has perhaps generated a degree of ire among other, more critically inclined, planning scholars who otherwise might substantially appear to share many ideas with her. But this does not mean that Healey's own attitude has been mild, forgiving or evasive with regards to the lacks and faults of the planning practices which she has investigated and analysed. From the beginning, Healey was always a sympathetic critic, but a critic nevertheless. A minoritarian 'loyal opposition' commentator who always believed in the positive potentials of planning, but always also – to paraphrase one of her great intellectual contemporaries, Stuart Hall – 'without guarantees' that this potential will be delivered upon and which, therefore, demands constant critical vigilance from planning practitioners and scholars alike regarding not just the 'what', but also the 'how' of unfolding planning practice (cf. Hall 1986[1983]).

Towards a Democratic Planning to Come

Patsy Healey has always argued the fundamentally normative nature of planning practice (see Part II in this volume as well as, for instance, Healey 1974b) and she has not been afraid to enter the, often heated, debates regarding what should be considered relevant and desirable directions for planning practice. Although her ideas on this issue have, to some degree, gone through various phases of development across her career, a key emphasis has always been on democracy. For instance in 1992 (p. 144) she writes that: 'the challenge for planning in the contemporary era lies at the heart of our efforts to reinterpret a progressive meaning for democracy in Western societies' (Healey, 1992: 144). Thus, a central 'concern' for her has been 'to affirm the relation between planning and democracy, and to propose a form of planning appropriate to late-twentieth-century developed Western countries, with their demands for citizen power and their plurality of conflicting interests' (Healey and Gilroy, 1990: 28). In a particularly lucid description of this ideal, Healey elaborates how the idea of 'a people-centred democracy, with progressive rather than regressive tendencies built into it' will always be 'incomplete' and 'emergent':

> It is a direction to be struggled for, in the wider effort to 'will into being' societies in which a daily life conception of the world becomes the measure of how we should co-exist in an increasingly urbanized world. Such a conception would be grounded in a pluralistic, outward regarding attitude, yet acknowledging political community with those who share the spaces of our … daily lives. In such a world, many of our practices may damage not only each other but the conditions for life on our planet. In this struggle, competent experts, technocrats and bureaucrats are all needed. But their contributions should be grounded in an awareness of the multiple ways we diverse many live our lives and how we engage in collective action in civil society, not hitched to the old worlds of narrow elites or specialist politicians and officials, disconnected from the rest of us. (Healey, 2012b: 35)

For Healey, 'the potential for progressive governance practices which could help to sustain and promote more people-centred, open-minded, pluralist polities' (Healey 2012b: 20) lies in an 'ontology of mutuality' which further 'demands efforts not just in social mobilization, but in experimentation and careful learning from situated experiences', but also requires 'continual attention to resisting subversion into the well-established pathways of elite technocracy and corporatism, or into exclusionary populism' (Healey, 2012b: 34). This recurring concern with, on the one hand the democratisation of planning and, on the other hand a regard for planning as a practice which itself can contribute to the further democratisation of society, bears interesting parallels to the French philosopher Jacques Derrida's (2004) thinking on what he calls 'democracy to come'.

In Derrida's sense, 'to come' has nothing to do with the future in itself, but rather with future-oriented demands made in the present, demands that we need more democracy, now – for a better future (cf. Rancière, 2009). For Derrida, to demand democracy, in the present, is to argue that 'there is not yet any democracy worthy of its name' and that 'Democracy remains to come: to engender or to regenerate' (2004: 327). But, as Derrida highlights, this is precisely the crux of the democratic tradition, that it is always about a critique of the present, of the not-there-yet, making the democratic tradition, in practice, a 'inheritance of a promise' (Derrida, 2004: 327). Thus, to call passionately for democracy is to inscribe oneself into a tradition of a longing, of a lack in the present, of a demand for an always to-come. But this is at the same time a future-orientation that makes a difference *in the present*, which demands attention to

> the absolute and unconditional urgency of the here and now that does not wait and on the structure of the promise, a promise that is kept in memory, that is handed down … inherited, claimed and taken up … Not something that is certain to happen tomorrow, not the democracy (national or international, state or trans-state) of the *future*, but a democracy that must have the structure of a promise – *and thus the memory of that which carries the future, the to-come, here and now*. (Derrida, 2004: 330f., emphasis in original)

Thus, the invocation of democracy to come is a statement that, in the present, democracy is not all it could be. But it also indirectly entails the recognition that neither will it ever be. 'Democracy', understood thus, is hence not some form of blissful harmonic end-state, but rather a reaching, a direction of struggle. It is the articulation of a promise that generates a particular orientation towards the future which at the same time entails a demand for ethical justice in the present – a taking to task for present wrongs (cf. also Patton, 2007: 773). What thinking and speaking about 'democracy to come' boils down to for Derrida, is the necessity of grasping the historical irony of the democratic tradition and inheritance, the inevitable betrayal of its promise in every concrete historical attempt at engendering it in practice, but at the same time to nevertheless adhere to this ethos, to argue that 'and yet it is necessary to believe it' and to stand for 'I believe in it, I promise, I am in on the promise ... I am taking action or am at least enduring, now you do the same' (Derrida, 2004: 336).

We suggest that Derrida's figure of 'democracy to come' eloquently captures Patsy Healey's ideas, both of democracy in general and the promise of, and calling for, a democratic planning 'to come', in particular. In the light of the above, it becomes auspiciously apparent that in a way, Healey's thinking is profoundly utopian. 'Utopian thinking' is today generally used as a denominator to describe thought about, and hope for, a better future (Bradley and Hedrén, 2014).[3] In 2015, Patsy Healey writes that 'I have come to understand the planning project as about collective endeavours to shape place qualities to promote better trajectories than might otherwise occur' (2015a). The idea of 'making better places' is crucial, as her 2010 book title illustrates. As Healey reflected already in 1974, to her, '[i]t is possibly this particular quality of planning, that it is associated with a desire for hopeful visions of the future, that makes it on the one hand so complex a phenomenon in the intellectual and practical sense and consequently, such an absorbing object of study' (Healey, 1974a: 230).

Even more directly she bemoaned how '[o]nce founded upon radical and utopian thinking, planning has become institutionalised and enmeshed in methodology', thus losing its 'transforming fervor' (Healey, 1974b: 602). But as intimated above, Healey has never given up on the possibility that it may become so once again, but has rather actively kept alive 'the hope that 'progress', a 'project of becoming', is still possible' (Healey, 1992: 152). So this utopian inclination is where Healey's recurrent interest in planning and 'change' appears to have its keystone, an interest in 'not just ... the relation between 'knowledge and action', as we so often assert in the planning field, but in *that* knowledge and *those* ideas which help to alter trajectories' (Healey, 2015b).

As noted by Bradley and Hedrén (2014: 7–8) contemporary 'utopian' thinking generally does not entail any suggested blueprints for the future, but instead most often proceeds by way of experimentation, 'exploring certain principles and engaging with process, change and critique'. So when we suggest that Healey's thinking can be understood as utopian we by no means imply that she in any way or manner appears to call for a transcendent, ideal world, or suggest specific routes towards this, but rather we understand utopianism as denoting a pointing out of a severe lack in the present and a possible trajectory towards the future that might remedy this lack (cf. Levitas, 2001). In relation to Healey's work, the utopianism it evinces of is anything but a belief in transcendence in the shape of some unattainable ideal society, but rather a shrewd analysis of present problems, and an articulation of a desire for a different future in the form of a more just, open and equitable society with room for the flourishing of all kinds of different ways of life. In this understanding, utopianism thus becomes the corollary of Healey's stated 'capacity to distance herself from an object and feel committed to its improvement' (Hajer, 2009: 138), which perhaps also forms the basis for the fascinating underlying 'tension between critical analysis and normative prescription' throughout Patsy's work, noticed by Huxley (2009: 141).

For Patsy Healey, '[t]he task of the planning enterprise' is thus 'to critically interrogate the governance practices that currently exist and to help ... develop different approaches where these are seen to be failing' (Healey, 2003: 116; cf. also Steele, in Part III.1). Nevertheless, she is keenly aware that this will not just come about by itself. It requires continuous effort and serious struggle. On a careful reading of Healey's work across the decades, the importance of conflict and struggle surfaces as a critical theme. From her perspective, as she puts it in 2006, 'transformation will and *should* involve struggle, as it would re-distribute access to resource flows, regulatory practices and the formulation of policy ideas' (Healey, 2006: 315, emphasis added). In her early days, Healey was a critic of the 'consensus view' of society 'under which planning legislation has been nurtured', (Healey, 1974b: 603). She attacked this idea as a false and dangerous delusion and rhetorically asks, 'is it not helpful to reveal differences and to expose conflicts?' as 'only then can the various groups

3 On the merits of utopian thinking in relation to spatial planning, see Fainstein (2009) and Friedmann (2000).

among us decide whether and with whom it is worth compromising' (Healey, 1974b: 604). She writes that 'the awareness of conflicts and inconsistencies leads to the perception of the need for adjustment and from this perception to adaptive activity, to changes. If the introduction of planning increases the conflicts, increases the awareness of inconsistencies, then it can operate as a contributory generator of adaptive changes' (Healey, 1974a: 224).

Even if a focus on struggle becomes less pronounced over the years, it remains visible in her early terms – 'argumentative planning' and 'planning through debate' – for what the book publishers renamed 'collaborative planning'. She never appears to have completely abandoned this perspective: 'we each have a responsibility to struggle to improve the world we find ourselves in' (2015a). Particularly, she has repeatedly returned to highlight the role and responsibility of both planners and planning academics as 'public intellectuals' (Steele, in Part III.1). Discussing planning scholarship she has argued that 'as academics we are not on the sidelines … We are deeply involved in shaping governance capacity in our localities, and need to think carefully about the social, political and moral dilemmas such involvements generate for us' (Healey, 2010a). The role of the 'community of planning scholars' is thus 'to develop the critical understandings and moral implications of such struggles as they focus on places and their qualities, and to enrich the array of ideas in circulation about what shaping place futures might involve' (Healey, 2015b).

Further, such changes should not be forcefully imposed from above. Instead, it is in specific cases, 'filled with struggles and contradictions as they evolve' where we can find that 'micro-practices, pursued critically and carefully over time, can generate transformative change in place qualities and in political cultures, in a co-production of agency expectations and structuring dynamics. They provide empirical experiences which throw light on the potential for progressive governance practices which could help to sustain and promote more people-centred, open-minded, pluralist polities', enacting 'the normative ambition of moving towards political cultures which centre on the way 'we-the-people', in our various multiplicities, live and seek to live our lives' (Healey, 2012b: 20).

But even if the overarching task as identified by Healey is 'the invention of democratic processes' to collectively shape place qualities in better trajectories, she has, nevertheless, always been careful to point out that 'there are many democracies which might be invented'. For democracy to be worthy of its name in the eyes of Patsy Healey, '[l]earning and listening, respectful argumentation, are not enough' (Healey, 1992: 159). Instead, '[w]e need to develop skills in translation, in constructive critique, in collective invention and respectful action to be able to realise the potential of planning understood as collectively and inter-subjectively addressing and working out how to act in respect of common concerns about urban and regional environments. We need to re-work the store of techniques and practices evolved within the planning field to identify their potential' (Healey, 1992: 159).

Therefore, as Healey has constantly reiterated, what calling for 'democratic planning to come' (as well as planning for 'democracy to come') demands of us is not so much a stable blueprint of an optimal end-state that will never arrive, but rather a constant critical vigilance through the development of 'critical antennae' (Healey, 2012b) which will call for the constant 're-working' of how democracy is performed 'in practice'. Such a conceptualisation of democracy is thoroughly process-focused, it is – in the words of philosopher Bernard Stiegler a 'taking care of *movement*' (Stiegler, 2010: 80, emphasis in original; Metzger, 2014). Or, in Healey's own words, democracy becomes understood as 'a direction to be struggled for' rather than a truth to be asserted (Healey 2012b: 35), based on 'continuously searching for what could be and never being satisfied with what is' (Healey, 2014, pers. comm.).

Connections

The production of this volume has, in itself, involved and engaged many connections. We thank our individual chapter authors for so eagerly accepting our invitation to write, and for producing such valuable contributions in a timely manner. We are also very grateful to those who gave us their time to review and make critical comments on the individual papers. Our thanks to Mona Abdelwahab, Guy Baeten, Kang Cao, Helène Frichot, Catharina Gabrielsson, Maria Håkansson, Kathryn Hegarty, Amy Rader Olsson, Huw Thomas and Hans Westlund for their critical attention, insight and support.

We also thank Louise Forthun for permission to use a copy of her artwork, *Orange Building Site*, 1989–1990, on the cover of this book. Louise is one of Australia's foremost artists who has been painting the city for more than twenty years. She is inspired by the city as social and cultural construct. Louise's images are made with painstakingly hand cut paper stencils which are over sprayed with coloured pigment to produce intricate patterns of shards and slivers which are overlaid on a background of pure colour. Her paintings do not resolve easily into coherent images; neither figure nor ground is forefronted so that the images hover between the objective and non-objective, between discernible forms and fragments of the city and pure abstraction (Corbett Lyon, 2011, http://louiseforthun.com/).

The project could not have been realised without the purchase of copyright permissions to reproduce the selected papers from Patsy Healey. We are grateful to the Dean's Office of the KTH School of Architecture and the Built Environment and to the RMIT School of Global, Urban and Social Studies whose generous grants assisted with purchase of copyright permissions as well as funding the excellent language editing provided by Paul Jones for a number of chapters in the book. Thanks also to Hans Westlund and Göran Cars for getting us through a dire strait in the production process by way of critical assistance in swiftly shoring up the necessary funds on the KTH side.

Our particular thanks to Val Rose at Ashgate Publishing for support and oversight of the volume. It was Val's faith in the project which has enabled it to come to fruition in the way in which it has. We would also like to express our utmost gratitude to Patsy Healey for her, at first hesitant, but with time seemingly increasingly enthusiastic engagement in this project. Of course, Patsy's ideas, suggestions and thoughtful comments were critical to the development of this book.

Jonathan Metzger is also deeply grateful to Patsy Healey for her continuous support and encouragement, and for taking the time and patience to discuss texts and thoughts with me, from my first stumbling attempts at getting any kind of intellectual grasp on my own experiences as a planning practitioner, up until my more recent undertakings as a neophyte planning scholar. I also want to thank Jean Hillier for inviting me into this extremely exciting project. Working and talking with you has taught me so much about the world of planning studies that I otherwise would never have known, and has really helped me in stretching my thinking beyond its previous horizons.

Finally, Jean Hillier wishes to thank Patsy Healey for her inspiration, mentoring and friendship over the last 30 years or so. I could not have had a better role model, as an academic scholar, teacher and leader, but especially as a human being. I owe Patsy Healey an enormous debt of gratitude. This volume represents a small token towards a repayment which can never be complete.

References

Albrechts, L. (2009) 'From strategic spatial plans to spatial strategies', *Planning Theory & Practice*, 10(1): 142–5.

Albrechts L., Healey, P. and Kunzmann, K. (2003) 'Strategic spatial planning and regional governance in Europe', *Journal of the American Planners Association*, 69(2): 113–29.

Bradley, K., and Hedrén, J. (eds) (2014) *Green Utopianism: Perspectives, Politics and Micro-Practices*, New York: Routledge.

Campbell, H. (2009) 'Learning about scholarship ... on a Red London Omnibus', *Planning Theory & Practice*, 10(1): 147–9.

Cockburn, C. (1977) *The Local State*, Oxford: Pluto.

Department of Communities and Local Government (DCLG) (2011) *Localism Act 2011*. Norwich: The Stationery Office. www.legislation.fov.uk/ukpga/2011/20/pdfs/ukpga_20110020_en.pdf [accessed 8 April 2014].

Derrida, J. (2004) 'The last of the rogue states: The "democracy to come", opening in two turns', *The South Atlantic Quarterly*, 103(2): 323–41.

Dewey, J. (1927) *The Public and its Problems*, Athens, OH: Swallow Press.

Douglas, M. (1966) *Purity and Danger: An analysis of concepts of pollution and taboo*, New York: Frederick A. Praeger.

Dyrberg, T. (1997) *The Circular Structure of Power*, London: Verso.
Fainstein, S. (2009) 'Spatial justice and planning', *Justice spatiale | Spatial Justice*, 1/2009.
Fischer, F. and Forester, J. (eds) (1993) *The Argumentative Turn in Policy Analysis and Planning*, London: UCL Press.
Forester, J. (1985) *Critical Theory and Public Life*, Cambridge, MA: MIT Press.
Forester, J. (1989) *Planning in the Face of Power*, Berkeley: University of California Press.
Forester, J. (1997) 'Beyond dialogue to transformative learning: How deliberative rituals encourage political judgement in community planning processes'. In: D. Borri, A. Khakee and C. Lacirignola (eds), *Evaluating Theory-Practice and Urban-Rural Interplay in Planning*, Rotterdam: Springer, pp. 81–103.
Forester, J. (1999) *The Deliberative Practitioner*, Cambridge, MA: MIT Press.
Forester, J. (2009) 'How lucky we are: A glimpse at Patsy Healey's contributions', *Planning Theory & Practice*, 10(1): 134–6.
Frank, A.G. (1966) *The Development of Underdevelopment*, New York: Monthly Review Press.
Frank, A.G. (1967) *Capitalism and Underdevelopment in Latin America.* New York: Monthly Review Press.
Frank, A.G. (1969) *Latin America: Underdevelopment or Revolution*, New York: Monthly Review Press.
Friedmann, J. (1965) *Venezuela: From doctrine to dialogue*, Syracuse: Syracuse University Press.
Friedmann, J. (1967) 'A conceptual model for the analysis of planning behaviour', *Administrative Science Quarterly*, 12(2): 225–52.
Friedmann, J. (1972) *Urbanization, Planning and National Development*, Beverly Hills, CA: Sage.
Friedmann, J. (2000) 'The good city: In defense of utopian thinking', *International Journal of Urban and Regional Research*, 24(2): 460–72.
Geertz, C. (1983) *Local Knowledge*, New York: Basic Books.
Giddens, A. (1984) *The Constitution of Society*, Cambridge: Polity Press.
Graham, S. and Healey, P. (1999) 'Relational concepts of space and place: issues for planning theory and practice', *European Planning Studies*, 7(5): 623–46.
Hajer, M. (1995) *The Politics of Environmental Discourse*, Oxford: Oxford University Press.
Hajer, M. (2009) 'From Oxford onwards', *Planning Theory & Practice*, 10(1): 136–8.
Hall, S. (1986) 'The problem of Ideology-Marxism without guarantees', *Journal of Communication Inquiry*, 10(2): 28–44.
Harvey, D. (1973) *Social Justice and the City*, London: Edward Arnold.
Healey, P. (1974a) 'Planning and change', *Progress in Planning*, 2(3): 143–237.
Healey, P. (1974b) 'The problem of ideology', *The Planner*, 60(3): 602–4.
Healey, P. (1975) 'Urban planning in a Venezuelan city', *Town Planning Review*, 46(1): 63–82.
Healey, P. (1979) 'Networking as a Normative Principle with particular reference to local government and land use planning', *Local Government Studies*, 5(1): 55–68.
Healey, P. (1982) 'Understanding land use planning. The contribution of recent developments in political economy and policy studies'. In: P. Healey, G. McDougall and M.J. Thomas (eds), *Planning Theory: Prospects for the 1980s*, Oxford: Pergamon Press, pp. 180–93.
Healey, P. (1983) *Local Plans in British Land Use Planning*, Oxford: Pergamon Press.
Healey, P. (1985) 'The professionalisation of Planning in Britain: its form and consequences', *Town Planning Review*, 56(4): 492–507.
Healey, P. (1991a) 'Researching planning practice', *Town Planning Review*, 62(4): 447–59.
Healey, P. (1991b) 'Models of the development process: A review', *Journal of Property Research*, 8: 219–238.
Healey, P. (1992) 'Planning through debate', *Town Planning Review*, 63(2): 143–62.
Healey, P. (1994) *Strategic Spatial Planning as a Process of Argumentation.* Centre for Architecture and Planning Research Working Paper. Perth: Curtin University.
Healey, P. (1995) 'Integrating the concept of social diversity into public policy', In: J. Beall, L. McFarlane and R. Ebbs (eds), *Women in the City: Housing, services and the urban environment*, Paris: OECD, pp. 50–56.
Healey, P. (1997/2006) *Collaborative Planning: Shaping places in fragmented societies*, London: Macmillan.
Healey, P. (2003) 'Collaborative planning in perspective', *Planning Theory*, 2(2): 101–23.
Healey, P. (2004) 'From collaborative episodes of institutional transformations: Experiences in city region spatial planning in Europe', Paper presented to ACSP conference, 21–24 October. Portland, Oregon.

Healey, P. (2006) 'Transforming governance: challenges of institutional adaptation and a new politics of space', *European Planning Studies*, 14(3): 299–319.

Healey, P. (2007) *Urban Complexity and Spatial Strategies: Towards a relational planning for our times*, London: Routledge.

Healey, P. (2009) 'The pragmatic tradition in planning thought', *Journal of Planning Education & Research*, 28(3): 277–92.

Healey, P. (2010a) 'Re-thinking the relations between planning, state and market in unstable times'. In: P. Pinho and V. Oliveira (eds), *Planning in Times of Uncertainty*, Porto: FEUP edicoes, pp. 11–22.

Healey, P. (2010b) *Making Better Places: The planning project in the 21st Century*, London: Palgrave Macmillan.

Healey, P. (2011) 'Civic capacity, progressive localism and the role of planning', Nathaniel Lichfield Memorial Lecture, Newcastle University, 6 October 2011. www.ncl.ac.uk/guru/documents/lichfieldflyer.pdf [accessed 18 August 2013].

Healey, P. (2012a) 'The universal and the contingent: Some reflections on the transnational flow of planning ideas and practices', *Planning Theory*, 11(2): 188–207.

Healey, P. (2012b) 'Re-enchanting democracy as a mode of governance', *Critical Policy Studies*, 6(1): 19–39.

Healey, P. (2014) 'Learning the craft of research: A continuing process'. In: Silva, E., Healey, P., Harris, N. and Van Den Broeck, P. (eds), *The Routledge Handbook of Planning Research Methods*, New York: Routledge.

Healey, P. (2015a) 'Finding my way: A life of intellectual exploration and institution-building'. In: B. Haselberger, L. Saija and J. Knight (eds), *The Future of Planning: Personal stories in the evolution of planning thought*, London: Routledge (forthcoming).

Healey, P. (2015b) 'Interpretive understandings in the planning field'. In: M. Bevir and R. Rhodes (eds), *The Routledge Handbook of Interpretive Political Studies*, London: Routledge (forthcoming).

Healey, P. and Barrett, S. (eds) (1985) *Land Policy: Problems and alternatives*, Aldershot: Gower.

Healey, P. and Barrett, S. (1990) 'Structure and agency in land and property development processes: Some ideas for research', *Urban Studies*, 27(1): 89–104.

Healey, P., Davis, J., Wood, M. and Elson, M. (1982) *The Implementation of Development Plans*. Report for the Department of the Environment. Oxford: Oxford Polytechnic.

Healey, P. and Gilbert, A. (1985) *The Political Economy of Land: Urban development in Venezuela*, Aldershot: Gower.

Healey, P., and Gilroy, R. (1990) 'Towards a people-sensitive planning', *Planning Practice and Research*, 5(2): 21–9.

Healey, P. and Hillier, J. (1995) *Community Mobilisation in Swan Valley: Claims, discourses and rituals in local planning*. WP49. Department of Town and Country Planning. Newcastle-upon-Tyne, University of Newcastle.

Healey, P., Hillier, J. and Davoudi, S. (1988) *Gender Issues in Planning Education*. Education for Planning Discussion Paper. Bristol: EPA.

Healey, P., Khakee, A., Motte, A. and Needham, B. (1997) *Making Strategic Spatial Plans: Innovation in Europe*, London: UCL Press.

Healey, P., McDougall, G. and Thomas, M.J. (1982) 'Theoretical debates in planning: Towards a coherent dialogue'. In: P. Healey, G. McDougall and M.J. Thomas (eds), *Planning Theory: Prospects for the 1980s*, Oxford: Pergamon Press, pp. 5–22.

Healey, P., McNamara, P., Doak, A. and Elson, M. (1985) *The Implementation of Planning Policy and the Role of Development Plans*. (2 volumes). Report for the Department of the Environment, Oxford: Oxford Polytechnic.

Healey, P., McNamara, P., Elson, M. and Doak, J. (1988) *Land Use Planning and the Mediation of Urban Change*, Cambridge: Cambridge University Press.

Healey, P. and Underwood, J. (1978) 'Professional ideals and planning practice', *Progress in Planning*, 9(2): 73–127.

Healey, P. and Upton, R. (eds) (2010) *Crossing Borders: International exchange and planning practices*, Abingdon: Routledge.

Healey, P. and Williams, R. (1993) 'European planning systems: Diversity and convergence', *Urban Studies*, 30(4/5): 201–20.

Hillier, J. (2005) 'Straddling the poststructuralist abyss: Between transcendence and immanence?' *Planning Theory*, 4(3): 271–99.

Hillier, J. (2007) *Stretching Beyond the Horizon: A multiplanar theory of spatial planning and governance*, Aldershot: Ashgate.

Hillier, J. and Healey, P. (2008) 'Introduction to Part II', *Political Economy, Diversity and Pragmatism, Critical Essays in Planning Theory*, Volume 2, Aldershot: Ashgate, pp. 205–13.

Huxley, M. (2009) 'Theoretical debates, planning practice and spatial processes', *Planning Theory & Practice*, 10(1): 139–42.

Innes, J. (1985) *Federal Requirements and Local Planning Capacity*. Berkeley, CA: Institute of Urban and Regional Development, University of California, Berkeley.

Levitas, R. (2001) 'For utopia: The (limits of the) utopian function in late capitalist society'. In: B. Goodwin (ed.), *The Philosophy of Utopia*, London: Frank Cass, pp 25–43.

Lukes, S. (1974) *Power: A radical view*, London: MacMillan.

Massey, D. (1984) *Spatial Divisions of Labor: Social structures and the geography of production*, New York: Methuen.

Metzger, J. (2014) 'Spatial planning and/as caring for more-than-human place', *Environment and Planning A*, 46(5): 1001–11.

Moseley, M. (2003) *Rural Development: Principles and practice*, London: Sage.

O'Connor, J. (1973) *The Fiscal Crisis of the State*, New York: St. James' Press.

Offe, C. (1974) 'Structural problems of the capitalist state'. In: K. von Beyme (ed.), *German Political Studies*, Vol. 1, London: Sage, pp. 31–57.

Offe, C. (1975) 'The theory of the capitalist state and the problem of policy formation'. In: Lindberg, L., Alford, R., Crouch, C. and Offe, C. (eds) *Stress and Contradiction in Modern Capitalism*, Lexington: D.C. Heath, pp. 125–44.

Patton, P. (2007) 'Derrida, politics and democracy to come', *Philosophy Compass*, 2(6): 766–80.

Pløger, J. (2010) 'Urbanicity, (neo)vitalism and becoming'. In: J. Hillier and P. Healey (eds), *Conceptual Challenges for Spatial Planning. The Ashgate Research Companion to Planning Theory*, Farnham: Ashgate, pp. 319–41.

Rancière, J. (2009) 'Should democracy come? Ethics and politics in Derrida'. In: P. Cheah and S. Guerlac (eds), *Derrida and the Time of the Political*, Durham, NC: Duke University Press, pp. 274–88.

Thomas, H. and Healey, P. (1991) 'Preface'. In: H. Thomas and P. Healey (eds), *Dilemmas of Planning Practice*, Aldershot: Avebury, pp. x–xiv.

Vigar, G., Healey, P., Hull, A. and Davoudi, S. (2000) *Planning, Governance and Spatial Strategy in Britain*, London: Macmillan.

Watson, V. (2014) 'Co-production and collaboration in planning – the difference', *Planning Theory & Practice*, 15(1): 62–76.

PART II
Normative Perspectives

Introduction to Part II
Normative Perspectives

Jean Hillier

Patsy Healey has always argued that the planning project is strongly normative. Normativity incorporates issues of value, good, ought, rationality, obligation and so on (Jean Hillier, Chapter 5). Normativity distinguishes the value side of fact/value distinctions and the 'ought' aspect of is/ought distinctions. It is expressed in a 'double way' (Healey, 2015a): both in the individual and collective, personal and professional ethics of practices (expressed, for example, in professional Codes of Practice) and in the search for better outcomes (expressed, for example, through social and environmental justice). Normativity tends to reflect the cultural 'norms', or shared values, of social institutions (such as 'family values', values of the planning profession and so on). Normative statements make claims about how things should or ought to be: what is good or bad, right or wrong (e.g. walkability, mixed use zoning, setback distance, privacy etc.). In the terminology of standards, including those which permeate development assessment planning, normative means prescriptive: i.e. what ought to be done in order for something to comply with the standard. Normativity, therefore, is concerned both with substance ('doing the right thing') and process ('doing it right'). For Patsy Healey, these two dimensions of normativity are fundamentally co-constituted. Of course, what constitutes 'the right thing' or 'the good' and how one 'does' right or good – or, for that matter, what constitutes and how one does 'bad', 'wrong' or even 'evil' – are matters of morals and ethics, as the chapters in this Part by Jean Hillier and Francesco Lo Piccolo discuss.

Norms may be regarded as rational, instrumental necessities: an operational style of working. The ideas of planners in practice working with norms and value judgements, mediating between a plurality of wide-ranging, often conflicting, interests, have been central to Patsy Healey's academic work from the 1970s to the present. A key question for her is 'in whose interests are these operational styles deployed?' (see Healey, 1979). For Healey, planning is always-inevitably normative, though practitioners may be unconscious of the norms that they invoke or serve in their actions and inactions. As Healey (1974a; Healey and Zwirner, 1973) stresses, there is a need for practitioners and other actors to make themselves conscious of, and reflect on, the norms (or ideologies) they perform.

In her empirically-based 1978 paper with Jacky Underwood (sections reproduced in this Part), Patsy Healey demonstrates that, in preparing policies or strategies, planners may call upon common norms (such as a common concern with social policy) to build alliances with other professional actors and, in turn, to be influenced by what other actors expect of them (pp. 92–3). However, broad common norms may exist only rarely in today's practice, and planners may have to select amongst competing norms held by different groups of actors.

In this Part, Normative Perspectives, we have brought together discussion of normative ethical concepts, such as those relating to duty, rule-following, moral standards and so on (in Chapter 4 by Francesco Lo Piccolo and Chapter 5 by Jean Hillier), with broader issues of professionalisation (in Chapter 6 by Mee Kam Ng), the role of politics (all three chapters) and democratisation.

How planning is performed in practice has always been a crucial element of Patsy Healey's research, stemming from her own practice experience in London in the 1960s. She has embraced empirical work 'in the field' from her doctoral studies onwards. As she writes with Huw Thomas: 'accounts by practitioners of how they approach their work are invaluable' (1991: x). In that volume, (Thomas and Healey, 1991: xiii) Patsy Healey admits to being directly influenced by the critical pragmatism of John Forester, who brought the work of Jürgen Habermas to the attention of the planning world in the 1980s. Apart from a passing interest in consensus-building (see this volume, Part III.2), Healey was more interested in Forester's material on listening to, and critically interpreting, the work of practitioners. This was – and still is – the aspect of critical pragmatism which 'appreciates multiple and contingent or evolving forms of knowledge, local and scientific, initial opinion and considered judgment' (Forester, 2013: 6) which interests her. Healey was also influenced by Luigi Mazza's (1986) ideas that plans should be regarded as 'sets of technical and political claims, to be understood in the context of relations external to the text' (Healey, 1996c: 264). The political may be as important (if not more so) than the technical.

Patsy Healey had recognised this point already in her doctoral work in South America (1974b). She also emphasised (1974b: 207) that 'social order' is more than mere physical order, and engages moral order, long before Fischer (1990) argued that policy analysis involves both rational-technical and normative claims. In 1992, Healey commented that 'we now appreciate that technical knowledge is inevitably infused with biases reflecting particular interpretive predictions and normative values' (p. 9). Drawing upon Habermas' (1984) three principles of reasoning – rational-technical, moral and aesthetic-expressive – she claimed 'the domain of moral philosophy' as relevant to planning practice (1992, 1996c).

Patsy Healey's papers reproduced in this Part illustrate the long trajectory of her concern with ethical and political normativities, embracing both the planning profession and other practitioners in the planning project assemblage. The contributions by Francesco Lo Piccolo, Jean Hillier and Mee Kam Ng find resonance with, and draw inspiration from, Healey's work, summed up beautifully in Lo Piccolo's footnote revealing his joy at discovering 'the same intellectual yearning and ethical tensions which were inspiring some of my [own] research experiences and questions'.

In the remainder of this Introduction, I introduce the key elements which the various papers address: the planning profession, knowledge, normativity, politics and democratisation. I aim to demonstrate how far thinking has come since Davies' (1972) assertion that 'the planning milieu is one long Gun Fight at the OK Corral' (cited in Healey, 1991a: 15).

The Planning Profession

From some of her earliest published research, Patsy Healey has critically engaged with planning practitioners, asking what they do, how and why they do it (e.g. Healey, 1975a, 1976, 1985; Healey and Evans, 1974; Healey and Fudge, 1983; Healey and Underwood, 1976, 1977, 1978). By 1985 she had become quite critical of the institutionalised form of professionalisation in planning, suggesting that it 'had encouraged the homogenisation of planning expertise' (1985: 503) and inhibited the development of a flexible capability to respond to changing circumstances. Healey recognised the need for flexibility some 25 years ahead of a more widespread call for dealing with uncertainties in a flexible, adaptive manner (see, for example, Balducci et al., 2011).

She also observed (in 1985) that a 'current association' of planning with large state bureaucracies both reduced the demand for, and the prestige of, planners. As is well known, the efforts of Margaret Thatcher's Conservative government in Britain (1979–1990) to 'roll back the State' significantly damaged both demand and prestige in a way from which it had only just begun to recover when further attacked by Eric Pickles (Secretary of State for Communities and Local Government since 2010), who famously referred to planning departments as 'the last bastion of communism and sheer bloody-mindedness' (in anon, 2010). Unsurprisingly, between 2006/7 and 2009/10 some two-thirds of English local planning authorities reported a decline in the numbers of planning staff, with over 29% of authorities losing more than 10% of their staff (Ove Arup/DCLG, 2010). This trend can potentially lead to a sense of demotivation and depression amongst planners, though as Jackson (2013) has found, several practitioners, amongst whom I would include Patsy Healey, continue to 'chip away at stuff', trying to make a difference as best they can.

Describing herself as an 'anthropological observer' or 'voyeur' (Healey, 1991b: 450), Patsy Healey has empirically studied planners in practice in many jurisdictions around the world. In doing so, she has recognised two key issues: that much so-called 'objective' research is implicitly guided by researchers' values and is, therefore, inherently subjective (Healey, 1991b) and that ideas 'copy/pasted' unthinkingly across different countries are rarely successful (Healey, 2010b).

Influenced by John Friedmann's (1987) concept of planning as 'knowledge for action in the public domain', Patsy Healey and Rose Gilroy (1990: 26, reproduced in this Part) ask 'what knowledge and what values does and should a planner bring forward? What grounds for raising such knowledge and values does a planner have?'. They suggest that planners need to develop an 'epistemological consciousness', thinking through what constitutes 'knowledge', how it is valued and so on; questions which remain as relevant today as in 1990.

Healey's analysis of Ted Kitchen's professional 'day' as then-Acting County Planning Officer in Manchester (Healey, 1992) brilliantly portrays how what a 'savvy' planner, such as Kitchen, says and does, is structured by a framework of rules and conventions, but he nevertheless manages to move skilfully between differing assemblages and networks of knowledges and values. Actualisation of practice outcomes depends, as Healey (1992: 19) writes, 'on the ways in which opportunities and constraints are perceived, debated, and confronted in daily practice'.

In her analysis of strategy-making in Amsterdam, Milan and Cambridge (2007), Patsy Healey recognises the relational nature of planning practice and the importance of contextual, institutional specifics. Often, those identified as 'planners' (by themselves and/or other actors) do not sit in municipal planning offices, as they may have done some 20 years previously. Healey (2007: 177) found 'planning officials' in a range of governance agencies, in commercial companies and lobby groups, as consultants and as academics (see also, Healey, 2008a), even as politicians. Increasingly, a range of other actors is involved in planning decision making across the various scales of government, political, business and interest group networks and so on, in forms of network governance, as I discuss below. I end this section with a quote from Patsy Healey (1985: 5010) which remains as pertinent today as when it was written: 'no government can dismantle the planning system without devising an alternative mechanism to manage the coordination of spatial externality effects and the mediation of conflicts over land use change and development'.

Knowledge

The conceptualisation of planning as an activity – planning practice – has generally shifted over the last 30 years or so away from that of a government-centred activity pursued through formally constituted administrative structures and legally authoritative, enforceable plans and standards. It has become increasingly recognised that planning practice is actively constituted by a wide range of relational assemblages of human and non-human actors with differing values, knowledges and expectations.

Influenced by Friedmann's (1987) definition of planning (above) as involving the translation of knowledge to action, Patsy Healey (2008a: 861) asks what kinds of knowledge and what mechanisms of translation are involved: i.e. what is the relation of knowledge to action. She (Healey and Hillier, 2008: xvii) outlines a pragmatist-inspired view that understandings, values and 'goals' should not be derived from 'pure forms' or scientific 'laws', but rather from 'continuous socially situated activity in probing inquiry, collective sense-making and the testing out of ideas to see if they "work" and mobilise attention'. As such, she is concerned, not only with knowledge per se, but with learning: 'the social processes through which what people learn and get to know is validated and filtered; with whose experiences, capacities and "imaginaries" get to count' (2013a: 284).

Patsy Healey has always been intrigued by questions concerning what is distinctive about the knowledge mobilised in planning activities. Questions of what is its substance and where does it come from, recur throughout much of her work. As she points out, there are two dimensions to these questions: firstly, content, or what planners need to know or know about; and secondly, what counts as knowledge, what does it mean to 'know about' something? (Healey and Hillier, 2008: xvii). Broad ontological and epistemological questions such as these, which Healey continually raises, deserve more attention in planning theory and practice than they generally receive.

Patsy Healey (2007, 2008a) develops a four-'box' matrix of forms of knowledge in which, on one axis, she contrasts knowledge which is explicitly systematised (e.g. through codification, such as in Mee Kam Ng's example of the Basic Law in Hong Kong, in Chapter 6), with experiential or practical knowledge, and on the other axis, tacit knowledge is contrasted with explicit knowledge (such as best-practice guides or the Hong Kong planning standards and guidelines as explored by Mee Kam Ng). Recognising that this matrix is merely one possible way of reading events, Healey argues that it offers an interpretive epistemology to explore potentially generative encounters between multiple rationalities (2007: 248).

Healey and Underwood (1978) were concerned with 'the nature and problem of planning knowledge' in a profession that was undergoing significant restructuring in England. Their empirical research was grounded in the apparent contradiction between theoretical ideal procedural models of planning practice and what practitioners actually seemed to be doing. Whilst this theory/practice relation has hopefully changed in the last 35 years, with planning theories being increasingly grounded in practice, there seems relatively less change in relation to the authors' findings that 'the most important sources of planners' knowledge … appeared to be experience' (1978: 109). Healey and Underwood were also among the first to recognise that planning is concerned with persuasion – persuading an audience (of elected representatives, other officers, local residents and so on) to agree with something, whether a development assessment, policy or other recommendation – and the importance of communication in this. As Patsy Healey concluded, regarding Ted Kitchen's 'planner's day', it was his communicative skill which made his knowledge effective in action (1992: 19). In her chapter below, Mee Kam Ng also argues that communication lies at the core of the problems besetting planning practice in Hong Kong. She suggests that practitioners need to begin to 'tread in unknown territory' to work with different actors and to 'learn in public' how to engage in intelligent and respectful co-production of governance processes.

Recognising that knowledge is constructed in social practices, Patsy Healey became interested in the politics of knowledge production and the struggles over meaning and which knowledge counts in governance processes. Mee Kam Ng offers the interesting example from Hong Kong of the Confucianist idea of planners as intellectual elites and what this implies for practice. Healey (2007, 2008a) demonstrates that what passes as 'knowledge' is 'created' through processes of debate, encounter and challenge rather than simple acceptance of formalised, expert, scientific accounts.

Such processes are permeated with issues of ethics. Reflecting on her own empirical research experiences with planning practitioners, Patsy Healey writes about the importance of building trust and respect with the practitioners involved. Especially when working with confidential information, she reflects that: 'the socially responsible researcher must act morally' (1991b: 457). A decade and a half later, she similarly comments that academic research and consultancy 'raises important ethical issues for all academics who get involved in policy work, demanding careful attention to why their advice is being sought, how it may be used, and the consequences of such use' (2008: 875).

Normativity

Throughout her career, Patsy Healey has been interested in the normative questions inherently involved in research practices (e.g., 1991b) as well as in those of professional planning. With Jacky Underwood (1978), she explored norm-reinforcing and norm-challenging networks and behaviours and the inherent 'normative orientation of the planning idea' (Healey, 2012b: 200). Patsy Healey's investigations of the micro-practices of governance over the past 45 years have led her to question practitioners' and researchers' claims of objectivity (e.g. 1991b) and of 'universally valid' (2012b) ideas. As Hillier discusses in Chapter 5, Healey warns practitioners to be cautious about unthinkingly accepting norms and 'general principles', especially if they have been imported as best practice from other jurisdictions.

In her early work, Patsy Healey asked 'can we act without belief? … [C]an we act without accepting as given certain values or assumptions about the way the world is organised[?]' (1974a: 603). These issues reappear in Healey's theoretical and practical work throughout her career. The volume, co-edited with Huw Thomas, *Dilemmas of Planning Practice* (1991), for instance, focuses on the ethics of individual practitioners' conduct in their everyday struggles with the often-conflicting values of different actors, including themselves. In the concluding chapter, the authors suggest that practitioners should cultivate a self-reflective practice (after Schön, 1983), akin to having an invisible critical 'ethical monitor' looking over their shoulders, so they do not behave 'stupidly', as Hillier explains in her contribution below.

Healey and Gilroy (1990, reproduced in this Part) suggest that a 'people-sensitive' form of planning requires six critical forms of consciousness, including epistemological, self-reflective and ethical consciousnesses. Whilst Healey rejects universalist, transcendent norms of justice and so on, she has argued that some form of systemic, institutional design mechanism or rules (1997) are necessary in the specification of requirements and rights (1995: 269), if decisions are to be taken in an inclusionary, communicative, ethical manner. As such, respect is an issue to which Patsy Healey returns on several occasions (1991b, 1993, 1996, 2010a, 2011, 2012a; Healey and Gilroy, 1990; Healey and Hillier, 1996), including respect for different arguments, positions, feelings etc. ('adversaries' not 'enemies'), respect for the role of formal government, respect for those who do formal government work, and respectful interactions between them (2012a: 31, reproduced in this Part).

Inspired by the pragmatist tradition of 'the good' as a potentiality rather than some 'transcendent being or scientific laws of societal development or formalised ideology' (Healey, 2009: 281), Healey (2012b, 2015) argues that 'good' place qualities and 'better' (rather than 'good') ways of doing governance are contextual and contingent potentialities rather than being set in stone in some 'Tablet' (2012b: 201) of 'good governance' or best practice indicators. An ethics of purpose of planning practice for Patsy Healey involves 'encourag[ing] people of all kinds to flourish while addressing the tensions of conflicting points of view and lifestyles', an ethics which she practices in her capacity as Chair of the Glendale Gateway Trust in North East England (2013b). She has recently expanded the concept of flourishing to include non-human species (2012b) in recognition of the ways in which planning activities affect non-humans in adverse ways.

Both Lo Piccolo's and Hillier's chapters in this Part explore and develop the normative ideas discussed above. Lo Piccolo, in particular, is concerned with practising a normative ideal in a non-ideal world in which some actors are not democratically recognised, where power-plays are often violent, insincerity of communication is rife and where the underlying nature of conflicts may not be apparent. Drawing on the plays of William Shakespeare, Lo Piccolo warns practitioners to heed the fate of Polonius who, not acting with the clichéd integrity which he demands of his son, is played for a fool, to die by Hamlet's hand. Shakespeare employs the character of wise fool, such as Lear's fool, Touchstone in *As You Like It* and Feste in *Twelfth Night*, to offer insightful commentary on particular issues and to destabilise conventional wisdom. As Lo Piccolo (Chapter 4) and Metzger (2010) suggest, planning, too, may need voices to speak unwelcome truths, pointing out foolish (or 'stupid', Hillier) attitudes and behaviours.

Hillier finds several resonances between the normative work of Patsy Healey and that of French philosopher, Gilles Deleuze. In the non-predictabilities of planning practice, chance (contingency) is coupled with necessity, as Hillier illustrates through the event of a dice-throw, resonating with Healey's (1974b) game analogy. Addressing Healey's (1991a: 33) question of whether planners are to be distinguished by what they know and do (knowledge and skills) or by their approach to what they know and do (values and ethics), Hillier discusses the problem of judgement from a Deleuzean-inspired ethical understanding of law as immanent, spatial and posthuman. She concludes that, for both Healey and Deleuze, the aim of practitioners is no longer to control a field, such as planning, but to intervene and experiment.

Politics

Patsy Healey has always recognised the politicised nature of planning practice. Knowledges and representations of place are filtered and reframed, 'called into consciousness' (2002) for certain effect. Healey and Gilroy (1990) argue that planners need to develop a political consciousness of power relations operating in the networks and assemblages in which they work. As Allmendinger (2011) points out, for example, development assessment has become often highly politicised as the implications of 'development' in general, development plans and development applications in particular, become apparent to local communities and other (such as environmental) actors.

Strategy-formation is also inherently small 'p' political (Reade, 1982). As Healey's case examples of the Newcastle-upon-Tyne 'Going for Growth' city strategy (Healey, 2005) and the 'Grainger Town Project' (Healey et al., 2003) illustrate, strategy-making and implementation are fraught with conflict. In Newcastle's 'West End', for instance, citizens were driven to 'waving placards of protest and releasing red balloons of distress about plans to demolish their locales' (Healey, 2005: 191). In contrast to such grassroots 'activism', the Grainger Town Project highlighted 'the insidious way that powerful interests outside and inside a locality can impose their agendas unless there are countervailing grass-roots forces to challenge and limit their operations' (Healey et al., 2003: 86). If planning practitioners are to have their recommendations and ideas listened to, discussed and accepted by elected representatives, citizens' groups, business interests and other professionals, they need the ability to position their ideas and give them meaning in the context of a strategic frame, rather than leaving them 'floating in the ether' (Healey, 2007: 192). Planners need to mobilise their ideas and activities, to create moments of opportunity in which to articulate persuasively the strategic focus behind their particular positions and actions (Healey, 2007: 275, Figure 9.1). As Healey says, sitting back won't work. She points out that attention must be paid to using 'persuasive and seductive power' in order that other actors buy in to the planners' message.

As Mee Kam Ng concludes, it is no longer feasible for planners to hide behind the façade of technocracy. Although unable to 'make trouble on the streets' (Mee Kam Ng, Chapter 6; also Hillier, 2002: 234), planners may have windows of opportunity to 'gain the ear' of those with direct decision-making influence.

Democratisation

Francesco Lo Piccolo (Chapter 4) declares Patsy Healey as a source of inspiration for his interest in the concept of planning as a way to enhance democracy. Patsy would add that, in turn, democracy should be a way to enhance planning. From an original research concern with political economic analysis of the role of capital in structuring events, Patsy Healey has graduated towards a more agentic approach (influenced by Giddens' [1984] concept of structure/agency structuration) for 'democratic plan-making in a plural democracy' (Healey, 1995). In the 1990s, she was interested (1996a, 1996b), as were many other planning theorists, in the potential of consensus-building, influenced by John Forester's and Judith Innes' work with Jürgen Habermas' theory of communicative action (Forester, 1989; Innes, 1994, 1995), but this was rapidly tempered for Healey by a recognition that consensus-formation in practice is context-dependent, relative to prevailing governance culture and power dynamics. If consensus is 'achieved' (on whose terms?), it may well be 'a fragile, incomplete and contestable outcome, which may or may not have enduring effects in structuring subsequent relations' (Healey, 2003: 114). Patsy Healey, therefore, advocates an inclusive, democratic, collaborative approach in a 'collective public realm' (2002) which embraces what Dunn (2005: 138) describes as 'a stumbling, myopic blend of quarrelling and shared exploration of the inescapable issue of how to sustain everyday lives together as agreeably as possible' (cited in Healey, 2012a: 24).

Referencing her earlier work on networking as a normative principle for planning (1979), Patsy Healey (2008b, 2011, 2012a) has worked with ideas of network governance which employ the institutional context of practices. She (2011) is aware of a problematic legacy of Habermasian communicative action in many mainstream network governance ideas, however, as well as the inherent problems of adopting formulaic approaches 'rather than encouraging people to think about the issues they should address and the questions they should ask as they design and practice the work of place governance in specific contexts' (2011: 7). As Healey poignantly continues, 'the quality of governance lies in its practices, not its formal principles'. She is interested in 'developing a civic capacity to promote integrated, richly informed and pluralistic conceptions of place qualities and connectivities as these are evolving into the future' (2011: 1). This requires recognition of conflict and power struggles rather than attempting to smooth them over in a 'quick fix' (2011: 10). She favours a more communitarian (2012a: 29), associative form of governance which reconfigures relations between formal governance structures and those of civil society.

Patsy Healey 'walks the talk' through her work with the Glendale Gateway Trust. She herself embodies the set of 'qualities to foster' which she advocates (2011, Table 1; 2012a, Table 1) for more progressive place-focused governance:

- Recognition of a shared, but diverse and often conflictual political community, in which distribution fairness matters;
- Recognition and respect of multiple identities and ways of life, as well as commonalities;
- Respect for different arguments, positions, feelings etc. ('adversaries' not 'enemies');
- A practice of 'intelligent', multi-sided discussion of issues – learning – and reasoning 'in public';
- Recognition of the complex relations in which 'parts' are connected to wider 'wholes' in open, often diffuse, ways;

- Critical respect for the role of formal government, while recognising the complexity of interactions between the spheres of state and society;
- Respect for those who do formal governance work and perform 'public services';
- Rich, responsive and respectful interactions between all levels and sectors of formal government.

Patsy Healey's empirical experience with governance micro-practices in the shape of the Trust and its interactions with public and private sector institutions, exemplify what she calls 'democracy-in-action' (2012a). As she explained in *The Guardian* newspaper, her work with the Trust 'has made me realise that, in practice, urban and rural-specific knowledge is not as important as the idea of the place and how its features can be used to benefit the community' (Healey, 2013b).

Place-focused, people-centred practices are critical not only in North East England, but the UK as a whole and beyond, as well illustrated by Mee Kam Ng's chapter on Hong Kong. Mee Kam calls for the re-design of the planning system and its practice to provide more meaningful arenas for consideration of citizens' views.

Conclusions

Throughout her career, as illustrated by the papers reproduced in this Part, those referenced in this Introduction and many others, Patsy Healey has demonstrated a strong and unwavering concern with the question of what constitutes the planning project: what knowledges, whose knowledges, and how they might be brought into practice, democratically and ethically. Her work, both as an academic researcher and as a practitioner, in London and with the Glendale Gateway Trust, engages her understanding of planning as inherently a political practice undertaken in a relational network or assemblage of human and non-human actors and structures.

Patsy Healey recognises that what constitutes 'planning activity' is 'an unclear and continuously shifting concept' (Healey and Hillier, 2008: xiv). The foci, approaches and values of planning are contested and change over time and place. There cannot be a transcendent, universal of 'good practice' or 'justice' in planning. Rather, Healey claims that planning attempts 'to promote the "betterment", however understood, of the human condition[1] as experienced through time and as lived in a wider world of human and non-human relations and forces' (Healey and Hillier, 2008: xv).

The idea of 'betterment' is central to Healey's work, *Making Better Places* (2010a). In the conclusion to this book, she defines 'better', with regard to what she terms 'the planning project', as 'improving the conditions of life for people' (2010a: 226) in future generations as well as present ones and as 'about how we relate to all others who inhabit the world with us and to the broader natural forces that shape our planetary existence' (2010a: 226). A big ask! However, as she (2011) explains, this task involves not only those trained as 'planners', but the collective efforts of many people, in many different arenas.

As the chapters by Francesco Lo Piccolo, Jean Hillier and Mee Kam Ng demonstrate, there are no quick fixes, best practices or universal formulae which can be applied if inclusionary, ethical planning is to take place. Part of Patsy Healey's legacy is to engage us, as researchers, practitioners and educators, to be reflexive about who we are, what we do, why and how we do it.

References

Allmendinger, P. (2011) *New Labour and Planning*, Abingdon: Routledge.

Anon (2010) 'Pickles: City planning offices are "bastion of communism"', *The Birmingham Post*. 9 October. www.birminghampost.co.uk/news/local-news/eric-pickles-labels-council-planners-3926761 [accessed 18 March 2014].

Balducci, A., Boelens, L., Hillier, J. Nyseth, T. and Wilkinson, C. (2011) 'Introduction: Strategic spatial planning in uncertainty: theory and exploratory practice', *Town Planning Review*, 82(5): 481–501.

1 I would add non-human condition too.

Davies, J.G. (1972) *The Evangelistic Bureaucrat: A study of a planning exercise in Newcastle-upon-Tyne*, London: Tavistock.
Dunn, J. (2005) *Setting the People Free: The story of democracy*, London: Atlantic Books.
Fischer, F. (1990) *Technocracy and the Politics of Expertise*, London: Sage.
Forester, J. (1989) *Planning in the Face of Power*, Berkeley: University of California Press.
Forester, J. (2013) 'On the theory and practice of critical pragmatism: Deliberative practice and creative negotiations', *Planning Theory*, 12(1): 5–22.
Friedmann, J. (1987) *Planning in the Public Domain*, Princeton, NJ: Princeton University Press.
Habermas, J. (1984) *The Theory of Communicative Action Vol 1*, Boston, MA: Beacon Press.
Healey, P. (1974a) 'The problem of ideology', *The Planner*, 60(3): 602–4.
Healey, P. (1974b) 'Planning and change', *Progress in Planning*, 2(3): 143–237.
Healey, P. (1975) *Planners' Debates on Policy Work in a London Borough*. Oxford Working Paper No. 22. Oxford: Department of Town Planning, Oxford Polytechnic.
Healey, P. (1976) 'London Borough Planning Policies: Co-ordinated Strategy or Policy Patchwork?' CES Structure Planning Conference Papers, CES CP16, London: CES.
Healey, P. (1985) 'The professionalisation of planning in Britain: Its form and consequences', *Town Planning Review*, 56(4): 492–507.
Healey, P. (1991a) 'Debates in planning thought'. In: H. Thomas and P. Healey (eds), *Dilemmas of Planning Practice*, Aldershot: Avebury, pp. 11–33.
Healey, P. (1991b) 'Researching planning practice', *Town Planning Review*, 62(4): 447–59.
Healey, P. (1992) 'A planner's day: Knowledge and action in communicative perspective', *Journal of the American Planning Association*, 58(1): 9–20.
Healey, P. (1993) 'Planning through debate: the communicative turn in planning theory'. In: F. Fischer and J. Forester (eds), *The Argumentative Turn in Policy Analysis and Planning*, Durham, NC: Duke University Press, pp. 233–53.
Healey, P. (1995) 'Discourses of integration: Making frameworks for democratic urban planning'. In: P. Healey, S. Cameron, S. Davoudi, S. Graham and A. Madanipour (eds), *Managing Cities: The new urban context*, Chichester: Wiley, pp. 251–72.
Healey, P. (1996a) 'The communicative turn in planning theory and its implications for spatial strategy-making', *Environment & Planning B: Planning & Design*, 23(2): 217–34.
Healey, P. (1996b) 'Consensus-building across difficult divisions: New approaches to collaborative strategy-making', *Planning Practice & Research*, 11(2): 209–18.
Healey, P. (1996c) 'The communicative work of development plans'. In: S. Mandelbaum, L. Mazza and R. Burchell (eds), *Explorations in Planning Theory*, New Brunswick, NJ: Centre for Urban Policy Research, Rutgers, The State University of New Jersey, pp. 263–88.
Healey, P. (1997) *Collaborative Planning: Shaping places in fragmented societies*, Basingstoke: Macmillan.
Healey, P. (1998) 'Building institutional capacity through collaborative approaches to urban planning', *Environment & Planning A*, 30(9): 1531–46.
Healey, P. (2000) 'Planning theory and urban and regional dynamics: A comment on Yiftachel and Huxley', *International Journal of Urban and Regional Research*, 24(4): 917–21.
Healey, P. (2002) 'On creating the "city" as a collective resource', *Urban Studies*, 39(10): 1777–92.
Healey, P. (2003) 'Collaborative planning in perspective', *Planning Theory*, 2(2): 101–23.
Healey, P. (2005) 'Place, identity and governance: transforming discourses and practices'. In: J. Hillier and E. Rooksby (eds), *Habitus: A sense of place* (2nd edition), Aldershot: Ashgate, pp. 189–218.
Healey, P. (2007) *Urban Complexity and Spatial Strategies: Towards a relational planning for our times*, London: Routledge.
Healey, P. (2008a) 'Knowledge flows, spatial strategy making, and the roles of academics', *Environment and Planning C, Government and Policy*, 26(5): 861–81.
Healey, P. (2008b) 'Civic engagement, spatial planning and democracy as a way of life', *Planning Theory & Practice*, 9(3): 379–82.
Healey, P. (2009) 'The pragmatic tradition in planning thought', *Journal of Planning Education & Research*, 28(3): 277–292.

Healey, P. (2010a) *Making Better Places: The planning project in the Twenty-First Century*, Basingstoke: Palgrave Macmillan.
Healey, P. (2010b) 'Introduction: The transnational flow of knowledge and expertise in the planning field'. In: P. Healey and R. Upton (eds), *Crossing Borders: International exchange and planning practices*, London: Routledge, pp. 1–25.
Healey, P. (2011) 'Civic capacity, progressive localism and the role of planning' Nathaniel Lichfield Memorial Lecture, Newcastle University, 6 October 2011. www.ncl.ac.uk/guru/documents/lichfieldflyer.pdf [accessed 18 August 2013].
Healey, P. (2012a) 'Re-enchanting democracy as a mode of governance', *Critical Policy Studies*, 6(1): 19–39.
Healey, P. (2012b) 'The universal and the contingent: Some reflections on the transnational flow of planning ideas and practices', *Planning Theory*, 11(2): 188–207.
Healey, P. (2013a) 'Learning the city: Knowledge and translocal assemblage', *Planning Theory & Practice*, 14(2): 283–5.
Healey, P. (2013b) 'Working lives: the local professor', *The Guardian*. 9 January. http://www.theguardian.com/local-governmjent-network/2013/jan/08/working-lives-local-professor [accessed 8 March 2014].
Healey, P. (2015) 'Interpretive understandings in the planning field'. In: M. Bevir and R. Rhodes (eds), *The Routledge Handbook of Interpretive Political Studies*, London: Routledge (forthcoming).
Healey, P., de Magalhães, C., Madanipour, A. and Pendlebury, J. (2003) 'Place, identity and local politics: Analysing initiatives in deliberative governance'. In: M. Hajer and H. Wagenaar (eds), *Deliberative Policy Analysis*, Cambridge: Cambridge University Press, 60–87.
Healey, P. and Evans, V. (1974) *Approaches to a Research Problem: Theory and the Study of Planners*. Kingston Planning Workshop Paper No. 3. Kingston upon Thames: School of Planning, Kingston Polytechnic.
Healey, P. and Fudge, C. (1983) *Local Planning in Practice (3): Camden, Manchester*. SAUS Working Paper No. 35. Bristol: School of Advanced Urban Studies.
Healey, P. and Hillier, J. (1996) 'Communicative micropolitics: A story of claims and discourses', *International Planning Studies*, 1(2): 165–84.
Healey, P. and Hillier, J. (2008) 'Introduction'. In: J. Hillier and P. Healey (eds), *Foundations of the Planning Enterprise. Critical Essays in Planning Theory*, Vol. 1, Aldershot: Ashgate, ix–xxvii.
Healey, P. and Underwood, J. (1976) 'Planning Departments in the London Boroughs' *RTPI News*. Autumn.
Healey, P. and Underwood, J. (1977) *The Organization and Work of Planning Departments in the London Boroughs*, CES CP18, London: CES.
Healey, P. and Underwood, J. (1978) 'Professional ideals and planning practice', *Progress in Planning*, 9(2): 73–127.
Healey, P. and Zwirner, W. (eds) (1973) *Ideologies in Planning*. Planning Workshop Paper No. 2. Kingston upon Thames: School of Planning, Kingston Polytechnic.
Hillier, J. (2002) *Shadows of Power: An allegory of prudence in land use planning*, London: Routledge.
Innes, J. (1994) *Planning Through Consensus-Building: A new vision of the comprehensive planning ideal*, Berkeley, CA: UCA, IURD.
Innes, J. (1995) 'Planning theory's emerging paradigm: Communicative action and interactive practice', *Journal of Planning Education & Research*, 14(3): 183–90.
Jackson, J. (2013) '"I think planning is about chipping away at stuff": The voices and activities of public service planners'. Paper presented at SOAC Conference, Sydney. www.soacconference.com.au/wp-content/uploads/2013/12/jackson-Governance1.pdf [accessed 18 March 2014].
Mazza, L. (1986) 'Guistificazione e automania degli elementi di piano', *Urbanistica*, 82: 56–63.
Metzger, J. (2010) 'Strange spaces: A rationale for bringing art and artists into the planning process', *Planning Theory*, 10(3): 213–38.
Ove Arup/Department of Communities & Local Government (2010) *Planning Costs and Fees, Final Report*. https://www.gov.uk/government/uploads/system/uploads/attachment_data/file/6381/1769333.pdf [accessed 18 March 2014].
Reade, E. (1982) 'The theory of town and country planning'. In: P. Healey, G. McDougall and M.J. Thomas (eds), *Planning Theory: Prospects for the 1980s*, Oxford: Pergamon Press, pp. 43–58.

Schön, D. (1983) *The Reflective Practitioner: How professionals think in action*, New York: Basic Books.
Thomas, H. and Healey P. (1991) *Dilemmas of Planning Practice*, Aldershot: Avebury.

Chapter 1

Professional Ideals and Planning Practice: A Report on Research into Planners' Ideas in Practice in London Borough Planning Departments*

Patsy Healey and Jacky Underwood

Aims and Assumptions

The research reported here was begun in 1973 and, as with most projects which extend over a number of years, it has been reviewed from time to time, leading to some changes in emphasis. In particular, there has been a shift from a broad interest in planners' use of theory in practice to a concern with the interaction of planners' ideas about their roles and the organisational contexts within which they develop their roles. This chapter outlines the initial aims of the research, discusses the ways they were limited and interpreted as work progressed and summarises the main contextual assumptions made.

Initial Aims

As stated in an application for research funds made to the Centre for Environmental Studies in1974,[1] the broad object was to investigate the 'way planners in practice use and relate to theories and concepts'. The research was to examine the following questions:

1. What use do planners in action make of ideas, concepts, theories and ideologies?
2. Under what circumstances do they define them?
3. From where are such conceptions derived?
4. What is their relation to 'planning' as a 'government activity, a profession and a social movement' (Foley, 1963)?

It was suggested that the findings of such an investigation would

1. Provide a contribution to the academic understanding of the local authority planning situation in this country;
2. Provide evidence of the way in which theory relates to practice and consequently provide some vital information for those developing applied theory;
3. Provide an insight for practitioners concerned to interpret their situation, especially for those anxious to improve the quality of planning practice;
4. Provide valuable case material, particularly for teaching purposes.

Such aims touched on a potentially wide area of concern, including the relationship of knowledge and action, the nature of the knowledge claimed by any group of specialists and its application particularly within public

---

* Extract from *Progress in Planning*, 9(2): 73–137 (1978). This work was written jointly with the late Jacky Underwood. This extract consists of Chapter 1 on Aims and Assumptions and the final Chapter 6 Concluding Summary and Discussion. Editorial inserts for the purposes of this partial reproduction are inserted in square [ ] brackets.

1 Theory in practice: a research programme into how planners in practice use theory, submitted to CES, May 1974 […].

organisations, the characteristics of planners' knowledge, and the nature of planning as a government activity. The particular emphasis within this wide area was on the way the ideas (theories and concepts) of planners interacted with the context provided by 'practice', the organisational context. It was assumed that although planners had some degree of independence from this organisational context, the latter might be expected to have some influence on planners' ideas. The initial focus of the research programme, therefore, was on the individual planner as claimant to specialist knowledge and skills. He was seen as resolving dilemmas presented by the field to which this knowledge applied and by the problem of how to apply it within organisations. In other words, the emphasis was very much on planners as professionals, rather than planning as a government activity or social movement.

The nature of planners' knowledge and skills has, however, been a subject of controversy ever since town and country planning was established as an obligatory government activity [in the UK] through the 1947 Town and Country Planning Act. Until the 1960s, the dominant school of thought among planners emphasised planning as the physical development and design of settlements, represented to practitioners by Keeble's textbook (Keeble, 1969). This position conflicted in many respects with that expressed in various government statements in the 1940s about the nature and purpose of town and country planning (see Cullingworth, 1975; Schuster, 1950). These viewed land use planning as an expression of national economic, social and strategic objectives. During the 1960s, new ideas began to filter through to planners from the United States. These stressed the notion of planning as a general process of rational public decision-making (see Hebbert, 1977). Proponents of this approach, such as Friend and Jessop (1969), Faludi (1973), Chadwick (1971) and McLoughlin (1969) owed much to operations research, systems theory and theories of organisation behaviour. Such approaches are concerned with ways of reducing 'malfunctions' in organisations and stress ideal arrangements rather than observed ones. In particular they focus on the ideal of the decision based on a rational examination of comprehensively collected information, i.e. decisions scientifically arrived at. Because of this essentially idealist/prescriptive stance allied to a functional perspective, such authors neglect the structural dimensions of decision-making. In other words, they ignore the importance of particular configurations of power and influence within which individuals and organisations operate. It was in reaction to this approach that the research programme was formulated. It was felt that more empirical evidence was needed about what planners actually did and what ideas they called upon when operating within public sector organisations understood as political entities in the sense of both their internal composition and their location within a wider context of power and influence. It was hoped that this would help to counterbalance the output of ideal organisational structures and operational guidelines (see Faludi, 1973; DoE, 1971, 1972; Skitt, 1975). In effect, this research began as a critical contribution within what is often known as 'procedural planning theory' (Hightower, 1969). The concern of this approach was with the processes of planning and with planners as potentially significant actors in defining the operation of these processes, rather than the examination of the content of planning policies. Because of this emphasis on individuals and their interactions, while at the same time recognising the importance of the structural context within which these interactions occur, a major issue in the development of the work was the relation between the two.

The Focus of the Research

The research programme set out to investigate, via case studies of planners in 'practice', the ideas planners were referring to, the way they were using them in different sorts of organisational contexts and the influence of such contexts on the ideas used. London Borough planning departments were selected for the case studies partly because they were familiar and accessible to the researchers and partly because of their intrinsic interest as the bottom tier of a 10-year-old two-tier local government system.

Given the approach set out above, the purpose of the case studies was to investigate the interaction of planners' ideas and the organisations within which they work. Analysis focussed on two central problems, identifying the ideas used, their nature and origins in relation to the body of knowledge associated with planning, and the problem of interpreting planners' organisational situation. We sometimes referred to these as the 'ideas strand' and the 'organisational strand'. The weaving together of these two 'strands' was our major interest, and involved us in a consideration of the following issues which provide important boundaries to a local government planner's sphere of operations. The first of these was the nature of the planning system which local authorities are required to implement, and the political context (broadly understood) within which local administrations operate. The second was the relationship of planners (an occupational group in local government claiming specialist knowledge and hence professional status) to the institutions that make up 'the planning profession'. These two issues highlighted the need to give more attention to planning as a government activity.

It became increasingly clear that our original interactional approach would have to be located within a structural perspective. To attempt to make such a link between the 'interactions' of particular groups of people and the wider 'structures' of which these interactions form a part raises a general conceptual problem. As Hall et al. (1975) have found in their case studies of social policy, it is difficult to make adequate sense of case studies without taking account of the structure of society. On the other hand, the structural approaches available take inadequate account of the importance of individual interactions which such case studies demonstrate. Not wishing to enter into this methodological debate, for the purposes of this study we have assumed that it is valid to 'frame' an interactional perspective with structural considerations. These structural considerations had to be of a kind which would allow us to locate the organisations within which the planners we studied worked (London Borough planning departments) and the various institutions relevant to planners as an occupational group. Clearly these two aspects are interrelated in any model of the structure of society as a whole. However, for convenience and because the concern in what follows is to outline those assumptions which have been made about the structural context of the cases studied in detail, the two are treated separately here.

Taking as a starting point planners as local government officials, some assumptions are necessary as to the nature of local government, central government and the society within which central government exists insofar as they 'frame' planners' spheres of operation. In particular, the emphasis will be on the contribution of various structural factors to bounding the 'autonomy' of planners. Autonomy is used here to mean an individual's (or group's) capacity to define his work in his own terms (and hence introduce 'specialist' concepts in doing so). It is used in the dictionary sense of 'self-government', as in its use in the autonomy of local government in respect of central government.

The Structural Context Framing Planners' Spheres of Operation

Firstly, we assume a society in which control over the means of production and consequently wealth are unequally distributed, that society being therefore composed of a number of groups whose interests conflict and which have unequal resources to engage in these conflicts. Within such a context, government has increasingly taken an interventionist role as regulator, distributor and producer. In exercising this interventionist role, government has been influenced by some interests more than others, and the interests which it has seen appropriate to favour have varied over time. The result has been that in any field of public intervention, policies and procedures often reflect conflicting purposes (see Wilding and George (1975), on social welfare policy).

Within this context, public intervention in the use and development of land is a field in which regulation of the private market is the major activity, although the purposes of this regulation are disputed, as is the limitation of government intervention to a largely regulatory rather than promotional role. For the performance of this task, central government has enacted legislation defining rules for performing the task, and setting up institutions (local planning authorities grafted onto local government). It has established for them some obligatory tasks (the control of development, the preparation of structure plans) and a whole range of discretionary tasks (e.g. local plans, conservation area plans), and in connection with each of these tasks it has laid down certain procedures aimed at ensuring that its agents (local planning authorities) comply with the spirit of central government intentions.

However, in interpreting and applying central government laws, tasks and procedures, local planning authorities are allowed considerable autonomy by central government. This is partly because of the general principle that local government should have some autonomy and partly because of the contradictions and uncertainty in central government's own policy (see, for some comment on these in relation to land use policy, Hall et al., 1973; Harrison, 1972; Neutze, 1975). For the purposes of the present study it was assumed that local planning authorities had a significant amount of autonomy, and hence how that autonomy was used was important in the interpretation and application of the planning system.[2] It does not necessarily follow that the occupational group employed to administer this system has any autonomy. From many studies of the community power variety (see for example, Clark, 1974; Rhodes, 1975), it is now well established that how a local authority exercises its autonomy must be related to the particular structure of power and influence within which that authority exists.

These vary in scale (national interest groups may dominate in some areas, local in others) and in type (unions, developers, industrialists and amenity groups can be dominant influences which local councillors represent, reflect or react to). In some areas, the electorate, via pressure groups, is a major influence on the councillors and officers of a local authority. Elsewhere, councillors may persist in control of local authorities with very little evidence of representativeness or responsiveness. Similarly, in some local authorities, officers rather than councillors dominate in defining policies and procedures (see for examples, Chamberlayne, 1976; Dearlove, 1973; Newton, 1976). Evidence on the influence of officers has led to contradictory arguments. Some studies have allocated a major role to officers (see Davies, 1972, on planners). Others have indicated a minor role (see Ambrose and Colenutt, 1975). Newton (1976, Chapter 7) suggests that the relationship is a fairly equal one. The first type of study led Pahl to formulate what has become known as the managerialist thesis, i.e. that local government officials (and other urban managers) play a dominant role in allocating resources within the urban system. Subsequently, Pahl himself changed his position (Pahl, 1975, Chapter 13), and the validity of his thesis has been the subject of some debate (see Harloe, 1975). Meanwhile, in response to Pahl's original encouragement, a few studies of local government officers have taken place, notably those of Norman (1975).

Our conclusion in respect of this debate is that it is reasonable to assume that local government officials do have some influence on how local autonomy is used but that this influence varies. In some cases, officials can be identified as the dominant force in identifying issues, defining policy and implementing procedures and programmes. Elsewhere, their influence is more subtle, but significant, for example in 'filtering' demands for action by constituents and councillors (see Hall et al., 1975, for this in central government), in managing the agenda for discussion by councillors and constituents and generally in contributing to the organisation of issues into and out of discussion arenas, which Bachrach and Baratz (1970) have referred to as the mobilisation of bias. Beyond these influences, there are also the many ways in which the administrators of policies displace an organisation's goals by their day to day interpretation of their tasks (see Hill, 1972, Chapter 5). It may be that with the increasing politicisation of local government[3] the power of officials is being reduced. As will be seen, our case studies provide some evidence of this. Nevertheless, at the present stage in our understanding of local government, there seem to be many reasons for studying a particular group of officials, especially planners around whose work there is in any case so much controversy.

The Influence of Planners

Though local government officers on the whole can be assumed to have some influence on how local government autonomy is interpreted, some groups of officers are likely to have more influence than others. Evidence regarding the relative influence of planners provides contrasting examples. Some planning case studies illustrate how planners' ideas have foundered within the internal power struggles of their authorities (see Muchnick, 1970), a well-known conflict in this field being that between road-building engineers and environment-preserving planners. Other examples can be found of powerful planning departments (such as the London County Council before 1965, or Newcastle City in the 1960s), and it is 'the planners' who are seen by the outside world as responsible for many unacceptable outcomes. Traditionally, the most powerful departments in local authorities have been those of the Town Clerk, the Treasurer and the Chief Education Officer. We therefore viewed local authority administrations as a collection of different groups competing

2 This assumption may possibly have been more appropriate to the early 1970s than to the present period of economic recession with consequent greatly reduced development pressure.

3 Councillors in urban areas in the UK have for many years organised themselves on national party lines. This seems to be happening also in the new Shire counties and county districts created in the 1974 reorganisation of local government.

for the power to define tasks for themselves and for control over the administration, each group variously constrained by and manipulating the wider political environment.[4]

Within this context, planners' claims to influence rest on two grounds, firstly their role in the administration of the planning functions of a local planning authority, and secondly, their claim to 'professional' status. This claim to possessing 'specialist' knowledge and skills justifies members of planning departments when seeking to exercise responsibility for particular tasks within the local authority and hence in demanding requisite resources (usually staff) and status (pay, position, departmental status, etc.). The image of 'professional' status is sustained [in the UK] by a 'professional' institute and special courses for those wishing to become institute members. The opinions or approaches advocated by accepted 'pillars' of the 'professional' establishment can be used as support by groups of planners and the Institute itself has from time to time acted as a pressure group to promote the interests of its members (e.g. advocating that planning should be a separate local authority department or making representations to central government on various matters). The whole paraphernalia of professionalism locates planners firmly within the established middle class (although the implication, often drawn, that planners automatically serve the interests of the middle class, cannot be assumed).

However, we have seen the basis of planners' claims to professional status as contentious on two grounds. Firstly, most 'qualified' planners are employees and most of these work in local government. Thus they are not in the position of independent specialists, controlled by the rules and ethics of their peers, which is the ideal to which occupational groups seeking professional status aspire (see Johnson, 1972).

The kind of tension which results when 'professional' values conflict with employer's values, common to occupational groups in which most members are employees, is compounded by the changes and controversy over the nature and purpose of planning *within* the 'professional' channels. As one planning theorist has commented, planners have become confused over what they are planning, who is supposed to plan, and for what purpose (Prost, 1976). As a result, it cannot be assumed that the ideas planners will refer to in defining their task within the autonomy available to them will be the same from department to department, from group to group or from planner to planner. One of the issues for investigation has therefore been to identify the various sets of ideas, to examine under what circumstances a particular set has been seen as relevant, and how one set rather than another came to be accepted by a group of planners in such a way that it could become the view they were collectively putting over to an outside audience.

Limitations of the Research

The assumptions discussed above about the nature of the organisational context which constitutes planners' 'practice', and about the problematical nature of planners' claims to possessing 'specialist' knowledge and 'professional' status provide the basis on which a number of detailed concepts are developed in Chapter 3. At this point, we should note that because of our emphasis on the ideas of planners in relation to the organisations within which they work, we are not able to discuss directly either the outcomes of the activities in which the planners we studied were involved or the interests these served. Nor did we focus explicitly on the relationship between central government, the Greater London Council and the London Boroughs.

Our material provides many insights on these issues and we have attempted to draw these out where possible. However, the emphasis of our investigations has been very much on the planners' view of their world and their ambitions within it. The justification we offer for such an emphasis is that planning is a controversial activity and the nature of planners' knowledge is sufficiently ambiguous to make a study of it worthwhile.

[Chapter 2 presents the methods used in the research. The focus was on planners at work in the London Boroughs. It drew on three strands of work: a study of how development plans were being produced in a selection of boroughs, a participant-observation study of a single planning department (published separately (Underwood 1980)), and a survey of all London Borough planning departments undertaken through interviews with planners at various levels and loci in the organisation. The empirical work was undertaken between 1974 and 1977. Chapter 3 introduces the concepts and questions which focused the research. A first set focused on

4 See Pettigrew (1973) and Bachrach and Baratz (1970) for this perspective on organisations.

planning knowledge. This considered the general problem of the relation of knowledge to action, the kinds of knowledge which planners use, and the origins of this knowledge. A second set focused on the organisational dimension of planners' work, covering the 'action space' available to planners, the networks they used and created to pursue their work, and the reference groups whose values and norms were respected by the planners we studied. Chapter 4 draws on the empirical work selecting three 'case histories'. These are all concerned with planners' policy work, and in particular the preparation of 'Borough Plans'. 'The first case describes the preparation of a Borough Plan in a small low status planning office in an outer London Borough. The second describes the same task in a large but poorly managed inner London Borough planning department which was accorded little status within the council' (p. 95 of original text). The third case, an inner borough and the subject of the participant-observation study, is of a situation where the planning function was more sophisticated and influential than in the other two.

Chapter 5 draws on the whole body of empirical material to address the questions set in Chapter 2. As regards the knowledge planners refer to, we found a range of conceptualisations in use. Some of this knowledge was implicit, but sometimes there were tensions within departments about the appropriate understanding of the planning task and the knowledge to use. We identified the following conceptualisations of the planning task: as the performance of statutory duties; as ad hoc land management; as urban design; as the provision of organised information; as the systematic consideration of policy issues; as co-ordinated land management; as social management. The most important sources of knowledge, as well as of norms, appeared to be experience in planning work. Professional channels and planning education were identified by planners as of lesser importance, though these influences may have reached them more indirectly. In some instances, for example the third case presented in Chapter 4, a planning team was energetically developing new ideas. The knowledge planners emphasised related to the audiences they wished to convince. For many, these were located in the work situation, though some sought to influence the wider professional context. As regards the organisational dimension of planners' work, the 'action space' available to planning teams was variable from borough to borough. It was affected in particular by the attitudes of councillors, but also by the extent of inter-departmental coordination. In some departments, planners tried to expand this space by offering some kind of service in exchange for influence (for example, information); by demanding co-operation, and by forming alliances. These efforts in turn sometimes led them to re-evaluate their own role and knowledge, reaching out to new ideas in the professional or political field. However, the less pro-active planning teams tended to look to each other as their main reference group. Yet there were several possible reference groups to which planners could refer in their work, and sometimes significant clashes between them. Professional values could challenge councillors' political commitments; the 'people of the borough' could be seen as having different concerns than the politicians. The diversity of views within professional circles about the nature of the planning task and the limited attention to how to operationalise different views in practical contexts meant that 'few guidelines were available for the often uncertain practitioner' (p. 116). This uncertainty also raised questions about the claims planners made to specialist status. It was also evident that planners were neither 'neutral nor consistent in the interest [*sic*] they are likely to promote' (p. 116).]

Concluding Summary and Discussion

The purpose of the research reported here was to investigate the notions about planning to which practising planners referred, in order to discover how these notions were influenced by and influenced the 'practices' or organisational situations in which they were applied and to comment on the nature of planners' knowledge and skills, given their claims to 'professional' status. The assumption underlying the research, as outlined in Chapter 1, was that, although the sphere of influence of planners as an occupational group implementing an area of government policy was limited by a number of structural factors, within that sphere planners might be expected to have some autonomy in determining the principles by which their influence was exercised. A major concern which developed in the research was the extent to which the ideas to which planners referred in determining the principles by which they sought to guide and promote their activities should be seen as dependent or independent of the structural factors determining the organisation's context within which they worked.

Conceptualisations of Planning

As discussed in Chapter 5, we found a range of ideas about the nature and purpose of planning (or 'conceptualisations' of planning) among the various groups of planners studied. Knowledge of these ideas in these cases was predominantly acquired through contact with planners and planning debate in the course of work, although all were seen as relevant ideas for planners as an occupational group to use. We have suggested that, although many planners are not aware of the range of ideas about planning held among planning practitioners nor are they necessarily clear about the particular ideas to which they are referring, 'professional channels' nevertheless convey alternative approaches to planning practitioners, and, conversely, practitioners feedback modifications to approaches into the collection of ideas recognised as existing within the profession.[5] Some of the approaches which we identified were ones which have been dominant traditions or paradigms among planners (urban design, planning as the formulation and review of policies in general). Others (such as co-ordinated land management) represented relatively new ideals circulating within the profession.

These approaches to planning held by practitioners should not necessarily be seen as descriptions of what was actually done, although sometimes this was the case. They were the basis of planners' aspirations to perform particular roles and acquire a certain status within their organisational context. Thus approaches were changed when they did not appear to provide a valid basis for claims to roles and status. The range of approaches within the professional channels allowed practitioners to make such changes while still presenting them as valid 'planning' ideas.

To summarise, we found planners to be engaged in an ongoing process of expressing and modifying their conceptualisations of planning in order to legitimate claims to a wider role than was available to them. Because those we observed had little experience of the wider roles, these conceptualisations were more significant as images or symbols of themselves necessary both for their self-justification in such ambiguous situations and for their justification to others, than as the basis for operational knowledge and skills. Consequently, successful claims to wider roles led to difficulties in acquiring relevant operational knowledge and skills.

The Influence of Planners

There was tension between the influence planners sought to exercise, reflecting their self-image as professionals, and the influence which they were in a position to exercise. This was particularly evident in the area of 'policy work', with which all three cases discussed here were associated. The planners studied were claiming the right, explicitly or implicitly, to influence both the way local authority resources were allocated and the evolving pattern of land use and development in their areas. Although we did not focus on the outcomes of planners' actions (see Chapter 1), the cases studied provided us with much evidence that the influence of planners on the former was marginal, though varying […]. The London Boroughs Survey suggested that the planning department with a major influence on local authority resource allocation is unusual in London, and that it is to be found only where there is a demand for 'policy expertise' which the planners are offering but no-one else can provide at the time […] and/or where land use and development problems are major policy issues, recognised by the council as a whole. With the recent decline in economic activity and reduction in public expenditure, the incidence of this sort of problem has declined. Even when there were major town centre and housing redevelopment schemes, architects, treasurers and estates departments might be more centrally involved than planners.

Recognising this limited potential for influence on council affairs as a whole, we found several planning departments putting more emphasis than previously on their established association with land and development problems. Without a study of outcomes, there is no real evidence from our research on the extent of planners' influence in this arena. In the 1960s, when development pressure was substantial in some parts of London (and when officers were often allowed more autonomy by councillors), only a few planning departments seem to have exercised a powerful role in influencing the pattern of development in their areas. Elsewhere, those planners who had influence were more likely to be reacting to pressures on them (see Ambrose and Colenutt,

5 By 'professional channels' here we mean a broader arena than that represented by the professional institute (see Chapter 3).

1975, on Southwark). This points to our conclusion that not only are the boundaries of planners' action space and hence potential for influence more limited than planners would wish, but also that in attempting to increase their influence, they adopt principles which are likely to be acceptable to those they seek to influence. In other words, their principles come to be influenced by the interests of the dominant pressures within the organisational context. In the cases discussed here, these dominant interests as identified by planners were those represented by certain key councillors. Underlying this was the definition of the role of the planning function as defined by central government through statute, regulation and advice. To quote from a recent draft Borough Plan written statement:

> In an inner London Borough whose area is intensively developed and where many of the significant social and economic forces are only marginally influenced by council controls and policies, the range of planning alternatives is very limited. (London Borough of Camden, 1976)

To conclude then, because of their limited powers and because, in the exercise of these powers, the principles adopted are unstable and subject to the influence of the interests which surround them, the planners in these cases cannot be considered as exercising a major influence on the events in which they were involved. Further, their approach to planning, although referred back to professional channels as a form of legitimation, can hardly be treated as a variable independent of the organisational context.

The Interests Planners Tend to Serve

One implication from this conclusion, as discussed at the end of Chapter 5, is that planners as an occupational group cannot be assumed to be serving a single interest. One way of identifying the interests any group of planners are serving would be to identify the dominant pressures on them and the interests these represent. However, how such pressures are interpreted by the planners concerned, into a 'relevant' approach to planning and into the work they do, may reflect some distortion of these pressures. Each conceptualisation of planning defined in the first section of Chapter 5 is likely to emphasise some values more than others. An urban design approach maximises the interests of amenity and heritage lobbies, or, in the days when urban design involved the promotion as well as the conservation of development, the interests of certain developers and related interests (see Ambrose and Colenutt, 1975). At the opposite extreme, a social policy/management approach should maximise the interests of disadvantaged groups in society. The information provision, policy and co-ordinated land-management approaches are often presented as technically neutral, but the emphasis on efficient and effective management not associated with clear social welfare goals would tend to discriminate between interests, and more generally may be associated with managerialist conceptions of government which have been attacked both by the right (such as Lindblom, 1965) and the left (see Benington, 1976).

Finally, it must be recognised that individual planners, planning teams and planning departments may be working with or against the dominant pressures on them, depending on their approach to planning and general ideology at the time. [In the first case presented in] Chapter 5, the policy team built up alliances with Conservative councillors and corporate planners under the banner of better corporate management. In some other boroughs, planners and councillors working towards social welfare goals are in alliance against developers. However, the problems faced by planners trying to work against the dominant pressures on them are compounded by uncertainty as to what interests the planning function can be made to serve. While it is clear how the task of interpreting and applying the land-use planning system can be used to serve business and development interests (maximising private profits from land), and amenity interests (via encouraging good design and conservation), it is not really clear how far the system can be used to serve social welfare interests. Nor is it known with any specificity how existing planning practice unintentionally affects social welfare.[6] Planners elsewhere in London who are considering social policy/management approaches are faced

6 Harvey (1973) represents a pioneering work in this field and more recently the government sponsored Inner Area studies (DoE, 1977) have provided much evidence on the effects of various policies on Inner City residents. See also the recent CES Conference on Urban Change and Conflict (at York University (1977)). Much more thought needs to be given to this problem, however.

with similar dilemmas. Their uncertainty leaves them as open to 'manipulation' by other interests as their colleagues adopting approaches which seem to be neutral.

The Ambiguity of Planners' Claims to Professional Status

Not only were the planners encountered in our studies uncertain about the interests they should, could or did further, but many questioned the legitimacy of their claim to professional status. Our own approach, viewing planning as a field of government intervention and planners as an occupational group created by government, raised questions about the validity of these claims (see Chapter 1). Our observations have provided examples of planners more concerned with essentially ideological discussion, about what should constitute a valid planning approach in a particular context and hence what problems should be considered and in what terms, than with the application of technical expertise to defined problems. Is this apparent lack of basis in technical expertise as a justification for a planner's role a phenomenon specific to these particular groups of planners, or is it a fundamental problem for all planners?

A study of this kind carried out in the 1960s would probably have found most London Borough planning teams struggling to perform their statutory duties, with a few 'leaders' adhering to an urban design approach trained in architectural or engineering knowledge and skills which might or might not have any relation to statutory duties. Then, the problem as discussed within the profession was that there were not enough skilled professionals. Today, the 'professional' problem is of a different order. What we have been describing is the variety of possible approaches to planning contained within the professional discussions, with uncertainty as to the operational implications of each approach. This, combined with the organisational pressures of practice, led to instability in approaches. This instability in turn has meant that rather than build up expertise around a particular conceptualisation, planners have been shifting their approach (known to planners colloquially as changing with the fashion). However, it is not clear that the 'expertise' that could be associated with the various conceptualisations discussed here relates in any way to the type of knowledge and skills associated with 'professionals' or 'technical experts'.

The literature on the professions in general varies in how it treats the nature of professional expertise. To take two examples, Wilensky (1964) suggests that the basis for a claim to exclusive technical competence should be neither too general and vague or too narrow and specific. Eliot (1972) defines professions as based on the exercise of skills derived from an 'extensive and theoretical body of knowledge which requires more than routine application'. Any claim to 'expert' status, whether linked to professional institutions or not, would seem to require an association with knowledge and skills which have some *specificity* (not too general and vague, a clear object), some *universality* (applicable to a 'universe' of problems) and some *technical intricacy* (more complex than an experienced well-trained mind can handle). The knowledge and skills that were displayed by the planners studied here hardly satisfied such requirements. They were mostly graduates (with presumably well-trained minds) and were referring to ideas and knowledge acquired mainly through exposure to the ideas of those they worked with (complemented to a limited extent by courses in planning), and through exposure to the tasks related to implementing and applying the land-use planning system in local government. It is little wonder then, that the planners observed had such problems in establishing successful claims to particular expertise in areas beyond those associated with their statutory role. To return to Johnson's analysis of occupational group control, planners have been operating as if they were part of a collegiate profession, when in fact they are involved in a process of state mediation. As with other such occupational groups, they therefore have multiple reference groups, and, now that urban design work is of much less importance, they have little in the way of *particular* expertise.

Planners as Implementors of the Land-use Planning System

In effect, this study demonstrates what others have noted before,[7] namely the contradiction between the notion of planning as 'professional expertise' on offer to a wide range of clients in the private and public sector, and planners as employees taken on to implement an area of government policy. It has often been argued

7 See, for example, Foley (1960) and McLoughlin (1973).

that planners needed professional status to be able to 'compete' within professionalised local government administrations (see Cherry, 1974). The professions with which planning was to compete – architecture, engineering, surveying, law, medicine, accountancy – had all been developed based on substantial private sector work. Planners however, are in a different category, more akin to housing officials and social workers, as personnel created to implement a particular area of government policy. For such occupational groups, 'technical expertise' is probably of less importance than a knowledge of various purposes of government in establishing such policy, knowledge of the means and effects of intervention by government and experience in defining and applying detailed policies and programmes. For land-use planning, this would involve much more emphasis on substantive understanding of government policy on land-use, development and related issues and on the role of land as a resource in economic and social terms.[8] It would involve awareness of the changes and contradictions within government policies and of the way in which different interests have been furthered by such policies. Even then, it has been argued that the application of government policy over land is more a matter of public debate over values that of elaboration by specialists (Roberts, 1976). To repeat the quotation from Evelyn Sharp, former Permanent Secretary to the Ministry of Housing and Local Government, which Cullingworth places on the frontispiece of his well-known introduction to Town and Country Planning in Britain (Cullingworth, 1972):

> Planning policies depend more on political than technical objectives; and they are always in a state of evolution. Planning is essentially a service rather than a science in its own right.

The contradictions between what planners claim to do (on the basis of a skill-based occupational group model) and what they actually do (as an occupational group created by government policy) are complicated by the idealism of planning thought. Planning has persistently been portrayed as a mechanism for achieving social reform of some kind, whether it be by urban design, by rational decision-making procedures, by comprehensive urban governance or by social planning. As professionals, planners have been marketing themselves as experts in achieving ideals, rather than as exercisers of particular skills which would achieve specific products (a building, better health, win a case at law). They have often been employed to achieve ideals, though subsequently criticised when the ideal takes practical form (see for example the history of the ideals and outcomes of central area redevelopment). It is in this context (of the symbolic, rather than operational, quality of conceptualisations of planning) that it can be suggested, then, that planners have acquired financial and occupational status on the basis of a variety of ideals, rather than recognised expertise. Although planners may change their ideals, the underlying contradictions – ideals versus expertise as a basis for claiming professional status; a professional model of occupational control versus state mediation models – remain. Possibly the relevant contemporary comparison of an emerging profession is with the community workers, where idealism associated with goals of social reform is strong but roles and practices are unclear and ambiguous.[9]

The Neglect Of Land-Use Planning As A Government Activity

The contradictions we have identified are between the three manifestations of planning as identified by Foley (1963) – the government activity, the profession and the social movement (see Chapter 1). Our observations lead us to conclude that although planners' practice is a government activity, they still tend to approach it in the terms of a professionalised skill-based occupational group, anxious for more status than the area of government activity necessarily legitimates, and backed by an idealism that provides the moral basis to claims for more status. As a result, there appears to be a comparative neglect in planning thinking both among the planners observed here and in the wider 'professional channels' about matters concerned with government policy in respect of land-use. If there is such discussion, it tends to focus on demarcation disputes (e.g. in relation to the Community Land Act). Now and again, in our interviews or in other areas of discussion with London planners, the point was made that 'planning is supposed to control the private sector', yet the implications

8 This is by no means a new suggestion; see the Schuster Report (1950).

9 For examples of thought in this field see Evens (1974) and Jones and Mayo (1974).

of this were rarely incorporated into subsequent comments. In other words, such a perspective does not fit easily into the established conceptualisations of planning. As a result, those planners whose tasks are directly related to implementing the land-use planning system have little guidance. (One reason for the neglect of development control is the lack of concern in planning thought with planning as government intervention in the use and development of land.) Individuals and individual departments are left to work out how far and with what justification they can extend central government's acceptance of 'valid planning grounds' in relation to development control and local plans, and how to interpret the ambiguities within central government advice. Discussion within the professional channels gives little guidance to practitioners on the nature of the relationship of government to the private sector in the field of land-use planning. Enthusiasm for public participation and public agency co-ordination has obscured rather than clarified this area. In other words, the idealism of planning thought and the association of planning with the application of technical expertise have inhibited the development of critical awareness about the nature of the purposes and procedures underlying the tasks planners are supposed to perform, and left practitioners with little guidance and much confusion.

This lack or critical awareness in turn inhibits the development of an appreciation of how the government-established task can be interpreted and applied to serve different interests. (A comparison can be made here with housing policy, which had traditionally been the subject of political debate, and hence there is more awareness that different policies serve different interests.) A further consequence has been the lack of evaluative research on the operation of planning policies. There have been Hall's major study (Hall et al., 1973), more recent studies of planning law (McAuslan, 1975; Roberts, 1976) and a number of isolated pieces of research.[10] However, Eversley and Moody (1976), reviewing the planning research effort in the past 10 years, can still list 'fairly fundamental questions' about the operation of the planning system and planning policy which have not yet received much investigation. Finally, the understanding of government intervention in land is hindered by the fragmentation of administration and legislation is this area, in which a variety of public sector agencies are involved, with a range of inadequately related policy vehicles intended to co-ordinate them (see, for one area of land-use planning, McLoughlin and Thompson, 1975).

Towards a Reorientation

Although our study was intended as an analytical investigation, our concluding hypotheses and the implications we draw from them lead us to make certain suggestions to planners (an occupational group to which we both belong). Given the range of ideas about planning circulating within the profession and the vagueness about the nature of many of them,[11] we suggest that further clarification of ideas and interests represented in planning discussions is an urgent task. More fundamentally, our approach and our conclusions lead us to advocate reorientation of planning thinking towards interpreting planning as government intervention in the use and development of land, and planners as an occupational group created to implement that area of intervention.[12] Such a reorientation will not, of course, substantially increase planners' influence on events, since, as we have suggested, this is limited by factors related to the structure of society as a whole. Greater understanding of government activity may nevertheless help to make practitioners more effective in exercising what influence is open to them, through greater awareness of the constraints and conflicts within which they work.

This implies moving away from interactional perspectives on planners, their ideas and their activities towards a structural view of the planning system, and its outcomes, an issue already raised in Chapter 1. The professional model of the planner assumes the individual expert interacting with his client. Discussions with practitioners seeking to increase their status within their organisation, or to get co-ordination with various groups and agencies, tend to focus on strategies for influencing others, for example, the bargaining networks discussed in Chapters 3 and 5. This kind of emphasis has been developed into the status of a planning approach

10 Recent areas of concern have been General Improvement Areas, high density housing, amenity policy, residential environment standards, Green Belt policy, AONB policy and development control.

11 See Oxford Polytechnic (1976), the discussions of the Radical Institute Group and the Conference of Socialist Planners, and the discussions about a recent RTPI Discussion Paper (RTPI Working Group, 1976) for some idea of the range and quality of current planning thought.

12 See the contributions by Allen, Cuthbert, Healey and Reade in Oxford Polytechnic (1976) for a similar conclusion.

by Friend et al. (1974) in the UK, and Argyris and Schon (1974) in the US. An awareness of this level of organisational and inter-agency politics is of great value for anyone associated with complex government organisations and has recently been presented as especially relevant to the contemporary problems of British local government (Harris, 1976). However, our contention is that without an understanding of interest conflict within society and its reflection in government policy, the nature of inter- and intra-agency conflicts cannot be adequately understood. Put another way, when a practitioner is confronted with the criticism that 'planning is a failure', without an external standard of some kind such as that provided by some form of structural perspective, there is no means of assessing how much of this criticism applies to the practitioner, [and] how much to the general characteristics of society as a whole.

Final Comment

Although under attack from each other, from academic commentators such as ourselves, from status competitors in local government and from the public at large, representatives of the planning profession nevertheless continue to produce 'idealistic' statements aimed at establishing the task of planning and the role of practitioners as more acceptable and more important than it is currently seen to be (see RTPI Working Group, 1976, for a recent example). Meanwhile, there are several indications that central government is engaged in a major, if incremental, evaluation of planning policies and procedures. Thus representatives of the profession continue to ignore the contradictions between their claim to recognition as professionals (marketing ideals more than the expertise) and the role that is legitimated for most members of the profession by government. This myopia can be interpreted sociologically as a typical example of a new occupational group trying to claim higher status in the social structure for its members. However, self-mystification would seem to be unhelpful to the body of planning practitioners, unhelpful to society, which pays for planners and consumes their services, and ultimately likely to be counter-productive to the search for greater status. So long as idealism and professionalism rather than a concern with the nature and operation of planning as an activity of government dominate planning thought, then the ideas of practitioners who have to make continual resolutions of the contradictions in these three 'activities' of planning are likely to remain as varied and confused as we have found them in this study. Because of these contradictions, the disaffection of both practitioners and those to whom they provide their services with planning and planning thought is likely to increase.

This study has focussed on what members of a particular occupational group implementing a particular area of government policy understood as their role and function. It has concluded that their understanding was only indirectly related to their legitimate role as implementers of a national system of government intervention in the use and development of land. There are, of course contradictions in government policy, but unless the purpose and value of government intervention in land in a mixed economy such as the UK can be demonstrated, and unless planners can offer some particular expertise with which to assist the operation of such intervention (perhaps a well-trained mind, a knowledge about that area of policy, its implementation and its connections with other policy areas, and an understanding of the detailed expression of strategic questions), then planners' claims to the several thousand well-paid jobs in local planning authorities are on shaky ground. There is no doubt a good case for government intervention in land and development as a distinct area of policy and implementation. Similarly, many planning practitioners and planning school students are well able to define and implement such policy creatively. But to maintain credibility in the eyes of central government, local government and the public at large, it is time that thinking among planners was focussed more directly on land use policy, its implementation and the interests and ends this can serve. This might provide planning practitioners with a more stable and substantial grounding in 'planning knowledge' than the planners studied here had available to them.

References

[*Only those cited in this edited text are included.*]
Ambrose, P. and B. Colenutt (1975) *The Property Machine*, Harmondsworth: Penguin.
Argyris, C. and D. Schon (1974) *Theory in Practice*, San Francisco: Jossey-Bass.

Bachrach, P. and M.S. Baratz (1970) *Power and Poverty: Theory and Practice*, New York: Oxford University Press.

Benington, J. (1976) *Local Government Becomes Big Business*, London: CDP Information and Intelligence Unit.

Chamberlyne, P. (1975) The coming of a new type of member, *Municipal Review*, 554: 6–7.

Chadwick, G. (1971) *A Systems View of Planning*, Oxford: Pergamon Press.

Cherry, G. (1974) *The Evolution of British Town Planning*, Leighton Buzzard: Leonard Hill.

Clark, T.N. (1974) Community autonomy in the national system. In: Clark (ed.), *Comparative Community Politics*, Beverley Hills, CA: Sage.

Cullingworth, J.B. (1972) *Town and Country Planning in Britain*, (revised 4th edition), London: Allen and Unwin.

Cullingworth, J.B. (1975). *Environmental Planning: Volume 1: Reconstruction and Land Use Planning 1939–1947*, London: HMSO.

Dearlove, J. (1973) *The Politics of Policy in Local Government*, Cambridge: Cambridge University Press.

Department of the Environment (DoE) (1971) *Management Networks: A Study for Structure Plans*, London: HMSO.

Department of the Environment (DoE) (1972) *The New Local Authorities: Management and Structure* (Bains Report), London: HMSO.

Department of the Environment (DoE) (1977) *Inner Area Studies summaries of consultants' final reports*, London: HMSO.

Eliot, P. (1972) *The Sociology of the Professions*, London: Tavistock.

Evens, P. (ed.) (1974) *Community Work: Theory and Practice*, Oxford: Alistair Schornach Ltd.

Eversley, D. (1973) *The Planner in Society*, London: Faber.

Eversley, D.E.C. and M. Moody (1976) *The Growth of Planning Research Since the Early 1960s*, London: SSRC.

Faludi, A. (1973) *Planning Theory*, Oxford: Pergamon.

Foley, D. (1960) *Controlling London's Growth*, Berkeley: California University Press.

Foley, D.L. (1963) British town planning: One ideology or three? *British Journal of Sociology*, II: 211–31.

Friend, J.K. and W.N. Jessop (1969) *Local Government and Strategic Choice*, London: Tavistock.

Friend, J.K., J.H. Power and C.J.L. Yewleit (1974) *Public Planning: The Intercorporate Dimension*, London: Tavistock.

Hall, P., H. Gracey, R. Drewett and R. Thomas (1973) *The Containment of Urban England*, London: Allen & Unwin.

Hall, P., R. Land and A. Webb (1975) *Change, Choice and Conflict in Social Policy*, London: Heinemann.

Harloe, M. (ed.) (1975) *Proceedings of the Conference on Urban Change and Conflict*, CES CP14, Centre for Environmental Studies, London.

Harris, R.J.P. (1976) Inter-authority decision-making: some implications for local government, Local Government Studies, 2: 17–26.

Harrison, M.D. (1972) Development control: the influence of political, legal and ideological factors, *Town Planning Review*, 43: 255–74.

Harvey, D. (1973) *Social Justice and the City*, London: Edward Arnold.

Hebbert, M. (1977) The Evaluation of British Town and Country Planning, PhD Thesis, University of Reading.

Hightower, H.C. (1969) Planning theory as contemporary professional education, *Journal of the American Institute of Planners*, 35: 326–9.

Hill, M.J. (1972) *The Sociology of Public Administration*, London: Weidenfeld & Nicolson.

Johnson, T.J. (1972) *Professions and Power*, London: Macmillan.

Jones, D. and M. Mayo (eds) (1974) *Community Work: One*, London: Routledge & Kegan Paul.

Keeble, L. (1969) *Principles and Practice of Town and Country Planning*, London: Estates Gazette, (first published 1952).

Lindblom, C.E. (1965) *The Intelligence of Democracy*, New York: Free Press.

London Borough of Camden (1976) *A Plan for Camden. Written Statement: the Reason Why*, London Borough of Camden.

McAuslan, P. (1975) *Land, Law and Planning*, London: Weidenfeld & Nicolson.
McLoughlin, J.B. (1969) *Urban and Regional Planning: A Systems Approach*, London: Faber.
McLoughlin, J.B. (1973) The future of the planning profession. In: Cowan, P. (ed.) *The Future of Planning*, London: Heinemann.
McLoughlin, J.B. and R. Thompson (1975) Structure planning and other policy vehicles. In: Drake et al. (eds) *Aspects of Structure Planning in Britain*, CES RP20, Centre for Environmental Studies, London.
Neutze, M. (1975) Urban land policy in five western countries, *Journal of Social Policy* 4: 225–42.
Newton, K. (1976) *Second City Politica*, Oxford: Clarendon Press.
Norman, P. (1975) *Housing Allocation Procedures in Slum Clearance Areas: A Comparison of Departmental Styles*, Discussion paper in Social Research, No. 11, University of Glasgow.
Oxford Polytechnic (1976) *The Content of a Planning Theory Course*, Oxford Working Paper no 25, edited at the Department of Town Planning, Oxford Polytechnic.
Pahl, R.E. (1975) *Whose City?* (2nd edition), Harmondsworth: Penguin.
Pettigrew, A.M. (1973) *The Politics of Organisational Decision-Making*, London: Tavistock.
Prost, R. (1976) Teaching planning theory, *Planning Outlook* 19: 30–49.
Rhodes, R.A.W. (1975) The lost world of British local politics, *Local Government Studies* l: 39–59.
Roberts, N.A. (1976) *The Reform of Planning Law*, London: Macmillan.
Royal Town Planning Institute (RTPI) Working Group (1976) *Planning and the Future*, London: RTPI.
Schuster, G. (1950) *Report of the Committee on Qualifications of Planners 1950*, London: HMSO.
Skitt, J. (ed.) (1975) *Practical Corporate Planning in Local Government*, Leighton Buzzard: Leonard Hill.
Underwood, J. (1980) *Town Planners in Search of a Role*, School of Advanced Urban Studies Occasional Paper No 6; University of Bristol.
Wilding, P. and V. George (1975) Social values and social policy, *Journal of Social Policy*, 4(4): 373–90.
Wilensky, H. (1964) The professionalisation of everyone, *American Journal of Sociology*, 10(2): 137–58.

Chapter 2
Towards a People-Sensitive Planning*

Patsy Healey and Rose Gilroy

Preface

That planning, as an idea and practice, should belong to citizens is a long-standing ideal within the planning tradition. However planning may have been used and misused, advocates for planning have offered it as a mechanism for the democratic management of our collective concerns with respect to land, property, location, and environment.[1] But how planning thinkers and practitioners have imagined the relation between citizens and planning activity has been strongly influenced by prevailing ideologies about social and political organisation. Howard and Unwin in the nineteenth and early twentieth century in Britain sought to liberate people from the discriminations of class and an elitist society. But they failed to notice the oppressions of patriarchy. Faludi, and Mannheim before him, believed in planning as a tool of scientific management at the command of citizens. Their concern was to challenge the totalitarian oppressions of European populist and communist regimes of the mid-twentieth century. In retrospect, both placed too much faith in the power of scientific analysis and management as instruments of democratic purposes.

By the 1970s, the rediscovery of the political nature of planning merged with the realisation that citizens, especially in more affluent societies, have a plurality of interests. Sometimes dimensions of interest may coalesce to produce broad categories, such as the systematically disadvantaged (in the labour market, in access to housing, in access to education and health). These dimensions may be intersected by major social divisions (such as gender and race). But attempts to predefine the groupings and categories which make sense to citizens (as in planners' categories – of age, income group, class, interest group, gender, ethnic group) may all too readily force citizens into categories they do not feel an identity with.

The major democratic movement of the late twentieth century in any case challenges the capacity and desirability of politicians, bureaucrats, and technocrats to act 'for' citizens guided by some superior knowledge of what 'people' want and 'the public interest'. People no longer trust experts to define their interests for them. As 'consumers' people are encouraged to demand that public agencies deliver goods and services appropriate to their particular concerns. As 'citizens', people demand not just accountability, but active involvement in the framing and implementing of public policy.

In the 1970s in Britain, this pressure led to all kinds of experiments in planning processes, from 'public participation' procedures in plan-making, to community development and community action in inner-city programmes, and tenant management in public housing schemes.[2] At the same time, citizens who were property developers energetically pushed themselves not only into the negotiations over individual planning schemes, but into policy-making at structure and local plan level (see Rydin 1986). This was recognised by planners in the explicit discussion of bargaining and negotiation, with planners being encouraged to develop 'negotiation skills'. Such practices have continued energetically in the 1980s, as development has become increasingly market-led. While in theory, new-right philosophy offered to realise citizens' demands for greater control over the state through the economic logic of consumer preferences, in practice 'business' has been allowed to play an increasing role in defining and implementing public action. At the same time, new quangos such as Urban Development Corporations and the proposed Housing Action Trusts originally had no formal

* Previously published in *Planning Practice and Research*, 5(2): 21–9 (1990).

1 Hall (1988) illustrates this clearly. See also such diverse figures as Howard (Beevers, 1988), Faludi (1973) and Friedmann (1987).

2 A review of all community development and public participation experience of the 1970s, along with the more directly political initiatives of the 1980s, such as area decentralisation and the inner-city 'task forces', would be very timely.

mechanism for community representation, though this has been altered in response to tenant feeling. However, citizens concerned with the environment and quality of life have refused to be excluded. 'Community' and 'environment' are in the political rhetoric of all parties in Britain, and, to help address the conflicts within and between 'business' and the 'community', the idea of planning has been politically refurbished even by the current Tory administration.

But the kind of planning which citizens, as 'business' and 'community', demand for the 1990s is not a return to the benevolent state paternalism of the 1950s; nor even the competitive politics of pluralism where articulate groups fight to persuade government agencies to adopt or at least accommodate their viewpoint. There is rather a search for considered ways of debating the issues of economic, social, and environmental opportunities, challenges, and threats in a way which can establish priorities and principles. This demand has been strengthened and consolidated by the growing concern with environmental sustainability. Many people are searching for a philosophy and principles to constrain the politics of self-interest, and seek debates which interrelate scientific knowledge and moral values. At one level, this means the search for a form of 'planning as debate'.[3] Citizens, as 'business and community', also want to be actively involved in discussion of their particular projects for change and of the effects of other people's projects on their activities, neighbourhoods, and properties.

If this pressure is to be recognised within planning practice, a form of planning must be found which allows citizens to enter into the process of planning itself. This has many implications, which are currently being debated within planning theory.[4]

Planning is increasingly presented as a practice of 'intermediation' between potentially conflicting interests, searching for routes through divergent interests, developing terrains of consensus, but challenging attempts by dominant political, economic, and social groups to impose their definitions on situations. This intermediation, while informed by systematised technical knowledge, primarily operates through the exchange of knowledge and values *between people* and the development of ideas *among people.* Planning method thus becomes *interactive,* using discursive communication as a key tool. Yet until very recently such communication was ignored within the planning discussion of method. It was assumed to be inevitable but unimportant. It was therefore effectively 'invisible'.

'Interactive planning'[5] has many forms and operates at different levels, from the set-piece dramas of a public inquiry, or a formal council session, to committee meetings, telephone discussions, working sessions in small groups, and one-to-one negotiations between a public official and a citizen (a developer perhaps, or someone in search of a vice or a service provision). This chapter will focus on one-to-one communication in informal contexts. It is this kind of interaction which has had least attention in the planning field, both because it is 'invisible' or unrecorded and difficult to research and because it brings us up against our own personalities and communication competences.[6] Yet many contemporary planners would admit that they spend more time in this way than in any other mode. And the way planners perform in such interpersonal interaction is critical to the capacity of the planning community to offer expertise in a citizen-respecting interactive way which is critical to the new 'landscapes of democracy' which citizens now strive for.[7] This is not an easy task given our contemporary multi-cultural society, as any engagement with racial issues shows.[8] What demands does it make of planners?

This chapter presents some examples of normally invisible or unrecorded interactions between people, one of whom is a planner or equivalent public official. It then makes some suggestions' about the dimensions of a 'people-sensitive' planning which incorporates an appreciation of such communications.

3 For a development of this idea elsewhere see Healey (1989a).

4 Notable contributions include those of Friedmann (1987) and Forester (1989).

5 Friedmann (1973) uses the term 'transactive planning'.

6 It has in fact taken one of the authors more time than it should to recognise the significance of many of the ideas discussed here. Part of the problem has been personal 'intellectual blinkers'. Part has been a personal tendency to compartmentalise the academic/technical from the personal/emotional. Feminist debate and the intellectual forces of 'postmodernism' have helped in recognising this and reconsidering its effects.

7 For a wider discussion on new modes of democracy see Held (1987).

8 See *Planning Practice and Research* special issue, 1989, Vol. 4, No. 1.

These examples and ideas are drawn from three areas of work in which one of the authors is currently engaged. The first is research on the changing constitution of the development industry. This work takes an institutional perspective on the relations between 'structure' (the forces which drive the decision-making of firms, groups, and individuals) and 'agency' (the way individuals negotiate and develop networks of relationships to pursue their interests) (see Healey and Barrett, 1990). The second is a study of the way knowledge and values are conveyed in the daily working lives of planners, developing some of John Forester's ideas (see Forester, 1985; 1988; 1989). The third, was the work of the RTPI's Women and Planning Working Party whose report *Planning for Choice and Opportunity* was published in 1989. Viewing planning issues from the perspective of women and the division of gender is an object lesson in how established principles, values, and categories can become so deeply entrenched that half the population is led to undervalue their perceptions, experiences, and practices. It also illustrates the very significant effort that is needed to translate principles into action, since the action needed, whether with respect to design layouts, forms of discussion, housing management practices, transport provision, or office management, requires people to change not only their values and assumptions, but the way they relate to others.

The second author's background adds two complementary strands: experience in public sector housing and voluntary work as a counsellor. Central government's plethora of legislation to chain down local government and housing in particular has resulted in the housing service examining its basic philosophies. The old style 'we're here because we're here' is no longer acceptable and is being replaced with the consumer-orientated 'we provide a public service. That service is only valuable if valued by its customers'. Education and training in the housing world are increasingly concerned with examining culture with a view to opening up local government to citizens. This means acquiring customer skills and determining ways of getting into meaningful dialogue with customers. Meaningful dialogue – learning the language of the client – is at the heart of effective counselling. To counsel is not to give advice or push the client down a particular path, but to let the client see himself or herself fully and through this discovery achieve personal growth. As local government offices look for ways of including citizens in decision-making, they must adopt many counselling skills – active listening, non-judgemental acceptance, and the ability to empathise. How can people fully play a part in the decision-making process unless we 'enable' them to do so. This means letting citizens into our language and officers entering that of citizens. How can we work together if we have no understanding of each other?

In all of these arenas, the forces which shape and constrain possibilities for action are clearly evident in the interactions of agents. Structure is made active in communications and decisions. In other words, a focus on the fine detail of communication is not an *alternative* to an account of the forces which organise economic, social, and political activity. Rather it reveals an important dimension of the inter-relationship between the structure of society and the way people think, talk, and act within it.

People Talking: Planners in Conversation

A people-sensitive planning involves explicit attention to the way knowledge, ideas, and values are conveyed in interactive relationships. Interaction involves 'talk' and non-verbal exchange, rather than the 'texts', 'calculi', and 'designs' upon which planning method tends to focus. In texts, calculi, and designs, knowledge, ideas, and values can be filtered, to cut out what is considered inappropriate or inadmissible.[9] Texts are documents, or records, constructed for particular purposes. Conversations, and particularly those which are not on some public stage, such as a public inquiry, a council meeting, or a televised discussion, are 'filtered' by the expectations about roles, relationships, and appropriate behaviour which people have about each other when they meet or make a telephone call. But within the relationship, they cannot avoid confronting each other as human beings, with many other roles and interests than those which may be the intended focus of the conversation. Conversations, 'talk', compared with texts, calculus, and design, are thus inherently more multi-dimensional in their potential meanings and reference points.

9 There have recently been some very interesting comments on the way plans as texts may be 'deconstructed' (see for example, Mazza, 1986).

Obviously in a chapter such as this, the conversations have been filtered by the very process of recording them. The record tends to focus on words, reducing the layers of nonverbal communication and tone of voice. In writing up a record for publication, the researcher presents an extract only. Quoting from research reports imposes a further selection process. Such filtering is essential if those formerly 'inadmissible' activities are to be admitted to the lexicon of planning processes and methods. But part of understanding interactive processes is a recognition of such multiple layers of filtering.

Five cameos are presented. The first is from original research. It illustrates a planning office discussion between a planner and the architect for a developer. The second shows a social worker–client relationship. In this case, noted by John Forester, in order even to begin to deliver the service the client sought, the social worker had to empathise with the client. The third case, again from original research, is back in the planning office, and a telephone conversation between a highly aggrieved citizen and a senior planning officer. The fourth episode presents the point of view of a citizen, this time a perceptive squatter seeking to persuade a service official to reconnect the squat's electricity supply. This case is taken from a novel, in which no social-scientific barriers are imposed on the dimensions of a relationship. The final case is presented in a more summarised form, providing a record of meetings held in a deliberately discursive manner in order to provide an input to the radical Greater London Development Plan of the early 1980s.

1: Negotiating Gains

This case focuses on a meeting between a senior planner and the architectural advisers to a major international high-technology company. The company is purchasing 20 acres of council-owned school playing fields for its regional headquarters. In the planner's view, the site is the best urban development opportunity in the region, adjacent to the motorway network, and halfway between the city centre and the regional airport. It has a long frontage to the main road and is of high environmental quality.

> ACPO: I think we need to do two things. We need to look at this and then we need to talk about how this is going to be handled.
>
> Arch: [Lays out the plans, which are mainly elevations, tastefully coloured with a green wash.] That would be the Nell Lane elevation. It's the same amount of storeys, the same planning configuration; what we're proposing now is that the gable end – masonry with bits between – that we would have a panel-type system … All of it is the same type of building … with more masonry tiles and more lightweight infill.
>
> ACPO: Continuing with the idea of some landscaping on that balcony, so that you actually get some landscaping up there. Arch: Well, what this scheme means is windows that are not projecting flush with the wall. They've [i.e., the company's architects] asked us to have a look at some more tiles, which may have that option, but that's just a flat face in this drawing, so there's no balconies, or anything like that.
>
> ACPO: So that may still be to come?
>
> Arch: It may be, but this is the scheme that they [the company's representatives] saw yesterday and liked. They weren't keen on balconies.
>
> ACPO: You've certainly lightened the appearance of that frontage as compared to what was before. It looks substantially lighter …
>
> [They discuss further details.]
>
> ACPO: [Reviewing the overall position.] Well, straight away, let's say without any argument that I can see quite a sizeable improvement. It does add some variety. Everything else is unchanged. The plan form is the same, the phasing is unchanged. You're not making any changes to the landscaping. It is simply, I don't mean to be perjorative in saying that, it is that you've actually looked at the elevations and you've lightened them.

> [Architect agreed.] Now, [the company's local representative] was quite keen [on these new designs]. He really was saying that as far as he was concerned, this was quite a noticeable improvement. You only have to put that [the plans] up against the previous ones to know that that's true. [Your boss] said to me that he was actually quite excited about this. Are you now at the stage as a practice where you are happy to see your name put to this?
>
> Arch: [Confirms this.]
>
> ACPO: OK. What we then need to spend a minute on is where do we go from here. Because the fact of the matter is that that [the new plans] is quite a significant change from the scheme that the Committee has actually approved. Too significant for us to say as officers, well we'll accept that as an amendment. What we therefore need *to* talk about is the way in which we're going to deal with this, given that it will require a fresh consent [i.e., a new full planning permission]. What I would suggest … [thinking] … the company want to make an early start on site, don't they?
>
> Arch: They're still looking at October time [note: this conversation was in the previous March].
>
> ACPO: So we've got a bit of time. You see, what strikes me is that we might be best looking at this informally first of all, because I think what we might be able to do is talk to interested parties – elected members and local people – show them what we've now got, say that in our view these are an improvement and that we would be quite prepared to take them as a substitution for the submitted scheme and deal with it through officer powers. If we get objections to these elevations, then of course we aren't able to do that. But if that's the position, then we'd say to you, well, fair enough, bang in the formal application fairly quickly. But it might be useful, if you think you've got the time, to let us go through an informal consultation stage first, because what we might then be able to do is actually save some time, and deal with it as officers.
>
> Arch: [Hesitant] Would that take the form of a public meeting again?
>
> ACPO: No, I don't think so. I think what we would do is to ask you to write to us explaining the changes you want to make, enclosing the drawings, and seeking our advice about the way in which this should be handled. What I think I would then want to do would be to talk to the politicians within the city, and to people like the [local] Residents' Association, informally and say: Look, this is what we've now got. We think this is quite an improvement. We think we are quite prepared to deal with this through officer powers as a substitution. What do you think? … And if they say: Yes, we think that's an improvement, we'd like to see it there, then I think we might be able to advise you that we can deal with it that way. But if they say: Yes, I think it's an improvement, but we really think we need to consult the public more formally, we could then come back to you and say: I'm sorry, you'll have to submit a formal, fresh application. We wouldn't have a public meeting though. I think we're clearly talking only about a detailed change here to the elevations and I think we could deal with that in the normal way. But the method I'm suggesting to you might actually save time, rather than make it worse.
>
> (From Healey, 1989b)

This case is very similar to one described by Forester (1988, pp. 213–14). In his analysis, Forester is interested in the power relations of such an interaction.

> The developer has initiative, capital and land. The planner, without formal power, nevertheless has numerous sources of influence. (p. 230)

In the case described above, the planner is clearly the dominant partner, setting the agenda, structuring the next round of action, developing the reasoning behind what is proposed. However, both parties are anxious to agree. The ACPO offers his position ('a sizeable improvement'), asks for the Architect to do the same. He guides the Architect through the steps that have to be decided, provides advice and reassurance, and presents himself as willing to be helpful throughout. However, it is not just a case of getting the two of them to agree. As Forester notes, a two-party conversation:

> quickly comes to be seen as a multi-party, multi-use negotiation which involves a host of public sector bodies ... The planner plays an active professional role in addition to serving as a rough proxy for the interests of others on the local Planning Board, in official agencies, in elected offices. (p. 232).

Forester suggests we may view both parties involved in such a conversation as engaged in a form of 'trading between issues'. In this trade, the planner's 'commodities' derive from the uncertainty and ambiguity of the institutional situation the developer faces. Which way will the residents and councillors decide? How long will it take? The planner is also 'testing' the amendments to the design against his perception of the likely reactions of residents and councillors, and ensuring that the course of action proposed does not step outside agreed procedures. The planner in this situation, needs skills in appreciating the architect's context, focusing the interests of those he 'represents' in relation to specific issues. He is also helping to 'organise the agenda', to shift the architect's attention towards the issues of concern to planning.[10]

2: To Be Angry is to Be Involved

> First I thought I could at least be polite, that I'd be dealing with the poorest and the most downtrodden of society, that even if I didn't have the power to do much, I could be polite. But then I saw that some people were just so personally obnoxious that it was the most I could do to be business-like. Being polite to them was more than I could do. Then, some people just expected the agency to give them hell, and they acted like it. There was one woman – she was just impossible to deal with. She just yelled and screamed and pounded her fists on my desk – and nothing I could say did anything. There wasn't anything I could do; I'd try to talk to her, but she'd yell and demand this and that – she was just irate. Then once I couldn't take it anymore. I threw my casebook down on the floor, slammed my fist, and yelled right back at her. What happened? She had a big smile on her face, and in the first calm and steady voice I'd ever heard out of her, she said, 'Well, there! You'll be all right yet!' I was astonished. It seemed I hadn't really been paying attention to her, taking her seriously, really listening to her, until then. (Quoted in Forester, 1989, p. 112)

This episode is included by Forester in a chapter entitled 'Listening'. We do not know what the social worker's own problem really was. Was she herself frustrated because she could do so little to help her clients, or was it a case of personal antipathy? Whatever the reason, until that moment, the social worker concerned had thought she should remain within the language and rules of being an objective service worker, being 'business-like'. But some people could not begin to relate to such behaviour, finding it alien and hostile. By allowing her own feelings to be expressed, she entered a dimension of communication which the 'client' in this case could understand. Forester argues that planners must learn to 'enter into' the languages and 'assumptive worlds'[11] of those with whom they interact. The skills required are not unlike those needed for counselling work.[12] A good counsellor 'listens' not only for facts but for feelings, to gauge the emotional temperature, gain trust, and find out how to help. This may require the emotions to be released in outbursts, in order that those involved can begin to relate as mutually-respecting adults. It is all too easy in relationships to replicate parent–child patterns of authority and of resistance to it. Forester suggests this requires planners to develop their capacities to:

- Be attentive, 'demonstrating an attitude of caring involvement, inquiry and wonder' (page 111). This requires an active approach to listening, conveyed through body language.
- Ask good questions, 'digging into what matters ... to reveal and clarify unspoken significances, values and concerns ... A planner who probes more and presumes less can learn a good deal about the interests and fears of neighbourhood residents and developers alike' (page 111).

10 Forester develops this point in an earlier article (Forester, 1985).

11 The phrase 'assumptive worlds' has been widely used. For a note on the origins of this term see Barrett and Fudge (1981, page 278, note 45).

12 See in particular transactional analysis, a form of psychotherapy. It involves examining a single transaction 'wherein I do something and you do something back and determining which part of the multiple natured individual is being activated' (Harris, 1973, page 12).

- Explore meanings and implications. This serves to identify 'fundamental ambiguities of intention and obligation' (page 111).
- Help to correct the self-deception of others (for example, to dispel a myth about a planning regulation) and the fear many people have of authorities and bureaucratic systems.
- Provide respect, treating the other in a relationship seriously, rather than as a 'tool or an object or a numbered client' (page 112). This involves the terms one uses to address people, the way one listens, the arrangement of seating in a room, etc.

3: Honest Explanation for Bureaucratic Failure

This case is in one of Britain's Metropolitan Districts in 1988. Financial cutbacks have meant staff shortages. The development control workload has fallen behind. One officer's case load is in arrears as a result of sickness. An applicant has been waiting two months for a decision on her project and has rung the Council to enquire about progress. She wants to extend her day nursery and needs to coordinate planning permission, her builder, and her financial sponsor. She has discovered that there has been little progress on her application. So she has rung the chief officer. The conversation has been recorded from the officer's side so we have to infer what the applicant said (and thought!).

> First of all, can I say I'm very sorry for the situation in which we've placed you. I'm not going to try and make excuses about that because that doesn't help you at all. All I would say is that nobody's done anything deliberate here. We've been caught up by staff shortages and staff absence from sickness. But nevertheless the position that we are in is that we are going to give your application priority and we will try and get it dealt with as quickly as we reasonably can ...
>
> As I understand, what you did was to amend the original submission by adding a garage and some downstairs toilets, and it's these that we would need to consult on ... Well, we've checked through the file and I think there's some more of that [consultation] still to be done, which ... well ... I have to tell you the truth, I'm not going to lie to you. There's some more of that straight away. If that doesn't raise any particular objections, and I've no particular reason to believe that it will, we will then issue the consent using our own powers without it needing to go to the committee. We can do that fairly quickly, to issue a decision. If it does raise objections, the elected members, the committees of the council, have to consider it in that situation. So the reason why I couldn't give you a straightforward answer to your question: 'how long will it take' is that it depends upon whether there are any objections ...
>
> I would think, within a fortnight ... I understand you have a deadline and I had a feeling that it would be because of matters to do with your builders ...
>
> Alright, can you leave it that I've looked at this. I clearly would not have wanted it to be like this. It is unfortunate that it's affected by sickness. We will therefore give dealing with it in the way I've just described priority and we'll try and get it done as quickly as we can ...
>
> Yes, they looked at the wrong list. Again, I'm not going to lie to you. That's the truth. They looked at the wrong list. [End of conversation.]
>
> [Officer to researcher] What can I say? She tells me how unfair it is and it is unfair.
>
> (From Healey, 1989b)

The planning officer in this case was caught in a 'structural dilemma', with three dimensions. He believed that planning should provide a helpful service to all citizens, assisting them to coordinate their own projects and safeguard their values. His office had failed to perform to this criterion, because resource revenues were such that he had inadequate staff to do so. But he could not bypass procedures, as the planning service had to operate according to the values and procedures agreed by councillors. Within these constraints, he nevertheless decided to 'fast-track' this application, responding to the pressure of 'voice', in this case a citizen's asserted complaint. Yet what did this mean for all the other cases which might be caught in the backlog? The officer here was very conscious of these dilemmas, and revealed most of them to the applicant. Clearly, the applicant had a right to be aggrieved and the planning office and the council were 'in the wrong'. The planning officer did not change the principles agreed by his Council for dealing with planning applications. By acknowledging the office's faults, he allowed the applicant to 'score a point' and the applicant in turn could respond to being treated courteously. The officer, however, while 'feeling guilty' on behalf of the service, sought to follow Fisher and Ury's advice and 'separate the people from the problem' (1983, Part 11). Thus he could explain what went wrong without being disabled by personal guilt, which often leads people to distort events which are embarrassing for them. How easy is it to do this?

4: Persuasive Manipulation

Alice seeks to persuade a council to delay demolition of the property she and various others have squatted in, and to get public services connected to the property:

> Alice ran lightly up the steps and, knowing her way from having been in so many similar buildings, went straight to the first floor where she knew she was in the right place, because there was a room where ten or so people waited. Unpaid bills, new accounts, threats of disconnections: a patient little crowd of petitioners. From this room opened two doors, and Alice sat herself so as to be able to see into both rooms. As the doors opened to emit one customer, and admit another, Alice examined the faces of these new arbiters, sitting behind their respective desks. Women. One she knew, after a single glance, she must avoid. The letter-of-the-law, that woman, judged Alice, seeing a certain self-satisfaction in competence. A thin face and lips, neatly waved fair hair, a smile Alice had no intention of earning. But the other woman, yes, she would do. She was large, and her thick tight dress held her solid and secure, performing the function of a corset, but from this fortress of a dress emerged a large soft rather girlish face and large soft hands. Alice adjusted her seat, and in due course found herself sitting in front of this motherly lady who, Alice knew, several times a day stretched things a little because she was sorry for people.
>
> Alice told her story, and described – knowing exactly what she was doing – the large solid house which inexplicably was going to be pulled down so that yet another nasty block of flats could be built. Then she produced her official-looking Council envelope, with the letter inside.
>
> This official, Mrs Whitfield, only glanced at the letter, and said, 'Yes, but the house is on the agenda [for demolition] that's all, it hasn't been decided'. She turned up a card in the cabinet beside her, and said, 'No.43 I know it. 43 and 45. I walk past them every day to the Underground. They make me feel sick'. She looked, embarrassed, at Alice, and even blushed.
>
> 'We have already begun to clean 43 up. And the dustmen are coming tomorrow to take it all away'.
>
> 'You want me to get the power switched on now before knowing what the Council decides?'
>
> 'I'm sure it's going to be all right', said Alice, smiling. She was sure. Mrs Whitfield saw this, felt it, and nodded.
>
> (From Lessing 1986, pp. 58–9)

The 'client' in this instance deliberately manipulates feelings to construct a personal individual understanding between the service worker and herself. Alice perceives a 'motherly' woman who can be manipulated by playing certain games. Alice plays 'the good child' role and earns 'the mother's' approval.

The service worker's preparedness to understand and empathise, which Forester seeks to encourage, is here hijacked to stretch the rules of the service agency. Both become partners in the project of 'saving a house'. As the novel develops, it emerges that Alice is ambivalent in this situation, seeking support for the resurrection of the house, though she despises the values of the 'system' represented by the service worker.

However, she uses her understanding of the system's procedures and values to manipulate it. In this instance, Alice was able to find a basis for developing trust between the two of them. Nevertheless, the service worker faced an ethical dilemma, as she was bending established procedures to Alice's benefit.

5: Learning to Understand Each Other

This case arises from the Community Areas policy of the now-abolished Greater London Council. The political regime of the GLC in the early 1980s sought to promote policy initiatives and practices which actively involved citizens, recognising the plurality of their interests and the variety of divisions of disadvantage. Through the Community Areas policy, grants were available to groups within the areas for a range of purposes. This involved officers who had never dealt with grants before in negotiation with groups.

> The commitment of the majority of officers working on grants grew as they had more contact with the groups. Rather than acting as 'independent assessors', officers began to provide a wide range of help to the groups in negotiating the Byzantine legal and financial arrangements within the GLC. Bureaucrats became practical internal advocates for the groups. Initially, their help depended on the strong political backing for the programme, but it developed its own dynamic as the programme wore on. This practical support became more significant as the grants were challenged in the Courts, leading to more complex procedures. By the time of abolition [of the GLC], many officers on the community areas team eschewed their own job hunting to continue working [overtime] to sort out money before the last day, and this was greatly appreciated by the groups.
>
> As a consequence, grant negotiations came to include a much more explicit communicative element. At first, both officers and groups were quite wary of each other, and there were strong elements of covert strategic action on both sides. However, as officers were able to be more open with groups about problems with their applications, groups became more able to express emotional support for officers. Thus, even when officers did make mistakes which adversely affected groups, the groups began to interpret these more generously, as understandable mistakes in a fraught environment, rather than as unwarranted exercises of irresponsible bureaucratic power. (Allen, 1986, p. 44)

This case illustrates the way the isolation of the officer in the third example and the ethically uncertain position of the service worker in the fourth might be overcome. Under the umbrella of strong political support for the groups, officers learnt how to 'join in' with the groups to facilitate their objectives and act as supportive 'intermediators' in the complex machinery of bureaucracy. Administrators and experts become skilled advisers assisting citizens to identify their own interests, represent their demands, and make their own decisions. If officers came to understand the needs and values of the groups, the groups in turn came to appreciate the 'structural dilemmas' of the officers. Such practices are 'of-course common between planners and representatives of the development industry, because astute developers and their consultants recognise the value of such mutually reinforcing relationships'. Community development workers are usually expected to work in this way. The critical ethical issue in such situations is how to balance engaging with the group, and other values and interests the 'experts' may be required to bear) in mind. The planning officer in the first and fourth case provide examples of how this may be done.

Dimensions of a People-sensitive Planning

Once citizens, whether as 'business' or 'community' become involved in policy-making and in policy implementation, planning work comes out of the studio and the technical office and becomes increasingly discursive in its practices. Many professions (such as general practitioner doctors and social workers) have long had practices of this kind, and their interactive dimensions are only now being fully appreciated. For planners, the development of an adequate understanding of interactive practice is complicated by two factors. Firstly, they are charged with the duty of administering a regulative system rather than providing a service to people. Secondly, in this and in the promotional and advisory work in which planners engage, planners interact with people in highly varied positions and roles. A development control case worker may meet in a day enraged local residents, a local small business seeking minor expansion, a volume housebuilder explaining the potential for substantial development, a local group seeking a site for a religious function, and a land scout acting as agent for a major national property company. The senior officer in the third case quoted had spent time during the working day with secretaries, with his immediate senior, with colleagues working on aspects of office reorganisation, with a developer whose application had been refused by councillors largely on political grounds, with consultants to another developer discussing design details of a major job-generating development, in a formal Council Committee Meeting, chairing a committee of officers trying to carry through the politicians' requirement for neighbourhood decentralisation, and with junior colleagues in difficulty over a project.

Faced with this diversity of interactions, planners confront a range of multi-dependent, ethical and contextual issues. Firstly, within each interaction, or conversation, knowledge and values are exchanged. What knowledge and values does and should a planner bring forward? What grounds for raising such knowledge and values does a planner have? This requires planners to be aware of the nature of the knowledge they deploy, how it relates to the knowledge used by others in the relationship, and what kind of 'validity claims' planners can and should bring forward when introducing particular items of information or understanding. To cope with this, planners need to develop an 'epistemological consciousness'.

Secondly, every interaction is a power relationship, partly 'structured' by the roles and rules which set the context for the conversation, and partly by the 'power of personality' which everyone possesses. We routinely use 'conversational' strategies in our interactive relationships, often unconsciously. As Forester argues, this helps to reinforce established rules about relationships and hence established power structures. Alice, in the fourth example, used values and behaviour patterns learnt as a child to manipulate bureaucracy. GLC officers learnt new conversational strategies, encouraged by politicians and community groups. Forester believes that awareness of the power relations of our conversations should help us to resist embedded power and create new ways of doing things. This points up very clearly the need for an ethical as well as a political consciousness. We must be able to 'ground' not merely our knowledge claims but the ethics of our choices about conversational strategies in the politics of the interactive situation.

Thirdly, planners need to develop the capability to enter into the 'assumptive worlds' of those with whom they interact, and to make their own 'world view' available to those with whom they relate. This means understanding not only what is said, but why it is said, and what this means to the speaker. This requires 'getting behind' the position of 'the other' and looking at the situation from his or her point of view. This was the breakthrough made by both the social worker and the client in example 2, and the key to success in the cooperative alliance which developed in example 5. This in part requires a socio-psychological consciousness, about what matters to people, and the way they may organise their lives – but it also requires a 'structural' consciousness, which allows planners to see the way the forces of social organisation shape people's attitudes, expectations, and interests. The service worker in the fourth example was clearly a kindly, understanding person but she ought to have asserted the rules within which she was working. A council official in this same story moves into the squat with her unemployed partner as the only way to save enough money to buy a place of their own in expensive London. The novel charts subtly the clash of world view between the 'bourgeois' aspirations of the council official and the would-be terrorists and others with whom the squat is shared.

This raises a fourth point. A critical capacity for planners is the ability to relate an understanding of people, including ourselves, as unique individuals working out our lives, with the social structures which frame our existence and which we in turn reinforce and change by what we do. As Allen notes, we need to understand:

> the links between our everyday subjective understandings and relations with people, on the one hand, and with broader social structures, on the other hand. How does our participation in these relationships link with broader social structures? (Allen, 1988, p. 4)

Using the sociological terminology of Giddens (1984), we need to develop the capacity to understand the relationships between 'structure' and 'agency'.

This implies that in our understanding of 'social structures' and in the way we relate as people to those we interact with, we should avoid three 'traps'. The first is an individualist one. Two manifestations of this have been developed in planning thought. The first, strongly advocated in public policy in Britain in the 1980s, is the doctrine of consumer preferences. This suggests that all that we need to know is what people say they want and value, or, less interactively, what they actually spend their money on. Yet what we say we want and value at one point in time is the product of the relationships we find ourselves in. Within these, we may be manipulated to want different things. But we may also seek actively to renegotiate our wants and values, as occurred in the final case discussed above. The second individualist interpretation assumes that how people relate can best be understood in psychological terms, with any problems resolved by forms of therapy. Yet no therapy could have removed the 'structural dilemmas' under which the planner in the third example laboured.

The second 'trap' is a structural one, which sees individual behaviour as driven by social forces which leave little scope for individual choice. Such an attitude downgrades the role of 'conversational strategies' and understanding 'assumptive worlds'. It has the effect of 'disempowering' individuals from trying to change things, unless in combination in some massive social movement.[13] It also encourages planners to view the world in terms of structural categories, such as 'workers' and 'capitalists', or 'developers' and 'community groups', 'male' and 'female', 'white' and 'black'. In everyday life, we are always both 'using labels' and recognising the uniqueness of individuals and their specific actions. Such labelling blinkers our ability to 'listen' to what people are trying to say, and builds up barriers which may reinforce the very divisions which we may be trying to change. When people complained of being pigeonholed in this way, such structural thinkers might accuse them of 'false consciousness', in not appreciating the importance of the impact of social structure on their ways of thinking. Such arrogance sits ill with the role of a planner. More fundamentally, in our complex societies, the structuring forces which shape our lives affect us in diverse ways, and in diverse spheres of our activity. Sometimes it is the dimension of class which is important; sometimes the challenge to patriarchy; sometimes the challenge to the bigotries of racial or religious prejudice. We each of us choose which to assert when, and thereby change the structuring forces themselves. An approach to interaction which fails to recognise this is very likely to be a 'people-insensitive' planning.

The third 'trap' is much more difficult to avoid, because categorising people into types or groups is a strategy we all use to organise the diversity of our experience of people. The notion of society organised into 'groups' who compete and conflict through the pluralist 'politics of voice' permeates much planning practice at the present time. Planners feel themselves as part of a group, a 'community of planners', or 'the planning officers of council x or consultancy y'. In many relationships, interactions are between a planner, as a representative of the council, and another representative, of a pressure group, a residents' group, or business group. Gans (1969), who in the late 1960s argued for a planning which was 'client oriented', suggested that planners should identify their client groups and conduct surveys to find out what their clients wanted – a mixture of pluralist politics and consumer preferences. His American compatriot, Davidoff (1965) developed this pluralist line by arguing that planners should attach themselves to groups whose values they shared and help them 'advocate' for their interests. As Allen argues, such a pluralist approach

> sees social groups only in terms of their 'competing' strategic objectives [and] denies the significance of communicative action as the basis for action-binding, coordinating decisions. (1986, p. 62)

In other words, in treating society as a plurality of groups, we should avoid collapsing our understanding of the structuring forces which generate groupings of various kinds with our appreciation of individual activity within groups into a typology of groups with exclusive membership.

A critical ingredient of a people-sensitive planning is therefore to develop forms of understanding and communication which can relate an understanding of the structuring forces which shape our societies to the specific experience of the individual people (including ourselves) with whom we relate in our roles as planners. To summarise, it requires:

- An epistemological consciousness: in what way do we know what we contribute as knowledge, why do we think it is valid, what knowledge are others contributing, what should we accept as valid 'knowledge' when competing claims are made?

13 This disempowering consequence of some Marxist contributions of the 1970s is well-criticised by McDougall (1982).

- An ethical consciousness: what values do we introduce, where do they come from, how legitimate are such values and our choices about conversational strategies in interactive situations?
- A political consciousness, of the power relations embedded in the roles which bring us to communicate and the behaviours we think are appropriate in the relationships we develop as planners.
- A socio-psychological consciousness, of the ways our personalities affect other people and they us; of the range of dimensions of the communication available to us, and the legitimacy of our use of these in different contexts.
- A structural consciousness, of the forces which shape our societies and the way these may be manifest in our roles, relationships, and conversations. The more we know of the world, the greater our vision of the world is, this should open us out as people and create greater tolerance, less judgemental attitudes, and greater self-awareness.
- A self-reflective consciousness, which allows us as planners to monitor our use of knowledge, values, skills, and behaviour, both so that we can change inappropriate behaviours and, perhaps more importantly for democratic planning, make available to others our reasons for what we do and say. This does not of course mean that we become 'perfect beings'. But it should mean that we are more aware of our limitations, as planners and as individual people.

People, Democracy, and Planning

This chapter has grown out of two over-riding concerns. The first has been to affirm the relation between planning and democracy, and to propose a form of planning appropriate to late twentieth century developed Western countries, with their demands for citizen power and their plurality of conflicting interests. Some people, including the present Tory administration (see the account by Gamble, 1988), are fundamentally unsympathetic to this enterprise. Democracy, understood in this way, appears unmanageable and a cost on economic growth, except in the form of consumer preferences expressed in the market place. A fundamentalist environmentalism could also seek to restrict democratic processes in the interest of safeguarding environmental qualities. Some people may consider a people-sensitive planning an ideal, but inappropriate for our times, or that we lack the intellectual concepts, the skills, methods, and ethics.

This chapter argues in contrast that we now have both the intellectual foundations and the citizen demand to realise the ideals of our forebears. The 1970s in Britain saw a great deal of experimentation in community action and people's involvement in the management of collective enterprises from environmental reclamation to tenant management. In the 1980s, business and community have both struggled to explain to central and local politicians what they need and what kind of relationships 'make sense' to them. Meanwhile, the philosophical contributions of Wittgenstein on language and Habermas on 'communicative action' have filtered broadly across the intellectual spectrum to assist our understanding of the nature of communication and the relation between 'structure' and 'agency' in communicative acts. An important requirement for a democratic planning in the contemporary world is to 'empower' people to persuade politicians, bureaucrats, and experts to 'listen' to their demands, and to criticise what the state is doing on their behalf. The second over-riding concern of this chapter is to help to turn public service bureaucracy 'inside out'. A people-sensitive planning will be fostered where planners see themselves as providing a service to citizens, and the economy. This requires that planners need to learn to listen and to empathise with their various clients, without losing a critical attitude to the potential conflicts between what people want and value.

Patsy Healey is Professor and Rose Gilroy lecturer in the Department of Town and Country Planning, Newcastle University.

Acknowledgements

An earlier version of this chapter was given at the International Conference on Planning Education, Birmingham, September 1989. Comments from Franco Corsico, Alessandra de Cugis, John Forester, and Huw Thomas are much appreciated.

References

Allen, J. 1986, 'Smoke over the Winter Palace: the politics of resistance and London's Community Areas', Paper to the second international Planning Theory in Practice Conference, Torino, Italy, September.

Barrett, S.N. and Fudge, C. (eds) 1981, *Policy and Action* (Methuen, London).

Beevers, R. 1988, *The Garden City Utopia: A Critical Biography of Ebenezer Howard* (Macmillan, London).

Davidoff, P. 1965, 'Advocacy and pluralism in planning' *Journal of the American Institute of Planners* 31(4): 331–8.

Faludi, A. 1973, *Planning Theory* (Pergamon, Oxford).

Fisher, R. and Ury W. 1983, *Getting to Yes* (Penguin, Middx).

Forester, J. 1985, 'Critical theory and planning practice', in J. Forester (ed.),*Critical Theory and Public Life* (MIT Press, Cambridge, MA) pp. 202–27.

Forester, J. 1988, 'Sources of influence in planning practice and their implications for development negotiations', in P. Knox (ed.), *The Design Professions and the Built Environment* (Croom Helm, London) pp. 208–35.

Forester, J. 1989, *Planning in the Face of Power* (University of California Press, Berkeley, CA).

Friedmann, J. 1973, *Retracking America* (Anchor Books, New York).

Friedmann, J. 1987, *Planning In the Public Domain* (Princeton University Press, Princeton, NJ).

Gamble, A. 1988, *The Free Economy and the Strong State: The Politics of Thatcherism* (Macmillan, London).

Gans, H.J. 1969, 'Planning for people, not buildings' *Environment and Planning* 1(1): 33–46.

Giddens, A. 1984 *The Constitution of Society* (Polity Press, Cambridge).

Hall, P. 1988 *Cities of Tomorrow* (Blackwell, Oxford).

Harris, T.A. 1973, *I'm OK – You're OK* (Pan Books, London).

Healey, P. 1989a *Planning for the 1990s* Working Paper No. 7, Department of Town and Country Planning, University of Newcastle.

Healey, P. 1989b, 'A day's work', Paper to AESOP Congress, Tours, France, November.

Healey, P. and Barrett, S.M. 1990, 'Structure and agency in land and property development processes' *Urban Studies* 27(1): 89–104.

Held, D. 1987 *Models of Democracy* (Polity Press, Cambridge).

Lessing, D., 1986, *The Good Terrorist* (Grafton, London).

McDougall, G. 1982, 'Theory and practice: a critique of the political economy approach to planning', in P. Healey, G. McDougall and M. Thomas (eds), *Planning Theory: Prospects for the 1980s* (Pergamon, Oxford) pp. 258–71.

Mazza, L. 1986, 'Giustificazione e autonomia degli elementi di piano' *Urbanistica* 82: 56–63.

Rydin, Y. 1986 *Housing Land Policy* (Gower, Aldershot).

Chapter 3
Re-enchanting Democracy as a Mode of Governance*

Patsy Healey

Introduction

In our contemporary Western democracies, politics is played out as much in the finegrain of public policymaking and its translations into policy practices as it is in the grander dramas of national political debates and electoral campaigns. In my own field of urban and regional planning – collective efforts to shape place qualities and connectivities – this is strikingly evident (Healey 2010, Corburn 2010, Owens and Cowell 2011). Such qualities and connectivities are an important dimension of life experience and opportunity for citizens and businesses. Yet qualities which are judged valuable now and in the future are not readily promoted either by the 'self-regulation' of market processes, or by the functional divisions of public policy institutions fostered by mid-twentieth-century welfare states. In the highly-urbanised situations and globalised cultural and economic referents which influence how people think and act these days, all kinds of concerns bump up against each other – those of residents with diverse lifestyles, businesses of various sizes and spatial reach, service delivery agencies, those of groups promoting heritage, and acting as spokespeople for all kinds of environmental dimensions. As Massey (2005) remarks, they are 'thrown together' in shared spaces. Any substantial proposal for change is likely to generate all kinds of tensions and conflicts, bringing to the surface knowledge and values in a plurality of claims and voices. The institutional sites provided by the arenas of formal planning systems have often become key arenas where these conflicts play out (Owens and Cowell 2011).

As substantive issues about place qualities are explored and focused, these public realm arenas also reflect, reveal and help to generate the qualities of prevailing political cultures (Healey 2007, 2010). Because of the multiplicity of claims for attention which can surface, with many sectors and scales of concern in play in complex power dynamics, the spatial planning field has a long history of normative interest in richly participative democratic processes. In promoting such processes, proponents have recognised the importance, in our contemporary twenty-first-century world, of grounding political practices and policymaking processes in daily life concerns and the worlds of practical action.[1] This has encouraged a research focus in the spatial planning and area development field on the micro-practices of democracy-in-action.[2] For it is in these micro-practices that discriminations, exclusions and oppressions become manifest in the lives of actual people – in accessing services, finding ways to make a living, getting recognition in the workplace, sharing the public realm of streets, parks and playgrounds, etc.

A focus on the micro-politics of policy-in-action (Barrett and Fudge 1981) does not necessarily imply merely a concentration on the worlds of agency. What can be very revealing in arguments over place qualities is the way, as Wagenaar (2011, p. 21) puts it, the 'general' and the 'particular' are folded into each other. Nor does it imply merely a filling out of the landscape of a static political culture. It is rather a dynamic landscape, rich in emergent political forms and substantive conceptions of place qualities, continually interacting with images and practices drawn into the future from the past. Small struggles which hardly seem to make a difference may sometimes, if built on in the flow of time, lead to major changes in political cultures. Within the planning field, stories from places such as Seattle (Sirianni 2009) and Vancouver (Healey 2010) attest to the way new political cultures

* Previously published in *Critical Policy Studies*, 6(1): 19–39 (2012).

1 See, for example, the work of Briggs (2008), Sirianni (2009), Corburn (2010), Friedmann (2011) and Wagenaar (2011).

2 Examples range from studies of national government discourse formation (Hajer 2005), to those focused on environmental conservation and resource management issues (Goldstein 2010, Innes and Booher 2010), and on area development and neighbourhood politics (Healey et al. 2003, Corburn 2005, Gonzalez and Healey 2005, Healey 2005, Hillier 2000, Moulaert et al. 2007, Wagenaar 2007, Sirianni 2007). See also Briggs (2008).

can be 'grown' through participative practices centred on neighbourhood organising as well as social group organising. In Vancouver, a city-wide practice of producing neighbourhood design guidelines in participative ways (Punter 2003) gradually evolved so that the practices of the development industry also changed, shaped by the emerging political culture. As a quite new city form and scale emerged over time, citizens as well as tourists came to admire their city and its planning approach. Cases such as this, filled with struggles and contradictions as they evolve, show that micro-practices, pursued critically and carefully over time, can generate transformative change in place qualities and in political cultures, in a co-production of agency expectations and structuring dynamics. They provide empirical experiences which throw light on the potential for progressive governance practices which could help to sustain and promote more people-centred, open-minded, pluralist polities.

From this vantage point, then, in this article, my aim is to explore further what this co-production could involve, if focused by the normative ambition of moving towards political cultures which centre on the way 'we-the-people', in our various multiplicities, live and seek to live our lives. In the next two sections, I expand on the experience of place-focused governance, and provide two illustrative cases of neighbourhood organising. I then review the broader context of the struggles between such forms and the elite, technocratic forms of governance which came to dominate in the twentieth century, and the potential of the new governance forms developing in the 'interstices', 'voids' and broader relations which now surround the old forms of 'welfare states'. I argue that, in current times in countries such as the UK, some of these forms of 'network governance' hold a potential for people-centred practices to emerge, though their realisation is contingent upon struggles to resist other trajectories. I then present seven hypotheses about general qualities to foster and tendencies to resist as specific governance practices and political cultures evolve.

Spatial Planning and Urban Policy as Fields of Practice

Contested and ambiguous, it is never easy to find a general term for identifying a place focus in public policy. I will use two terms, which also in some Western European countries have been used to denote distinct public policy fields. 'Spatial' planning centres on the promotion of place qualities and issues of spatial connectivity. Yet 'places' cannot be confined to administrative jurisdictions. Nor are they separate bits of a physical or social 'jigsaw' which can be neatly slotted together to create a wider picture. They are formed by complex, multi-scalar processes of overlapping webs of material activity and social processes of identity formation and recognition (Healey 2004a, Massey 2005). 'Urban' policy is sometimes focused on a similar agenda, the place qualities of cities. For many years in the UK, this policy field was associated with 'problems' in urban areas, particularly linked to 'areas' of concentrated poverty and other forms of marginalisation. Some analysts have recently focused on an 'area development' or 'neighbourhood' agenda in the search for socially innovative governance practices. More recently, there has been a shift of attention to cities as singular entities – in competition with each other, in a discourse of 'economic competitiveness', or as units in environmental systems which generate 'footprints' of variable dimensions.[3] Yet what 'urban areas' are taken to be, and how persistent and spatially concentrated inequities arise within them, cannot be contained merely by a study of the people and areas affected. They too are formed by complex, multidimensional and multi-scalar webs of relations (Healey 2007). These characteristics are enough to claim that both fields are full of 'wicked problems', which continually escape political management and defeat the problem-solving capacity of formal government.

These fields also share two other characteristics. The first is the challenge of addressing a 'spatial' or 'area' focus in the organisation of public policy, especially in institutional designs inherited from mid-twentieth-century welfare states. These created divisions between functional programs – health, education, different sectors of business, arts and culture, etc. Around these programs, 'policy communities' (Rhodes 1997, Hoppe 2010) developed, drawing together government departments and levels, pressure groups and supporting consultancies into a nexus of relationships, modes of thought and practices. Exactly how the spatial connections between these programs were to be coordinated when delivered in particular places often remained under-specified, or relegated to a separate 'policy field', expected to co-exist with other programs,

3 See the literatures on urban governance and environmental policy. For an analysis of competing discourses in the spatial planning field in the UK in the 1990s, see Vigar et al. (2000).

or to be dealt with by local administrations. Spatial planning and urban policy as policy fields were thus left to deal with the 'spill-overs' and 'boundary limitations' created by other programs, a coordinative role within government organisations. Experience in performing this work in different Western countries in the second part of the twentieth century highlights the difficulties of this project and how the challenge of place-focused coordination varies with institutional configurations within the state sphere and the political cultures or 'traditions' which accrete around them.[4]

It is not just the coordinative challenge that makes a place-focus in public policy so difficult. It also demands sensitivity to the socio-spatial diversity of people's daily life experiences, since it is through such experiences that people get to think about what they value and resent about places and their qualities. Such diversity is not just about the many social groups to which we owe allegiances or into which we get pigeonholed by some agency. Each of us has multiple identities through which we acknowledge allegiances to several groups, often with concerns which are not actually compatible with each other (Connolly 1987, 2005, Sen 2006). Thus we may make inconsistent demands on many of the state's functional policy communities, pulling policy in several directions at once. This has long been evident in the field of spatial planning activity. Formal planning systems are given the task of coordinating other government programs as they 'land' on particular places and sites. At the same time, they are charged with addressing the many different and conflicting concerns people have about places and their qualities, whether these are streets, neighbourhoods, zones or nodes of an urban area or of large cities and regions.

The arenas of planning systems have long been recognised as providing institutional sites where all kinds of issues citizens are concerned about are given expression in an effort to get heard (Owens and Cowell 2011). In these arenas, usually created to give legitimacy to state interference in private property rights, affected citizens bring forward concerns which seem neglected by politicians and the organisation of formal government activity (e.g. environmental issues; fear of health effects; sudden surges in local demographics). Although sometimes deferring to the authority of 'science' or 'experts', citizens also bring in their so-called 'local' experiential knowledge (Corburn 2005). In the micro-practices of spatial planning systems and urban policy programs, 'front-line professionals' are continually faced with negotiating, mediating and balancing the complex diversity of claims and counterclaims that surface vigorously in these encounters.[5] The 'profession' of community development often presents itself as providing specialist expertise in developing this interface (Gilchrist 2009). However, many front-line professionals involved in spatial planning and urban policy work are also actively engaged in 'community development' in one form or another. It is no surprise that all kinds of initiatives to encourage 'people/citizens/communities' to 'participate/engage' more positively in this balancing act have been incorporated into the design of planning systems and in urban policy initiatives. Nor is it a surprise that strong pressures have built up within formal government systems to resist the potentially disruptive effects of such participation, especially in the field of spatial planning, where property rights and investment values are at stake.[6]

But attempts to contain these disruptions generate wider impacts in political cultures. Limitations on the promise of being heard and valued turns citizens into distrustful cynics (Stoker 2006), while frontline public sector workers and professionals charged with contradictory objectives and behaviour targets become disoriented and demoralised. Instead of reaching across boundaries between the spheres of the state and civil society, the result can be a reinforcement of boundaries, widening the so-called 'disconnect' between the supposedly authoritative 'them' and the rest of 'us'. This fuels not just apparently apathetic responses to invitations from the state sphere to 'participate', to 'get engaged' with public policy issues. It also generates all kinds of protest against 'authority', by individuals, spontaneously formed groups and wider movements and special interest associations (Stoker 2006, 2010, Hoppe 2010).

4 The concept of 'culture' has many different definitions, but is here taken to refer to the historical accumulation of modes of thought and practice in which members of a political community are embedded and help to shape. Bevir et al. (2003) refer to this phenomena as 'tradition'.

5 See Healey and Underwood (1979), Forester (1999), Laws and Forester (2007), Durose (2009), Sager (2009).

6 See Rydin (1999) for a review of public participation in the planning field in the UK, and Burby (2003) for the United States.

Within the *urban policy* field (in the UK and Europe generally), this ongoing tension has led to substantial experimentation in building 'partnerships' of various kinds at the interface generated by state interventions in the development and management of deprived and/or rundown parts of cities (Pierre 1998, Sullivan and Skelcher 2002). These are often called 'multi-actor' partnerships because they connect up public agencies and cross-citizen/state boundaries as they pull together those with different kinds of stake around a place focus. These may be formed as a result of local activism and protest, sometimes manifest as violent disorder, which forces recognition of more respectful, informed and better coordinated responses from public agencies. Area development policies visibly affect the streets where people live, shop and co-exist as neighbours. It is difficult to leave people out as active voices in such contexts. Some of these new governance arenas and practices have had transformative impacts at local levels, as residents have found voice and learned more (and become more critical) about the workings of governance processes. In this way, such initiatives have helped to shape local political cultures, although the direction of such 'shaping' may lead to cynicism and even more lack of trust in the political class and policymaking professionals.

In the *spatial planning* field, this is also the case, but there are other stakeholders as well – land and property owners, builders and developers, providers of infrastructures, defenders of wildlife concerns and heritage conservation, advocates of the promotion of economic competitiveness, of environmental sustainability and/or the general provision of a fairer distribution of housing opportunities. Many of these stakeholders are accustomed to conducting struggles over the management of spatial organisation and opportunity at the national level and in the courts, since property rights and substantial financial values are at stake. So the micro-practices of site development and struggles over area management become connected to the arenas of national government, nationally-orchestrated public inquiry processes and the legal system. Many countries have seen attempts to redesign planning system institutions in attempts to provide space for these multidimensional struggles and yet contain them so that they are not too costly and do not inhibit the provision of land, buildings and infrastructures for the changing needs of a changing society. Just as important have been the less formalised practices of networking and informal partnership which often develop where boundaries need to be crossed.[7] Yet these are often hidden from view, forms of 'network governance'[8] which do valuable coordinative work but yet are not transparent and easily open to public scrutiny. Such networks may be 'horizontal' in reach within the state, and between the state and key economic actors, but yet act as introverted policy communities, lacking a sensibility to the diverse concerns of 'we citizens' in all our multiplicities.

So both fields – spatial planning and urban policy – show vividly the tensions faced by modern states, as their authority is challenged and their practices are continually called into question by critical, cynical and distrustful publics and political cultures. Honed to deliver the apparent basic needs of citizens and defend nation states in a 'cold war' context, the organisation of twentieth-century welfare states no longer provides the problem-solving and delivery capacity for the new agendas of a more demanding citizenry, more conscious of our multiplicities, and more concerned about different issues, both in relation to how we live in the world at large, and in the places of our daily lives. Nor do they deliver the coordinative capacity sought by business communities faced with increasingly complex and dynamic contexts, linked to global environmental and geopolitical threats as well as more localised systems.

It is in this context that all kinds of experiments are underway in constructing new practices and arenas. As many argue, a dominant line of 're-configuration' has centred on introducing so-called 'neoliberal' conceptions of a public sphere driven by market logics and 'new public management' principles (Leitner et al. 2007, Jessop 2008, Diefenbach 2009, Sager 2012). This challenges the 'corporatist' arrangements and technical policy communities into which the practice of formal government policy development and delivery

7 There is an important contrast here between 'partnership' governance forms which have emerged endogenously, and situations where such arrangements have been 'imposed' by a higher tier of government. In the UK, the tendency has been for 'impositions', attached to funding incentives, which have often been counterproductive to the development of endogenous evolutions.

8 I use the term 'governance' to refer to the arrangements by which a polity is governed (see Le Galès 2002, Hoppe 2010). In using the term 'network governance', I refer to shifts towards more horizontal, networked forms in which the distinctions between formal departments and levels of government, and between state, society and economy become much more blurred (Sorensen and Torfing 2007, Hajer 2009, Innes and Booher 2010, Hoppe 2010).

seemed to many to have stabilised, unable to respond to an increasingly unstable context. Co-existing uneasily with this agenda is at least a rhetorical concern for more participatory governance practices which reach out to citizens' daily life concerns. Yet this often subsides into a kind of re-active populism, layering one new initiative on top of another, increasing the potential for 'delivery failures'. Under the radar of such rhetorics and reactions, governance forms are also being 'invented', in practices located in the 'in-between' spaces between the overlapping and fluid spheres of the state and civil society.[9]

So what can we learn from this experimentation about what a movement towards a more participatory political culture might involve? By this, I do not mean a culture where we are all busy doing governance work along with all our other activities in life (see also, Stoker 2006). Instead, what the concept of participatory democracy means, in my usage, is a people-centred polity with a culture grounded in what Dunn (2005, p. 138) refers to as 'a stumbling, myopic blend of quarrelling and shared exploration of the inescapable issue of how to sustain everyday lives together as agreeably as possible', in ways which recognise the plurality of ways we live our lives (Connolly 1987, 2005, Sen 2006). Such a people-centred, pluralistic polity seeks to reduce the tensions produced by self-centred egoism (Dunn 2005), gross inequalities and the marginalisation of some groups into 'unacceptable others'. It encourages people to develop, as one dimension of their identities, a sense of a shared political community, while still maintaining awareness of the diversity and conflicts which differentiate and divide people.[10]

In such a conception of political life, those who do governance work, whether elected politicians and officials, or those doing administrative or advisory work for governance purposes, come to be judged by the extent to which they convey that they live among us and not apart from the flow of our lives. From this perspective, political ideas and cultures, policy programs and governance practices are evaluated in terms of their appreciation of our lived experiences in all their multiplicities, not by over-generalised abstract principles. The spatial planning and urban policy literatures are rich in accounts of experiments in new government forms. But many of these merely look at specific episodes, often constructed through ethnographic methodologies and narrative presentations. There has been less attention to providing historical narratives within which transformative impacts beyond individual experiences can be seen more clearly.[11] In the following section, I provide sketches of two situations which illustrate how citizen-centred governance capacity was built through time. Both are grounded in place-focused neighbourhood organising.

Transforming Governance Cultures in People-centred Ways

Those struggling for political change tend to focus on capturing control of key arenas and on the installation of new institutional designs and/or policy programs through which to redirect and re-regulate the way resources flow. Yet it is in the micro-dynamics of practices that such ambitions are brought into being, brought to life and experienced by both those involved in their 'implementation' and by those who are supposed to experience their outcomes, as many examples show.[12] In these practical realisations, government initiatives aiming at 'reform' and 'change' are frequently resisted and distorted by the force of established organisational cultures and practices. But they are also sites where new meanings and practices may be forged. Whatever the impact of initiatives intended to be transformative, their long-term consequences depend both upon these practical interpretations and experiences and on how these then embed themselves in wider governance processes and

9 My broad division of society as consisting of the overlapping spheres of state, economy and civil society draws on Urry (1981). I do not use these as distinct categories with clear boundaries, but as a heuristic lens through which to focus on areas of social dynamics. As many are commenting, what is interesting in the present evolution of new governance forms is the search for new ways to link between (intermediate, cross boundaries) the spheres (see Cornwall and Coelho 2007, Holston 2008, Jessop 2008, Hajer 2009).

10 Here I draw on conceptions of deliberative democracy and the public sphere as developed by Habermas (1996, chap 8), Dryzek (2000), Fischer (2003), Fung (2005) and Connolly (2005).

11 Work in the interpretive policy analysis field is beginning to provide such accounts, to supplement the classic narratives of Meyerson and Banfield (1955), Altshuler (1965), and, more recently Flyvbjerg (1998) (see Throgmorton 1996, Corburn 2005, 2010, Healey 2007, Sirianni 2009, Phelps 2012).

12 See examples already cited in note 2.

become part of the culture of the wider polity. These levels – of initiatives, processes and cultures – co-evolve, but along different timescales and subject to different pressures (Healey 2004b).

The first case shows the continued persistence of an old working-class community in Newcastle, in north east England, and the governance capacities built up through struggles for improvements in daily life conditions and for a social democratic politics. Relative to many other neighbourhoods with such a general history, the Walker area has a very rich civil society of active groups and networks, both informal and formal, as well as individuals who do small acts of 'peace-keeping' work. But the dominant attitudes to collective life are relatively inward-looking, even though the area has experienced substantial economic and political changes. Can the capacity for informal collective action survive as a resource to address future challenges? The second case is of a poor inner city neighbourhood in Kobe, Japan, neglected for many years by formal government. Here activists have struggled to create a voice for the area and insert that voice in local government arenas, becoming a 'leading edge' of the slow transformation of government focus towards more attention to citizen's concerns. Their work shows active agency, building up collective knowledge and relational resources (intellectual and social capital), through generating the mobilisation energy to sustain informal governance activity and to campaign to change formal government policy development and delivery practice.

In north east England, a social democratic, working class, political culture lives on in the minds of citizens, community activists and many involved in formal governance initiatives. In Walker,[13] throughout the second part of the twentieth century, the city council has initiated many innovative governance ideas. It has energetically embraced most national programs to promote physical redevelopment, economic development, environmental sustainability and neighbourhood improvement. Newcastle politicians have been important nationally when the Labour Party has been in power. Meanwhile, the local economy has shifted, especially in Newcastle as the centre of Tyneside, into a role dependent not on heavy industry as in the past but on public services, education and administration, and the service sector generally. From the 1990s, a professional middle class has expanded as the local economy has shifted from dominant old industries to public sector institutions. Associated with this, urban regeneration initiatives have helped to 'gentrify' areas in and around the city centre. Meanwhile, poorer neighbourhoods around the city centre, the old heartlands of Labour support, have seen out-migration of those who could, and in-migration of people from outside the UK, searching for available housing.

For many years, as these changes evolved, the neighbourhood of Walker retained its character as a stable but poor working-class community. Work opportunities for men were dominated by nearby shipyards, though by the end of the century, very little such work was left. Community organising, to provide facilities and support for local people, has built up over many years, supported by volunteer effort, churches, the unions and sometimes by local and national politicians and officials. Its experience contrasts with some other Newcastle neighbourhoods, notably the West End, where conflicts between different groups have often broken out into intimidation and violence. Analysis of social networks in Walker suggests a relatively high density and interaction between them compared to many other poorer neighbourhoods in other parts of the city and UK urban areas generally.[14] In a recent review, around 40 agencies – from small groups to the city council, churches, and health and police authorities – could be identified. Many of the community groups received small sums from the city council. Linkages between the groups, both formal and informal, were quite strong. Through these, and their articulation with political representatives both local and national, grass roots governance capacity for small-scale coordinating, problem-solving and initiative-taking has developed strongly.

Much of this social capital built on a history of working-class activism, which came to focus on issues of access to and the condition of housing and neighbourhood services. This has been coupled with concerns for work opportunities for younger generations and issues about street-level safety and security. Generational and lifestyle change has seen the heavy drinking and domestic violence which has long been part of the northern working-class culture spill out onto the streets, mixed up with drug-dealing. Recently, the social mix of the area has changed, partly as a result of the in-migration of people from elsewhere, visibly and often culturally different from older residents, and partly from generational change with younger people and their lifestyles

13 See Madanipour and Bevan (1999), Madanipour and Merridew (2004), Healey (2005) and the report by Corrighan (2011).

14 See Corrighan (2011).

upsetting older people. Local and national governments have encouraged such 'dispersal' to relieve pressures in the crowded south east of the country. Incomers have taken up housing left vacant by residents who moved away to areas perceived as 'better'. Tensions between incomers and established residents sometimes generate street level violence, though not on a large scale. While some established residents are prepared to take action to deal with behaviour which gets too violent or disturbing for other residents, the incomers are not (yet) linked in to the richly networked neighbourhood fabric.

The situation became more complex in the 2000s when a major national initiative which aimed to renew housing stock which no-one then wanted to rent or buy 'arrived' in the area. Opportunities for more private housing were to be created by a large-scale re-development and renewal project with the objective of changing market perceptions of the area. If property values then rose, private capital generated by building new houses on sites created by demolishing 'unwanted' stock could be used to cross-subsidise the renewal work on the rest of the stock. Initially, this national Housing Market Renewal program was welcomed by people in the area, with vigorous local deliberation about which houses to demolish and how to address re-housing and general area improvement issues. But after initial progress, the program was brought to an abrupt halt in 2008 by the global financial crisis and the collapse of the UK housing market. In 2010, a new government cancelled the program 'midway', leaving many projects at a halfway stage of development, with no public funding to deal with vacated sites and fulfil promises made to residents. As a result, many residents were forcibly reminded of the dangers of relying on capitalist entrepreneurship and of the failure of public agencies to prevent the outcome which they, as residents, had to live with. These days, the public agencies, 'them', are seen as a problem and no longer a protective resource. Along with other public agencies, the city council is faced with working to re-build trust while faced with major budget cuts imposed by the new national government. The challenge now is whether the historical richness of the area's 'civil society' infrastructure can adapt to these changes, reducing the potential for the kind of destructive conflicts seen in other neighbourhoods in Newcastle and some other British cities. Is Walker's civic governance capacity just a remnant of a social-democratic past, or does it show the qualities of adaptive resilience which will help to promote better living conditions for future generations, however hard the economic future that may have to be faced?

My second example comes from Kobe, in Japan.[15] It illustrates a situation where a civil society initiative grew over time to contribute to, and find a position in, a democratic widening of local governance processes. It shows how resilient governance capacity can be built up from grassroots initiative, when combined with a wider movement for political and organisational change. In Japan, government activity operates in an even more centralised way than in England. Nevertheless, the Kobe experience has served as an exemplar for a more people-centred approach to governance activity, in a country long dominated by a primary focus on economic growth and on construction investment. Kobe in the 1960s witnessed enormous transport construction projects as part of the country's modernisation drive. As in many other cities across the world, these projects drove through inner city neighbourhoods, in many of which industrial activity, often heavily polluting, was located cheek-by-jowl with worker housing. Japanese citizens were expected to tolerate poor living conditions for the sake of a better future which a focus on economic development would bring. In the 1960s, however, the country experienced major environmental health hazards from this development approach, leading to vigorous protest from citizens. At national level, this was addressed by the imposition of tough emission standards. But there was still little attention to local living conditions (Sorensen 2002).

In this context, some activists began to promote local development initiatives. Initially primarily organised to protest and demand improvements from the local council, some of these developed into local community development (*machizukuri*) agencies. These were groups of neighbours who came together to provide social welfare services to each other, raise funds for better community facilities, and produce their own guidelines for how development should occur, bearing in mind what different neighbours were concerned about. One of the very first *machizukuri* councils emerged in Kobe in 1963 as a voluntary initiative to protest against the impact of road traffic growth and the quality of the public realm. This was promoted by labour union leaders and did not last beyond the late 1970s, but another initiative, in the Mano area in Kobe, continues to this day. Its origin lay in concerns about the impact of air pollution on children's health. It built on an earlier neighbourhood activist group that had been working on local improvements since the 1960s. External

15 Some of the text for this case comes from Healey (2010, pp. 85–92). See also Sorensen and Funck (2007).

expertise had been sought from volunteer academics and students. These earlier experiences helped both citizens and municipal politicians and officials to learn a new practice together.

A *machizukuri* study group was formally established there in the 1980s, involving 27 community members (including neighbourhood organisations, shopkeeper and industry organisations and others), four academics and four city government officials (Sorensen 2002, p. 272). The group included women, traditionally excluded from political organisations.[16] In 1981, Kobe municipality took the pioneering step of enacting the first *machizukuri* ordinance, a set of policies and guidelines about how the local administration would relate to voluntary groups such as that in Mano. This gave more status to *machizukuri* councils. The local administration set up an 'expert despatch system' to provide technical help to these councils, and allowed formal agreements to be made between the city council and the *machizukuri* councils over local development and environmental management issues (Watanabe 2007, p. 51). By 1995, 12 *machizukuri* councils were registered under Kobe's 1981 ordinance, and a further 16 had some degree of recognition (Funck 2007).

Then, in 1995, a massive earthquake struck the city. Over 4000 people died, highways collapsed and a great deal of injury was caused to people, place and property. In Kobe and across the nation, this disaster focused a spotlight on urban living conditions and how to manage them. One lesson was that the 'construction state' had not been as technically good at construction as it claimed, and the story of the collapsed highways unravelled to reveal poor technique and corrupt contracting practices. A second lesson was that civil society across the nation responded, as it has since done in the 2011 triple disasters in North East Japan, with a surge of volunteer effort to provide assistance to the stricken area. Thirdly, the national government was slow to organise a relief effort. Finally, organisations such as Kobe's *machizukuri* groups were the most effective agencies organising initial relief and were able to act as key players in the longer-term reconstruction effort. The networks they had built up in their local areas, and their connections within and beyond the municipality, provided a mobilisation and organisation capacity that helped reduce the spread of fire, rescued the elderly and infirm, and provided shelter. The *machizukuri* groups in areas such as Mano then did significant work in developing reconstruction plans and guidelines, drawing on the practices that had built up within the city since the late 1970s.

In Kobe, then, a people-centred local governance practice was available when most needed. As the reconstruction effort has been completed, the groups have continued to act to pay attention to the quality of local environments and promote conditions for improved liveability, in a fluid relation with the work of the municipal council, although many of the original residents have moved away.[17] As a result, many people in Kobe have come to participate in active ways in local place management and development, although tensions between civil society groups and the city council remain. Some argue that, in Japan today, it is at this community development level that the real innovations in Japanese political culture are taking place.

Towards More Participatory Political Cultures

In the cases sketched above, residents, community groups, small-scale business operations, politicians and government staff stretched across boundaries between formal government and people concerned about their daily life living conditions. In these efforts, they sought to transform and sustain rich and knowledgeable people-centred interactions, both to provide governance outcomes which would assist people in their daily life concerns, and to promote an awareness of the wider public realm of what being 'thrown together' in particular places involved in those contexts. As micro-practices, their activities illustrate the daily, yearly and intergenerational 'grind' of 'living' a relationship which connects our daily life worlds – as residents, citizens, workers, socialisers, 'volunteers', business enterprise owners and managers, etc. – to governance activity, and particularly that which surrounds and performs 'formal' government work. If we were to 'travel' within the worlds of formal government activity, there too we would find micro-practices in which government

16 It is the women and children who, until recently in Japanese society, lived most of their daily lives in their home neighbourhoods, men being much more socially involved in their workplace networks (Funck 2007).

17 This is partly due to generational change, and partly because many people left after the earthquake and did not return.

programs are devised, announced and performed. It is through these focusing and filtering processes that policy ideas and modes of performing government activity take shape and get packaged up, to 'travel' out into an array of sites where government activity is carried out and into the surrounding political culture.[18] Looked at in this way, what get to be called 'power structures' can be seen as particular sets of interactive relations and arenas, often themselves full of inconsistencies and conflicts, which have, at a particular moment, come to be in a position to exercise some control over substantial material resources and how they are accumulated and distributed. From these positions, their justifications and claims to legitimacy get a privileged 'sound' through the foghorns and babble of the wider political culture in which they exist.

As is by now widely-recognised in the democracies of the richer developed world, an elite class of national politicians, policy experts and often major corporations and large non-governmental organisations (NGOs) have tended to dominate how policy directions get articulated and then to shape how the material resources and regulatory power available to formal government are used. From their micro-locations in Kobe, citizen activists knew they were struggling to change such a practice and culture. In Walker, citizens had thought that the social democratic state for which they had struggled so hard in the first part of the twentieth century would deliver a governance process which understood their lives and needs. Now they feel let down by the formal government. Yet both cases suggest potential for a much richer and more citizen-respecting relation between formal government activity and civil society energy.

This, however, demands a changing ethos and practice from those working within the public sector. The tendency of governments and an elite nexus of politicians, senior officials, representatives of powerful business groups and other lobby groups to pull away from 'we-the-citizens' between elections has long been recognised. For example, Dewey (1991 [1927]) in the first part of last century saw the tension as between elite and participative forms of democracy. Habermas (1984, 1996), who drew significantly upon Dewey, expressed the tension as a duality between a system world and a life world. More recently in European debate, the talk has been about a 'democratic deficit' (Stoker 2006, 2010) or a 'diminished' form of democracy (Skocpol 2003).

Elite democratic practices may have won out in the twentieth century in the West, but we have been living with its crumbling weaknesses for the past few decades. Protest movements of the 1960s and since challenged the 'technocrats' as the source of authority, and/or the way in which political elites were co-habiting in their networks with business driven nexuses. In the terminology of late 1960s radicalism, which still reverberates in working-class communities such as Walker in Newcastle, the welfare state came to be seen as hijacked to the interests of large, corporate companies. These days, the criticism (often from business as well as the civil society sphere) is of a self-serving political class, which includes politicians but also all kinds of lobbyists, special interest groups and think tanks (Crouch 2004, Stoker 2006). Meanwhile, the delivery of government programs has been carried out by 'policy communities' of interested stakeholders (Rhodes 1997). Uncertain and fragmented, this political nexus picks up people's concerns as transmuted through a sensationalist media, and converts them into interventions without risking the complex and often politically unsettling processes of multifaceted 'reasoning in public'. In other words, political elites are no longer informed elites but uncertain, re-active populists, oriented not to problem-solving but to political survival, blown by the winds of 'mediatised' public concerns as they travel in the leaky rafts of the latest policy ideas being promoted by think tanks and policy consultancies, with few principles and little 'firm evidence' to guide them (Crouch 2004, Stone 2007, Hajer 2009).

Politicians, public officials and their advisers in recent years have all recognised the corrosive effect this has had on the quality of political cultures and especially the public realm of political discourse. Citizens feel alienated and detached from formal government and formal government activities repeatedly experience legitimacy and accountability crises in program design and delivery. It is this sense of a corroded public realm which motivates the re-thinking of what 'democracy' can mean. This re-thinking is evolving in several directions. A so-called 'neoliberal' strand centres on reducing the scale and reach of formal government and directing its processes according to principles borrowed from how markets and businesses are supposed to work. Citizens are expected to learn the 'game' of acting as 'service' customers, making demands on provision and assuring quality by informed choice. But many of us find it difficult to think narrowly of

18 This 'travelling' metaphor is taken from actor–network theory, especially Callon et al. (2009).

our 'utilities' in this way (Sen 2009). Drawn into micro-practices initiated by government agencies with a formal duty to involve us citizens in policy formation, monitoring and evaluation, we often realise that all our involvement is doing is to help frontline officials achieve their targets, which thereby keeps the funding in place for the service they are providing. From direct experience and from the media, Walker residents sensed this influence in their interactions with city council officials. Such practices feed, rather than reduce, cynicism and demoralisation (see Sager 2009, Diefenbach 2009, Durose 2009). More systemically, this strand of thinking promotes an individualist ontology and attention to self-interest, rather than recognising mutuality and the value of building good relations as a way to address the challenge of co-existence in the shared spaces of urban life. In Walker and Kobe, and the many other similar cases where neighbours mobilise to address collective problems, an ontology of mutual co-existence motivates many of the activists who get involved.

A more communitarian strand recognises this mutuality but shifts governance activity almost wholesale into civil society and the capacity for 'self-regulating systems' to evolve.[19] Such ideas get some support from the literature on 'insurgent' movements in urban areas where formal government has been too weak or too corrupt to respond to the demands of citizens for better material conditions (Cornwall and Coelho 2007, Holston 2008, Briggs 2008). Both Walker and Kobe could provide 'celebratory cases' for such a direction. But when adopted by politicians and administrations, that is, the representatives of formal government, this rhetorical invocation of self-governing capacity seems more like an off-loading of responsibility and a prop to arguments for funding cuts.

A third strand of re-thinking democracy takes an ontology of mutuality more seriously. It recognises that, while people do have identities as self-interested persons, at the same time we build a sense of ourselves through recognition of, and interaction with others around us – among family and friends, but also among the 'neighbours' with whom we share urban life. This is the terrain occupied by those interested in the potential of forms of 'deliberative democracy' and 'network governance'. It accepts that civil society needs formal government in many ways (Sirianni 2009). Proponents advocate two qualities in particular. The first is the significance of public argumentation, 'reasoning in public', in ways which recognise the diversity of perspectives, the complexity of many issues faced and the uncertainty of knowledge. Performing 'reasoning in public' in a way which achieves some respect in a wider polity is a hard challenge but yet creates a kind of trust which can become a resource for dealing with complex issues (see Wagenaar 2007, Hajer 2009, Callon et al. 2009, Sen 2009). Secondly, they underline the importance of attending to the micro-practices of the networks of relations which connect people, and in particular, stretch between, and draw back into co-relation, the spheres of civil society and the state. Hajer (2009), echoing the arguments of those who see structural forces demanding a new kind of governance, suggests that such examples may be forerunners of new governance landscapes and political cultures. In such potentially emergent polities, the *formal* institutions of representative democracy are but one dimension, not the dominant one, of a polity described in terms of the various networks which connect arenas and stakeholders in all kinds of combinations, in dispersed governance forms, rather than in the form of organisational hierarchies and voting procedures (Sorensen and Torfing 2007). In the case of Kobe, this sense of being part of a wider political community was an important dimension of local activism, in contrast to the Walker case, where a sense of such a wider identity had somehow been subcontracted to the social democratic state and then lost.

These strands are often presented as conflicting and incompatible approaches. Yet it may be more productive to see them as each highlighting a possible way forward in the effort to pull our polities back to a more engaged and interactive relation between the spheres of state and civil society. Although a 'neoliberal' direction may have the potential to increase inequalities among social groups, and a communitarian self-regulating approach may lead to marginalisation or exclusion of minority groups, these outcomes are not inevitable. This suggests that the search for an inclusive, knowledgeable, pluralistic, people-centred way of doing and living a democratic polity may be located as much in the evolving practices emerging at the intersection between these different strands, rather than the pursuit of an idealised model developed from any one of the strands. Instead of a search for a universal formula of some kind through which to foster richer

19 Such ideas draw on discussions about the formation of 'social capital' (Lewis 2010), and on Luhmann's (1982) conception of self-regulating systems.

interactions between citizens and governance practices, this conclusion leads to a search for *qualities* to foster and *capacities* to build within specific governance contexts.

This takes the argument back to the significance of attending to the dynamics of governance micro-practices and the particularities of their contexts. If it is in the micro-practices of governance activity that people get to learn what building more people-centred, pluralistic and richly intelligent political cultures in which diverse voices get to be listened to respectfully can mean as enacted, then what qualities and capacities might help to move in such a direction?

Building a Participatory Polity

Experiences from the spatial planning and urban policy fields are rich in suggestions of what these might be. Drawing on these, as well as well as the conceptual debates on rethinking democracy,[20] I put forward the following propositions or hypotheses about what to look for and encourage to help shape emerging governance processes and cultures in more progressive directions. Table 3.1 summarises these into seven 'qualities', counterpoised against qualities which should be resisted. They should be understood as 'qualities' of a polity in which the culture of the polis is richly informed by, responsive to and actively interacts, in fair and respectful ways, with the plurality of worlds of citizens.[21]

The first involves a recognition that people in Western democracies are concerned about the quality of their governance, both materially (how they experience its particular material aspects) and with respect to the sense of collective identity and pride which is engendered by feeling part of a polity they respect and value. We/they are not apathetic. We have opinions and care about the political communities in which we live. But we do not feel that we are listened to, despite the new technologies which allow us to send all kinds of messages all over the place. We may feel 'delivered to', but we sense that our experience as citizens has, in Skocpol's (2003) words, somehow been 'diminished'. We feel 'left out'. People do not necessarily want to engage in governance work, but we want to feel confident that we are respected and that our daily life worlds as we live them (not as the media sensationalises them) stretch into the consciousness and sensitivities of those who engage in work for the political community. This sense of being part of a political community also provides the basis for the feelings of solidarity and fairness upon which support for such values as distributive justice and environmental care builds.

Table 3.1 Towards more people-centred governance cultures/polities

Qualities to foster	Qualities to resist
Recognition of a shared, but diverse and often conflictual political community, in which distributional fairness matters	Dualistic conflict Factional stereotypes
Recognition of multiple identities and ways of life, as well as commonalities	Divisions into exclusive groups and singular identities
Respect for different arguments, positions, feelings, etc. ('adversaries' not 'enemies')	Competitive persuasion Struggles between 'enemies' Destructively violent conflict

continued ...

20 There is a substantial literature around concepts of deliberative democracy, and participative democracy. I have found the work of Connolly (1987, 2005) and Sen (2006, 2009) particularly useful in developing my own thinking in relation to the arguments in this literature. See also Wagenaar (2011).

21 The format for Table 3.1 was inspired by Fisher and Ury (1983). Sorensen and Torfing (2009) provide a somewhat similar exercise.

Table 3.1 (*concluded*)

Qualities to foster	Qualities to resist
A practice of 'intelligent', multi-sided discussion of issues – learning and reasoning 'in public'	Polarised extremes Fixed positions and interests
Critical respect for the role of formal government, while recognising the complexity of interactions between the spheres of state and society.	'Us' vs. 'Them' polarities Sharp divisions between 'public' and 'private' spheres
Respect for those who do formal government work and perform 'public services'	Continual negative critique (of incompetent and unresponsive bureaucrats or public service staff)
Rich, responsive and respectful interactions between all levels and sectors of formal government	Hierarchical paternalism and struggles over 'organisational turf'

Secondly, a polity which cultivates people-centred governance needs the capacity to combine a sense of commonality and common destiny with an appreciation of the diversity of identities we all have. In our debates about democracy and in our experiments in participatory democracy, we all too often (especially when engaging in a governance activity) fall into the trap of creating categories of people and assigning them a set of interests, as if we had singular identities. This danger is one of the key insights of interpretive forms of policy analysis (Wagenaar 2011). Such stereotyping encourages what Sen (2006) refers to as federal monoculturalism, or pluralist monoculturalism. Yet, as noted above, most of us have, and enjoy having, multiple identities. We are residents here, parents there, employees on work days, friends in the evenings, and sports enthusiast at weekends. We go to a clinic for an ailment, and may use a faith for our spiritual and moral refreshment. In these different worlds, we not only have different identifications but develop specific networks and cultural associations. These may enrich us, but quite often they clash. A richly participative democracy needs to cultivate the qualities of what Sen (2006) calls 'pluralist multiculturalism'. A particular challenge in this context is to work out what distributive justice in specific situations actually involves. In Walker, for example, which groups of houses should get the very small amounts of investment now available for stock improvement? Working through this challenge collectively is likely to lead to an outcome which is perceived as more fair than merely imposing a general rule or technical solution on those involved (Sen 2009).

Thirdly, the cultivating of practices of multisided public discussion, through media from street protest, to televised presentation, online chatter and face-to-face encounters of various kinds, in which participants are 'listened to' rather than 'positioned', does not imply the smoothing away of differences and the sharp divisions which generate some of the most intense conflicts in our societies. Conflict adds grit and edge which challenges the complacency of routine and habit. A vigorous public realm is surely one which recognises that people have very many different perspectives and concerns, and that the gain of some will often be a loss for others. The cultivation of debate and public reasoning is as much to expose these differences and conflicts as it is to reduce conflicts which arise from misunderstanding and failures of recognition. A people-centred approach to governance is not an easier path to managing our mutual co-existence. But it is a richer one for generating a polity in which conflicts can be understood and the grounds for them respected, even as choices have to be made from time to time about which 'side' to take and whose values to promote.[22]

Fourthly, and as part of this respect for citizens, we all need to give more attention to the kind of 'intelligence' we use when addressing issues of public concern. It is 'easy politics' to sensationalise and polarise, and to pigeonhole attitudes into ideological categories. This encourages a public culture where people 'take positions' and then battle over who should win. It is more difficult but yet more necessary in our complex societies, to develop understanding of the multisided nature of many issues and the difficulties faced by those involved in governance interventions on our behalf. Hajer (2009) looks at examples of different ways

22 Here I draw on arguments about the value of 'agonistic' versus 'antagonistic' approaches to conflict, see Connolly (2005), Sorensen and Torfing (2007), Hillier (2007) who all comment on Chantal Mouffe's formulation.

in which very complex issues got to be presented and re-interpreted in the unfolding of debates 'in public'.[23] We all need encouragement to 'learn in public', through discussions and interactions, in ways which allow issues to be unfolded and their many dimensions appreciated.[24] This does not mean leaving the debates only to those skilled in formal public debate, such as lawyers and politicians and their persuasive skills, or to scientists with their particular forms of testing and evidence. All of us can and do 'reason' about issues. We 'rationalise'. But, as anthropologists tell us, we vary in the grounds upon which we construct our reasoning and in what counts for us as evidence. This is what Dewey (1991) meant by 'cultivating the intelligence of a polity' through teaching us all how to be better 'scientists'. It is also what Lindblom (1990) argues when he encourages critical probing of policy arguments, and what Habermas (1996) searches for in his emphasis on the nurturing of the capacity for 'reasoning in public' as a key element of a rich and critical democratic life.

Fifthly, the energy of civil society does not replace the need for a state sphere, and for well-crafted, capable and respected institutions of formal government (Sirianni 2009). The 'withering away' of 'the state', as imagined in Marx's *Communist Manifesto*, would leave us bereft of ways of re-distributing resources more fairly, of restraining the build-up of world-damaging armaments, or of responding to changes in our climatic conditions. It would make it difficult to establish legal principles to safeguard human rights in the face of the actions of systematically unfair or violent treatment by some communities.[25] It would be difficult to link and resolve conflicts between the different 'communities' to which our individual and diverse collective identities acknowledge allegiance. Formal government and accompanying political mechanisms to render government institutions accountable and hence legitimate are needed at federal, national, state/province/region and local levels, along with supra-national linkages and global arrangements. However, these exist within, and need to reflect and interact with a political culture in which multiple civil society voices and movements can find expression, and connect to the arenas of formal government. It is through this multiplicity that the hegemonies of dominant forces (such as those of corporate capitalism, or of a powerful political ideology) can be continually challenged. This implies that political cultures and the practices of government institutions co-evolve, shaped by complex processes of mobilisation, networking and persuasion. It is through such interactions that the streets of daily life and the corridors of government activity move nearer together. Many analysts of governance activity worry about the problems of accountability presented by the existence of the diverse networks through which these interactions take place. These seem to undermine the hierarchy of legitimacy which is portrayed in the nesting of one formal level of government within another, each with its formal arenas for making key choices related to the public realm. Yet it has always been the case that the discussions and choices which take place in formal arenas are framed by all kinds of other debates and negotiations, only some of which are transparent to the publics on whose behalf choices are being made. The solution is not to discipline all networks back into the shape of formal organisations. Instead, if governance activity is becoming more diverse and often less formalised, as suggested in the 'network governance' idea, the ethics of those who do governance work and the mechanisms for holding them to account demand careful scrutiny as regards both institutional designs and the practices of a polity as a whole.

Sixthly, encouraging greater attention to the energy and voice of civil society movements for change does not imply a continual hostile critique of government institutions. This undermines the capability of government and encourages continued loss of respect and legitimacy for the state sphere. It discourages people from becoming politicians and public officials, and disheartens all those who work in 'the public sector' (see Lyons 2006). As citizens, we are surely right to judge adversely those politicians who seem just to be 'in the game' for self-advancement (power-seeking for its own sake), and officials who follow rule-books or the orders of superiors, without critical attention to the justification for the rules and orders, or their relevance to the circumstances they are dealing with. But as citizens, we also benefit from encouraging capable people committed to advancing the 'public good' through working in government institutions and the public sector. As several are now arguing,

23 Hajer's examples are: how to react to the murder of a film-maker known for his anti-Islamic views in a city renowned for its tolerance, how to decide what to build after the 9/11 bombing of the Twin Towers in the World Trade Center in New York, how to re-build confidence in British food standards after the 'mad cow' experience. See also, Innes and Booher (2010), for their collective experience of 'collaborative' governance forms.

24 See Laws and Forester (2007) for a discussion of the dimensions and forms of such interactions.

25 This is a key concern underlying Habermas' work (Specter 2010).

this demands an effort in re-valuing the work of public bureaucracies, and re-enchanting public administration with a sense of its social mission and moral responsibility (du Gay 2000, Sager 2009).

Seventhly, there can be no doubt that such an interactive, inclusive approach to rethinking the relations between civil society and the state, in which more participative forms of governance become 'normal', benefits from strong *local* governance institutions, both of formal government and less formalised ones (see Sirianni 2009). Richly participatory forms of governance seem to evolve more readily in places where local administrations have considerable autonomy and access to resources they can raise themselves. But 'localism' by itself can also end up exclusionary, unresponsive and incompetent. Instead, more attention is needed in formal institutional design and re-design to how tasks, resources and responsibilities are distributed between levels of government and how civil society voices and networks can reach into every level. This demands an analysis of networks which moves beyond simple categories of thin/thick and weak/strong, to a more nuanced and ethnographic investigation of network patterns and flows through which meanings are constructed and activities performed (Wagenaar 2011).

The above list is partly a possible checklist of working hypotheses for those seeking to build more people-centred governance practices, in which people get to feel that their governance activity 'knows and cares' about who they are, and in which those performing such activity sense at least some understanding and respect from those they are doing governance work 'with and for'. It also summarises issues which are the focus of active struggle in the polities of western democracies and elsewhere at present. But it should not be read as a 'good practice formula' to describe an ideal type or a reality which we can arrive at by appropriate steps. It points towards a 'culture of practice' (a way of being/doing; not just a way of saying). Every polity has its own particular configuration, and its own specific history and geography. It has its own 'moments' and 'windows of opportunity' for something different to happen (Tarrow 1994). How these moments and windows are grasped depends a lot on those who have the judgment to see them and the practical skill and energy to use them. Politicians, in their role as democratic representatives, have no monopoly on this capacity in complex urbanised societies, which is why they too may find the practices of participatory democracy, once they have got used to them, more congenial than those of a technocratic, professionalised elite democracy or a reactive populism.

Concluding Comment

In this article, I have argued that people-centred governance, or participatory democracy, describes the practices of polities which struggle to evolve into richly-intelligent, inclusively-fair and multiple-pluralist political communities. These practices arise not just from the creation of formal institutions, but emerge in interaction with the attitudes, values and sensibilities in the society as a whole. The governance processes and political culture of polities co-evolve. For a people-centred polity to emerge, an effort in 'intelligent' and respectful co-production is needed, as Dewey (1991) argued nearly a century ago. It demands efforts not just in social mobilisation, but in experimentation and careful learning from situated experiences. It also requires continual attention to resisting subversion back into the well-established pathways of elite technocracy and corporatism, or into exclusionary populism.

In North America and Western Europe, we have been vigorously experimenting for some time with participative practices. The fields of spatial planning and urban policy and planning have been important loci for such experimentation. This has resulted not just in new kinds of governance process, but in re-thinking the substantive foci of governance attention and making a real and valued difference to people's daily life experiences. As analysts of governance processes, 'we researchers' need to see such experiments as not just particular ways of 'managing' yet another type of problem, but as part of political struggles, as contributions to re-configuring governance institutions and breathing new life and energy into our polities, as part of an effort to 'will into being' some new kind of political community. As Jessop (2008) argues, many experiments will fail, or not work as intended. Yet learning takes place as much from failure as success. As researchers, we need to address carefully exactly why and how 'they didn't work', to identify where the constraints lay, and where strategic transformative energy might have been better targeted. We also have a role in offering concepts and vocabularies to help those who are experimenting practically to push forward their initiatives. The fields of spatial planning and urban policy provide a good focus for such a research endeavour.

But we must also remember that the idea of a people-centred democracy, with progressive rather than regressive tendencies built into it, will always be 'incomplete' and emergent. It is a direction to be struggled for, in the wider effort to 'will into being' societies in which a daily life conception of the world becomes the measure of how we should co-exist in an increasingly urbanised world. Such a conception would be grounded in a pluralistic, outward regarding attitude, yet acknowledging political community with those who share the spaces of our 755 daily lives. In such a world, many of our practices may damage not only each other but the conditions for life on our planet. In this struggle, competent experts, technocrats and bureaucrats are all needed. But their contribution should be grounded in an awareness of the multiple ways we diverse many live our lives and how we engage in collective action in civil society, not hitched to the old worlds of narrow elites or specialist politicians and officials, disconnected from the rest of us.

Acknowledgements

This article is adapted from a talk given at Roskilde University, Denmark, in December 2009, at the Conference Governance Networks: Democracy, Policy Innovations and Global Regulation, and from a paper given at the Interpreting Democratic Governance Conference organised by the Local Governance Research Unit of De Montfort University, UK, on 23–24 September 2010. My thanks to David Booher and John Forester for their very helpful comments on an early version of the paper, and to the CPS referees and Henk Wagenaar for their guidance. Thanks also to Jeff Corrighan for permission to make use of work undertaken for NCRN, Newcastle.

References

Altshuler, A., 1965. *The city planning process: a political analysis*. Ithaca, NY: Cornell University Press.
Barrett, S. and Fudge, C., 1981. *Policy and action*. London: Methuen.
Bevir, M., Rhodes, R.A.W., and Weller, P., 2003. Traditions of governance: interpreting the changing role of the public sector in comparative and historical perspective. *Public administration*, 81, 1–17.
Briggs, X. da Souza, 2008. *Democracy as problem-solving*. Boston, MA: MIT Press.
Burby, R.J., 2003. Making plans that matter: citizen involvement and government action. *Journal of the American Planning Association*, 69(1), 33–49.
Callon, M., Lascoumes, P., and Barthe, Y., 2009. *Acting in an uncertain world: an essay on technical democracy*. Cambridge, MA: MIT Press.
Connolly, W.E., 1987. *Politics and ambiguity*. Madison, WI: University of Wisconsin Press.
Connolly, W.E., 2005. *Pluralism*. Durham, NC: Duke University Press.
Corburn, J., 2005. *Street science: community knowledge and environmental health justice*. Cambridge, MA: MIT Press.
Corburn, J., 2010. *Toward the healthy city: people, places and the politics of urban planning*. Cambridge, MA: MIT Press.
Cornwall, A. and Coelho, V.S.P., eds, 2007. *Spaces for change? The politics of citizen participation in new democratic arenas*. London: Zed Books.
Corrighan, J., 2011. Understanding neighbourhood dynamics: a framework. Available from: http://www.newcastlecrn.org.uk.
Crouch, C., 2004. *Post-democracy*. Cambridge: Polity Press.
Dewey, J., 1991 (1927). *The public and its problems*. Athens, OH: Swallow Press/Ohio University Press.
Diefenbach, T., 2009. New public management in public sector organizations: the dark sides of managerialist 'enlightenment'. *Public administration*, 87(4), 892–909.
du Gay, P., 2000. *In praise of bureaucracy*. London: Sage.
Dunn, J., 2005. *Setting the people free: the story of democracy*. London: Atlantic Books.
Durose, C., 2009. Front-line workers and 'local knowledge': neighbourhood stories in contemporary UK local governance. *Public administration*, 87(1), 35–49.

Dryzek, J., 2000. *Deliberative democracy and beyond.* Oxford: Oxford University Press.

Fischer, F., 2003. *Reframing public policy: discursive politics and deliberative practices.* Oxford: Oxford University Press.

Fisher, R. and Ury, W., 1983. *Getting to yes: negotiating agreement without giving in.* Harmondsworth: Penguin.

Flyvbjerg, B., 1998. *Rationality and power.* Chicago, IL: University of Chicago Press.

Forester, J., 1999. *The deliberative practitioner: encouraging participatory planning processes.* London: MIT Press.

Friedmann, J., 2011. *Insurgencies: essays in planning theory.* London: Routledge.

Funck, C., 2007. Machizukuri, civil society, and the transformation of Japanese city planning: cases from Kobe. *In*: A. Sorensen and C. Funck, eds. *Living cities in Japan: citizens' movements, machizukuri and local environments.* London: Routledge, 137–56.

Fung, A., 2005. Deliberation before revolution: towards an ethics of deliberative democracy in an unjust world. *Political theory*, 33(2), 397–419.

Gilchrist, A., 2009. *The well-connected community: a networking approach to community development.* Bristol: Polity Press.

Goldstein, B., 2010. Epistemic mediation: aligning expertise across boundaries within an endangered species habitat conservation plan. *Planning theory and practice*, 11(4), 523–47.

Gonzalez, S. and Healey, P., 2005. A sociological institutionalist approach to the study of innovation in governance capacity. *Urban studies*, 42(11), 2055–70.

Habermas, J., 1984. *The theory of communicative action: reason and the rationalisation of society.* Vol. 1. Cambridge: Polity Press.

Habermas, J., 1996. *Between facts and norms.* Oxford: Polity Press.

Hajer, M., 2005. Rebuilding Ground Zero: the politics of performance. *Planning theory and practice*, 6 (4), 445–64.

Hajer, M., 2009. *Authoritative governance: policy-making in the age of mediatization.* Oxford: Oxford University Press.

Healey, P., 2004a. The treatment of space and place in the new strategic spatial planning in Europe. *International journal of urban and regional research*, 28 (1), 45–67.

Healey, P., 2004b. Creativity and urban governance. *Policy studies*, 25(2), 87–102.

Healey, P., 2005. Place, identity and governance: transforming discourses and practices. *In*: J. Hillier and E. Rooksby, eds. *Habitus: a sense of place.* Farnham: Ashgate Publishing, 189–218.

Healey, P., 2007. *Urban complexity and spatial strategies: towards a relational planning for our times.* London: Routledge.

Healey, P., 2010. *Making better places: the planning project in the twenty-first century.* Basingstoke: Palgrave Macmillan.

Healey, P. and Underwood, J., 1979. Professional ideals and planning practice. *Progress in planning*, 9(2), 73–127.

Healey, P. et al., 2003. Place, identity and local politics: analysing partnership initiatives. *In*: M. Hajer and H. Wagenaar, eds. *Deliberative policy analysis: understanding governance in the network society.* Cambridge: Cambridge University Press, 60–87.

Hillier, J., 2000. Going round the back: complex networks and informal action in local planning processes. *Environment and planning A*, 32(1), 33–54.

Hillier, J., 2007. *Stretching beyond the horizon: a multiplanar theory of spatial planning and governance.* Farnham: Ashgate.

Holston, J., 2008. *Insurgent citizenship: disjunctions of democracy and modernity in Brazil.* Princeton, NJ: Princeton University Press.

Hoppe, R., 2010. *The governance of problems: puzzling, powering and participation.* Bristol: Policy Press.

Innes, J.E. and Booher, D.E., 2010. *Beyond collaboration: planning and policy in an age of complexity.* London: Routledge.

Jessop, B., 2008. *State power.* Cambridge: Polity Press.

Laws, D. and Forester, J., 2007. Learning in practice: public policy mediation. *Critical policy analysis*, 1(4), 342–70.
Le Galès, P., 2002. *European cities: social conflicts and governance*. Oxford: Oxford University Press.
Leitner, H., Peck, J., and Sheppard, E.S., eds, 2007. *Contesting neoliberalism: urban frontiers*. New York: The Guilford Press.
Lewis, J.M., 2010. *Connecting and cooperating: social capital and public policy*. Sydney: UNSW Press.
Lindblom, C.E., 1990. *Inquiry and change: the troubled attempt to understand and shape society*. New Haven, CT: Yale University Press.
Luhmann, N., 1982. *The differentiation of society*. New York: Columbia University Press.
Lyons, M., 2006. *National prosperity, local choice and civic engagement: a new partnership between central and local government for the 21st century*. London: The Stationary Office.
Madanipour, A. and Bevan, M., 1999. *Walker, Newcastle upon Tyne: a neighbourhood in transition*. Newcastle: CREUE, University of Newcastle, 67.
Madanipour, A. and Merridew, T., 2004. *Neighbourhood governance, social exclusion and urban renaissance: a case study of Walker, Newcastle upon Tyne*. Newcastle, Global Urban Research Institute, University of Newcastle upon Tyne, 85.
Massey, D., 2005. *For space*. London: Sage.
Meyerson, M. and Banfield, E., 1955. *Politics, planning and the public interest*. New York: Free Press.
Moulaert, F. et al., 2007. Introduction to special issue: social innovation and governance in European cities: urban development between path dependency and radical innovation. *European urban and regional studies*, 14(3), 195–209.
Owens, S. and Cowell, R., 2011. *Land and limits: interpreting sustainability in the planning process*. 2nd ed. London: Routledge.
Phelps, N., 2012. *An anatomy of sprawl: planning and politics in Britain*. London: Routledge.
Pierre, J., ed., 1998. *Partnerships in urban governance: European and American experience*. Basingstoke: Palgrave Macmillan.
Punter, J., 2003. *The Vancouver achievement*. Vancouver: UBC Press.
Rhodes, R.A.W., 1997. *Understanding governance: policy networks, governance, reflexivity and accountability*. Buckingham: Open University Press.
Rydin, Y., 1999. Public participation in planning. *In*: J.B. Cullingworth, ed. *British planning policy: 50 years of regional and urban policy*. London: Athlone Press, 184–97.
Sager, T., 2009. Planners' role: torn between dialogical ideals and neo-liberal realities. *European planning studies*, 17(1), 65–84.
Sager, T., 2012. *Reviving critical planning theory*. London: Routledge.
Sen, A., 2006. *Identity and violence: the illusion of destiny*. New York: W.W. Norton and Co.
Sen, A., 2009. *The idea of justice*. London: Allen Lane/Penguin.
Sirianni, C., 2007. Neighbourhood planning a collaborative democratic design. *Journal of the American planning association*, 73(4), 373–87.
Sirianni, C., 2009. *Investing in democracy: engaging citizens in collaborative governance*. Washington, DC: Brookings Institution.
Skocpol, T., 2003. *Diminished democracy: from membership to management in American civic life*. Norman, OK: University of Oklahoma Press.
Sorensen, A., 2002. *The making of urban Japan*. New York: Routledge.
Sorensen, A. and Funck, C., eds, 2007. *Making livable places: citizens' movements, Machizukuri and living environments in Japan*. London: Routledge.
Sorensen, E. and Torfing, J., eds, 2007. *Theories of democratic network governance*. Basingstoke: Palgrave Macmillan.
Specter,M.G., 2010. *Habermas: an intellectual biography*. Cambridge: Cambridge University Press.
Stoker, G., 2006. *Why politics matters: making democracy work*. Basingstoke: Palgrave Macmillan.
Stoker, G., 2010. The rise of political disenchantment. *In*: C. Hay, ed. *New directions in political science*. Basingstoke, Hampshire: Palgrave Macmillan, 43–63.

Stone, D., 2007. Recycling bins, garbage cans or think tanks? Three myths regarding policy analysis institutes. *Public administration*, 85(2), 259–78.
Sullivan, H. and Skelcher, C., 2002. *Working across boundaries: collaboration in public services*. Basingstoke: Palgrave Macmillan.
Tarrow, S., 1994. *Power in movement*. Cambridge: Cambridge University Press.
Throgmorton, J., 1996. *Planning as persuasive story-telling*. Chicago, IL: University of Chicago Press.
Urry, J., 1981. *The anatomy of civil society: the economy, civil society and the state*. Basingstoke: Macmillan.
Vigar, G. et al., 2000. *Planning, governance and spatial strategy in Britain*. Basingstoke: Palgrave Macmillan.
Wagenaar, H., 2007. Governance, complexity, and democratic participation: how citizens and public officials harness the complexities of neighbourhood decline. *The American review of public administration*, 37(1), 17–50.
Wagenaar, H., 2011. *Meaning in action: interpretation and dialogue in policy analysis*. New York: M.E. Sharpe.
Watanabe, S.-I.J., 2007. *Toshi keikaku* vs *machizukuri*: emerging paradigm of civil society in Japan. *In*: E. Sorensen and C. Funck, eds. *Living cities in Japan: citizens' movements, machizukuri and local environments*. London: Routledge, 39–55.

Chapter 4
Neither Cordelia nor Polonius: Ethical Implications for Planning in the Face of Blatant and Hidden Imbalances of Power, Status and Rights

Francesco Lo Piccolo

An (Autobiographical) Introduction and an (Instrumental) Use of Metaphors and Layers of Time

Let me start with a few autobiographical notes and a confession: I met Patsy for the first time before having read any of her research works. It was the summer of 1993 in Lodz and the occasion was the AESOP Summer School for Doctoral Students and Young Academics: with a background of architectural studies. I was starting my PhD planning research project with – as frequently happens to first year PhD students – lots of questions as well as doubts about the basis of my future research activity. At a time when Italian immigration was entering into a new phase (with little research in the field, a regressive political reaction to it and no awareness of planning responsibilities in the face of the newly changed social structure of the population), my intuitions and core research question were related to the issue of ethnic minorities and the role of planning in guaranteeing and recognizing substantial citizenship rights when formal and institutional rights do not eventuate. In short, at the base of my research interest was the brave, and for me, the new theoretical concept of planning as a way to enhance democracy. In just those three days of the Summer School in Lodz, I had several talks and fruitful discussions with Patsy: the first of a series, which I have enjoyed in the following years. That first encounter opened for me the way to a broad set of research networks and some profound long-term friendships,[1] as well as a new journey in planning literature, starting from Patsy and Rose Gilroy's article on a people-sensitive planning (Healey and Gilroy, 1990).

If I look now at most of my previous as well as my present research activities, I recognize the persistence of that initial question: how can planning be sensitive in the face of minorities' needs and wishes, especially when such minorities are not 'recognized', and not consequently involved in political encounters?

Patsy's work offered me a wide set of normative perspectives and intellectual stimuli, with all the challenges of applicability and transferability in my very different research contexts. Given my context-based research field and the place where I live, I experienced difficulties in applying normative principles that are nurtured by promises on behalf of deliberative democracy. They are somehow inappropriate for those contexts that are in some ways 'extreme'[2] but not irrelevant, and exposed, for example, to violence and to hidden and illegal abuse, as in the case of organized crime.[3] So, these are the questions of applicability and transferability of Patsy's normative principles that I want to explore in the following paragraphs. Applicability and transferability are also a matter of systematically using reflections deriving from 'extreme', particular and locally determined cases. As suggested by Brand and Gaffikin (2007:284), 'hard cases can illuminate the challenges and contradictions involved in a proposition, without laying claim to being typical'. Moreover, reflections deriving from extreme cases and contexts can be helpful in order to individuate and discuss some more general phenomena that are much more evident – and consequently analysable – in 'extreme' conditions.

1 As in the cases of Huw Thomas and Judith Allen, to whom I owe a debt for many of my research experiences and intellectual reflections.

2 Here the expression 'extreme contexts' refers to conditions in which organized crime, corruption and 'disorder' (as a result of 'conflicting orders') are structurally concentrated.

3 The expression 'organized crime' or 'criminal association' is used in Italy to refer to those illegal groups – Mafia in Sicily, Camorra in Campania and 'ndrangheta in Calabria – that are locally space-based and with strong territorial identity, links and roots, although their networking activity is on a wider, global scale.

In just one chapter it is a challenging task to develop all this, considering Patsy's work, my experiences and all the planning debates and researches that have developed in the last twenty years. Here I will mainly refer to some of Patsy's articles and book chapters on the one hand, and to some of my research experiences that have been described in some previous articles (Lo Piccolo, 2008, 2009; Bonafede and Lo Piccolo, 2010), trying to re-connect them and to develop some forward-looking research perspectives. In order to do so, I will use two Shakespearean characters, Cordelia and Polonius, as metaphors of some theoretical dilemmas in the theory-practice relationship.

Why use Shakespeare's tragedies of Hamlet and King Lear to talk about democracy, when both plays are located in a time of un-democratic political systems? In my view, they are instrumentally appropriate in order to clarify the difference between formal and substantial democracy, as my research experiences are mostly based on substantially un-democratic contexts. In other words, I am deliberately using two metaphors and three layers of time – the fictive feudal time of King Lear and Hamlet, the Tudor time of Shakespeare and our current time – in order to reflect on planning and democracy, as here I want to stress the ethical as well as operational implications for planners when they face huge imbalances of power, status and citizenship rights. The chapter draws parallels between Shakespearian personae and contemporary planning practice to highlight problematic aspects of this practice, particularly as it plays out in contexts where substantial democracy is sorely lacking.

A Normative Ideal: (Questions of) Applicability and Transferability

From the beginning of my research, my research questions have been supported by the reading of Healey and Gilroy (1990), as they highlight the ethical commitment of a pluralist approach to needs and requests. Planners, faced with diversity of groups and individuals, in their interactions confront a range of multi-dependent, ethical and contextual issues, which demand sensitivity to the socio-spatial diversity of people's daily life experiences (Healey, 2012:21). This assumption, and the normative ideal that nurtures it, was and still is, both very convincing and challenging, but the reading of that article does not entirely answer my questions. In fact, Healey and Gilroy refer to people and more often to citizens, but what about those who are not citizens? There is a brief reference to this in Healey and Gilroy (1990:27), talking about the disownment of minorities, but it is briefly developed. Also in the following and more recent of Patsy's works, the most frequently used expression is 'affected citizens' (Healey, 2012:22). Healey and Gilroy (1990) is fundamental as it highlights the need to develop sensitivity in the face of the other, of getting behind the position of the other, and looking at the situation from his or her point of view. But who decides who is 'the other'? Which sorts of 'others' are recognized, and consequently accepted? And which sort of 'others' are excluded?

In brief, at the base of my research work were and are, on one hand issues of equity and justice, and on the other, the issue of guaranteeing rights of citizenship, in particular the ethical implications of planning activities and research in ensuring citizenship rights, especially for those who are not substantially recognized as citizens. Who is or is not a citizen is, in fact, a serious issue, defining principles of inclusion and exclusion within the territorial boundaries of the national or local government (Harvey and Potter, 2009). In this category I do not just consider, for example, stateless persons, refugees or illegal immigrants with no political rights, but all the weak groups in the society that do not have any real opportunity for claiming their rights and even improving their status.

If we assume the notion of citizenship as the theoretical and political sphere where the inclusive or conflictual relations between 'different' subjects are defined (Bobbio, 1990), my research has been focussed on the relation between inclusive and exclusive forms of citizenship, and the recognition of them in urban planning. According to Zolo (1994:4), the notion of citizenship puts into relation individual subjective rights and pre-juridical reasons and conditions for political inclusion or exclusion. Consequently, the notion of 'substantial' citizenship can be considered as an indicator of the functional level of democratic institutions in providing fair and equal access to public life and activities, including the urban ones (Zolo, 1994:4). In other words, to analyse democratic institutions and their activities, planning included, under the perspective of substantial citizenship allows us to measure and evaluate the level of democracy of political systems *ex parte populi*, as it adopts a double perspective: the entitlement of rights and the endowment of rights (Barbalet, 1988).

The instrumental interpretation of the notion of citizenship has led some authors to the conclusion of the necessity of overcoming a 'restricted conception' of citizenship, as explained by Held (1987, 1989), Bobbio (1990) and Kymlicka (1995) in order to gain a more inclusive recognition of a cosmopolitan, rather than a mere universal, recognition of rights. This extensive interpretation of substantial citizenship is the one I am referring to in the following sections of the chapter.

In my perspective, influenced also by the work of Harvey (1996, 2000, 2003), Friedmann (1999, 2011), Holston (1999) and Sandercock (1998, 2003) amongst others, this extensive interpretation of substantial citizenship is strictly connected with, and complementary to, planning: the juridical recognition of citizenship guarantees the right to forms of representation and participation, related to the processes of transformation of the cities; at the same time, planning initiatives and their results are important in influencing citizenship's boundaries, in order to nurture (a more substantial) democracy (Lo Piccolo, 2010). In brief, and in line with Patsy's words (Healey, 2012), I have been considering planning as a way to enhance and re-enchant democracy,[4] due to the long history of planning's normative interest in richly participative democratic processes (ibid.:20).

So, three core issues are tightly intertwined: the controversial issue of citizenship in its multiple facets – the formal that is institutional, the substantial and the denied; the role of planning and its ethical implications; and the very idea of democracy in its dichotomy – institutional versus substantial. Behind and at the same time, at the base of this research interest, two other issues and questions arise, which are influenced by the local contexts of my research activity: how to deal with these issues in political and social contexts where the notion of substantial democracy faces several obstacles in its way? And how to set a progressive planning agenda where – also for the same reasons and *milieu* – the role of planning and planners is very weak and with a low level of efficacy, or it is just 'on the side' of power?

From Healey and Underwood (1978) and then Healey and Gilroy (1990) to Healey (2012) there is a normative political and ethical *fil rouge*. The emphasis – when analysing the processes of planning rather than examining their contents – is on individuals and their interactions, without underestimating the importance of the structural contexts and the consequent configurations of power and influence within which these interactions occur. It is also about planning and planners' activity as a complex set of inter-relationships or social networks, where planners interact with people in highly varied positions and roles and at the same time, where planners themselves perform in highly varied positions and roles. It also concerns all the ethical consequential dilemmas of these multiple roles (Lo Piccolo, 2009). Also of interest are: the potentials of micro-practices in planning activities, and urban dynamics as well; a recognition and appreciation of the multiple identities and ways of life, based on the pluralist multiculturalism approach described by Sen (2006, 2009); the importance of discourse in formal and, moreover, in informal contexts through the analysis of normally invisible or unrecorded dialogues with different actors – formal and informal, institutional and not, technical and not, with different levels and nature of knowledge, not to talk about their interests, cultures and values, and 'stretching across boundaries between formal government and people concerned about their daily life living conditions' (Healey, 2012:28). Healey's multi-textured view holds several other elements: the ethical commitment of respecting different arguments, positions and feelings, and a claim for establishing practices of intelligent, multi-sided discussions in order to cultivate 'the intelligence of a polity' according to the principles developed by Dewey (1991), and as deeply discussed in Healey (2009); under the influence of Forester (1989), the normative imperative of developing appropriate and just planning skills, being attentive, asking good questions, exploring meanings and implications, helping to correct the self-deception of others, providing respect, treating the other in a relationship seriously; and finally, the planners' mission of developing skills and sensitivity in order to nurture the capacity of 'reasoning in public' as a key element of a rich and critical democratic life, as developed in Habermas (1981, 1989, 1996).

This is a normative framework that has inspired all my research work. However, what are the conditions and, indeed, the pre-conditions for a fair and substantial public reasoning? The problems, and my own research mission concern:

---

4 I had not read Healey (2012) before having been asked to write this chapter. And I was surprised at finding in Patsy's article the same intellectual yearning and ethical tensions that were inspiring some of my research, although my fields of observation and analyses have been so different – culturally, politically and geographically – from most of those observed and considered in Healey (2012) as many other of her research works.

1. how to practice this normative ideal in those contexts or situations in which the 'reasoning in public' arena is a mere deceitful pretence, or where, on some occasions, there is no 'reasoning in public' at all; and
2. how to practise this normative ideal when some, either individuals or groups, are excluded from the 'reasoning in public' process, due to their citizenship or social or cultural status.

The metaphor of the deceitful 'reasoning in public' in Act I, Scene I of King Lear will be used in order to highlight risks and weaknesses that can occur in some substantially undemocratic contexts. Moreover, the consequences of the opening scene of King Lear will be considered in order to describe the effects of the loss of substantial citizenship rights. This is a crucial point that is not often deeply considered in Healey's reflections. In fact, in Healey's work a recurrent word is 'citizens', which in some contexts does not entirely include all the individuals and groups in the 'world's stage', that is the public planning arena, although Healey (2012:23) recently highlights the need for innovative problem-solving and delivery capacity for the new agendas of a more demanding citizenry, more conscious of our multiplicities. But what if you do not possess any power? Or others/the institutions do not recognize your right to any power? The 'denied citizenship' still remains an open question.

From this question three problematic issues are:

1. how to gain citizenship status in discursive practices for those who are not recognized as citizens, and the consequent ethical dilemma of imperfect (and consequently not entirely inclusive) systems;
2. how to develop this in contexts where insincerity is the rule, and where the gap and imbalance in role between the public arena and the closed rooms of power and the most powerful elites is so huge; and
3. what might be the role for planners and their technical knowledge/expertise when the nature of conflicts is latent or hidden as a consequence of the structural characteristics of the above mentioned contexts.

The metaphor of Polonius, unsuccessfully dealing with hidden conflicts in Hamlet, will be used in the following in order to highlight the frailty of technical expertise in contexts of 'unspoken controversies'. In some of my previous research works, I faced the problem of how to assume the inclusive process of communicative ethics (Forester, 1989, 1999; Healey, 1997) in contexts and in local societies that are unwilling to adopt that model. The experiences I analysed (Lo Piccolo, 2008, 2009; Bonafede and Lo Piccolo, 2010) show the attempts at facing this issue in a 'substantially un-democratic' context, where the institutional sites provided by the arenas of formal planning rarely become key arenas where the conflicts play out.

Contrary to that which Healey (2012:34) properly highlights about the benefits of strong local governance institutions in developing interactive, inclusive approaches to rethinking the relations between civil society and the state, the contexts I am referring to differ in two main aspects: a.) the weakness of local institutions; b.) the interference, and in some cases predominance, of 'third parties' – and not just organized crime, but also family and/or religion, or a mix of the all three! – in the relations between civil society and the state. In fact there are some orders, processes and actions that are beyond or out of the state's control. Most policy and planning analyses consider the state area of intervention, underestimating the role of other institutions, such as family, religion or illegal crime organizations. The Shakespearean metaphors I will use are helpful in order to include those and their role in the theoretical debate.

The effects on planning as well as on urban dynamics are significant: in areas that have a strong presence of organized crime, their powers are capable of bending local governments to support real estate investments, encouraging phenomena like unauthorized building, corruption and political patronage. The most insidious causes of these phenomena reside in the capacities to:

- defend, reconcile and promote particular private interests, deliberately jeopardizing the public ones; and
- de-legitimize the public administration, the politicians and professionals, by hook – corruption – or by crook – violence/intimidation – or guaranteeing a general condition of inertia/inaction/ inefficiency/ mistrust (De Leo, 2011, 2013).

From her work prior to 2012, Patsy is fully aware of how imperfect the interactive processes are and how they have to be continuously monitored and revised in a dialogical and communicative approach as well as in an antagonistic dimension, and while not ignoring or refusing the institutional dimension. However, what are the conditions for the cultivation of debate and public reasoning in different political and cultural contexts? An analysis of the opening scene of King Lear will provide some insights.

King Lear: The Disownment of Citizenship in Insincere (Planning) Discourses

The story opens (Act I, Scene I) with the elderly King Lear deciding to give up his power and divide his realm amongst his three daughters, Cordelia, Regan, and Goneril, on the basis of which one of them loves him the most. Goneril and Regan, corrupt and deceitful, lie to their father with sappy and excessive declarations of affection, in gushing, insincere speeches. Cordelia, however, refuses to engage in Lear's game, and replies simply that she loves him as a daughter should. Her lacklustre but firm retort, despite its sincerity, enrages Lear and he disowns Cordelia completely. When Lear's dear friend and advisor, the Earl of Kent tries to speak on Cordelia's behalf, Lear banishes him too from the kingdom.

Lear is a planning tragedy, and in my eyes is *the* most outstanding planning tragedy in the entire story of western literature. First, the plot and all the following tragic consequences, starts with a land division (zoning): lands, that is, space, are not just properties (that is, economic values) but power and status. In fact, much more than in other Shakespeare's plays, characters and their identity, status and power are named and identified with the lands they own, administer and rule – Albany, Cornwall, France, Burgundy, Kent, Gloucester. Formally, the lands' assignment (that is, planning decisions) is based on and justified by common principles and shared values, which are argued and expressed in public discourses. As Healey (2012) highlights, decisions are established as consequence of discursive inter-relationships, in formal, and more significantly, informal contexts and unrecorded dialogues, as in the case of the unrecorded dialogues between Goneril and Reagan. Consequently, the interaction among people in Scene I is a practice of intermediation between potentially conflicting interests, using discursive communication as a key tool (Healey and Gilroy, 1990:21). Substantially, most of the public discourses, including those of Lear's daughters Reagan and Goneril, but also those of their husbands Albany and Cornwall, of Gloucester's bastard and evil son Edmund, and even those of some partially involved stakeholders, such as Burgundy and France, are insincere and false, as they are oriented to power and to individual interests. Decisions are formally taken according to equity and rigour in (abstract) principles and rules, but they are substantially revealed, and considered, as the most unjust and inequitable by all the subjects involved, whether they benefit or are damaged by them. The results of public discourses are not just a land allotment, but a new established system of power, status and citizenship, which impedes some of the involved subjects from participating in the following phases of the institutional decision making process. Cordelia immediately loses her status and rights, and the same happens to the 'technician' Kent. In short, the tragedy of King Lear reveals that planning is not just about assigning pieces of land, but recognizing or denying citizenship, as a result of sincere or insincere speeches in public reasoning.

From Lefebvre (1968) to Mitchell (2003), many scholars (see Friedmann, 1992, 1999; Harvey 2000, 2003; Purcell, 2003) have discussed how the right to the city must be defended if we want to live in a diverse, just society, as the very idea of citizenship rights is fundamental to protecting the ideals of liberal democracy. Consequently, citizenship is the theoretical and political sphere where conflictive relations between different subjects are defined. Whatever the explicit or implicit aims of any planning decision – whether favouring functional, economic or aesthetic reasons – every planning initiative contributes to a redefinition of the boundaries of citizenship, consequently shaping spatial or non-spatial forms of social control (Lefebvre, 1974; Yiftachel, 1998; Hillier, 2002). This continuous process of redefinition by means of planning initiatives and decisions may occur directly or indirectly, with intentional or unexpected, witting (often) or unwitting (rarely) effects. The land assignment has immediate and evident effects on the status and the citizenship recognition of Cordelia and Kent, but Lear, too, unexpectedly and unwittingly loses his status as a consequence of his 'planning decisions'. A redefinition of the boundaries of citizenship that are, in other words, the rights of using the city will always occur, as an effect of the redistribution of resources deriving from planning decisions, and whether it is a matter of economies, spaces or rights. As Healey

(2012:20) highlights, 'this has encouraged a research focus in the spatial planning and area development field on the micro-practices of democracy-in-action. For it is in these micro-practices that discriminations, exclusions and oppressions become manifest'.

The possible 'extensive' or 'restrictive' interpretations of the concept of citizenship – as they are described by Kymlicka (1995) and Held (1989) – have some relevant consequences, because they show, theoretically and practically speaking, how citizenship may include or exclude (Lo Piccolo, 2010). The recognition of the differences and the guarantee of an effective equity imply a reconsideration of the idea of citizenship. Nowadays, the characteristic of citizenship status, both legal and political, prevails due to a paradox in history and a process of involution.[5] If compared with the political civil struggles and juridical recognitions of the 1970s, democratic rights and justice claims are much more threatened by neo-liberal capitalist globalisation and its effects upon the development and planning of cities (Routledge, 2010). In other words, the recognition of citizenship by institutions has recently ignored most of the capacity or, indeed, the potential willingness to encounter and engage with differences.

While a static concept of citizenship, considered as a guarantee of the acquired rights, is widely prevalent, a dynamic concept of citizenship, considered as an activity and a political practice where recognition, defence and plural articulation of the rights are taken into consideration (Kymlicka, 1995) is rarely promoted. Citizenship as a status thus creates new geographies, distinguishing between centres and peripheries of the right holders (Roche, 1992); and in this way juridical inclusion/exclusion has repercussions for spatial inclusion/exclusion. Coming back to King Lear, the three layers of time overlap, with implications for citizenship and status. Cordelia's rise and fall present a clear example of the disownment of citizenship in planning processes. Based on her speech, Cordelia loses her status and consequently, she has no rights; she is not a citizen anymore, and presents as a metaphor of the marginalization of some groups into unacceptable 'others' within the political arena and due to the 'reasoning in public' effect. The same destiny (from presence to absence) is reserved also for the technician Kent. In the case of Cordelia, I am referring to the loss of status of all the minority groups or their advocates who are ignored in every political decision-making process, and the consequent disownment of their citizenship rights. In the second case, of Kent I am referring to the loss of the role, that is, the substantial disclaimer of the institutional validity of technical expertise when technical expertise faces power, reclaiming at the same time critical thinking and justice: Kent appears, in other words, as he who is punished for speaking truth to power, as an aggressively radical planner.

Here is the problem of truth and insincerity in public planning discourses. According to Young (2000), weaker parties can sometimes achieve their political goals when the democratic process is open and fair, and when there is sustained public discussion in which they have a chance to persuade fellow citizens of the justice and wisdom of their cause. Under the influence of Forester (1989), Healey and Gilroy (1990) highlight the importance of a people-sensitive planning, which is based on an epistemological, ethical, political, socio-psychological, structural and self-reflective consciousness. Most of Patsy's subsequent works (Healey, 1992a, 1993, 1996, 1997) are devoted to developing such an approach, method and sensitivity. If, according to Healey and Gilroy (1990:27) 'we must be able to 'ground' not merely our knowledge claims but the ethics of our choices about conversational strategies in the politics of the interactive situation', the failure of the reasoning in public in the first scene of King Lear shows the absence of a theory of truth or of some truth-criteria on which planning discursive inter-relationships can rely.

Healey and Gilroy (1990) highlight, as one of the critical ingredients of a people-sensitive planning, the importance of an ethical consciousness, but how to nurture and defend such consciousness still remains an open question. In the Habermasian theoretical framework they refer to, rationality and ethics rely on accuracy,

5 Bobbio (1990), describing the increase in the number of the spheres of rights, highlights how now more than ever rights cannot be ascribed to an abstract category – the generic man – but rather to different and peculiar categories – the specific man – according to the characteristics and the privileges of the various social statuses. As status, e.g. family status, the status of the free citizen, the status of the legitimate child etc., was in the past. Today citizenship is an instrument for the differentiation – and consequently the separation and the discrimination – of some subjects within the social body from others; hence, the status of citizenship is a privilege. This phenomenon is more evident in the urban dimension, especially if we consider the inability of the contemporary welfare-state to cope with 'a more demanding citizenry, more conscious of our multiplicities' (Healey, 2012:23).

integrity, accountability and sincerity of what participants say in communicatively rational discussions, and on a planner's duty of responsibly constructing dialogues and knowledge according to truth and rigour. This requires planners to be aware of the nature of the knowledge they use, how it relates to the knowledge used by others, and the kind of 'validity claims' brought forward (Healey and Gilroy, 1990:26). Discourses have to be 'fair' and 'right', and based on an ethical dimension of justice as well as honesty. However, as often happens, Goneril and Reagan's discourses are not truthful, and are based on rhetorical devices that are instrumental and entirely false. Cordelia is honest, but her speech is not persuasive. Lear's choice consequently lacks justice. It is profoundly unjust and unethical.

In fact, if we analyse the 'reasoning in public' process in Lear, some unsolved problems still remain. Cordelia refuses to make a big public fuss about her love for Lear; her honesty and integrity contrast with her sisters' selfish insincerity, which sets and perverts the context for the conversation. As highlighted by Healey and Gilroy (1990:27), 'every interaction is a power relationship, partly 'structured' by the roles and rules which set the context for the conversation, and partly by the 'power of personality' which everyone possesses'. Cordelia's character is not one-dimensional; she is more than just 'the good daughter' that her father would desire. Cordelia refuses to go along with Lear's little love test because she is simply too principled, indeed, too authentic to take part in something so socially low and fake; or, she appears to be completely alien to strategic action. A lot of Cordelia's motivation comes not from frustration with her father, but from anger at her sisters' insincerity, that is, from her speaking truth in the face of power.

The ethical dilemma posed by the effects of insincere speeches in public reasoning requires further insights; it is still a highly debated issue. Although Healey (1992a, 1996, 1997) considers Habermasian communicative ethics a valuable conceptual resource from which follows the ethical challenge of mediating conflicts in inclusionary processes, Hillier (2003) analysed the Habermasian concept of rational consensus formation and its fallibility, describing the possibilities of the permanence of conflict, non-reciprocity, domination and the related power games that are inherent in planning decision-making. In my experience, the argumentative turn fails if the discursive approach is not based on a truthful and open level of confrontation: it is not just a matter of imbalance of powers; it is a matter of making conflictive positions clear and visible, according to agonistic theory (Hillier, 2003; Gunder, 2003). In other words, in Lear as in many of my research experiences, public planning discourses are neither honest nor authentic and sincere; but this happens in many other situations! There is a further problem, and one peculiar to some undemocratic contexts. As a consequence of Cordelia's public refusal to take part in an insincere discourse that hides conflicts of interests as well as of values, she is abruptly and violently interrupted by the undemocratic power of Lear. The result is that Cordelia loses her status and consequently her rights and, therefore, there is no longer room for representation, as in the metaphorical figure of *homo sacer* used by Agamben (1998) in order to describe those circumstances when hegemonic power suspends the law with the intention of excluding, depriving of rights and marginalizing within the juridical order. This is exactly what happened, for example, in the case analysed by Bonafede and Lo Piccolo (2010), which explored the ambiguous role that participative practices tend to assume in fragile and substantially undemocratic contexts.

In my metaphor, we could say that Cordelia is not the planner, as she could be considered a disempowered citizen, like asylum seekers or refugees, but also, the technical expertise that is, of planners in our case, is subjected to a similar treatment. In fact, Lear's advisors are silent, becoming, sooner or later, victims of the conflicts of power, as in the case of Gloucester,[6] or they are banished after reclaiming their intellectual independence, losing their status and rights of citizenship, as in the case of Kent.[7]

6 When the loyal Gloucester realizes that Lear's daughters have turned against their father and are planning to murder the King, he decides to help Lear in spite of the danger. Regan and Goneril discover Gloucester has warned Lear of their plot, and Cornwall, Regan's husband, accuses Gloucester of treason and gouges out his eyes.

7 Kent is notable for speaking honestly, which is a rather rare position in King Lear. Kent is not afraid to be blunt when he tells Lear he is a fool for believing that Goneril and Regan care about him or, when he tells Cornwall that he does not like the look of his face. Kent, although banished by Lear, remains to try to protect the unwitting King from the evils of his two remaining daughters. He disguises himself and takes a job as Lear's servant, named Caius.

From Cordelia to Polonius

In such circumstances, conflicts can arise or just remain hidden, when, as in most of my experiences, there is no moral shock, there is no room for representation, 'local voices' are disappearing, powerlessness is not contested by the powerless, and protests are latent. It is not just a matter of imbalance of powers (Hillier, 2002; Hoch, 1994), or of progressive and radical urban movements involved in resisting the assaults on their life worlds and working for better living conditions (Routledge, 2010). Rather, it is a matter of making conflictive positions clear and visible in order to have the opportunity to deal with them. Whether on the side of collaborative planning or of the agonist (Hillier, 2003; Gunder, 2003; Pløger, 2004), what about those contexts where conflicts are eluded, implicit or drained, that is, devoid of their oppositional power? The stories that I dealt with (Lo Piccolo, 2008, 2009; Bonafede and Lo Piccolo, 2010) show such a context where deprived and minority groups such as immigrants and local poor, children and young people have passively accepted policies that confirm their inferior status or the absence of public policies in favour of de-regulating and privatizing the public realm, and without expressing any particular form of protest and/or conflict.

In a review of the role of planning practices in its conflictive dimensions, some authors see conflict as a disruptive force that causes imbalances in a system of interrelated parties, others view conflict as a potentially positive force that can promote change, integration and adaptability (Turner, 1991) and some others (Mouffe, 1999; Friedmann, 1992 and 2011; Harvey, 2000; Holston, 1999) suggest that innovation arises in a situation of social conflict. Subsequently, depending on the levels of community, political awareness and empowerment, as well as on the manipulative strength of dominant interests, planning may take deliberative/interactive forms, which clearly express its political role as a social learning practice. Alternatively, it may assume traditional, deterministic models or, at best, tokenistic and paternalist participatory forms that often exclude minority groups.

But if we look at the role, effectiveness and legitimacy of planning knowledge in dealing with conflicts, in many cases we could notice that anyway, technical knowledge is weak and frail. If we look back at Healey and Underwood (1978), most of their cases are affected by conflicting purposes: the major role of interest conflicts within society and its reflection in government policy (as well as in planning policies) is one of the conclusions of the Healey and Underwood report (p. 123). At the same time the power of officials, planners included, varies and can be reduced according to some structural variables of the local contexts.[8] In their conclusions, Healey and Underwood (1978) highlight that the influence of planners on political decision making is marginal and limited, especially in conflictive contexts. Many planning theorists have discussed limits and weaknesses of the technical dimension of planning knowledge, and the evolution of Patsy's work offers a noticeable contribution in this regard. Nevertheless, the neo-liberal stream has determined a 'revival' of technical knowledge and technocrats, globally and locally. So, it has become again an issue to be debated, although – yesterday as today – when conflicts arise, planning technical knowledge is the first to die. This is the role and fate of the planner when speaking truth to power, as in the case of Kent: a role model for the democratically committed planner who even when disavowed keeps an allegiance to justice and democracy, but this is also the fate of the institutional bureaucrat in the enforcement of his/her mandate role.

Hamlet: A Series of Murders, But Who Is the First to Die?

Here comes Polonius, listed as a member of the Danish Privy Council, chief counsellor of the King and the father of Laertes and Ophelia. If you read, on the one hand Healey and Underwood's (1978) analysis of planners' attitudes and ways of acting and, on the other, the huge literary reviews on Polonius' character, then you will immediately come to the conclusion that either Polonius is *the* planner or that planners are in many occasions in their working life Polonius!

8 'Planners' claims to influence rest on two grounds: first, their role in the administration of the planning functions of a local planning authority, and second, their claim to 'professional' status. This claim of possessing 'specialist' knowledge and skills justifies members of planning departments when seeking to exercise responsibility for particular tasks within the local authority' (Healey and Underwood, 1978:82).

Self-assured, cynical, self-centred, flatterer, long-winded, sly, devious, false, shrewd, immoral, a sermoniser, meddling, political, arrogant, despicable, vain, hypocritical, manipulative, verbose, insincere, self-absorbed – Polonius is also described as a windbag by some and a rambler of wisdom by others. It has also been suggested that Polonius only acts like a 'foolish prating knave' (Hamlet, 3.4, 195) in order to keep his position and popularity safe and to keep anyone from discovering his plots for social advancement. While he is a blowhard, and he does spout aphorisms that were clichés, even in the sixteenth century, his clichés constitute sound advice[9] and his observations reveal the nature of a weak but ruthless politician:[10] in other words, the technical advisor of the established power.

The analogy and similarities between Polonius' character and planners' habits and acts as described in Healey and Underwood (1978) are truly amazing; and most of my personal observations of planners' daily activities, methodologically inspired by Healey (1992b), arrive at the same conclusions. Polonius is a man bred in courts, exercised in business,[11] stored with observation, confident in his knowledge,[12] proud of his eloquence.[13] Such a man excels in general principles,[14] but fails in particular application: he is knowing in retrospect and ignorant in foresight. While he depends upon his memory and can draw from his depositaries of knowledge, he utters weighty sentences, and gives useful counsel. However, as the mind in its enfeebled state cannot long be kept busy and intent, the old man is subject to the dereliction of his faculties; he loses the order of his ideas and entangles himself in his own thoughts, until he recovers the leading principle and his former train of thought.[15] Not only are the boundaries of Polonius' space for action and, hence, potential for influence more limited than he would wish; in attempting to increase his influence, Polonius adopts principles that are likely to be acceptable to those he seeks to influence (Healey and Underwood, 1978:119) and, consequently, Polonius is the first to die.

It is interesting to me that Polonius is the first to die in the long series of murders in Hamlet. The murdering is evidently the result of several conflicts of power, all of which are hidden and latent for most of the play. Even the prequel of the tragedy of the murder of the King of Denmark, Hamlet the father, poisoned by his brother Claudius, is a hidden conflict that is absent from the reasoning in public and is revealed only in 'un-recorded' dialogues (e.g. those between Claudius and Polonius, Claudius and Gertrude, or Hamlet and the Ghost). Highlighting this metaphor and its consequent reflections are also inspired by the research method used in Healey and Gilroy (1990) as well as in many other works of Patsy, such as Healey (1992b), which focuses on one-to-one communication in informal contexts, giving examples of normally invisible or unrecorded interactions. In fact, the invisible (or dark) side of planning interactions is particularly challenging when analysing not just imbalances in power, but also that dimension of hidden conflicts which is frequently forgotten or underestimated in our reflections and evaluations.

Coming back to the hidden conflicts in Hamlet and the consequent series of murders, the list of the few survivors is also worth attention. To reiterate, the only survivors are: Fortinbras, prince of Norway, the new established power when the conflict ends; Horatio, listed as a poor scholar, friend and confidant of Hamlet – the story teller; the grave-diggers, listed as first and second clown – the outsiders; and, more significantly for me, the players, representing neither the power nor the technical knowledge, but the creative and independent knowledge of art.

---

9 In Act 1, Scene 3, Polonius gives advice to his son Laertes who is leaving for France, in the form of a list of sententious maxims.

10 Polonius may be elderly and demented but he must have been at least politically adept. He admits that he is not a man of great prestige, and yet, he has risen to be counsellor to King Claudius. Presumably, he counselled King Hamlet, the father, as well.

11 Planners are not independent specialists (Healey and Underwood, 1978:83) and are engaged in processes of bargaining (ibid.:92).

12 Planners might seek to use this claim to specialist knowledge and professional status as a resource in the competition for power and influence within local authority administrations (Healey and Underwood, 1978:88).

13 Planning knowledge would appear to consist more of symbolic statements than 'technical expertise' (Healey and Underwood, 1978:111).

14 Planners are seeking to maintain support based on common norms (Healey and Underwood, 1978:111).

15 There is also a general tendency to respond to constraints by limiting the approach to planning to that associated with planning by statute and by tradition (Healey and Underwood, 1978:112).

All the others, the eight principal characters, are murdered, but the first to die is Polonius. This is not by chance and, in my view, there is an interesting metaphor in this as Polonius appears as the technical knowledge in the midst of many hidden or latent conflicts. There are many other interpretations of Polonius' death, but in any of them there is anyway a general agreement: it is not by chance,[16] and there is a deep design in being the first of the series.[17] I am not interested, of course, in a review of the different interpretations, but here I highlight my use of this metaphor. In a context where most of the conflicts are latent or hidden, the first victim is not power itself, but technical expertise. Hamlet had slain Polonius instead of Claudius and during all the tragedy the direct or blatant conflict is eluded. This attitude of avoiding conflicts or hiding them leaves technical knowledge in a very weak and ambiguous role:[18] its only very powerful role lies merely in the functioning of apologetics and excuses for decisions taken on other grounds.

Polonius' acts and character fit very well with planners' attitude to serve the interests of those they are working with, the reference group, as it is described in Healey and Underwood (1978:114). At the same time, Polonius' presumption of neutrality is merely fake and inconsistent. Healey and Underwood (1978:116) highlight that planners are neither neutral nor consistent in the interest they are likely to promote, and this observation contrasts with the assumption of technical neutrality. So, the metaphor of Polonius brings also some ethical dilemmas. The ethical dilemmas I faced for many years in the planning research field (Lo Piccolo and Thomas, 2008) are already present in Healey and Underwood (1978) and concern planning practice: for what purposes is 'planning knowledge' used by planners? For what audiences are planners producing such knowledge? Who draws the boundaries of planning knowledge, and in what way? Polonius uses his knowledge only in order to serve the power or to help it in maintaining its status of privilege, and despite his efforts, Polonius is the first to die, due to the weakness of technical reasoning.

Pruitt and Rubin (1986) classify five strategies for dealing with conflict: contention, yielding, problem solving, withdrawal and inaction. The last two are adopted with the deliberate intention of exploiting the status quo in order to avoid conflict and to maintain the supremacy of the elites. The conflictive dimension is in some cases implicit or latent, hidden or concealed; it involves opposite interests but also different values. This contribution argues that as long as conflicts are latent or as long as the blatant dimension of conflicts is only about conflictive interests, a professional technical approach, which wants to be formally neutral and impartial, will lose. If the conflictive dimension of values over interests of public planning is more explicit in confrontation so, the chance of being the first to die is more unlikely.

The guiding principle in cases of this sort is that although it is not possible to share the same values, in order to allow government action it is possible and necessary to open a space of interaction, of overt opposition instead of hiding. The players used by Hamlet (Act 3, Scene 2) in order to unmask King Claudius and to reveal his crime – that is, in order to make the conflict blatant – are an appropriate metaphor for this potentially successful approach. In fact, the metaphor of the play within the play is particularly appropriate for distinguishing that which is 'open' and that which is 'secret' or 'hidden' in public reasoning, suggesting potential innovative strategies. If 'any proposal for change generates tensions and conflicts, bringing to the surface knowledge and values in a plurality of claims and voices in shared spaces' (Healey, 2012:20), it is necessary also to consider unmasking and interpreting latent conflicts in order to then understand how to find

16 'No other incident shows so deep a design, or is so appropriate for its purpose. Hamlet, acting blindly through impulse, slays the wrong one; the result is guilt. … Through reflection, Hamlet is unable to slay the right one; thus he cannot perform the great injunction laid upon his soul. Such is his dilemma; if he acts, it is through impulse, and he falls into guilt; if he reflects, he cannot act. … Through the death of Polonius, Hamlet has committed the very crime which he was seeking to punish; the son of a father murdered has himself murdered a father. … So this incident offers the profoundest illustration of Hamlet's character, and, at the same time, furnishes the motive of his death' (Denton, 1877).

17 Having slayed Polonius and being sent to England by Claudius in order to be executed at his arrival, Hamlet finds the orders and changes them to order Rosencrantz and Guildenstern killed. Ophelia reacts to her father's death with utter madness, falls in a stream and drowns. Claudius tries again to eliminate Hamlet, arranging a sword duel between Hamlet and Laertes, who is enraged over Polonius' and Ophelia's death. During the duel, a secretly poisoned drink is offered to Hamlet; he declines and instead, his mother, Gertrude drinks it. Laertes deceitfully scratches Hamlet with the poisoned sword to ensure Hamlet's death. Hamlet unknowingly then switches swords with Laertes, and cuts and poisons him. Laertes, dying, admits of Claudius' treachery. Weakening, Hamlet fatally stabs Claudius, and Hamlet begins his death speech.

18 The planners' uncertainty leaves them as open to manipulation by other interests as their colleagues adopt approaches which seem to be neutral (Healey and Underwood, 1978:120).

any common value that will allow us to change, in positive ways, the state of affairs, even facing the risks, conflicts and unexpected results of an open reasoning-in-public process.

Re-thinking Democracy: Neither Cordelia nor Polonius

In her recent work, Patsy Healey warns us that a people-centred polity, 'requires continual attention to resisting subversion back into the well-established pathways of elite technocracy and corporatism' (Healey, 2012:34). She also suggests (ibid.:30) that nowadays there are different strands for re-thinking democracy in the face of a corroded public realm. Amongst them, the strands of 'deliberative democracy' and 'network governance' take an ontology of mutuality more seriously than others, and have two main qualities: 1.) the significance of public argumentation, that is, 'reasoning in public' in ways that recognize the diversity of perspectives, the complexity of many issues faced and the uncertainty of knowledge; and 2.) the importance of micro-practices and networks of relations that connect people, drawing back into-correlation the spheres of civil society and the state.

My reflections are concerned with how to transfer and apply such a normative ethos. In my 'extreme' territories, blatant as well as latent conflicts place themselves and reproduce themselves within the gap between public objectives and the distortion of private interests, in the forms of corruption, patronage and un-democratic (hidden) control.

Healey (2012:33) is fully aware that, 'it has always been the case that the discussions and choices which take places in formal arenas are framed by all kinds of other debates and negotiations, only some of which are transparent to the publics on whose behalf choices are being made'; more generally, we all know that political discussions in closed rooms have frequently taken place, everywhere and anytime. However, when the 'closed rooms' activity is the very frequent or the only rule, technical knowledge become very weak and useless, and reasoning in public is a mere deceitful pretence, especially at the local level of governance.

One solution can be to move towards a more diverse and often, a less formalized governance activity, as suggested in the 'network governance' idea (Healey, 2012), though some pre-conditions are necessary:

1. The existence of the space of democracy, according to Arendt (2005). In order to have a space where Cordelia – even after having lost her rights of citizenships and with a weak status – will be allowed to talk and to be listened to, in order to enable and foster those qualities that Healey (2012:31) individuates as some of the essential requirements for moving towards more people-centred governance and polities.[19]
2. The awareness of the necessity of playing multiple roles in multiple scenes (Lo Piccolo, 2009), as Kent or Edgar or Hamlet know, and accordingly act.
3. The adoption of a normative perspective in order to make latent or hidden conflict as visible and as publicly debated as possible, instead of just relying on merely technical knowledge and expertise.[20]

In a blatant or a hidden conflictive situation of opposing interests, the role of values is fundamental; I agree with Watson (2006:44) that, 'introducing questions of value into deliberative processes is not necessarily contrary to a recognition of multiple and conflicting rationalities'. It is not easy to bring back the 'irreconcilable' opposition between different and divergent values to a level of democratically lawful and operationally possible discussion. The reference parameter can be found in the ambit of rights and their acknowledgement, in their spatial declination and how this translates into being 'operational': in other words, citizenship rights (Lo Piccolo, 2010; Lo Piccolo and Thomas, 2001) in their substantially extensive interpretation as described and explained by Kymlicka (1995) and Held (1987, 1989). This chapter focuses on the predominant value of citizenship rights as a key issue in the effort of moving from a professional-technical knowledge to a socio-political and ethical dimension of planning knowledge.

19 Recognition of multiple identities; respect for different arguments, positions and feelings; 'intelligent' multi-sided discussion of issues-learning and reasoning in public; rich, responsive and respectful interactions.

20 The limits of a presumption of planners' technical neutrality is described in Healey and Underwood (1978) as well as the contradictions between what planners claim to do and what they actually do (p. 121).

When considering the issues of citizenship and status, we should highlight the political role that planning may have for minority groups and individuals, representing a potential political arena for them through the micro-practices of democracy-in-action: 'small struggles which hardly seem to make a difference may sometimes, if built on in the flow of time, lead to major changes in political cultures' (Healey, 2012:20). The appropriation of spaces and the construction of new uses and practices in them, in 'institutional' or 'insurgent' forms, provide a guarantee of the claims of those rights of citizenship that often are denied at the political and/or juridical levels. Local actions of participation, formal or informal practices of coexistence in urban space, as well as inclusionary planning initiatives, become all-significant for a redefinition of the category of citizenship with respect to the substantial changes (and plural articulation) of the social corpus in the contemporary city (Lo Piccolo, 2010). According to Healey (2012:28), 'this, however, demands a changing ethos and practice from those working in the public sector'.

The Habermasian (1981, 1989) substantive norm, which is at the base of Patsy Healey's work, sometimes fails not just because of the gap between its theory and practice (Gunder, 2003) or because it represents a restrictive model that does not apply to most decision-making processes (Hillier, 2003). In some cases, such as those I refer to (Lo Piccolo, 2009; Bonafede and Lo Piccolo, 2010), it also fails due to the absence of a basic level of access to citizenship rights that are the indispensable ingredients – as I would say, pre-conditions – for Arendt's (and hence Habermas') space of democracy, which Cordelia lacks and which is the physical and metaphorical arena where 'reasoning in public' practices can really take place and flourish. According to Arendt (2005), the public space of democracy can be defined as the ambit where all the discursive issues can show their many-sidedness and people can freely show their own plurality by acting on and expressing their plural opinions. Thus, the public space of democracy coincides with the political space of freedom. The sense, but not the aim, of politics is the freedom of plurality. Therefore, understanding a political situation means acknowledging a large framework of different viewpoints and positions from which the situation can be considered and judged.

I highlighted previously (Bonafede and Lo Piccolo, 2010) the need for the existence of political spaces where conflicts can fully take place and agreements emerge through discursive and dialogic approaches, using some concepts as developed by Hannah Arendt (1958, 2005). In this approach, a substantial precondition for any form of 'public reasoning' is the existence of political spaces such as Arendt's idea of a discursive and active 'infra-space': the tangible space of the *agora* and the metaphorical space of democracy. This exists among equal individuals freely debating. Wherever this active and conversational freedom is lacking, a proper political space does not exist (Arendt, 2005).

Accordingly, citizenship rights do not only represent a framework of reference that is predominantly theoretical, or an abstract body of principles to which to rhetorically appeal, but are a litmus of daily analyses/assessments (Zolo, 1994), debates, choices and deliberations that in our ambit of thought and intervention are all specific, minute, practical and tangible. This commitment, however, assumes recognition of an ethical dimension to the planning discipline that is neither obvious nor automatic. Healey (2012:28–9) suggests a changing ethos and practice from those working within the public sectors, in the face of the current weaknesses of elite democratic practices and the protests and/or challenges against the 'technocrats' as the source of authority. In my experience, planners cannot act as Polonius, otherwise they will die first or they will be blinded, as in the case of Gloucester. They can learn lessons from time to time, from Kent, the disavowed, democratically committed planner who does not renounce to speak truth to power; Edgar in his role of 'Poor Tom', the insurgent planner who acts outside the formal framework of institutionally established procedures; or Hamlet instrumentally using the creativity of the players, that is, involving alternative subjects and skills that are different from the technically orthodox approaches.

If Healey and Gilroy (1990:26) properly highlight that, 'planners interact with people in highly varied positions and roles', I would add that planners themselves should learn to act in different roles and positions at the same time, with all the ethical dilemmas arising from playing multiple roles in multiple dramas (Lo Piccolo, 2009) and developing that 'epistemological consciousness' (Healey and Gilroy, 1990:26), which is necessary in order to cope with the diversity of interactions and multi-dependent, ethical and contextual issues. In potentially emergent polities and methodological approaches such as those which enhance the plurality and multiplicity of roles, we may assume at the same time that the formal institutions of representative democracy are but one dimension, and not the dominant one (Healey, 2012:30). The role of informal networks is consequently a major issue, but in which ways do the formal and the informal levels cope or clash, even frequently, with the other, especially in very latent conflictive and substantially un-democratic contexts?

According to Arendt (1958), without the existence of a common space – both physical and metaphorical – initiatives fail or do not even take place. In planning terms, without a common and sincere place of debate, there is no space that is also efficacious for informal networks, because of the patronizing attitude of the political actors and the subjugated condition of all the other subjects. It is not just the old, classical contrast between elite and participative forms of democracy that I am suggesting, but a search for making conflictive positions clear and possibly, sincere. Although Kent belongs to the elite, he is forced to set up informal networks outside of the formal arena and of the 'reasoning in public' process. The same strategy is adopted by Hamlet in his 'use' of the players, in order to defeat insincerity. On the other hand, without a public space of freely truthful confrontation, Cordelia will never regain her status – which was lost as a consequence of the disregard of her distinctive and different character as well as her authenticity in the 'reasoning in public' game. And Polonius will be the first to die, if he pretends to maintain an ethical dimension. Or, if he is on the side of power, after being used, he will still die sooner or later.

Acknowledgements

I am grateful to Jean Hillier and two referees for their constructive comments on the earlier draft of this chapter. Ideas, comments and field research were shared with Giulia Bonafede. A special thank you to Judith Allen, who offered fruitful insights and comments, and who has also been a great companion in enjoying many Shakespeare plays in the last twenty years.

References

Agamben, G. (1998) *Homo Sacer: Sovereign Power and Bare Life*, Stanford: Stanford University Press.

Arendt, H. (1958) *The Human Condition*, Chicago, IL: The University of Chicago Press.

Arendt, H. (2005) *The Promise of Politics*, New York: Schocken Books.

Barbalet, J.M. (1988) *Citizenship. Rights, Struggle and Class Inequality*, Milton Keynes: Open University Press.

Bobbio, N. (1990) *L'età dei diritti*, Turin: Einaudi.

Bonafede, G. and Lo Piccolo, F. (2010) 'Participative planning processes in the absence of the (public) space of democracy', *Planning Practice & Research*, 25(3): 353–75.

Brand, R. & Gaffikin, F. (2007) 'Collaborative planning in an uncollaborative world', *Planning Theory*, 6(3): 282–313.

De Leo, D. (2011) 'Public sphere and illegal settlements: a case from the Naples metro-region'. In: M. Cremaschi and F. Ekardt (eds), *Changing Places, Urbanity, Citizenship, and Ideology in the New European Neighbourhoods*. Amsterdam: Teknè, pp. 199–220.

De Leo, D. (2013) 'Trading with enemies? The trading zone approach in successful planning processes in Sicily'. In: A. Balducci and R. Mäntysalo (eds), *Urban Planning as a Trading Zone*, Dordrecht: Springer, pp. 125–42.

Denton, J.S. (1877) *The System of Shakespeare's Dramas*, St. Louis: G. T. Jones and Company.

Dewey, J. (1991) (1927) *The Public and its Problems*, Athens, OH: Swallow Press/Ohio University Press.

Forester, J. (1989) *Planning in the Face of Power*, Berkeley: University of California Press.

Forester, J. (1999) *The Deliberative Practitioner: Encouraging participatory planning processes*, Cambridge, MA: MIT Press.

Friedmann, J. (1992) *Empowerment. The politics of alternative development*, Cambridge, MA: Blackwell.

Friedmann, J. (1999) 'Claiming rights: Citizenship and the spaces of democracy', *Plurimondi*. 1(2): 287–303.

Friedmann, J. (2011) *Insurgencies: Essays in planning theory*, London: Routledge.

Gunder, M. (2003) 'Passionate planning for the others' desire: an agonistic response to the dark side of planning', *Progress in Planning*, 60(3): 235–319.

Habermas, J. (1981) *The Theory of Communicative Action: Reason and the rationalization of society*, Boston, MA: Beacon Press.

Habermas, J. (1989) *The Theory of Communicative Action. Lifeworld and System: A critique of functionalist reason*, Boston, MA: Beacon Press.

Habermas, J. (1996) *Between Facts and Norms*, Oxford: Polity Press.

Harvey, D. (1996) *Justice, Nature, and the Geography of Difference*, Cambridge, Mass.: Blackwell.

Harvey, D. (2000) *Spaces of Hope*, Berkeley: University of California Press.

Harvey, D. (2003) 'The right to the city', *International Journal of Urban and Regional Research*, 27(4): 939–41.

Harvey, D. and Potter, C. (2009) 'The right to the Just City'. In: P. Marcuse, J. Connolly, J. Novy, I. Olivo, C. Potter and J. Steil (eds), *Searching for the Just City. Debates in urban theory and practice*, Abingdon, New York: Routledge, pp. 40–51.

Healey, P. (1992a) 'Planning through debate: The communicative turn in planning theory', *Town Planning Review*, 63(2):143–62.

Healey, P. (1992b) 'A planner's day: Knowledge and action in communicative perspective', *Journal of the American Planning Association*, 58(1): 9–20.

Healey, P. (1993) 'The communicative work of development plans', *Environment and Planning B: Planning and Design*, 20(1): 83–104.

Healey, P. (1996) 'The communicative turn in planning theory and its implications for spatial strategy formations', *Environment and Planning B: Planning and Design*, 23(2): 217–34.

Healey, P. (1997) *Collaborative Planning: Shaping Places in Fragmented Societies*, London: Macmillan.

Healey, P. (2009) 'The pragmatic tradition in planning thought', *Journal of Planning Education and Research*, 28(3): 277–92.

Healey, P. (2012) 'Re-enchanting democracy as a mode of governance', *Critical Policy Studies*, 6(1):19–39.

Healey, P. and Gilroy, R. (1990) 'Towards a people-sensitive planning', *Planning Practice and Research*, 52(2): 21–9.

Healey, P. and Underwood, J. (1978) 'Professional ideals and planning practice', *Progress in Planning*, 9(2): 73–127.

Held, D. (1987) *Models of Democracy*, Cambridge: Polity Press.

Held, D. (1989) *Political Theory and the Modern State*, Cambridge: Polity Press.

Hillier, J. (2002) *Shadows of Power. An Allegory of Prudence in Land-Use Planning*, London: Routledge.

Hillier, J. (2003) 'Agon'izing over consensus: Why Habermasian ideals cannot be 'real'', *Planning Theory*, 2(1): 37–59.

Hoch, C. (1994) *What planners do: Power, politics, and persuasion*, Chicago: Planners Press.

Holston, J. (ed) (1999) *Cities and citizenship*, Durham, NC: Duke University Press.

Kymlicka, W. (1995) *Multicultural Citizenship*, Oxford: Oxford University Press.

Lefebvre, H. (1968) *Le droit à la ville*, Paris: Anthropos.

Lefebvre, H. (1974) *La production de l'espace*, Paris: Anthropos.

Lo Piccolo, F. (2008) 'Planning research 'with' minorities in Palermo: Negotiating ethics and commitments in a participatory process', *Planning Practice & Research*, 23(2): 187–209.

Lo Piccolo, F. (2009) 'Multiple roles in multiple dramas: Ethical challenges in undertaking participatory planning research'. In: F. Lo Piccolo and H. Thomas (eds), *Ethics and Planning Research*, Farnham: Ashgate, pp. 233–54.

Lo Piccolo, F. (2010) 'The planning research agenda: Plural cities, equity and rights of citizenship', *Town Planning Review*, 81(6): i–vi.

Lo Piccolo, F. and Thomas, H. (2001) 'Legal discourse, the individual and the claim for equality in British planning', *Planning Theory & Practice*, 2(2): 187–201.

Lo Piccolo, F. and Thomas, H. (2008) 'Research ethics in planning: A framework for discussion', *Planning Theory*, 7(1): 7–23.

Mitchell, D. (2003) *The Right to the City. Social Justice and the Fight for Public Space*, New York, London: The Guilford Press.

Mouffe, C. (1999) 'Deliberative democracy or agonistic pluralism', *Social Research*, 66(4): 745–58.

Pløger, J. (2004) 'Strife: Urban planning and agonism', *Planning Theory*, 3(1): 71–92.
Pruitt, D.G. and Rubin, J.Z. (1986) *Social Conflict: Escalation, stalemate, and settlement*, New York: McGraw-Hill.
Purcell, M. (2003) 'Citizenship and the right to the global city: Reimagining the capitalist world order', *International Journal of Urban and Regional Research*, 27(3): 564–90.
Roche, M. (1992) *Rethinking Citizenship. Welfare, Ideology and Change in Modern Society*, Cambridge: Polity Press.
Routledge, P. (2010) 'Introduction: Cities, justice and conflict', *Urban Studies*, 47(6): 1165–77.
Sandercock, L. (1998) *Towards Cosmopolis: Planning for multicultural cities*, Chichester: John Wiley & Sons.
Sandercock, L. (2003) *Cosmopolis II: Mongrel cities of the 21st century*, London, New York: Continuum.
Sen, A. (2006) *Identity and Violence: The illusion of destiny*, New York: W.W. Norton and Co.
Sen, A. (2009) *The Idea of Justice*, London: Allen Lane/Penguin.
Turner, J. H. (1991) *The Structure of Sociological Theory*, 5th edn. Belmont, CA: Wadsworth.
Watson, V. (2006) 'Deep difference: Diversity, planning and ethics', *Planning Theory*, 5(1): 31–50.
Yiftachel, O. (1998) 'Planning and social control: Exploring the 'dark side'', *Journal of Planning Literature*, 12(2): 395–406.
Young, I. M. (2000) *Inclusion and Democracy*, Oxford, New York: Oxford University Press.
Zolo, D. (1994) 'La strategia della cittadinanza'. In: D. Zolo (ed.), *La cittadinanza. Appartenenza, identità, diritti*, Roma-Bari: Laterza, pp. 3–46.

Chapter 5
Throwing Dice: Between Contingency and Necessity in Spatial Planning

Jean Hillier

> If the role of planning theory is to provide conceptual resources for efforts to make places more just, livable and sustainable, contributions are needed which offer normative ideas.
>
> (Healey, 2000: 920)

Preamble

Some years ago, Patsy Healey and I were discussing a paper on ethics. Patsy commented that the problem with the paper was that 'they are still seeking God'. At the time, I didn't think too much about this, but later came to realise that Patsy's own ethics are reflexive, contingent and immanent rather than transcendent universals. Such a view of ethics – a pragmatics of 'acting well' – has much in common with that of French philosopher, Gilles Deleuze. Inspired by Patsy to not seek some form of transcendent universalist God, I offer this meditation on ethics and spatial planning.

Introduction

Institutions involved in practices of spatial planning are founded on norms which define duties and responsibilities, require actors to behave morally and generate effects which have ethical implications (Sayer, 2007: 261). 'Normativity' is a fairly recent term which includes issues of value, good, ought, rationality, obligation and so on. Normativity distinguishes the value side of fact/value distinctions and the 'ought' of is/ought distinctions (Finlay, 2010). It distinguishes between 'correct or incorrect, just or unjust, appropriate or inappropriate, right or wrong and the like' (Rouse, 2007: 3). Questions of norms and values are encountered in ethical, legal and political senses in professional disciplines such as spatial planning.

In planning, moral norms embody assumptions about what 'well-being', the 'good city' and so on comprise. There are states towards which planning aims, but inevitably fails to achieve. In normative terms, for example, a key objective of planning practice is to enable people and places to flourish (Healey, 2010a, 2011). In practice, however, particular planning policy and development decisions may be driven by other objectives, with other, often unanticipated effects on people's 'flourishing' capacities.

Patsy Healey (2000a: 917) regards spatial planning as inherently normative. Throughout her career, she has been interested in interrogating complex relations between concepts, research methodologies and analyses ('how do they come to think like this' [2014]), exploring norm-reinforcing networks (1979; Healey and Underwood, 1978, reproduced in this Part) and the ways in which practising planners deal with conflicting norms (Healey and Underwood, 1978). Her investigations of governance micro-practices in the United Kingdom and beyond always consider norms of values, governance and place qualities. For Healey, the spatial planning idea inevitably 'carries with it a normative orientation' (2012b: 200). However, her engagement with the unexpected and unintended consequences of planning has led her to question the 'universal validity' (2012b) of planning both as an idea and as a set of essentialised practices. She warns us to be cautious about unthinkingly accepting norms and 'general principles which could provide a basis for acting in and evaluating specific instances of planning activity' (2012b: 192), especially if these norms have been 'imported' as best practice from other jurisdictions.

Drawing on her readings of American pragmatism (particularly Charles Peirce, William James and John Dewey), Healey (2009) recognises the normativity of actors as non-unitary subjects whose identities are relationally formed in social contexts through processes of encounter and transformation. A pragmatist emphasis on discovery through experience and experiment, together with her recent hands-on-experiences with civil society governance activity, is reflected in her understanding of the importance of planners and plans allowing flexibility for the unexpected (2008a) faced with the dynamic complexities of place (2007a; 2012b). Healey argues that people-centred governance – and spatial planning as an activity of people-centred governance (Healey and Gilroy, 1990 reproduced in this Part; Healey, 2012a) – demands efforts of social mobilisation, of 'experimentation and careful learning from situated experiences' (Healey, 2012a: 34). She recognises that strategies as experiments will always be incomplete and emergent (2007; 2012b) and that 'many experiments will fail, or not work as intended' (2012a: 35), especially if they are inadequately contextualised, failing to pay attention to the particularities of the situations they are presumed to address.

Patsy Healey's work is less concerned with representing past patterns of experience and more with investigating the potentialities of spatial planning as experience in order to explore alternative connections amongst the differential elements in an organised way. I suggest that Patsy Healey's thinking increasingly resonates with the philosophy of poststructuralist scholars (see also Huxley, 2009), such as Gilles Deleuze, in that they recognise and attempt to work with key notions of immanence rather than the transcendence of universal norms, and emerging, transversal values rather than the implementation of canonical laws (see Braidotti and Pisters, 2012: 2). Between contingency or chance and necessity, or as Patsy Healey writes, 'what can be fixed ... and what can be left to emerge' (2005: 6).

In the paper discussing the trajectory of pragmatist thought and its relevance for spatial planning, Healey (2009: 282) mentions the contribution of Richard Rorty and his comment that 'James and Dewey ... are waiting at the end of the road which, for example, Foucault and Deleuze are currently travelling' (Rorty, 1982: xviii). At first sight, Patsy Healey and Gilles Deleuze may be unlikely chapter-fellows, yet I suggest that there are several resonances between them. Resonances may be found in Healey's interest 'in the way the dynamic fluidity of evolving relational webs intersects with the "fixes" that develop as certain ways of thinking and doing become consolidated into accepted practices' (2007b: 15) and in her statement regarding strategic spatial planning that 'all conscious interventions to shape urban dynamics need to be evaluated not as steps towards given ends ... Instead they are risky experiments' (Healey, 2007a: 31). I will develop this idea of evaluation, rather than judgement, below.

In 1990, Patsy Healey and Rose Gilroy suggested that spatial planning required six critical forms of consciousness (paper reproduced in this Part): epistemological, ethical, political, socio-psychological, structural and self-reflection. I engage in particular with ethical consciousness, though the other five forms are never far from the surface. Spatial planning practice inevitably has constitutive ethical and moral dimensions, as actors have to have some notion of how they are supposed to behave, their responsibilities, duties and so on (Sayer, 2007).

I discuss questions of contingency and necessity in relation to spatial planning and ask, if God plays dice, do spatial planners play dice too? I suggest that planning decision-making engages both contingency and necessity, a key aspect of which is how planners 'play' the 'planning game'. In this regard, I explore issues of normativity and judgement and 'problems' of decision-making in a contingent, ever-changing world. I draw out resonances between Healey's and Deleuze's conceptualisation of ethics and morals, highlighting relation and praxis; the relevance of an immanent ethics grounded in problem contextualisation and critical, reflexive thinking. A brief illumination of unthinking 'stupidity' in relation to planning practice emphasises this last point and leads to a call, in conclusion, for an ethical insurrection in spatial planning.

Do Spatial Planners Play Dice?

Spatial planning comprises a set of what I will argue are insufficiently situated practices (Haraway, 1988), affected by and affecting a wide range of assemblages of force relations. A diversity of local planning practices has evolved, inter- and intra-nationally, as 'local institutional capacity confronts wider opportunities and pressures' (Healey and Williams, 1993: 717). Not only are planning practices epistemologically different from each other (difference between or distinction from: not-x), they are inevitably different-in-themselves (ontological difference).

A city, such as London, and its planning system, are distinct from other cities and their systems, such as Tokyo. There is an identifiable and understandable difference between them. London and its planning system is not-Tokyo and vice versa. This is an empirical difference-between; a relation between multiple terms where we have objects which differ from one another (London and Tokyo and their planning systems) and an external mind or consciousness which is distinguishing between them. This is difference epistemically conceived.

However, London and its planning system, as objects, are themselves constituted by differences irrespective of whether any external minds are distinguishing them from other objects. London and its planning system differentiate or distinguish themselves through internal forces, rather than being externally distinguished. This is difference-in-itself; a powerful intensive difference of potentiality where there is an internal self-differing within differentials (dx).[1] Difference is thus not a thing, but a process. Knowledge, therefore, cannot be fixed prior to an engagement with difference.

'The course, potentialities and limits of order(ing) and the whole of which it forms can neither be predicted in advance nor definitively established: they continuously change, adapt and transform in ways unexpected and changing' (de Vries, 2013: 139). Planning practice never was, and never can be, one. It is always-already a differential field which productively engages in a world that is always in need of more than limited, situated knowledge and limited intervention. Planning interventions are 'risky experiments' (Healey, 2007a: 31) which are always-already different. Or, as Deleuze and Parnet suggest: '[e]verything is played in uncertain games' (1987: 147).

The notion of spatial planning as a game is not new (see, for example, Long, 1958; Keyes, 1969; Oatley, 1995; Pacudan, 1997; Healey, 1974, 1998a; Lubell et al., 2010; Galster, 2012; Tato and Vallejo, 2012; Garvin, 2013). There are several board and electronic games based on urban and also rural planning simulations (Poplin, 2011), including *SimCity* (1989), *Urban Plan* (2001), *City Creator* (2011), *Suburbia* (2012), *Grepolis* (2013) and *Roll for the Galaxy* (2013), the last of which is a dice game for two to five players featuring more than 120 custom dice, in which players attempt to work with chance (Lehmann et al., 2010). As Patsy Healey perceptively writes, 'it is the nature of the "game" itself which is as much the problem as the playing of it. Stakeholders come to place-making issues accustomed to the particular games of their own social worlds' (1998a: 1542).

Whilst Healey might not agree that either planners (Temple and Boon, 2002) or politicians (Callies, 1999) have the dice loaded against them, she would accept the role that contingency, chance or the undetermined play in the unfolding of events. As the physicist, Stephen Hawking synthesised: 'the universe does not behave according to our pre-conceived ideas. It continues to surprise us … . Einstein was doubly wrong when he said that God does not play dice. Not only does God definitely play dice, but He sometimes confuses us by throwing them where they can't be seen' (1999: np). This is a sentiment also expressed by Gilles Deleuze: 'God makes the world by calculating, but his calculations never work out exactly, and this inexactitude or injustice in the result, this irreducible inequality, forms the condition of the world. The world "happens" while God calculates' (1994: 222).

Spatial planners, then, are left to cope with the outcomes and externalities of the dice throws. When they adopt policy directions or gazette a strategic plan, they inevitably take a chance on what might eventualise. Or, as a definitional example in the Merriam-Webster online dictionary suggests, they 'stake something on a contingency', speculate or gamble. 'The mayor is gambling that the new policies will help rather than hurt the city' (Merriam-Webster, 2013: np).

It could be argued that politicians and planners 'gamble', or take 'risky bets' (Healey, 2004: 91), that in (un)anticipated events, or dice-throws, the dice will land favourably over the lifetime of a strategic spatial plan. The multiplicity of dice include those of economic situations (e.g. changes in financial interest rates affecting new residential and major project development), political situations (e.g. any regime change remains favourable to the plan's objectives), social situations (e.g. public opinion remains favourable to the plan's objectives) and environmental situations (e.g. major environmental 'disruption' will not occur within the

1 See Bryant (2009) for further explanation of epistemological and ontological difference.

plan's lifetime). Some dice will be unseen, while combinations of others may necessitate substantial revisions of strategic thinking and acting.

If planners cannot even locate the dice on occasions, where might they turn for theoretically-grounded support or assistance? Deleuze's ontology of difference (1983; 1990; 1994) engages with the aleatory, or chance. Deleuze (1990; 1994) discusses the issue of problems (perhaps strategic planning problems?) in relation to throwing dice: 'the singular points are on the die; the questions are the dice themselves; the imperative is to throw. Ideas are the problematic combinations which result from the throws' (1994: 198). Deleuze perceives a problem as a field of differential relations and their accompanying potentialities. Problems, then, are generative of 'ideas' or ways forward.

'Normal games' attempt to 'abolish chance' or to fix it at certain points (Deleuze, 1990: 71) by playing with pre-established categorical rules, as in most of the games mentioned above (and as Abbott [2005] reports planners as doing) and by calculating 'win-lose' probabilities over several metaphorical throws, 'choosing the best hypothesis which determines a win' (Deleuze, 1994: 198). Such 'normal games' contrast with what Deleuze (1990: 69–76) calls an 'ideal game', in which chance can change the rules and cannot simply be captured through probability over a succession of throws: 'every play ... puts those rules into question even if it obeys them' (Williams, 2008: 122). In Deleuze's ideal game, rules are transient, there is never a final winner or loser. Ideal games, nevertheless, possess the capacity to disturb 'reality, morality and the economy of the world' (Deleuze, 1990: 60).

Deleuze stresses the distinction between chance and the arbitrary or unconstrained. The chance of each dice throw resulting in a particular outcome is limited (one in 36 for two dice). For Deleuze the throw retains a coherent structure and process. There are two moments in a dice throw (Deleuze, 1983: 23–4): the dice thrown in the air or spinning on the table before they settle (perhaps akin to a debate over legislation or a particular policy-direction?) and the fall of the dice when they stop rolling and show a particular combination (the enactment of legislation or sign off of policy). The throw is the moment of chance; relationships between individual dice, the dice and the air, the players and so on, are contingent. Although players can attempt to cheat chance by loading the dice in their favour by some means (such as performing a ritual or special throw, or surreptitiously weighting a die/persuading or bribing planning decision-makers perhaps), nothing can be guaranteed.

When the dice have fallen, this is the moment of necessity.[2] A particular relationship of the dice to each other has actualised and must be lived through. There is no turning back. Contingent events, such as the dice throw, become necessary; becoming becomes being. 'Necessity and chance, being and becoming are simultaneously affirmed in their constitutive difference' (Conway, 1998: 7).

The issue is how to react to the above. For Deleuze (1983: 26–7), a bad player 'makes use of causality and probability to produce a combination that he (*sic*) sees as desirable'. Bad players deny the role of chance. They aim for a desired combination or outcome and calculate a set of probabilities based on potential combinations, but 'all that will ever be obtained are more or less probable relative numbers' (Deleuze, 1983: 27). Good players experiment. They acknowledge the necessity of chance and 'explore the virtual rather than ... cling to the actual' (May, 2005: 65), testing out relations, recognising the limitations of particular constraints and attempting to work with enabling constraints where possible. The game, for Deleuze, is 'to multiply the means of recognising and experiencing the multiplicity of ways by which things occur' (Fuller and Goriunova, 2012: 162).

There is a tension between Deleuze's ideal and normal games. Both are rule-structured to some extent, with assumptions about how to act (throw the dice on the table, don't weight the dice etc.). The question for planning practitioners, however, is who determines what type of game is being played? Who situates the game?

It is at this juncture that Deleuzian ethics may be of assistance. Deleuze (1983: 28–30) considers the dice throw, not in terms of probability, but, following Nietzsche (1961), acknowledges every throw as a unique event, as if the first throw were being performed anew each time. Chance is thus reaffirmed with each throw, rather than diminished or abolished.

2 See Deleuze's (1983) discussion of Nietzsche's affirmation of chance and necessity as yoked together, contrasting with Mallarmé's insistence on their total opposition.

The future of a city or a planning system is thus radically contingent. It is effectively unpredictable inasmuch as it contains the potentialities for unseen shifts in actors and rules which destabilise and change it. Similarly, chance encounters and interactions between systems can lead to destabilisation and transformation. Yet once the planning 'dice' have fallen, it is necessary for those in the planning assemblage (planning practitioners, engineers, surveyors, environmentalists, local residents, politicians, non-human animals and nature, etc. etc.) to cope with what actualises. Chance, 'coupled with necessity, is itself born of and reconstitutes the open, and in turn is disturbed and perpetuated by the action of chance, of change upon change' (Fuller and Goriunova, 2012: 162). The future, therefore, lies between contingency and necessity.

On Normativity, Ethics and Morals

Normativity refers to 'imperatives, duties, obligations, permissions and principles which do not describe the way the world is but rather prescribe the way it ought to be' (Jun, 2011: 91). Norms give reasons for actions. Originally a legal concept relating to the law and, especially, to the force of law, normativity expanded to become a moral concept and canon for 'correct thinking' (normative rationality). It is concerned with what is right or ought to be done (Korsgaard, 1996; Finlay, 2010), rather than what is good or valued. We should, however, recognise that norms are inherently political and, as such, some reflexive procedure is necessary to maintain constant vigilance of what norms we follow. As I discuss below, the right is co-extensive with issues of morality and the good with ethics.

Planning practice involves 'not only the learning of technical skills, but also the adoption of values and norms' (Schein, 2004: 10). Practitioners struggle with normative questions of resource distribution: who gets what, when and how (Laswell, 1958). Normativity derives from processes of practical reasoning. For Deleuze, normativity is an immanent process, rather than one relating to abstract universals. This means that the process itself must account for both the production of the norm as well as its possible alteration: 'One will indeed find normative criteria that govern, for instance, the application of the power of the State [e.g. norms embedded in planning legislation], but one will also find the means for the critique and modification of these norms [e.g. public consultation, democratic elections and so on]' (Smith, 2003: 308). Normativity, then, involves the ability to critique and transform existing norms and to create something new (Deleuze, 1994: 139). Existing norms in themselves do not permit the creation of the new, as it would be always-already foreseen. As Canguilhem writes, 'any normality limited to maintaining itself, hostile to any variation in the themes that express it, and incapable of adapting to new situations is a normality devoid of normative intention. When confronted with any apparently normal situation, it is therefore important to ask whether the norms that it embodies are creative norms, norms with a forward thrust, or, on the contrary, conservative norms, norms whose thrust is toward the past' (1994: 352).

Norms are inherently related to judgement and decision-making in that they prescribe what 'ought' to be according to some transcendent code. Deleuze (1998) critiques transcendent judgement, in the same way as he critiques normativity, for suppressing the production of the new. Rather than judging actions according to transcendent or universal norms or values of true and false, good and evil, Deleuze suggests that we should 'evaluate' them by determining the mode of existence which underlies them; what they can do. In one of his lectures, he gives the example, of Orson Welles' movie 'F for Fake', in which a forger of Vermeer's works of art was able to pass off the forgeries as real Vermeers. Deleuze explains that the forger managed to convince the experts by studying and making use of the criteria which the experts had established as constituting a true Vermeer. The experts judge the forged painting according to their criteria and decide, 'These are true Vermeers, they meet all the criteria' (Deleuze, 1984, my translation). Deleuze argues that the mistake which experts commit is to not recognise that what makes Vermeer a great artist is the power to transcend the 'true': i.e. of creative transformation (*puissance*). The forger simply paints according to some static template of 'true' criteria and the expert simply judges according to this static template, both of them unable to change or to comprehend creative transformation. For Deleuze, critique and creation are immanent. '[It] is not a matter of judging … in the name of a higher authority, which would be the good, the true; it is a matter, on the contrary, of evaluating every being, every action and passion, even every value, in relation to the life which they involve. … [I]mmanent evaluation, instead of judgment as transcendent value … ' (Deleuze, 1989: 141).

A judgement, such as a planning policy decision, is concerned with both the current situation to be changed (what is happening now) and also with creating new situations in the future (what will happen). But the specific moments in the future to which decisions refer are impossible to limit. There is always an unverifiable component that reverberates in the present in a zone of indeterminacy (Moore, 2012: 144). Judgements/decisions should be immanent processes, working creatively with the flux of live occasions.

In planning practice, however, the tendency is to think in a linear way; of cause and effect. Such thinking tends to be rather mechanistic and overlooks the multiplicity of potentialities which may lead to new decisions and so on. As Lampert (2012: 187) explains: 'a decision has to re-insert itself into a world that has been changing since its last engagement'.

Two simple examples:

Deleuze's (1993: 70) – and perhaps my or your – hesitation between staying in (at home or in the library) and working or going out to a nightclub. Judgement would normally objectify the two options as if they are weights on a scale, presuming that the options remain the same and that the decision-maker (Deleuze, me, you) also remains the same and makes a rational judgement between the options. However, whilst the decision is being taken, both the decision-maker and the options are themselves changing. So, if the decision is finally taken and acted upon some 60 or so minutes later, it may be taken for a reason not originally anticipated (the decision-maker became increasingly bored or 'stuck' with the paper). The nightclub option will also have changed (in terms of patrons, music, ambience etc.).

The second example concerns a planning decision for a major project in the north-east of England. The local authority planners spent some three to four years preparing a proposal for the Council, based on available information at the time. Council decided to go ahead with a three-phase development, which it preferred to undertake in partnership with the appropriate Regional Development Agency. It then took approximately another year for the Agency to decide to participate in the project. The Masterplan for the site was then prepared, taking a further 12 months. With the Masterplan agreed, phase one of the project, consolidating the site, could begin, some five years after the original project was mooted. Due to unforeseen difficulties, however, phase one took far longer than anticipated. It was still incomplete when the national Conservative Liberal Democrat coalition government was elected and disbanded Regional Development Agencies. Phases two and three of the project did not materialise. Decision-makers changed, options changed and the project fell into abeyance.

Whether as individuals deciding to stay in or go clubbing, or as planning practitioners judging to change a proposed major project, we have to think in advance about the undecidable, unknowable to-come. Deleuzean evaluation would attempt to decide or judge based on immanent criteria which emerge through encounters and interactions of variable rationalisations and subjectivations. Deleuzean evaluation would establish a broad set of criteria for which to aim (such as participating in an exciting event or regenerating an urban area) and identify what short-term or smaller-scale actions can do to progress the broad aim (such as going to an area of nightclubs or consolidating the site). These actions are then evaluated. Deleuze argues that evaluation is an ethical process. Instead of relating a human or non-human to a set of transcendent values, Deleuze would relate them to the mode of existence that they imply. In other words, ethics is concerned with what a human or non-human can do, rather than with what they are. Ethics is concerned with creativity and potentiality while morality looks at what 'is' and then claims 'what ought to be'.

Ethics: 'of what are you capable, what can you do?' (Kristensen, 2012: 19)

Patsy Healey has always regarded planning as both an intellectual and an ethical challenge. She links the Aristotelian (1934) tradition of ethics as questions of *eudaimonia*, or flourishing, with recent ethical concerns about how to determine the rules that should govern individual and collective behaviour. Healey's work always has ethics as a core consideration: from concern with ethical issues of responsibility with her research colleagues and research respondents, to the issues which drive her forward, exploring how spatial planning practitioners act and why, and with what procedural and distributional implications for those for and with whom they plan. Healey terms these 'ethics of conduct' and 'ethics of purposes' (2010a: 215–21).

Healey's seminal work on professional ethics with Huw Thomas (1991) has the following question at its core: 'Are planners to be distinguished by what they know and do (knowledge and skills) or by their approach to what they know and do (values and ethics)?' (Healey, 1991: 33). Several of the ethical dilemmas discussed in this volume – advocacy/neutrality, loyalty and duty – remain pertinent today. Healey's paper with Rose Gilroy (1990, reproduced in this Part) is concerned with the interaction of scientific knowledge and moral values in practice towards a 'people-sensitive planning', a notion which recurs throughout her trajectory (see, for example, 1998b; 1997/2006; 2000b; 2007b; 2011)

Patsy Healey draws our attention to the normative principles and responsibilities of planning as practised by individual practitioners (Healey and Underwood, 1978, this Part), in networks (Healey, 1979), developing a 'relationship of equals' (Healey, 1987: 283). Her response to the structural, political economy-inspired critique of collaborative planning, led by Huxley and Yiftachel (2000; Yiftachel and Huxley, 2000), was to point out that planning is 'inherently both normative and analytic' (Healey, 2000a: 917). It cannot, therefore, ignore values: 'the significant recognition is not merely about the way structuring forces enter into the fine-grain of practices, and that structuring forces have their own everyday practice forms[, i]t is also that ways of thinking, ideas, discourses … carry structuring power' (Healey, 2000a: 919). Thinking normatively thus offers a way of 'actively critiquing and challenging policy processes in the making' (ibid.).

Concerned with increasing the ability within planning communities to recognise and deal with what has the potential to develop and what to destroy democratic capacitation (Healey and Hillier, 1995: 37), Healey has offered us several ethically-based, pragmatist-inspired principles and directions for planners to consider. These range from developing 'an ethical consciousness' (Healey and Gilroy, 1990) and 'an idealism based on fundamental moral or aesthetic principle' in her early work on 'planning through debate' (1992: 147) and collaborative planning, to being prepared 'to stand up for values they care about', 'acting with honesty and integrity' (2010a: 220–21) and the importance of distributional fairness, recognition of and respect for others (2012a: 31, Table 1).

Patsy Healey's work is an integral part of a long history of research into the values and actions of planners, pioneered by Peter Marcuse (1976) and Elizabeth Howe (Howe and Kaufman, 1979, 1981; Howe, 1990, 1992, 1994) and Martin Wachs (1982, 1985, 1989) and Carol Barrett (1984, 1989, 2002) in the United States to Huw Thomas (Thomas, 1994; Thomas and Healey, 1991; Lo Piccolo and Thomas, 2008, 2009) and Heather Campbell (2002, 2006, Campbell and Marshall, 1999, 2000, 2002, 2005, 2006, 2012a, 2012b) in the United Kingdom, Sue Hendler (1990a, 1990b, 1991, 1994, 1995, 2002, 2005) in Canada, Wendy Steele (2009) and John Jackson (2013) in Australia, Vanessa Watson (2003, 2013) and Tanja Winckler (2011) in South Africa and Bent Flyvbjerg (1998, 2013) in Scandinavia.[3]

Most of the above work refers to ethics in its generally accepted Western sense; as rules of conduct recognised with regard to a particular issue or action. Ethics, then, are principles of 'right' or 'correct' action, commonly associated with professional Codes of Practice. Morals, in this interpretation, are the general principles on which ethics are based. They are more abstract than ethics, more subjective and personal: what an actor might refer to as 'common sense'.

While many of the above contributions are concerned with Healey's 'ethics of conduct' as matters of individual subjectivity, she, herself, is inspired by the pragmatist tradition in which 'the good' is regarded as a potentiality, not 'some transcendent being or scientific laws of societal development or formalised ideology' (Healey, 2009: 281) and where 'what is taken as "truth for now" is neither fixed nor reflective of some transcendent "truth"' (Healey, 2012b:194), but rather evolves in the dynamic contingencies of experimentation and experience of human and non-human actors. Healey has been particularly influenced by Dewey's (1954) view of 'life as experienced in a world of flux and contingency', requiring planning practitioners to 'navigate a dangerous and uncertain future' (both quotations Healey, 2008b: 389).

Semetsky (2003: 224; 2006: xxiv) argues a 'warranted assertibility of continuity between Dewey's thought and that of Deleuze'. Both Dewey and Deleuze were concerned with the importance of the experimental. They also reject the binary opposition between universals and particulars. As does Patsy Healey, who appreciatively cites Bernstein's (1983: 219) view that 'judgment is communal and intersubjective … It is not a faculty of Man in his universality, but of human individuals in their particularity and plurality'.

Deleuze (1998) makes a distinction between law as *loi*, distinguished by judgement, justification, institutionalism and statutory legalism, and law as *droit*, common law, case law or jurisprudence as an

3 My apologies if I have omitted anyone or any publication significant in this trajectory.

immanent practice.[4] Deleuze critically regards law, as *loi*, as associated with restriction, prohibition and limitation. Law (*loi*) is that which society utilises to function according to its pre-given structure. It striates, codifies and categorises things and 'distributes space according to the organisation of its own structure' (Philippopoulos-Mihalopoulos, 2012: 95), such that practitioners find it difficult to think otherwise, outside the proverbial box. Deleuze argues for a positive, creative practice of law as *droit* which extends human relations through 'an open practice of assessing singular situations and inventing ways of promoting rights, equality and freedom' (Bogue, 2012: 107). Deleuze also points out the key role of politics in the preparation, enactment and implementation of legislation. To adequately adjudicate (or evaluate) for Deleuze, then, involves the whole assemblage of operations through which a case develops and individuates and becomes an occasion for law to exercise (de Sutter and McGee, 2012: 6).

Deleuze prefers a 'clinical' approach (in the medical sense, see Smith, 2005) in which, presented with an assemblage of (socially constructed) 'facts' or 'symptoms' and a store of 'texts' (previous cases), the adjudicator reflectively diagnoses the just disposition of the case and argues its merits or 'treatment' (de Sutter and McGee, 2012: 207). As Smith (2005: 183) explains, 'no doctor would treat a fever or headache as a definite symptom of specific illness; they are rather indeterminate symptoms common to a number of diseases, and the doctor must interpret and decipher the symptoms in order to arrive at a diagnosis'. In a clinical approach, there is no determinate judgement of a ready-made problem which subsumes cases under rules, but rather careful, reflective formulation of the problem and its evaluation.

The ethical framework which results is more fluid, situational and embedded rather than being rigid, doctrinaire and a priori. Existing laws may be modified to turn what is new into an intelligible legal concept (Lefebvre, 2008). The legal 'solution' then gives determination, direction, specificity and meaning to the 'problem' that makes it pertinent. 'Problems are the transcendental condition for (attentive) judgments that are not straightforward applications of rules' (Lefebvre, 2008: 213).

Deleuze offers a philosophical way to open up a more positive, affirmative practice of law, which should be far more in tune with, for example, multi-ethnic societies, especially in the many parts of the world where colonial law (*loi*) was superimposed on Indigenous, customary law (or lore). I suggest that Patsy Healey's conceptualisation of 'the way the law works' (2010b: 8) is not remote from a Deleuzean-inspired ethical understanding of law as immanent, spatial and posthuman (human and non-human), where ethics refers to the system of principles in which morals are grounded and morality to practical differentiation between good/right and bad/wrong.

Morals: 'something superior to Being which plays the role of the Good' (Deleuze, 1980: np)

This definition of morals derives from Kant's (1964, 2004) notions of duty, obligation and principles of deontological, rule-based conduct, compared with an Aristotelian-derived ethic of practical reasoning based on the concept of virtue, which was developed into pragmatic ethics by philosophers such as John Dewey and Martha Nussbaum (Dewey, 1922, 1927, 1932; Nussbaum, 1990, 1994, Nussbaum and Sen, 1993).

In line with this latter definition, Deleuze's understanding of morality always implies something superior to Being which plays the role of the Good (Deleuze, 1980: np), such as a moral code which judges actions by relating them to a set of universal values. He (1980) argues that morality is a matter of essence and values; essence being potential reasonability and values being the realisation of essence. Moreover, morality is the system of judgement of which values are the fundamental element: 'those who have the taste for morality are those who have the taste for judgement' (Deleuze, 1980: np). Judgement for Deleuze is, as we have seen, an oppressive and stultifying force operating 'at the level of the body' (Deleuze, 1988: 130) through the auspices

4 The common law system of jurisprudence, in which courtroom decisions are informed by those of previous cases, is employed in the United Kingdom, Australia, New Zealand, Malaysia, India, the USA and Canada. In countries of continental Europe, South America and in Japan, civil law systems are derived from Roman law which relies on State legislature to create law and ensure its standardised operation.

Moore (2007: 34) defines jurisprudence as 'the mode of practising the law, where the law is engaged with anew in each and every situation'. Deleuze (1995) uses the term as 'a contextualised, positive and problem-specific thought located in the middle of concrete situations' (Philippopoulos-Mihalopoulos: 2102: 110, n10).

of the State, the church and other authorities who regard bodies as already-constituted units of assessment rather than bodies-in-the-making or people and things to-come. Transcendent distinctions between Good and Evil for Deleuze (1988, 1990a) are 'the illusions of a moralistic world-view that does nothing but reduce our power to act' (Roffe, 2005: np).

Similarly, we can identify a problem with utilitarian consequentialist ideas which often underpin planning decision-making. If one is to judge an action according to its anticipated consequences, as an outcome, it follows that it must be possible to track the relation between the action and its consequences. However, such thinking works *as if* the action has already occurred and anticipates a direct relationship between cause and effect, as in the two examples I discuss above. It evacuates uncertainty as it requires static conditions and does not allow for the likely and important potentiality that circumstances change, new human and non-human actors become involved and/or the action itself transforms what actualises into something non-anticipated. Each conceptual 'solution' is thus a creation which cannot be separated from the production of new unknowns (Stengers, 2006). As Bryant (2011: 27) argues, it is when moral principles fail in situations of uncertainty that one encounters the problem of the ethical; of need to rethink and recompose situations from an ethical perspective. Ethics, then, become a moment of thinking uncertainty.

Deleuze's understanding of ethics does not seek to articulate some transcendent 'truth' as such. It rather creates a perspective on what one can do. Deleuze (1994) argues against the unthinking, strict following of laws or rules because it diverts us from the creative experimentation which is more useful for processes of coping with change. He (1995a: 11) rejects Kant's notions of duty and so on, grounded in universality and necessity, as being determined external to experience. Instead, Deleuze proposes that necessity and chance are immanent conditions of what we can do; i.e. the principles of what we can do are constituted within doing itself.

Universals would effectively prevent immanence and movement (Deleuze, 1995b). They are 'abstractions that explain nothing' (Smith, 2003: 307). As Smith (ibid.) explains: 'what one finds in any given socio-political assemblage is not a universal "Reason", but variable processes of rationalization; not universalizable "subjects", but variable processes of subjectivation; not the "whole", the "one" or "objects", but rather knots of totalization, forces of unification, and processes of objectification'.

Deleuze clearly prefers ethics to morals. There is no judgement as such in ethics; rather what I have explained above as 'evaluation'. Instead of relating someone or something to a set of transcendent values, Deleuze would evaluate them through facilitative principles of how they 'occupy' the existence that they imply: 'you seek the enveloped modes of existence, and not the transcendent values. It is the operation of immanence' (Deleuze, 1980: np). Ethics shifts the focus from a unitary, rationality-driven consciousness to an ontological process which implies relational accountability, situatedness and ethical conduct (Braidotti, 2013: 343). Ethics is concerned with what someone or something can *do*, rather than with what they *are*: 'what are my/its/their capabilities and capacities?'

The above, to me, renders Deleuze's dice-throw an ethical process. The point he makes is not a directive as to what to do, but an orientation as to how to cope with the contingencies one faces: 'a process and not a goal' (Deleuze and Guattari, 1984: 133). In terms of planning 'problems', this implies that we should not be thinking of necessary paths or actions, but as 'a relation between necessary actual causal relations and expressive relations regarding sense, value or significance, where free actors can intervene' (Williams, 2008: 164). There is, therefore, neither pure necessity nor pure chance or contingency, but interferences and intertwinings between them. There are no absolutes of right and wrong, best and worst, but rather different degrees of potentiality.

Deleuze (1994) also advocates that we trace the nature of problems; the force relations of their conditions of possibility and so on, in order to ascertain what is the 'true' problem with which we are faced. We can then think through the conditions of possibility for thinking and doing otherwise (see Hillier, 2007, 2011). This problem-based model of ethical composition represents an attempt to embrace inventiveness and creativity in order to cope with the emergence of a 'problem' and the (re)composition of a collective undertaking in response (Bryant, 2011: 22). Planning thus becomes a process of the 'groping possibility to evaluate problems and solutions in the very process of their construction' (Stengers, 2006: np). Healey and Thomas (1991: 198) suggest that practitioners cultivate a self-reflective practice, critically thinking through, as far as possible, the knowledges referred to, how they are used, their validity and their potential impacts. The key to this practice is critical, reflexive thinking.

Stupidity: Do Planners Sometimes Behave Stupidly?

The term stupidity is derived from the Latin word *stupure*, to be insensible, and is related to 'stupor'. It means non-thinking, without reflexivity or substantive reasoning, therefore, rather than its more colloquial usage of 'dumb' or 'error' (Ronell, 2002). Adorno and Horkheimer (2002) regard stupidity as the result of thought's arrest: it is a transcendental structure of thought, an illusion internal to thought, not a disability. Stupidity is thus not related to a lack of intelligence, but rather a failure of courage to think critically and to situate practice. As such, it has an ethical dimension; of complicity with the status quo, habitual or accepted ways of thinking and acting.

Examples of 'stupidity' in spatial planning could include unthinkingly copy/pasting 'best practice' or ideas imported from elsewhere without evaluating their appropriateness for local situations. Participatory budgeting, for instance, has been adopted internationally as a 'solution' detached from the problematic field where it developed and treated as a 'thing' in isolation from social, cultural and economic context. The outcome is often that of failure. As Healey (2012: 190) comments, 'the technique could not just be extracted from its context of invention, uprooted and "planted" somewhere else'.

Mobilities of discourses and materialities, including policy ideas and practices, have long had significant consequences for spatial planning (see, for example, Peck and Theodore, 2010; McCann, 2011; McCann and Ward, 2011; Peck, 2011), from provision of parks and gardens and Ebenezer Howard's Garden City onwards. International transfer of cargo cult-like, so-called 'best practice' initiatives, such as New Urbanism, imported from the USA (McCann and Ward, 2010), or the 'Barcelona model' of culture-led regeneration (Gonzales, 2011) may result in inappropriate developments, such as Tuscan-style, eaveless housing in hot climates, as in Perth, Western Australia, and attempts to impose US-derived, New Urbanist walkability in Karratha, Western Australia (Shire of Roebourne, 2010), where summer temperatures average c.36° Celsius and regularly rise to over 45°C.

Intra-nationally, similar examples of 'non-thought' can occur, as in metropolitan-located consultants or designers preparing influential reports for rural local planning authorities on a 'fly-in/fly-out' basis. Examples include failure to account for the impact of ice freezing and the weight of snow on infrastructure and buildings in the Australian Alps (Beza, 2013). The Coroner's Report into the fatal landslip at Thredbo, Australia, in 1997, identified a major highway and key pipeline being laid in inappropriate locations, together with unsuitable building design and construction (Hand, 2000; Skertchly and Skertchly, 2001–2002). Since lack of time and resources, plus pressure of performance targets, may inhibit practitioners' capacities to check documents fully, 'un-thinking' behaviour may be a symptom of a complex set of underlying issues.

Deleuze (1994) links stupidity to a particular way of approaching problems: an inability to constitute, comprehend or evaluate a problem for what it can do. Stupidity refers to both an inability to distinguish epistemologically between different manifestations of something, conflating issues and simplifying them into what are effectively 'false' problems and also an inability to self-differentiate. It is associated with rigidity in thinking, stifling creativity: 'stupidity is a thought closed in upon itself, happy in its self-sufficiency but with a falsely or absurdly complete view of the world' (Colebrook, 2013: 27). Repetition and reproduction of unquestioned doxa, clichés, empty rhetoric and jargon can trap practitioners in the form of their own judgement, closing off potential for Deleuzean evaluation.

Alvesson and Spicer (2012) term this 'functional stupidity' with regard to organisations. It involves a lack of recognition of the incompleteness and uncertainty of knowledge and a non-questioning of knowledge claims and norms, dominant goals and logics (Alvesson and Spicer, 2012: 1213). This can provide a sense of certainty and smooth organisational functioning, unchallenged by doubts and difficult questions. Functional stupidity could include practitioners placing unquestioning faith in complex models of, for instance, housing capacity or urban metabolism. Alternatively, it could focus on the 'efficient' achievement of a given end, without questioning the end. A 'business case' of numbers may be made, for instance, for a Public Finance Initiative/Public Private Partnership to construct a new hospital without questioning either whether a new hospital is actually needed in that location or what the numbers represent and the ethical implications of using them in the hospital decision.[5] Issues are framed in (ideologically) focused ways.

5 See Dillard and Ruchala (2005) with regard to the hospital example, and Hillier and Van Wezemael (2008) for an example of construction of a new school.

'How is stupidity (not error) possible?' (Deleuze, 1994: 151). Jenemann (2013: 40) suggests that, rather than a basic lack of capacity for thinking, stupidity may be caused by 'psychological repression'; i.e. 'the repressive weight of a society arrayed against critical and abstract thinking'. It may be easier and less stressful to put aside doubt and critique and be like everyone else rather than to be the lone questioner of a policy or proposal or to think outside the accepted norms.

Stupidity tends to be associated with unthinking adherence to some transcendent rule, law, or obligation. Yet, as Deleuze and Guattari argue, 'the rational is always the rationality of an irrational' (2009: 36). Stupidity, then, is 'a spectre that perpetually threatens thought' (Bryant, 2007: np). The issue is to uncover the practices which lead to unthinking stupidity. Why, for instance, do we pose certain problems badly? Are there ways of acting and relating to other humans and non-humans that invite stupidity? (Bryant, ibid.). Could it be related to inadequate understanding of contingency and necessity?

Deleuze's philosophy allows us to reconsider the way in which we think thinking. His aim was to 'awaken thought from its natural stupor' through 'the contingency of an encounter with that which forces thought to raise up and educate the absolute necessity of an act of thought or a passion to think' (Deleuze, 1994: 139). In other words, to use chance events to force us to critical reflexion and thinking the new. Between chance and necessity lies thought, the construction of new ways of understanding, acting and becoming.

Non-rational, non-thinking planning practice forgets to pay attention to the concrete particularities of situations. If we are to counter the 'alien implants' (Healey, 2012b: 191) of copy/pasted, 'universalised' 'solutions' to spatial planning issues, critical, reflexive, situated thinking is necessary.

Conclusions: The Power of Ethical Insurrection

> *I think what we need is an insurrection ... Yes, an insurrection, but an ethical insurrection, not in the usual, moralising sense of the word, because that would mean continuing along the same path.* (Saramago, 2001: np, emphasis in original)

In the quotation above, the Portuguese writer, José Saramago, expresses a wish for society to move away from traditional questions of 'ought' rather than 'is' or 'can do', as 'ought' refers to some transcendent notions of good or right as source of authority over decision-making. As such, Saramago's desire for a practice of ethics rather than morals resonates with that of Gilles Deleuze and, I believe, that of Patsy Healey, for an ethics grounded in principles of immanence and local contextual appropriateness. An ethics of immanence criticises transcendent moral 'rules' and codes as separating us from our capacity to act: our power of acting (what Spinoza termed *potentia* or what Deleuze terms *puissance*).

Following Foucault (1977, 1982), Deleuze and Guattari (1987: 106) make a crucial distinction between power in French as *pouvoir* – instituted power of domination or coercive power – and *puissance* – the capacity to form emergent unities which respect the heterogeneity of their components, immanent forces/capacities of becoming, power to *act* rather than to *dominate* another. We often misread or ignore the potential of our own power and what we can do (our *puissance*) as being the power of the God, the rules, the good and so on (*pouvoir*). This then subjugates our power to the transcendent and significantly constrains what we can do. Infinite calls to universal justice or absolute responsibility to the other (as in Levinas or Derrida) can never be satisfied within frames of reference in which systems and institutions remain fundamentally unjust and irresponsible. Saramago's 'ethical insurrection', therefore, involves a critical, reflexive rethinking and transformation of our frames as references and how we as individuals and groups (such as spatial planners) recompose ourselves and connect with the human and non-human worlds around us in terms of what kind of theoretical and practical connections we want to foster and sustain (Lorraine, 2011). This would be to situate practices (after Haraway, 1988) in locatable, critical knowledges which are acknowledged as being particular rather than universal.

'No theory can develop without eventually encountering a wall, and practice is necessary for piercing this wall' (Deleuze and Foucault, 1972: 3). The theory/practice relationship in spatial planning has been at the core of Patsy Healey's scholarship from the 1970s (see, for example, Healey and Underwood, 1978), through to the present day. She regards the contribution of planning thought as linking together normative philosophy, urban and regional dynamics and the 'fine-grain' of governance practices (2000, 2012a, 2012b). Her work stimulates the development of new perspectives and insights into ways of thinking and doing.

Planning is concerned with an infinite array of potentialities which can emerge in practice. For both Healey (1997/2006, 2007b, 2011) and Deleuze, the aim is no longer to master or control a field (such as spatial planning), but to intervene and experiment. Problems cannot be 'solved' once and for all, but are those issues with which planners experiment in proposing responses (Gibson, 2011). Experimentation in spatial planning requires theory and practice to affirm the uncertain, the unthought or contingent. Strategic spatial planning, in particular, thus becomes 'a long affair of experimentation, requiring a lasting prudence' (Deleuze, 1998: 125) as the dice roll. Suspended between contingency, chance and necessity, spatial planners need to learn to play ethically, critically challenging universal norms that have become empty signifiers – meaning little or even masking abuses of power – and thinking reflexively, responding creatively to signs and 'problems' that go beyond what is known or can be done in a given situation (Williams, 2003: 135).

I conclude this chapter with Patsy Healey's (1998b: 12) words: developing capacity on '"what kind of collective game to play" and how to play it, centred on a grass-rooted approach to place-making in a world of diverse and open relationships, offers the hope of providing a rich "soil" in which individual projects, partnerships and associations can "grow" in flexible and innovative ways'.

Acknowledgements

My thanks to both reviewers, especially for stimulating questions about dice throws.

References

Abbott, J. (2005) 'Understanding and managing the unknown', *Journal of Planning Education and Research*, 24(3): 237–51.

Adorno, T. and Horkheimer, M. (2002) [1947] 'The genesis of stupidity'. In: *Dialectic of Enlightenment*, trans. E. Jephcott, Stanford, CA: Stanford University Press.

Alvesson, M. and Spicer, A. (2012) 'A stupidity-based theory of organizations', *Journal of Management Studies*, 49(7):1194–220.

Aristotle (1934) [c350 BC] *Nichomachean Ethics*, 23 vols, ed. H. Rackham, London: Heinemann.

Barrett, C. (1984) 'Ethics in planning: You be the judge', *Planning*, November: 22–5.

Barrett, C. (1989) 'Fair perspectives on ethics', *APA Journal*, Autumn: 474–9.

Barrett, C. (2002) *Everyday Ethics for Practicing Planners*, Washington, DC: APA Planners Press.

Bernstein, R. (1983) *Beyond Objectivism and Relativism: Science, hermeneutics and praxis*, Pittsburgh: University of Pennsylvania Press.

Beza, B. (2013) 'Realising design projects in the mountains of Australia: A repositioned problem-solving approach'. In: A. Bento-Gonçalves and A. Viera (eds), *Mountains: Geology, topography and environmental concerns*, Hauppage, NY: Nova Publishing.

Bogue, R. (2012) 'Nature, law and chaosmopolitanism'. In: R. Braidotti and P. Pisters (eds), *Revisiting Normativity with Deleuze*, London: Bloomsbury, pp. 98–112.

Braidotti, R. (2013) 'Nomadic ethics', *Deleuze Studies*, 7(3): 342–59.

Bryant, L. (2007) 'The pedagogy of problems and the figure of stupidity' larvalsubjectswordpress.com/2007/02/25/the-pedagogy-of-problems-and-the-figure-of-stupidity/ [accessed 5 September 2013].

Bryant, L. (2009) 'Ontological difference: Difference making a difference' larvalsubjectswordpress.com/2009/02/12/ontological-difference-difference-making-a-difference [accessed 3 September 2013].

Bryant, L. (2011) 'The ethics of the event: Deleuze and ethics without Αρχή'. In: N. Jun and D.W. Smith (eds), *Deleuze and Ethics*, Edinburgh: Edinburgh University Press, pp. 21–43.

Callies, D. (1999) 'An American perspective on UK planning'. In: D. Cullingworth (ed.), *British Planning: 50 years of urban and regional policy*, London: Athlone Press. pp. 264–75.

Campbell, H, and Marshall, R. (1999) 'Ethical frameworks and planning theory', *IJURR*, 23(3): 464–78.

Campbell, H, and Marshall, R. (2000) 'Public involvement and planning: Looking beyond the one to the many', *International Planning Studies*, 5(3): 321–44.

Campbell, H. and Marshall, R. (2002) 'Utilitarianism's bad breath? A re-evaluation of the public interest justification for planning', *Planning Theory*, 1(2): 163–87.

Campbell, H. and Marshall, R. (2005) 'Professionalism and planning in Britain', *Town Planning Review*, 76(2): 191–215.

Campbell, H. and Marshall, R. (2006) 'Towards justice in planning: a reappraisal', *European Planning Studies*, 14(2): 239–52.

Campbell, H. (2002) 'Planning: an idea of value', *Town Planning Review*, 73(3): 271–99.

Campbell, H. (2006) 'Just planning: The art of situated ethical judgement', *Journal of Planning Education and Research*, 26: 92–106.

Campbell, H. (2012a) '"Planning ethics" and rediscovering the *idea of planning*', *Planning Theory*, 11(4): 379–99.

Campbell, H. (2012b) 'Planning to change the world: Between knowledge and action lies synthesis', *Journal of Planning Education and Research*, 32(2): 135–46.

Canguilhem, G. (1994) *A Vital Rationalist*. ed. F. Delaporte, trans. A. Goldhammer, NewYork: Zone Books.

City Creator (2011) www.citycreator.com [accessed 20 August 2013].

Colebrook, C. (2013) 'The war on stupidity', *parallax*, 19(3): 24–33.

Conway, D. (1998) 'Tumbling dice: Gilles Deleuze and the economy of repetition', *symploke*, 6(1): 7–25.

De Sutter, L. and McGee, K. (2012) 'Introduction'. In: L. De Sutter and K. McGee (eds), *Deleuze and Law*, Edinburgh: Edinburgh University Press, pp. 1–14.

De Vries, L.A. (2013) 'Politics on the line'. In: B. Evans and J. Reid (eds), *Deleuze and Fascism*, Abingdon: Routledge, pp. 126–47.

Deleuze, G. (1980) 'Cours Vincennes. Ontologie-Ethique', 21/12/1980, *Les cours de Gilles Deleuze*, www.webdeleuze.com/php/texte.php?cle=1908&groupe=Spinoza&langue=2 [accessed 20 March 2013].

Deleuze, G. (1982) 'Cinéma: une classification des signes et du temps', Lecture 26, 14/12/1982, Part 1 *La voix de Deleuze*. http://www2.univ-paris8.fr/deleuze/article.php3?id_article=164 [accessed 29 April 2013].

Deleuze, G. (1983) [1962] *Nietzsche and Philosophy*, trans. H. Tomlinson, London: Athlone Press.

Deleuze, G. (1984) 'Cinéma: vérité et temps – la puissance du faux', Lecture 66, 12/06/1984, Part 3. http://www2.univ-paris8.fr/deleuze/article.php3?id_article=344 [accessed 29 April 2013].

Deleuze, G. (1988) [1981] *Spinoza: Practical philosophy*, trans. R. Hurley, San Francisco, CA: City Light Books.

Deleuze, G. (1989) [1985] *Cinema 2: The time-image*, trans. H. Tomlinson and R. Galeta, Minneapolis: University of Minnesota Press.

Deleuze, G. (1990) [1969] *The Logic of Sense*, trans. M. Lester and C. Stivale, London: Athlone Press.

Deleuze, G. (1991) [1953] *Empiricism and Subjectivity: an essay on Hume's 'Theory of Human Nature'*, trans. C. Boundas, New York: Columbia University Press.

Deleuze, G. (1993) [1988] *The Fold: Leibniz and the baroque*, trans. T. Conley, London: Athlone Press.

Deleuze, G. (1994) [1968] *Difference and Repetition*, trans. P. Patton, London: Athlone Press.

Deleuze, G. (1995a) [1963] *Kant's Critical Philosophy – the doctrine of faculties*, trans. H. Tomlinson and B. Habberjam, London: Athlone Press.

Deleuze, G. (1995b) [1990] *Negotiations 1972–1990*, trans. M. Joughin, New York: Columbia University Press.

Deleuze, G. (1998) *Essays Critical and Clinical*, trans. D.W. Smith and M. Greco, London: Verso.

Deleuze, G. and Foucault, M. (1972) 'Les intellectuels et le pouvoir', *L'Arc*, 49: 3–10.

Deleuze, G. and Guattari, F. (1984) [1972] *Anti-Oedipus: Capitalism and schizophrenia*, trans. R. Hurley, M. Seem and H. Lane, London: Athlone Press.

Deleuze, G. and Guattari, F. (1987) [1980] *A Thousand Plateaus: capitalism and schizophrenia*, trans. B. Massumi, London: Athlone Press.

Deleuze, G. and Parnet, C. (1987) [1977] *Dialogues*, trans. H. Tomlinson and B. Habberjam, London: Athlone Press.

Deleuze, G. and Guattari, F. (2009) 'Capitalism: a very special delirium'. In: S. Lotringer (ed.), *Chaosophy: Texts and interviews 1972–1977*, trans. D. Sweet, J. Becker and T. Adkins, New York: Semiotext(e), pp. 35–52.

Dewey, J. (1922) *Human Nature & Conduct*, New York: Henry Holt & Co.
Dewey, J. (1932) *Ethics*, Carbondale: Southern Illinois University Press.
Dewey, J. (1954) [1927] *The Public and its Problems*, Athens: Ohio University Press.
Dillard, J. and Ruchala, L. (2005) 'The rules are no game: From institutional rationality to administrative evil', *Accounting, Auditing and Accountability Journal*, 18: 608–30.
Finlay, S. (2010) 'Recent work on normativity', *Analysis Reviews*, 70(2): 331–46.
Flyvbjerg, B. (2013) 'How planners deal with unacceptable knowledge: The dubious ethics of the American Planning Association', *Cities*, 32: 157–63.
Fuller, M. and Goriunova, O. (2012) 'Worse luck'. In: R. Braidotti and P. Pisters, P. (eds), *Revisiting Normativity with Deleuze*, London: Bloomsbury, pp. 159–71.
Galster, G. (2012) *Driving Detroit: The quest for respect in the motor city*, Pittsburgh: University of Pennsylvania Press.
Garvin, A. (2013) *The Planning Game: Lessons from great cities*, New York: W.W. Norton.
Gonzales, S. (2011) 'Bilbao and Barcelona "in motion": How urban regeneration "models" travel and mutate in the global flows of policy tourism', *Urban Studies*, 48: 1397–418.
Grepolis (2013) www.en.grepolis.com [accessed 20 August 2013].
Hand, D. (2000) *Report of the Inquest into the Deaths Arising from the Thredbo Landslide*, Sydney: New South Wales Attorney General.
Haraway, D. (1988) 'Situated knowledges: The science question in feminism and the privilege of partial perspective', *Feminist Studies*, 14(3): 575–99.
Hawking, S. (1999) 'Does God play dice?' www.hawking.org.uk/does-god-play-dice.html [accessed 21 August 2013].
Healey, P. (1974) 'Planning and change', *Progress in Planning*, 2(3): 143–237.
Healey, P. (1979) 'Networking as a normative principle with particular reference to local government and land use planning', *Local Government Studies*, 5(1): 55–68.
Healey, P. (1987) 'City fathers, mandarins and neighbours: Crossing old divides in new partnerships'. In: O. Kaltorp (ed.), *Cities in Transformation: Transformation in cities*, Aldershot: Avebury, pp. 266–88.
Healey, P. (1991) 'Debates in planning thought'. In: H. Thomas and P. Healey (eds), *Dilemmas of Planning Practice*, Aldershot: Avebury, pp. 11–33.
Healey, P. (1992) 'Planning through debate', *Town Planning Review*, 63(2): 143–62.
Healey, P. (1998a) 'Building institutional capacity through collaborative approaches to urban planning', *Environment and Planning A*, 30(9): 1531–46.
Healey, P. (1998b) 'Collaborative planning in a stakeholder society', *Town Planning Review*, 69(1): 1–20.
Healey, P. (2000a) 'Planning theory and urban and regional dynamics: A comment on Yiftachel and Huxley', *International Journal of Urban and Regional Research*, 24(4): 917–21.
Healey, P. (2000b) 'Planning in relational space and time: Responding to new urban realities'. In: G. Bridge and S. Watson (eds), *A Companion to the City*, Oxford: Blackwell, pp. 519–30.
Healey, P. (2004) 'Creativity and urban governance', *Policy Studies*, 25(2): 87–102.
Healey, P. (2006) [1997] *Collaborative Planning: Shaping places in fragmented societies*, Basingstoke: MacMillan.
Healey, P. (2007a) 'Re-thinking key dimensions of strategic spatial planning: Sustainability and complexity'. In: G. De Roo and G. Porter (eds), *Fuzzy Planning*, Aldershot: Ashgate, pp. 21–41.
Healey, P. (2007b) *Urban Complexity and Spatial Strategies: Towards a relational planning for our times*, London: Routledge.
Healey, P. (2008a) 'Knowledge flows, spatial strategy making, and the roles of academics', *Environment and Planning C, Government and Policy*, 26: 861–81.
Healey, P. (2008b) 'Civic engagement, spatial planning and democracy as a way of life', *Planning Theory and Practice*, 9(3): 379–82.
Healey, P. (2009) 'The pragmatic tradition in planning thought', *Journal of Planning Education & Research*, 28(3): 277–92.
Healey, P. (2010a) *Making Better Places*, Basingstoke: Palgrave MacMillan.

Healey, P. (2010b) 'Introduction: the transnational flow of knowledge and expertise in the planning field'. In: P. Healey and R. Upton (eds), *Crossing Borders*, Abingdon: Routledge, pp. 1–25.
Healey, P. (2011) 'Civic capacity, progressive localism and the role of planning'. Nathaniel Lichfield Memorial Lecture, Newcastle University, 6 October 2011. www.ncl.ac.uk/guru/documents/lichfieldflyer.pdf [accessed 18 August 2013].
Healey, P. (2012a) 'Re-enchanting democracy as a mode of governance', *Critical Policy Studies*, 6(1): 19–39.
Healey, P. (2012b) 'The universal and the contingent: Some reflections on the transnational flow of planning ideas and practices', *Planning Theory*, 11(2): 188–207.
Healey, P. (2013) Planning theory: the good city and its governance'. In: J. Wright (ed.), *International Encyclopaedia for Social & Behavioural Sciences*, 2nd ed. Entry no. ISB2 74027. Oxford, Elsevier. www.elsevier.com/isb2 [accessed 20 August 2013].
Healey, P. (2014) 'Learning the craft of research: A continuing process'. In: E. Silva, P. Healey, N. Harris and P. Van den Broeck (eds), *Research Inquiry in Regional and Spatial Planning*, London: Routledge.
Healey, P. and Gilroy, R. (1990) 'Towards a people-sensitive planning', *Planning Practice and Research*, 5(2): 21–9.
Healey, P. and Hillier, J. (1995) *Community Mobilization in Swan Valley: Claims, discourses and rituals in local planning*. Working Paper No. 49. Perth, Western Australia: Curtin University.
Healey, P. and Thomas, H. (1991) 'Knowledge, skills and their validation: Knowing why, as well as what and how'. In: H. Thomas and P. Healey (eds), *Dilemmas of Planning Practice*, Aldershot: Avebury, pp. 193–9.
Healey, P. and Williams, R. (1993) 'European urban planning systems: Diversity and convergence', *Urban Studies*, 30: 701–20.
Hendler, S. (1990a) 'Professional codes as bridges between planning and ethics: A case study', *Plan Canada*, 3: 22–9.
Hendler, S. (1990b) 'Moral theories in professional practice: Do they make a difference?', *Environments*, 20(3): 20–30.
Hendler, S. (1991) 'Ethics in planning: The views of students and practitioners', *JPER*, 10(2): 99–105.
Hendler, S. (1994) 'Feminist planning ethics', *Journal of Planning Literature*, 9(2): 115–27.
Hendler, S. (1995) *Planning Ethics*, New Brunswick, NJ: Centre for Urban Policy Research.
Hendler, S. (2002) 'It's the right thing to do – or is it? Contemporary issues in planning ethics', *Plan Canada*, 42(2): 9–11.
Hendler, S. (2005) 'Towards a feminist code of planning ethics', *Planning Theory & Practice*, 6(1): 53–69.
Hillier, J. (2007) *Stretching Beyond the Horizon: A multiplanar theory of spatial planning and governance*, Aldershot: Ashgate.
Hillier, J. (2011) 'Strategic navigation across multiple planes: Towards a Deleuzean-inspired methodology for strategic spatial planning', *Town Planning Review*, 82(5): 503–27.
Hillier, J. and Van Wezemael, J. (2008) '"Empty, swept and garnished": The public finance initiative case of Throckley Middle School', *Space and Polity*, 12(2): 157–81.
Howe, E. (1990) 'Normative ethics in planning', *Journal of Planning Literature*, 5(2): 123–50.
Howe, E. (1992) 'Professional roles and the public interest in planning', *Journal of Planning Literature*, 6(3): 230–48.
Howe, E. (1994) *Acting on Ethics in City Planning*, New Brunswick, NJ: Centre for Urban Policy Research.
Howe, E. and Kaufman, J. (1979) 'The ethics of contemporary American planners', *JAPA*, 45: 243–55.
Howe, E. and Kaufman, J. (1981) 'The values of contemporary American planners', *JAPA*, 47: 266–78.
Huxley, M. (2009) 'Theoretical debates, planning practice and spatial processes', *Planning Theory and Practice*, 10(1): 139–42.
Huxley, M. and Yiftachel, O. (2000) 'New paradigm or old myopia? Unsettling the communicative turn in planning theory', *Journal of Planning Education and Research*, 19: 333–42.
Jackson, J. (2013) 'What planners do and think: A Melbourne case study', State of the Australian Cities conference, November, Sydney.
Jenemann, D. (2013) 'Stupider and worse: The cultural politics of stupidity', *parallax*, 19(3): 34–49.
Jun, N. (2011) 'Deleuze, values, and normativity'. In: N. Jun and D.W. Smith (eds), *Deleuze and Ethics*, Edinburgh: Edinburgh University Press, pp. 89–107.

Kant, I. (1964) [1785] *Groundwork of the Metaphysic of Morals*, trans. T.K. Abbott, New York: Harper and Row.
Kant, I. (2004) [1780] *The Metaphysical Elements of Ethics*, trans. T.K. Abbott, Adelaide: The University of Adelaide.
Keyes, L. (1969) *The Rehabilitation Planning Game: A study in the diversity of neighbourhood*, Cambridge, MA: MIT Press.
Korsgaard, C. (1996) *Sources of Normativity*, Cambridge: Cambridge University Press.
Kristensen, A.R. (2012) 'Thinking and normativity in Deleuze's philosophy'. In: R. Braidotti and P. Pisters (eds), *Revisiting Normativity with Deleuze*, London: Bloomsbury, pp. 11–24.
Lampert, J. (2012) *Simultaneity and Delay*, New York: Continuum.
Laswell, H. (1958) *Politics: Who gets what, when, how*, New York: Meridian.
Lefebvre, A. (2008) *The Image of Law: Deleuze, Bergson, Spinoza*, Stanford: Stanford University Press.
Lehman, T., Huang, W-H., Lantznester, T., Johnson, C. and Kelleher, S. (2010) Boardgame Babylon. Recorded at Gateway 2010. http://traffic.libsyn.com/boardgamebabylon/bgb_v86.mp3 [accessed 20 August 2013].
Long, N. (1958) 'The local community as an ecology of games', *American Journal of Sociology*, 64(3): 251–61.
Lo Piccolo, F. and Thomas, H. (2008) 'Research ethics in planning: A framework for discussion', *Planning Theory*, 7(1): 7–23.
Lo Piccolo, F. and Thomas, H. (eds) (2009) *Ethics and Planning Research*, Farnham: Ashgate.
Lorraine, T. (2011) *Deleuze and Guattari's Immanent Ethics*, Albany, NY: SUNY Press.
Lubell, M., Henry, A.D. and McCoy, M. (2010) 'Collaborative institutions in an ecology of games', *American Journal of Political Science*, 54(2): 287–300.
Marcuse, P. (1976) 'Professional ethics and beyond: values in planning', *JAPA*, 42(3): 264–74.
May, T. (2005) *Gilles Deleuze: an introduction*, Cambridge: Cambridge University Press.
McCann, E. (2011) 'Urban policy mobilities and global circuits knowledge: Toward a research agenda', *Annals of the Association of American Geographers*, 101: 107–30.
McCann, E. and Ward, K. (2010) 'Relationality/territoriality: toward a conceptualization of cities in the world', *Geoforum*, 41: 175–84.
Merriam-Webster (2013) Dictionary. www.merriam-webster.com/dictionary [accessed 14 August 2013].
Moore, N. (2007) 'Icons of control: Deleuze, signs, law', *International Journal for the Semiotics of Law*, 20: 33–54.
Moore, N. (2012) 'The perception of the middle'. In: L. De Sutter and K. McGee (eds), *Deleuze and Law*, Edinburgh: Edinburgh University Press, pp. 132–50.
Nietzsche, F. (1961) [1883–1885] *Thus Spoke Zarathustra*, trans. R. Hollingdale, Harmondsworth: Penguin.
Nussbaum, M. (1990) *Love's Knowledge*, Oxford: Oxford University Press.
Nussbaum, M. (1994) *The Therapy of Desire*, Princeton, NJ: Princeton University Press.
Nussbaum, M. and Sen, A. (1993) *The Quality of Life*, Oxford: Clarendon Press.
Oatley, N. (1995) 'Competitive urban policy and the regeneration game', *Town Planning Review*, 66(1): 1–14.
Pacudan, R. (1997) 'Transport, energy and the environment in Singapore: An institutional perspective', *Cities*, 14(3): 165–8.
Peck, J. (2011) 'Geographies of policy: From transfer-diffusion to mobility-mutation', *Progress in Human Geography*, 35: 773–97.
Peck, J. and Theodore, N. (2010) 'Mobilizing policy: Models, methods and mutations', *Geoforum*, 41: 169–74.
Philippopoulos-Mihalopoulos, A. (2012) 'Law, space, bodies: The emergence of spatial justice'. In: L. De Sutter and K. McGee (eds), *Deleuze and Law*, Edinburgh: Edinburgh University Press, pp. 90–110.
Poplin, A. (2011) 'Games and serious games in urban planning: study cases', *Computational Science and its Applications* – ICCSA2011, Vol.6783, pp. 1–14.
Roffe, J. (2005) 'Gilles Deleuze 1925–1995', *Internet Encyclopaedia of Philosophy*, www.iep.utm.edu/deleuze [accessed 6 September 2010].
Roll for the Galaxy (2013) Rio Grande Games. www.riograndegames.com/games.html?id=442 [accessed 20 August 2013].
Ronell, A. (2002) *Stupidity*, Urbana: University of Illinois Press.

Rorty, R. (1982) *Consequences of Pragmatism*, Minneapolis: University of Minnesota Press.
Rouse, J. (2007) 'Social practices and normativity', *Philosophy of the Social Sciences*, 37(1): 1–11.
Saramago, J. (2001) 'Magna Terra' delivered in Guatemala. March–April. In: Fundaçao José Saramago, Lisbon. [viewed July 2013].
Sayer, A. (2007) 'Moral economy as critique', *New Political Economy*, 12(2): 261–70.
Schein, E. (2004) *Organizational Culture and Leadership*, San Francisco, CA: Jossey-Bass.
Semetsky, I. (2003) 'The problematics of human subjectivity: Gilles Deleuze and the Deweyan legacy', *Studies in Philosophy and Education*, 22: 211–25.
Semetsky, I. (2006) *Deleuze, Education and Becoming*, Rotterdam: Sense Publishers.
Shire of Roebourne (2010) *Karratha City of the North Plan*. Shire of Roebourne. www.roebourne.wa.gov.au/assets/karratha%20city%20of%20the%20north%20plan%20summary.pdf [accessed 28 August 2013].
SimCity (2013) [1989] http://www.simcitycentral.net/simcity6/ [accessed 20 August 2013].
Skertchly A. and Skertchly K. (2001–2002) 'Thredbo disaster Coroner's inquest', *Australian Journal of Emergency Management*, 16(4) www.em.gov.au/Documents/ThredboDisasterCoronersInquest.pdf [accessed 28 August 2013].
Smith, D.W. (2003) 'Deleuze and the liberal tradition: normativity, freedom and judgment', *Economy and Society*, 32(2): 299–324.
Smith, D.W. (2005) 'Critical, clinical'. In: C. Stivale (ed.), *Gilles Deleuze: Key concepts*, New York: Acumen, pp. 182–93.
Smith, D.W. (2007) 'Deleuze and the question of desire: Toward an immanent theory of ethics', *Parrhesia*, 2: 66–78.
Steele, W. (2009) Australian urban planners: Hybrid roles and professional dilemmas', *Urban Policy & Research*, 27(2): 189–203.
Stengers, I. (2006) 'Gilles Deleuze's last message' www.recalcitrance.com/deleuzelast.htm [accessed 9 May 2013].
Suburbia (2012) Bézier Games. www.boardgamegeek.com/boardgame/123260/suburbia [accessed 20 August 2013].
Tato, B. and Vallejo, J.L. (2012) 'Open source planning for augmented citizens', *Velocitat*, 28. www.tdd.elisava.net/cdeccion/28/tato_vallejo-en. [accessed 20 August 2013].
Temple, M. and Boon, A. (2002) 'Uncollaborative planning? Implications for the spatial economic development of congested cities', *Urban Design International*, 5(1): 27–35.
Thomas, H. (ed.) (1994) *Values and Planning*, Aldershot: Avebury.
Thomas, H. (2012) 'Values and the planning school', *Planning Theory*, 11(4): 400–417.
Thomas, H. and Healey, P. (1991) 'Introduction'. In: H. Thomas and P. Healey (eds), *Dilemmas of Planning Practice*, Aldershot: Avebury, pp. 1–7.
Thomas, H. and Healey, P. (1991) 'Preface'. In: H. Thomas and P. Healey (eds), *Dilemmas of Planning Practice*, Aldershot: Avebury, pp. x–xiv.
Thomas, H. and Healey, P. (eds) (1991) *Dilemmas of Planning Practice: Ethics, legitimacy and the validation of knowledge*, Aldershot: Avebury.
Upton, R. (2002) 'Planning praxis: Ethics, values and theory', *Town Planning Review*, 73(3): 253–69.
Urban Plan (2011) [2001] www.owensworld.com/flashgames/play.php?id=33 [accessed 20 August 2013].
Wachs, M. (1982) 'Ethical dilemmas in forecasting for public policy', *Public Administration Review*, 42(2): 562–7.
Wachs, M. (ed.) (1985) *Ethics in Planning*, New Brunswick, NJ: Centre for Urban Policy Research.
Wachs, M. (1989) 'When planners lie with numbers', *JAPA*, 55(4): 476–9.
Watson, V. (2003) 'Conflicting rationalities: Implications for planning theory and ethics', *Planning Theory & Practice*, 4(4): 395–407.
Watson, V. (2013) 'The ethics of planners and their professional bodies: response to Flyvbjerg', *Cities*, 32: 167–8.
Williams, J. (2003) *Gilles Deleuze's 'Difference and Repetition': A critical introduction and guide*, Edinburgh: Edinburgh University Press.

Williams, J. (2008) *Gilles Deleuze's 'Logic of Sense': a critical introduction and guide*, Edinburgh: Edinburgh University Press.

Winckler, T. (2011) 'On the liberal moral project of planning in South Africa', *Urban Forum*, 22: 135–48.

Yiftachel, O. and Huxley, M. (2000) 'Debating dominance and relevance: Notes on the communicative turn in planning theory', *International Journal for Urban and Regional Research*, 24(4): 907–13 and 922–4.

Chapter 6

Planning in Hong Kong, an Undemocratic, Post-colonial Chinese Capitalist Society: Negotiating the Roles of Technocrats, Traditional Public Intellectuals and Reflective Practitioners

Mee Kam Ng

Introduction: Planning at a Crossroads – from 'Efficient' Technocrats to ... ?

Twenty-first century Hong Kong is at a crossroad and it seems that urban planning as a profession needs to reposition itself. While the political nature of spatial planning and development is increasingly recognised by the civil society, the rather 'toothless' urban planning system does not empower the technically oriented urban planners to make substantive changes. Healey and Underwood's (1978) account of town and country planners in London in the 1970s could not be a more apt description of the plight of planners in Hong Kong today: they are 'a much-criticised group of local government officials', for what they have promoted, resisted and omitted to consider; they claim the right to influence the evolving pattern of land use and development in their areas but their influence has invariably been marginal; and their focus on 'technically neutral' and 'efficient and effective management' without clear social welfare goals gives the impression that planning is merely serving 'business and development interests' (1978, pp. 77, 119, 120). Their conclusions, surprisingly, remain relevant for a challenged profession: 'technical expertise' is less important than critical knowledge of 'the purposes, means and effects of government interventions' and 'experiences in implementing detailed policies and programmes'. Such critical knowledge allows planners to develop competencies so as to critically interpret 'planning as government intervention in the use and development of land' in a conflict and interests-ridden society; the implementation of effective policies and programmes can also empower planners, to effect government-designated tasks that may serve diverse interests (Healey and Underwood, 1978, pp. 121–3).

Planners in Hong Kong are yet to accept the fact that 'people no longer trust experts to define their interests for them' and instead, seek 'active involvement in the framing and implementing of public policy' (Healey and Gilroy, 1990, p. 21). As in the United Kingdom before the dawn of the new millennium, social activists in Hong Kong are 'searching for a philosophy and principles to constrain the politics of self-interest, and seek debates which interrelate scientific knowledge and moral values' (op. cit., p. 21). In the United Kingdom, as argued by Healey and Gilroy (1990), the intention has been to 'mediate' conflicting interests, 'searching for routes through divergent interests, developing terrains of consensus, but challenging attempts by dominant political, economic, and social groups to impose their definitions on situations' (op. cit., p. 22). Advocating a 'people-sensitive planning', they contend that planners should listen carefully and empathise critically with the demands raised by different stakeholders whose varying interests and power are shaped by the wider socio-economic structuring forces (op. cit., 1990). The ultimate goal is to empower both the public and the planners so that they will, in effect, demand a reasonable pathway towards genuine consensus building, even in a politically charged and fraught power context.

Whither the role of urban planners in Hong Kong? The ecology of the planning profession in the city may be different from other places. Most urban planners in Hong Kong work for the Government. There are urban planners who work in the private sector but they also rely on planning studies commissioned by the Government or on development applications made by the private sector to the Government-appointed Town Planning Board. Local advisory district councils funded by the Government may commission small and ad hoc planning studies but civil society in general lacks the resources to engage professional planning services. In any case, few professional planners work for community interests. A brief overview of the political economy of Hong Kong bequeathed by its history and the consequent institutional set up, which are discussed in the following section, demonstrates how the role of urban planning has been curtailed. This limited role has only recently surfaced as a substantial problem; the economic growth oriented, colonial government generally was regarded as competent in improving people's lives and livelihoods, and especially following the marked economic expansion of the 1970s. However, as the city faces political, economic and social transitions, the contradictions between what planners claim to do and what they actually do frustrate the general public, especially social activists (Healey and Underwood, 1978). The complications can be seen in the failed attempt by the Planning Department to fundamentally revamp the outdated Town Planning Ordinance that was first enacted in 1939, and in the various planning sagas in social, cultural and heritage conservation (Ng, 2013). In this chapter, I invite urban planners in Hong Kong to revisit their roles as public intellectuals within traditional Chinese culture and to reflect on the relationships between 'scientific knowledge' and 'moral values'. Yet at the same time, I challenge the nature of twenty-first century planners to go beyond the elitist nature of Chinese intellectual thought to embrace the collective wisdom of different stakeholders, especially those who are directly affected by the development process – and this, within a web of skewed power relationships – to transform imperfect Hong Kong into a better, more sustainable and just society.

Hong Kong as a Non-democratic, Post-colonial Chinese Capitalist Society

Twenty-first century Hong Kong, a compact city of seven million, built on less than a quarter of its 1,104km^2 land mass, conjures an image of dynamism, efficiency and professionalism. This image is very different from the post-WWII slum city of the 1950s following an influx of over 1.5 million refugees from the mainland (Pryor, 1983, p. 23). Then Hong Kong was a poor, corrupt and exploitative colonial and refugee society with serious urban problems such as large squatter settlements, housing shortage, lack of employment opportunities and so on. The discontent of the citizens and effects of the Cultural Revolution on the China mainland led eventually to the riots of 1966 and 1967 (Cheung, 2009). This probably explains why by 1973 government expenditure on police was 800 per cent of that on social welfare in the crisis-ridden financial centre (Hong Kong Research Project, 1974, pp. 5, 28). Only eight per cent of the land was urbanised, accommodating a population of four million. In order to pacify an unsettled population with better living conditions, urban planners (then a 'hidden' profession within the colonial administration's Works Department) joined with other built-environment professionals to contribute to the amazing development of new towns so as to expedite the implementation of the Ten Year Housing Development Strategy. The dominant spatial development strategy was to bulldoze coastal zones, reclaim land from the sea and replace villages with 'modern' high-rise new towns to properly house a growing population (Bristow, 1984; 1989). The process was met with minimal resistance and things were done 'efficiently'.

Today, planners have their own Planning Department, established in 1990, and the profession could not be more meticulous in engaging the general public in the planning process. Yet, planners have constantly met with local resistance in almost every planning-related issue, from compulsory purchase of rural villages, to harbour reclamation, to urban renewal projects. What has happened in these intervening decades? If planners can no longer afford to be 'efficient faceless technocrats' behind a 'bulldozing' development culture that has increasingly been criticised, what roles should and could planners play at this historical juncture?

The social discontent with the unfolding of spatial development processes has to do with the awakening of the general public to the flawed political institutional arrangements in Hong Kong. Underlying the facade of an ultra-modern, sky-scraper packed metropolis is the antiquated, undemocratic political system. The current Chief Executive of Hong Kong was elected by an Election Committee of merely 1,200 members from different economic, professional and religious sectors and political establishments that include the Legislative Council and district councils, Heung Yee Kuk (Rural Council), National People's Congress and Hong Kong representatives of the National Committee of the Chinese People's Consultative Conference (HKSAR Government, 2013, p. 20). The Chief Executive not only nominates the Government's principal officials to be appointed by the Central Government, they also appoint the Executive Council which gives advice to the administration and approves policy matters. In 2012 the election process for the Legislative Council was so designed that only 40 of the 70 Councillors were elected through geographically based universal suffrage, with 30 of the seats going to functional constituencies representing business and professional sectors, with voting by only some seven per cent of registered voters (HKSAR Government, 2009). Hence, the political system cannot be called democratic.

Nevertheless, Article 45 of the Basic Law[1] stipulates that an ultimate aim is 'the selection of the Chief Executive by universal suffrage upon nomination by a broadly representative nominating committee in accordance with democratic procedures' (HKSAR Government, 1997, p. 15). In order to pressure the Government to implement universal suffrage, two academics and one minster of religion initiated a nonviolent campaign to 'Occupy Central with Love and Peace'(OCLP), to be held in July 2014 if the Central Government of China failed to allow citizens of Hong Kong universal suffrage in the 2017 Chief Executive Election and the 2020 Legislative Council Election.[2] There have been intense debates surrounding the election processes, and one recent appeal was made by the Bar Association chairperson to urge the Government not to present political objections as legal objections in deliberating on the pathways towards the promised universal suffrage under the Basic Law by 2017 (Lau, 2014).

As China's first Special Administrative Region (SAR) under the ingenious 'One Country, Two Systems' policy from 1997, Hong Kong's colonial history has bequeathed the city a unique, east-meets-west culture, which, unfortunately, has continuously been eroded by the heightened presence of the political will of the Central Government. No wonder some activists have refused to acknowledge the status of Hong Kong after 1997 as post-colonial – and liken the HKSAR to a city 're-colonised' by her motherland (Vines, 1998). In any case, the post-colonial status has at least forced Hong Kongers to reflect on their future (Crush, 1994), especially when the city is integrating into a much larger entity of which it has little knowledge or experience (DeFilippis and Smith, 1997; Skeldon, 1999). The transition to post-colonial status stimulated Hong Kongers to enjoy 'the freedom of imagination' to re-imagine the city differently (Yeoh, 2001, p. 464), and to aspire to a more equitable society (Cheng, 1989; Lin, 1992; Lu, 1985). Unfortunately this re-imagination was out of step with the political and economic realities of the time, and hence has laid the seeds of present day conflicts.

The post-colonial, Chinese dominated Hong Kong has continued to be an unashamedly liberal economy with the Government cautiously handling socio-economic issues such as setting the minimum wage (instituted in 2011 and at a current rate in 2014 of only US$3.85/hour) and exhibiting extreme tardiness in placing the issue of comprehensive retirement benefits firmly on the political agenda. Indeed, the city's return to China with its rapidly transforming centrally planned economy through 'neo-liberal' socialist market reform has perpetuated, if not worsened, the city's pro-growth tradition of privileging market forces. In the Basic Law, one of the four major principles is to 'facilitate the development of the capitalist economy' (HKSAR Government, 2013b, p. 11). Despite the fact that for 20 years the city has been voted as 'the world's freest economy' by the Heritage Foundation (The Heritage Foundation, 2014), her brand of capitalist economy feeds on a peculiar relationship between the Government, which is the owner of all 'land and natural resources' (Article 7, Basic Law) – property-related revenue has counted for 20 to 30 per cent of the total government revenue (Census and Statistics Department, various years) – and a super-rich class of 'land lords' (Poon, 2005):

1 The Basic Law is a constitutional document for the Hong Kong Special Administrative Region (HKSAR). It is a legal document stipulating the basic policies for the implementation of important concepts such as 'one country, two systems', 'a high degree of autonomy' and 'Hong Kong People ruling Hong Kong', and that Hong Kong's pre-1997 capitalist system and life-style shall remain unchanged for 50 years. The Basic Law of the HKSAR ('the Basic Law') was adopted on 4 April 1990 by the Seventh National People's Congress (NPC) of the PRC and came into effect on 1 July 1997. The Basic Law consists of a total of nine chapters with 160 articles; Annex I is about the methods for the selection of the Chief Executive; Annex II is about the methods for the formation of the Legislative Council and its voting procedures; Annex III is about the national laws to be applied in the HKSAR; and the design of the regional flag and emblem of the HKSAR (HKSAR, 2013a).

2 The OCLP campaign took place after a series of student strikes and civic activism on 28 September 2014. Dubbed as the 'Umbrella Movement', participants of the Occupy Movement camped in several strategic areas in different parts of urban Hong Kong. The movement lasted 78 days before the occupied sites were eventually cleared on 15 December. The Movement allowed citizens to experience with the urban realm in a different way and ignited their spatial re-imagination of the use of urban places.

> Land has worked as an enriching agent for the ruling class in Hong Kong.[3] Existence of such a ruling class, which thrives on land and spatial monopoly and control of other forms of economic assets, is itself a symbol that contradicts the ideas of a free economy – a state that Hong Kong is ironically believed to be in … a combination of the working of the land system and a non-competitive business environment, with government acting as the behind-the-scene director, has created a few economic titans that threaten to grow into ugly monsters (pp. 150 and 103).

And citing the World Socialist Website (Poon, 2005, p.123), 'Tung's [Hong Kong SAR's first Chief Executive] administration has proven to be as autocratic, remote and subservient to Hong Kong's powerful business tycoons as the previous British colonial governors'.

The birth of the Hong Kong SAR in 1997 therefore has been marked by two fundamental contradictions: a society that is ripe for democratic governance has to operate within the manipulated constraints of the Basic Law; and the post-colonial yearning for a more equitable and just society has to face the continuation, if not a heightening, of the lop-sided support for land-related business interests. The civil society's awakening to these contradictions helps explain the various planning controversies and crises; the 'toothless' planning mechanisms are perceived as instruments of the pro-growth government to promote the specific interests of the 'land lords'. The following section examines the problematic planning system bequeathed by the colonial administration.

Failure to Transform a 'Toothless' Planning Regime and its Consequences

The two most important spatial events that have shaped the geography of post-WWII Hong Kong were the development of new towns after the 1970s and the construction of the Ten Airport Core Projects in the 1990s, the last decade of British colonial rule. These spatial development projects were outcomes of socio-economic and political circumstances, more so than the result of vigorous urban planning exercises. Hence, planners' contributions and participation were rather 'accidental'. Technocratic planning had been instrumental in rationalising Hong Kong's post-WWII ad hoc and 'chaotic' urbanism brought about by the 'transferred industrialisation and urbanisation' resulting from the capital flight and in-migration from communist China after 1949 (Lethbridge, 1980; Wang and Wong, 1997). The planning system, embedded within an institutional framework crafted to accommodate rapid economic growth, was effective in promoting the horizontal and vertical expansion and densification of the city. From the post-WWII decades to the city's return to Chinese rule in 1997, planners together with other built environment professionals were widely regarded as competent experts or technocrats who had magically transformed the post-war slum city into 'Asia's World City'.

The built-environment professionals were particularly instrumental in shaping urban development after the territory-wide riots of 1966 and 1967, which were driven in part by social discontent of a rather corrupt and 'lawless' administration, and in part by the Cultural Revolution across the border in China. When the Government appreciated the social discontent, it endeavoured to change the administration through a range of measures, including the launch of district level administration, and establishing the Independent Commission Against Corruption (ICAC). Spatially, they announced the Ten Year Housing Programme to construct public housing to pacify the poorly housed population with the self-contained and socially balanced new towns built in the 1970s and 1980s (Bristow, 1984; 1989). Hence, the built environment professionals played an important role in driving the whole development process. More than 30 km^2 of land[4] had been reclaimed for the development of various new towns (Lands Department, 1996). And the end of this era of rapid spatial growth was marked by the successful implementation of the 'Rose Garden' projects, ten mega-infrastructure

3 While the Government owns all the land, the Government leases the right to land uses through land auctions or tenders. From 1999 to 2013 an application system was developed for developers to trigger land auctions or tenders through placing minimum bids that were seen as acceptable by the government. This was seen as a measure that gave developers considerable control over land supply (Liu and Ng, 2013).

4 As at 2012, public and private housing that accommodate over 90 per cent of the population are built on just 41km^2 (HKSAR Government, 2012a).

projects concerning the airport development, announced in 1989 and completed in 1998. Interestingly, the implementation of the 'Rose Garden' projects did not draw on the formulation of the planners' Territorial Development Strategy in 1984 (Ng, 1992). Rather, the projects were carried out in order to boost confidence in the future of the city after the political turmoil in Beijing in June 1989.

However, a closer examination of the planning system reveals that planners have extremely limited roles in plan making and implementation. Unlike legislation such as the Buildings Ordinance that was first enacted in 1889 (Mclnnis, 2003), the Town Planning Ordinance was not enacted until 1939 and due to the outbreak of WWII and the Japanese Occupation it was not implemented until the 1950s. Long before formal planning measures, the city had used land leases to control development. After WWII, population in the colony increased 400 per cent, from 600,000 in 1945 to 2.4 million in 1956 (Pryor, 1983, p. 28). The Buildings Ordinance at that time was intended to specify plot ratios and densities in the colonial city that could accommodate the rapidly increasing population. Besides mandating these essential development parameters the right to implement new statutory land use plans, too, was given to the Buildings Authority. In 1959 the colonial Government of Hong Kong introduced a clause in the Buildings Ordinance, specifying that 'the Building Authority has discretion to approve any building work plans which contravene a relevant statutory land use plan' (Cooray, 1997, p. 141). Planning and plan implementation through the exercise of this discretionary power of the Buildings Ordinance has produced undesirable urban landscapes.

Critiques from the Hong Kong Branch of the Royal Town Planning Institute forty years ago (in 1973) are still valid in understanding the limited roles of urban planners in Hong Kong. In short, urban planners have been denied the right to control the urban form because the power to specify permitted plot ratios, coverage and building height etc., rests with the Planning Regulations of the Buildings Ordinance, special conditions of the land lease, or on administrative measures such as density zoning (HKB of RTPI, 1973, p. 42). As the Planning Regulations of the Building Ordinance stipulate the height, plot ratios and site coverage standards for residential buildings according to 'the class of the site that in turn depends on the number of streets not less than 4.5m wide that the site abuts' (Buildings Department, 2009, p. 1), developers naturally endeavour to build to the maxima permitted by the Government who welcomes such development – as does the land owner – in order to maximise their land-related revenues (Table 6.1). As a result, planners have little room to modify development intensity and very often are expected to approve full development by the land owners. If planners cannot really control the development intensity specified by the Building (Planning) Regulations, the related population estimates and land use budget for roads, government, institutional and community facilities, then have to be based upon the full development allowable under the Buildings Ordinance (HKB of RTPI, 1973).

In other words, urban planners have little power to truly plan the form of a specific site because the Buildings Ordinance and the First Schedule of its associated Building (Planning) Regulations have already outlined the 'maximum plot ratios and site coverage permitted for domestic and non-domestic buildings in relation to building height' (Planning Department, 2003, chapter 2, p. 3). As stated in the Planning Department's *Hong Kong Planning Standards and Guidelines* (2003, p. 3), the Buildings Ordinance 'is the only statute which explicitly sets density limits and provides for their enforcement'. In theory planners can enforce plot ratio controls through the Outline Zoning Plans prepared under the Town Planning Ordinance, particularly in areas with infrastructure constraints, incompatible land uses and environmental considerations, and a wide range of other characteristics (Planning Department, 2003, pp. 3, 10). However, in reality any move to lower the plot ratio or building height will trigger strong objections from the development industry.

Urban planners in Hong Kong tried to regain the power to plan and implement urban development when statutory land use planning was extended to the rural areas and a separate Planning Department was established in 1990. Rural Hong Kong faced dramatic changes as a result of China's open door policy in the late 1970s and the consequent economic integration with Hong Kong's economy from the 1980s. A lot of the rural land in the New Territories was transformed into container storage yards, dumping grounds for construction and other materials. The 'destruction' of the rural landscape forced the Government to establish the Planning Department in 1990 and extended the Town Planning Ordinance to the New Territories. The then new Planning Department also tried to push through a comprehensive review of the Town Planning Ordinance (hereafter, Comprehensive Review) in order to enhance its 'openness, fairness, certainty, efficiency, effectiveness, affordability and comprehensiveness' (PELB, 1991). Table 6.2 outlines the proposed changes in the 1991 Comprehensive Review, in comparison with the suggestions made by the Hong Kong Branch of

Table 6.1 **Maximum permitted site coverage and plot ratio in relation to building height for domestic buildings under First Schedule of Building (Planning) Regulations**

Height of building in metres	Maximum site coverage (%)			Maximum domestic plot ratio		
	Class A Site	*Class B Site*	*Class C Site*	*Class A Site*	*Class B Site*	*Class C Site*
Up to 15	66.6	75	80	3.3	3.75	4.0
Up to 18	60	67	72	3.6	4.0	4.3
Up to 21	56	62	67	3.9	4.3	4.7
Up to 24	52	58	63	4.2	4.6	5.0
Up to 27	49	55	59	4.4	4.9	5.3
Up to 30	46	52	55	4.6	5.2	5.5
Up to 36	42	47.5	50	5.0	5.7	6.0
Up to 43	39	44	47	5.4	6.1	6.5
Up to 49	37	41	44	5.9	6.5	7.0
Up to 55	35	39	42	6.3	7.0	7.5
Up to 61	34	38	41	6.8	7.6	8.0
Over 61	33.3	37.5	40	8.0	9.0	10.0

Note: Class A site = a site, not being a Class B site or Class C site, and that abuts on one street not less than 4.5 m wide. Class B site = a corner site that abuts on two streets neither of which is less than 4.5 m wide. Class C site = a corner site that abuts on three streets none of which is less than 4.5 m wide.
(*Source:* Planning Department, 2012, p.19)

the RTPI in 1973 and the eventual implementation of Phase I of the recommended changes in 2004, as well as the implications of the results. It is clear from the Table that planners would like to enhance the planning system through engaging the general public during the planning process, from planning study to monitoring development applications, and they argue for the consolidation of disparate development control mechanisms under an amended Town Planning Ordinance so the Planning Department will be empowered to implement plans. However, planners have failed to effect the proposed changes. The following gives a brief account of planners' losing battle and some of the implications for those who are discontented with the planning scene.

After the first round of consultation on the Comprehensive Review, it was quite clear that except for the developers, the public at large, indeed, supported wider consultation in the preparation of statutory plans and in the planning process (HK LegCo, 1992, p. 1355). The then Secretary for the Planning, Environment and Lands Bureau (PELB) also pointed to the 'misguided fear' of the proposed planning certificate system and explained that the proposal was simply to 'enable the Planning Authority to deal directly with the planning aspects of building works rather than indirectly through the Buildings Ordinance as at present' (op. cit., p. 1356). Another five years lapsed and instead of a promised Blue Bill (draft legislation), the 1996 consultative White Bill was put forward for consultation.[5] One legislative councillor argued that 'the proposals [in the White Bill] … still lack transparency with regard to the composition of the Town Planning Board, the decision-making process in planning and the appeal procedure … it is imperative for the Government to introduce … the Blue Bill into this Council for scrutiny asap' (HK LegCo, 1996, p. 150). Other legislators criticised the design of the whole mechanisms as biased towards the interests of 'the Government, the public utility organisations and the property developers' (op. cit., p. 155).

5 A White Bill is published for consultation whereas a Blue Bill is the final version of a draft legislation that has been checked and cleared within the Government (HKSAR, 2012b).

Table 6.2 Amendments made at the 2004 Town Planning (Amendment) Bill when compared to suggestions put forward in the 1973 HK Branch of RTPI Report and the 1991 Comprehensive Review of the TPO and the Implications

	1973 HK Branch of RTPI Report	1991 Comprehensive Review of the TPO	2004 Town Planning (Amendment) Bill	Implications
Planning study	The public to be consulted before and during the preparation of plans. Period of public exhibition: 3 months	A planning study to be prepared & publicly advertised/ displayed for 3 months for public comments	n.a.	Without involving the public, the planning study may not capture all the key issues of local concerns.
Submitting representation	n.a.	Any member of the public would be able to submit comments to be published	Any member of the public would be able to submit representations to be published	Encourages activists to participate but their failure to change the course of events also frustrates them.
Exhibition time	n.a.	Would be extended from 3 to 6 months	2 months for making representations; & 3 weeks for commenting on representations	Allows the public to learn more about the plans but the time is short for responses
Plan review	To be reviewed at least once every 5 years	n.a.	n.a.	Review may allow avenues to proposed changes.
Planning permission	Regulations to be introduced 'to replace the notes ... requiring material changes of use ... to be submitted to the planning committee'	Written consent or notice to owners required Planning applications published for public inspection & comments Consideration from 2 to 3 months	Written consent or notice to owners required Planning applications published for public inspection & comments Consideration within 2 months	A measure that allows access to information previously only available by applicants & Government.
Interim-develop. control	n.a.	Deferred consideration for sites that are subject of objections	n.a.	Cannot stop undesirable developments immediately
A register of all planning app.	n.a.	A register available for public inspection	A register available for public inspection	Allows better access to information
Established use certificate and Building density	'Established use certificates' to be issued by the planning authority and to include density and development control in the TPO	Planning certificates issued by the Planning Authority & Planning related provisions in the BO should be consolidated in the new Planning Ordinance	n.a.	Planners have limited right to modify development intensity and the built form specified in the Buildings Ordinance

Source: synthesized by author from the respective Reports.

However, on the contrary, the Hong Kong Institute of Architects, Hong Kong Institute of Engineers and the Hong Kong Institute of Surveyors put forward a joint paper criticising the White Bill (1997, pp. 1–7), arguing that the expense entailed in the proposals would delay the development process, create uncertainty, deprive landowners of property rights, impose another layer of bureaucracy, and provide excessive power to the Planning Department. Naturally, these Institutes suggested that planning permission should only be required 'in exceptional cases' and that 'no planning certificates [issued by planners] should be required' (op. cit., p. 7). In other words, these professional groupings objected to a return of responsibility for, and therefore authority over urban planning to the planning profession. Although the HKIP repeatedly urged the Administration (1997, 1998) to release the compiled public views on the White Bill and publicise the Blue Bill as early as possible, the requests were not entertained.

Because of these controversial and opposed stances, the Government continued to consult, review and address the concerns of different parties. When the Town Planning Bill was eventually raised in the Legislative Council, it was in 2002, 11 years after the publication of the 1991 Comprehensive Review. In this round, the Government's determination to introduce comprehensive changes seemed to have waned. Instead of enacting a comprehensive Bill, the Government proposed to amend the TPO in three stages and planned to introduce the relevant legislative proposals into the Legislative Council in March 2003 (PPLW, 2002, p. 6). The first stage amendment was eventually passed by the Legislative Council on 7 July 2004 but the prospect of placing development control and implementation mechanisms back in the hands of urban planners remains as uncertain as ever (HK LegCo, 2003, p. 2).

As the Building (Planning) Regulations more or less set the development framework for urban development in the city, and planners are deprived of opportunities to implement plans, one may argue that government planners in Hong Kong when compared to their western counterparts have little direct experience in understanding the needs and sentiments of the general public for whom they are planning (Hoch, 2006). Before the 1990s, these limitations had not prevented the contributions of the planners because the dominant development ideology of the city was pro-growth and the results were seen to be positive for urban development in general. However, the context of Hong Kong changed after the 1990s.

Hong Kong's reversion to Chinese sovereign rule in 1997 almost coincided with the bursting of the economic bubble[6] triggered by the Asian financial crisis in 1998. The subsequent economic recession (Martin, 2007) and China's courtship with the capitalists in Hong Kong served to dramatically escalate the many debates related to the skewed power relationships in the political economy of the city. As a result, the political nature of urban planning and spatial development long hidden from view when the city had undergone several hectic decades of almost non-stop spatial developments became a bone of contention in the city. However, the soul searching by the economically battered citizens was met with an economically reforming China whose leaders were eager to engage the capitalists to ensure their continued support of the HKSAR. The emerging and increasingly vocal civil society has tried to challenge a growth machine run by the public and private sectors that has paid little attention to the third sector. The societal awakening to the spatial injustice incurred through planning related mechanisms embedded in an institutional framework that privileges economically powerful property development-related stakeholders (Poon, 2011) has led to incessant controversy over development issues, from urban renewal to the development of the New Territories; often these are seen as efforts by the Government to resume people's lived spaces for property development, and plans for reclaiming land from Victoria Harbour are seen as a greedy land grab from a natural water body with great cultural and environmental values to Hong Kongers (Ng, 2013, 2006).

Should the proposals in the 1991 Comprehensive Review be adopted, planners and hence the government might be able to identify and understand the sentiments and concerns of the affected stakeholders before a planning study is conducted. Should the TPO be revamped as first proposed in 1991, they would empower planners to specify development intensity and urban forms appropriate for each unique urban context and allow them the power to freeze undesirable developments and implement plans; thus some of the controversial development schemes such as those involving mega wall-like residential buildings along the waterfront (Ng, 2014a) might be resolved in a spatially more humane and acceptable manner. The failure to transform the toothless urban planning system means that planners can play a very limited role when civil society objects to

6 In August 1997, the Hang Seng Index plunged from 16,800 to 6,600 (Wong, 2011).

the spatial consequences of the growth machine. It is, therefore, imperative for the urban planning profession to reflect on its role in the future development of Hong Kong.[7]

Whither the Role of Urban Planners?

As I have argued elsewhere, if 'planners remain complacent to serve as technicians and continue to loathe politics, their services would then be monopolized by the ruling strata' (Ng, 1992, p. 10). In a questionnaire survey conducted in 1996, about 40 per cent of the surveyed planners trusted the capacity of the public to know of their own needs and make decisions, and 46 per cent believed that citizen participation could 'increase effectiveness of planning' (Ng and Chui, 1999). This relatively pro-citizen engagement stance explained the Planning Department's subsequent move towards more transparent practices in the late 1990s. In 2007, the then second Chief Executive promised to begin 'a new decade of greatest public participation in formulating government policies by collaborating with civic groups and citizens' (Chan, Wu and Pang, 2007). Indeed, some of the controversial planning issues were eventually resolved through, among other factors, citizen engagement. However, it should be noted that the original plans were made to promote economic growth to the neglect of natural, social and cultural heritage, social networks and there has been no change to the hard core formal legislative and institutional processes (Ng, 2013, pp. 371–3).

If engagement might have placated an unsatisfied public in the 1990s, the awakening of the public to the skewed power relationships in the city and their discontent at the Government's policy bias towards the interests of the property developers have aroused negative sentiments among the community. For instance, the Government halted land auction and the construction of subsidised home owner flats in 2002, which many alleged was under pressure by developers as property prices plummeted after the Asian Financial Crisis. In 1999 the Government introduced the 'application list system' in land sales, giving developers the initiative to 'trigger' developments. As a result, the slowing down of land development has led to dwindling supply of housing units and soaring property prices, and has earned the city 'the title of the most unaffordable housing in the world for the fourth straight year' (Liu, 2014). Hong Kong's median home price was more than half a million US dollars for a flat of 40–50m^2. Many have attempted through the internet to publicize and verify how government planning and development have favoured the big tycoons[8]. As the Government has lost the trust of the people, it is perhaps no longer effective for urban planners to simply enhance their modes of engaging the public. They have to take sides.

Possible Ways Forward?

A two pronged approach may be appropriate, for the profession to resolve the current predicament. On one hand, planners need to move beyond their comfort zone as 'neutral professional experts', and subject their professional work to moral scrutiny. They have to reassess the political economy of the city to realise what they can do, and by reference to transforming the existing politically biased practices for the construction of a more equitable, socially just and environmentally friendly society. On the other hand, planners need to learn how to reconnect with their service targets and equip themselves with the necessary skills and attitudes so that their hard work can create emotional resonance and support in the community (Friedmann, 1987; Forester, 1989; Healey and Gilroy, 1990; Hoch, 2006).

7 The Public Affairs Committee of the Hong Kong Institute of Planners organised two workshops titled 'From Bitter Debates to Planners' Answers' in January 2014, attended by about 20 planners. Though the discussions had been constructive and thought-provoking, conclusions are yet to be reached after the workshops.

8 In recent years, many activists have used the internet to expose the biased policies or developments initiated by the Government. Examples include The Real Hong Kong News, http://therealnewshk.wordpress.com//?s=property&search=Go; the Local Research Community, https://zh-hk.facebook.com/localresearch; Land Justice League, https://zh-tw.facebook.com/landjusticehk; The Professional Commons, http://www.procommons.org.hk/.

Healey and Gilroy (1990, p. 15) argue that a people-sensitive planning is only possible when planners understand how 'the structuring forces which shape our societies' affect 'the specific experience of the individual people'. Such planners will need to have an epistemological consciousness, an ethical consciousness, a political consciousness, a socio-psychological consciousness, a structural consciousness as well as a self-reflective consciousness (op. cit., pp. 15–16). This may sound like a far cry from the technocratic breed of urban planners that has developed in colonial Hong Kong since the last century. Yet, some of these requirements of people-sensitive planners are surprisingly similar to the qualities intellectuals are expected to possess in traditional Chinese culture, and consolidated during the Spring and Autumn and Warring States periods (770 to 221 B.C.). Conventionally, intellectuals in the Chinese culture are often seen as those who understand the 'ways' （道） that can save the world （明道救世） (Yu, 2010b, p. 49). To Confucian scholars, the roles of intellectuals were to 'reconstruct social values; to establish meanings of life for the people; to continue lost teachings for past sages; and to establish peace for all future generations'.[9] What, then, are the values held to be ideal for traditional Chinese intellectual elites? The Book of Rites describes a world of Grand Unity (大同世界), a harmonious existence of a great state and settled and self-sufficient families (Holloway, 2009).

This utopian state is the imaginary of traditional Chinese elites. As argued by Hua and Guo, 'the most serious crisis of Chinese culture is that Confucianism represents the values of the intellectual elites … The Chinese lack a popular value system' (2007, p. 5). The knowledge of the ideal world (the 'way') is the inner drive for intellectuals to practise 'inward transcendence', a 'self-reflective' quality that helps intellectuals examine critically an unethical world that has gone astray (Yu, 2010a, p. 74). In other words, intellectuals in the Chinese traditions always maintain, in Healey and Gilroy's terms, 'a structural and political consciousness' (1990) to monitor the performance of the ruling class. 'One serves his prince according to what is right, and when he finds he cannot do so, he retires' (Du and Song, 1995, p. 48). 'Rather die of outspokenness than live in silence' highlights intellectuals' natural calling to 'remonstrate those in power', to speak truth to power (Yu, 2010a, p. 77). Mencius (c.250BC to 150BC) commanded intellectuals not 'to be tempted by money and ranks, shaken by poverty and hardship, nor succumbed to power and force'[10] as they should be the conscience of society.

In other words, Chinese intellectuals are not only required to focus on morality issues, monitoring those in power and challenging social injustice, they are also expected to possess an admirable 'socio-psychological consciousness': 'in his own conduct, he is humble; in serving his superiors, he is respectful; in nourishing the people, he is kind; in ordering the people, he is just' (Du and Song, 1995, p. 45). This emphasis on the importance of monitoring the virtues of those who rule rather than mobilising resources the lay people possess in enhancing governance perhaps explains why there has been a persistent lukewarm response to the call of community planning in Hong Kong. Even social activists in various controversial planning and development projects display a rather distinctive elite culture (Ng, 2014b). Hence, returning to the roles as traditional Chinese intellectuals to recognise the ills of a society bankrupted by imbalanced power relationships, to criticise injustice practices and to speak as the conscience of society, while essential, are no longer sufficient for urban planners to meet the challenges in twenty-first century Hong Kong. Urban planners have to learn anew how to communicate better, and collaborate with the directly affected stakeholders and from there develop a deeper understanding of how the structural forces shape the daily lives of individuals and families in local places (Healey and Gilroy, 1990).

Hence, a dialogical and co-learning relationship between the urban planners and their service targets are paramount. While planners in Hong Kong find it extremely difficult and challenging to understand the varying needs of different stakeholders in carrying out current mega projects, Schon (1982, p. 300) encourages planners to be reflective practitioners; embracing uncertainties in the planning process, inspiring the service targets, those whom are the beneficiaries of the their work to adopt a learning attitude; and actively connecting with the thoughts and feelings of different stakeholders, with a view to establishing long-term connections. This is a real challenge in Hong Kong given its peculiar political economy and an institutional set

9 Reference: modified from Premier Wen Jiabao's speech in Harvard University on 10 December 2013, http://www.news.harvard.edu/gazette/2003/12.11/10-wenspeech.html and http://www.360doc.com/relevant/178781079_more.shtml, accessed in January 2014. The phrases in Chinese are: '為天地立心，為生民立命，為往聖繼絕學，為萬世開太平'.

10 Original Chinese: '富貴不能淫，貧賤不能移，威武不能屈'.

up that marginalises urban planners in the whole planning and development process. Reclaiming their roles as intellectuals, they have to find a practical pathway to operationalise their 'ways'. Healey (2012) outlines the qualities that planners should promote in the course of communication or dialogue: constantly monitoring if beneficial distrbutional outcomes are achieved, and at the same time, fostering rich, complex, pluralistic, intelligent and respectful multi-scalar interactions:

- Recognition of a shared, but diverse and often conflictual political community, in which distributional fairness matters.
- Recognition of multiple identities and ways of life, as well as commonalities.
- Respect for different arguments, position, feelings, etc.
- A practice of "intelligent", multi-sided discussion of issues – learning and reasoning "in public".
- Critical respect for the role of formal government, while recognising the complexity of interactions between the spheres of state and society.
- Respect for those who do formal government work and perform "public services".
- Rich, responsive and respectful interactions between all levels and sectors of formal government' (op. cit., 31).

Many have synthesised means to achieve these principles through correcting distortions in comprehensibility, sincerity, legitimacy and truth in communications (Forester, 1989) or through transformative practices of interpersonal transaction and dialogue (Friedmann, 1987). Hoch (2006, p. 379) even suggests that 'we can use our sensitivity to emotional response to organise planning messages that will elicit the desired response [from our service targets]'. The idea that Hong Kong's foremost need is efficiency sustained a techno-centric practice of the built-environment professionals for several post-war decades. Growing planning controversies since the 1990s have demanded that planners move out of their comfort zone as technocrats to reflect on their roles as morally engaged Chinese public intellectuals and people-sensitive reflective practitioners. This is not easy as planners have to return not just to the elite-oriented Chinese tradition that has been eroded over the years by colonialism, capitalism, utilitarian and consumerist cultures; they will also need to tread into unknown territory to work with different stakeholders, to 'learn in public', and to acknowledge the possibilities of 'intelligent and respectful co-production' of political culture and governance processes (Dewey, 1991 cited in Healey, 2012) – all formidable challenges, indeed, in the skewed political economy of Hong Kong.

Concluding Remarks

If it is no longer feasible for urban planners to hide behind the facade of technocracy, and if urban planners are not supposed to be social activists who strive to resolve problems by making trouble on the street, perhaps urban planners can go back to Chinese cultural knowledge and reflect on their roles as intellectual elites and thereby courageously reclaim their traditional calling as moral advocates for justice and righteousness, and as a force to remonstrate against the ruling class and to speak truth to power and the people. In the context of Hong Kong, planners need to push for the further amendment of the Town Planning Ordinance and to guarantee a genuine engagement of the different stakeholders, especially those who are marginalised and have no resources to participate in the formal planning processes. However, a bigger challenge, from both a professional and cultural point of view, is to demand that urban planners work much closer with various stakeholders at the local scale and inspire a political culture and governance processes, and ones that first, do not lose sight of the 'dark side' of the unbalanced power structure and have the diligence and sensitivity to correct the faulty institutional set up in a capitalist society such as Hong Kong; and on the other hand, that continue with a faith in the importance and possibilities of building 'richly-intelligent, inclusively-fair and multiply-pluralist political communities' (Healey, 2012, p. 34). This terrain has huge potential and implications for a culture that hitherto has known very little the potential of its ordinary citizens.

Acknowledgements

This project was supported by the Research Grants Council of the Hong Kong Special Administrative Region, China (Project number CUHK750610). I would like to thank the two anonymous reviewers as well as the editors for their useful and encouraging comments. All the faults that remain are, of course, mine.

References

Bristow, R. (1989) *Hong Kong's New Towns: A Selective Review*, Hong Kong: Oxford University Press.

Bristow, R. (1984) *Land-use Planning in Hong Kong: History, Policies and Procedures*, Hong Kong: Oxford University Press.

Buildings Department (2009) *Practice Note for Authorized Persons, Registered Structural Engineers and Registered Geotechnical Engineers APP-124*, Hong Kong: Buildings Department.

Census and Statistics Department, Hong Kong Special Administrative Region Government (HKSAR) *Hong Kong Annual Digest of Statistics*, various years.

Chan, C., Wu, E. and Pang, D. (2007) 'Tsang pledges to consult masses', *The Standard: China's Business Newspaper*, 6 July 2007.

Cheng, J. (ed.) (1989) *Political Expedition: Views of the Hong Kong Association for Democracy and People's Livelihood*, Wide-angel Press, Hong Kong (in Chinese).

Cheung, G.K.W. (2009) *Hong Kong's Watershed: The 1967 Riots*, Hong Kong: Hong Kong University Press.

Cooray, A, (1997) 'Recent developments in planning law', *Law Lectures for Practitioners 1997*, pp. 135–60.

Crush, J. (1994) 'Post-colonialism, de-colonization, and geography', in A. Godlewska and N. Smith (eds), *Geography and Empire*, Oxford: Blackwell, pp. 333–50.

DeFilippis, J. and Smith, N. (1997) 'Post-colonialism? Reflections on Hong Kong and Mars', *Environment and Planning D: Society and Space*, pp. 505–7.

Dewey, J. (1991) *The Public and its Problems*, Athens: Swallow Press/Ohio University Press.

Du, G. and Song, G. (1995) 'Relating human rights to Chinese culture: The four paths of the Confucian Analects and the four principles of a new theory of benevolence', in M.C. Davis (ed.), *Human Rights and Chinese Values*, Hong Kong: Oxford University Press, pp. 35–56.

Einstein, A., *Albert Einstein Quotes*, available at: http://www.brainyquote.com/quotes/quotes/a/alberteins385842.html#YFzEeHrATLFEixJL.99, accessed in January 2014.

Forester, J. (1989) *Planning in the Face of Power*, Berkeley: University of California Press.

Friedmann, J. (1987) *Planning in the Public Domain: From knowledge to action*, Princeton, NJ: Princeton University Press.

Friedmann, J. (1973) *Retracking America: A theory of transactive planning*, New York: Anchor Press.

Healey, P. (2012) 'Reenchanting democracy as a mode of governance', *Critical Policy Studies*, 6(1): 19–39.

Healey, P. and Gilroy, R. (1990) 'Towards a people-sensitive planning', *Planning Practice and Research*, 52(2): 21–9.

Healey, P. and Underwood, J. (1978) 'Professional ideals and planning practice', *Progress in Planning*, 9(2): 73–127.

Hoch, C. (2006) 'Emotions and planning', *Planning Theory and Practice*, 7(4): 367–82.

Holloway, K. (2009) *Guodian*, Oxford University Press.

Hong Kong Branch of the Royal Town Planning Institute (HKB of RTPI) (1973) *A Review of the Hong Kong Town Planning Ordinance Chapter 131*, Hong Kong.

Hong Kong Institute of Architects, Hong Kong Institute of Engineers and the Hong Kong Institute of Surveyors (1997) *Joint position paper on the Consultation Paper on Town Planning Bill.*

Hong Kong Institute of Planners (HKIP) (1998) *HKIP Written Submission*, Public Affairs Committee, February 1998.

Hong Kong Institute of Planners (HKIP) (1997) *HKIP's Position Paper on the TPB*, December 1997.

Hong Kong Institute of Planners (HKIP) (1996) *HKIP's Views on the Town Planning Bill.*

Hong Kong Legislative Council (HK LegCo), 18 December 1996. *Official Records of Proceedings.*

Hong Kong Legislative Council (HK LegCo), 15 January 1992, *Official Records of Proceedings*, pp. 1307–59.

Hong Kong Research Project (1974) *Hong Kong: A Case to Answer*, Hong Kong Research Project Spokesman Books.

Hong Kong Special Administrative Government (HKSAR Government) (2013a), *The Basic Law*, Hong Kong: Information Services Department, available at: http://www.gov.hk/en/about/abouthk/factsheets/docs/basic_law.pdf, accessed in April 2014.

Hong Kong Special Administrative Region Government (HKSAR Government) (2013b) *Methods for Selecting the Chief Executive in 2017 and for Forming the Legislative Council in 2016: Consultation Document*, available at: http://www.2017.gov.hk/en/consult/document.html, accessed in January 2014.

Hong Kong Special Administrative Region Government (HKSAR Government) (2012) *Hong Kong Year Book 2012, Appendices*, available at: http://www.yearbook.gov.hk/2012/en/pdf/Appendices.pdf, accessed in January 2014.

Hong Kong Special Administrative Region Government (HKSAR Government) (2012) *How Legislation is Made in Hong Kong: A Drafter's View of the Process*, Hong Kong: Department of Justice.

Hong Kong Special Administrative Region Government (HKSAR Government) (2009) *Public Consultation on the Methods for Selecting the Chief Executive and for Forming the Legislative Council in 2012*, 18 November 2009.

Hong Kong Special Administrative Region Government (HKSAR Government) (1997) *The Basic Law of the Hong Kong Special Administrative Region of the People's Republic of China*, available at: http://www.basiclaw.gov.hk/en/basiclawtext/images/basiclaw_full_text_en.pdf, accessed in January 2014.

Hua, S. and Guo, S. (eds) (2007) *China in the Twenty-first Century: Challenges and opportunities*, New York: Palgrave MacMillan.

Lands Department (1996) *Survey and Mapping Office Series AR/9/RD* (3rd ed.).

Lau, S. (2014) 'Top justice officials back independence of Hong Kong courts', *South China Morning Post*, 13 January 2014.

Lethbridge, D. (1980) (ed.) *The Business Environment in Hong Kong*, Hong Kong: Oxford University Press.

Lin, X. (1992) *An Uncertain Future*, Yuan Jing Chu Ban Shi Ye Gong si, Taipei (in Chinese).

Liu, Y. (2014) 'Hong Kong ranks world's No.1 for "most unaffordable" housing', *South China Morning Post*, 22 January 2014.

Liu, Y. and Ng, J. (2013), 'Land supplication list scrapped in bid to control supply', *South China Morning Post*, 1 March 2013.

Lu, F. (1985) *Towards a Democratic Hong Kong Ruled by Hong Kongers*, Hong Kong: Beichen Xueshe (in Chinese).

Martin, M.F. (2007) *Hong Kong: Ten years after the handover*, CRS Report for Congress, available on: http://www.fas.org/sgp/crs/row/RL34071.pdf, accessed in January 2014.

McInnis, A. (2003) *Butterworths Hong Kong Building Law Handbook* (2nd ed.), Hong Kong: LexisNexis.

Ng, M.K. (2014a) 'The state of 'planning rights' in Hong Kong: A case study of "wall-like" buildings', *Town Planning Review*, 85(4): 489–511.

Ng, M.K. (2014b) 'Intellectuals and the production of space in the urban renewal process in Hong Kong and Taipei', *Planning Theory and Practice*, 15(1): 77–92.

Ng, M.K. (2013) 'Got the controversial urban planning job done? An institutional perspective', in J. Cheng, *Evaluating the Tsang Years 2005–2012*, City University Press, Hong Kong, pp. 347–73.

Ng, M.K. (2006), 'World-city formation under an executive-led government: The politics of harbour reclamation in Hong Kong', *Town Planning Review*, (77)3: 311–37.

Ng, M.K. (1992) 'The changing politics in Hong Kong: Wither the role of planners?', *Planning and Development*, 8(2): 2–12.

Ng, M.K. and Chui, E. (1999) 'From values to praxis: Problems and prospects of advocacy planning in Hong Kong', *Planning and Development*, 15(2): 70–96.

Panel on Planning, Lands and Works (PPLW), Hong Kong Legislative Council (2002) *Panel on Planning, Lands and Works; Minutes of meetings*. Accessible at: http://www.legco.gov.hk/yr02–03/english/panels/plw/minutes/pl021206.pdf, accessed in April 2012.

Planning, Environment and Lands Branch (PELB) (1991) *Comprehensive Review of the Town Planning Ordinance: Executive Summary*. Hong Kong: Government Printer.
Planning Department, HKSAR (2003) Residential densities, *Hong Kong Planning Standards and Guidelines, Chapter 2*, pp. 1–34.
Poon, A. (2011) *Land the Ruling Class in Hong Kong*, Richmond, BC, Canada.
Pryor, E. G., (1983) *Housing in Hong Kong*, Hong Kong: Oxford University Press.
Schon, D.A. (1982) *The Reflective Practitioner*, Hong Kong: Basic books.
Skeldon, R. (1997) 'Hong Kong: colonial city to global city to provincial city?', *Cities*, 14: 265–71.
The Heritage Foundation (2014) *2014 Index of Economic Freedom*, available at: http://www.heritage.org/index/country/hongkong, accessed in January 2014.
Vines, S. (1998) *Hong Kong: China's New Colony*, Hong Kong: Aurum Press Ltd.
Wang, G. and S.L. Wong (1997) (eds) *Dynamic Hong Kong: Business and Culture*, Centre of Asian Studies, The University of Hong Kong.
Wong, D. (2011) *Hang Seng Risk Adjusted Index Series*, available on: http://www.hsi.com.hk/HSI-Net/static/revamp/contents/en/dl_centre/presentations/P_HSRAIe_20110324.pdf, accessed in January 2014.
Yeoh, B.S.A. (2001) 'Postcolonial cities', *Progress in Human Geography*, 25(3): 456–68.
Yu, Y.S. (2010a) 'An examination of the history of Chinese intellectuals', in J. He (ed.), *Intellectual Thoughts of Yu Ying Shi*, Shanghai Old Text Press, pp. 67–82 (in Chinese).
Yu, Y.S. (2010b) 'A brief discussion of the similarities and differences between Chinese and Western intellectuals', in J. He (ed.) *Intellectual Thoughts of Yu Ying Shi*, Shanghai Old Text Press, pp. 45–56 (in Chinese).

PART III
Places and Practice

Part III.1
The Planning Development Nexus: How Places are Produced and Changed

Introduction to Part III.1

The Planning Development Nexus: How Places are Produced and Changed

Jonathan Metzger

> All cities require the production of space in the form of both buildings and sites for various activities. Yet the processes through which this production has been accomplished have been given little attention in the social scientific literature on urbanisation and urban development. The role of landownership, the organisation of the construction industry, the nature of the finance invested in urban development and the significance of intermediaries, from developers to property consultants, lie hidden or are given little more than passing reference in many historical accounts of urban development.
>
> (Healey and Barrett, 1990:89)

The above quote outlines a research programme and area of inquiry that most contemporary readers would not spontaneously associate with the works of Patsy Healey. Nonetheless, the study of the social mechanisms that in various ways affect or even partially drive how places are produced – quite physically – primarily through the activities of the land and property development sectors, has been a key interest for Healey, particularly in focus in her work in the 1980s and early 1990s. Her attention on the physical production of space and place hasn't primarily focused on urban design, but rather on the institutional settings – economic, social, political and cultural – that come to bear down on specific areas and locales to produce pressures or directions for change. The interplay of these forces direct investments in land and property in a market-based economy, leaving some places overheated with investment and soaring property prices, such as the Southeast of England in the 1980s–1990s (and still today). Meanwhile, as the flip-side effect of the same mechanisms, other areas become neglected or abandoned, the urban 'black holes of informational capitalism', as so aptly put Castells (1998), for instance parts of northern England during the same time-period – such as the deindustrialized Tyneside to which Patsy would arrive as a newly appointed Professor and Head of Department of Town and Country Planning at the University of Newcastle upon Tyne in 1988.

Already before this, since the mid-1980s, Healey had taken as her mission to study these processes, their emergent variegated effects on specific places and the range of both expected and potential responses these patterns in turn may come to generate. She identified a sore lack of empirical work recognizing the complexity of interactions within and around processes of land and property development, arguing that '[u]nderstanding these processes is … a necessary task, both to assist in the practical activities of urban management and property development and economic development, and to develop a critical capacity to evaluate such practices' (Healey and Barrett, 1990:90). More recently, she has reiterated this claim, lamenting that there still 'has been surprisingly little academic work on the land and property development sector, particularly from a more economic and sociological point of view' (Healey, 2010:19, n 17).

In her work on land and property development she has displayed a particularly keen interest in the relationship between public planning and the development industry, investigating the potential role of methods and devices for partially guiding or directing private investments in a fundamentally capitalist society marked by strong property rights, a high level of private land ownership and socio-spatially highly uneven concentrations of capital, as well as a strong political will towards promoting market-led spatial development, such as in Thatcherite Britain in the 1980s (cf. e.g. Healey and Nabarro, 1990; Healey et al., 1992; Healey 1992a, Thornley 1991). Consequently, this part of the book, *The Planning/Development Nexus*, focuses on issues of land and property development, relations between planning and land and property markets and the implementation impacts of planning instruments. It includes three reproduced papers by Healey as well as three original contributions, which all in different ways relate to issues of land and property development as seen from a wider institutional perspective and with an attention towards the micro-politics of local and situational agency and contingency.

Two of the reproduced papers (Healey and Barrett, 1990; Healey, 1992a) have been selected to both highlight the important and still influential major shifts that occurred in property development-related British planning practice at the time when they were written, such as 'market aware' local development plans, as well as Healey and colleagues' (such as Sue Barrett and Paul McNamara) introduction of novel intellectual resources for the understanding of these developments. First and foremost among these was perhaps Giddensian structuration theory, offered as a tool for grasping and strategically appropriating these shifts towards 'progressive' aims. The third reproduced paper (Healey, 2010) is a recent actualization of these research interests in which Healey draws upon her earlier research in the area to discuss the potential role and function of spatial planning in the face of the ongoing current global recession and related financial and property crises. It aptly illustrates that although Healey never was a fully-fledged Marxist per-se, she nevertheless accepts large parts of the overarching political-economy analysis of broad economic patterns and shifts. At the same time, she is persistent that this frame of analysis will never be enough in itself to understand which difference active practice actually can make to how the effects of these 'structuring forces' are channelled by private as well as public actors so as to bear down on the unfolding fate of specific locales or places.

The above intuition is also the focus of Benno Engels' Chapter 12 in this section, which examines the historical influence of long economic waves on planning practice. While arguing that long waves undoubtedly exist, the chapter raises the question to what degree changes in planning practice can be 'read off' the turning of more overarching tides in the economic fortunes of capitalist accumulation. Drawing upon a neo-Marxist SSA (Social Structures of Accumulation) approach, Engels analyses the effects of the three most recent such long waves on the planning systems of the UK, US and Germany. He reaches the conclusion that although both 'upswing' and 'downswing' phases of long waves see the introduction of major planning legislation, the extent or focus of planning intervention does not appear to be mechanistically related to particular moments in these cycles. Rather, just as Healey has argued (Healey, 2010), it appears as if it is in times of instability, present both in the deep slumps and high peaks of long waves as well as in the shifting of conjunctures, windows of opportunity exist to introduce quite far-reaching planning reforms and frameworks of spatial development. But there is no evidence of a direct correlation between the exact content of such reforms and frameworks and the correlated position in a long wave at the time of its introduction. Instead, Engels – following Healey – draws a conclusion about the critical role played by specific institutional frameworks, including differences in general and political cultures, in channelling exactly how such broad economic swings are responded to by planning practice, reaching the conclusion that 'economic long waves have conditioned, but not determined human and institutional agency' in the planning field (Engels, this volume).

Responding to Healey and Barrett's (1990) call for more research focus on the sources and vehicles of investment capital in the property industry, and how these affect the development of the industry, Paul McNamara provides a unique in-depth investigation of structural shifts in the UK property industry, 'changing almost everything about how the UK commercial property market operates' during the past decades. He further discusses how these changes may come to affect investors' interests in the fate of specific places, as well as their interactions with the planning system – while concluding that the overarching research agenda on land and property development processes outlined by Healey and Barrett in 1990 (Healey and Barrett, 1990, see also further below) most probably will 'remain as pertinent for the next thirty years or so' (McNamara, Chapter 11).

In Chapter 10, Wendy Steele tackles the role of planning in a depicted era of 'feral capitalism' and 'wild cities', in which global capitalism has run amok. In the chapter Steele musters empirical evidence from Australia detailing the mounting pressures to 'cut red tape' and 'speed up' planning and land allocation processes, asking – in the face of persistent such pressures for most part of the past three decades in the Western world – which are the viable alternatives for planning that are both sensitive and responsive to the interdependencies between economic development, environmental protection, and social equity? In such an environment, where can we find the 'seeds of urban change' and how can they be nurtured? If we take our cue from Healey, Steele suggests, this would imply fostering the capacity in territorial political communities for ongoing 'place-making' activities at the local scale through better understanding how places are produced and change, and critically rethinking the complex inter-relationship between planning, land, growth and development. This would then in extension contribute to a more informed critical questioning concerning for whom, for what purpose – and driven by what reform values – planning systems, plans and practices purport to 'perform'.

The remainder of this introduction to Part III.1 will pick up on some thorough-reaching themes and important developments in Patsy Healey's work on land and property development processes, particularly focusing on her interest in the interplay between wider economic structuring forces, how these come to affect particular places and locales, and the potential influence of active governance agency – such as planning initiatives – to guide or affect these developments. An interest which saw her focus move from a stricter political economy-based theoretical perspective in the early 1980s towards a more Giddensian structuration approach in the later part of that decade, albeit without completely shedding her political economy influences.

Combining Political Economy and Structuration Theory

Questions concerning how various types of societal 'structures' are formed and may evolve or change have been in Healey's focus ever since her PhD-studies in the 1970s (see e.g. Healey, 1974). As she brought this interest to bear upon the study of land and property development processes in Britain during the early years of the 1980s she was initially deeply influenced by a reinvigorated political economy approach that asserted itself across the social sciences at the time. This development was perhaps particularly forcefully in human geography – a discipline closely related with planning studies – much thanks to the seminal studies produced by David Harvey around this time (see e.g. Harvey 1973, 1982; for examples of Healey's more political economy oriented works refer to e.g. Healey, 1979, 1982, 1983). Other important influences for her during this period included Massey and Catalano (1978) as well as Ball (1983). Soon, though, Healey discovered that the still quite rigid analytical frameworks of neo-Marxist political economy failed to accommodate the complex nuances evinced by her empirical experiences of spatial development processes in South America as well as in Britain. These studies led her to conclude already in 1974 that 'structures' can never be understood without paying attention to the continuous but often also unstable processes and mechanisms of *structuration*, a term she later came to incorporate from the works of sociologist Anthony Giddens. Healey and Barrett (1990) carefully explore this intuition in a text that has since become quite influential and which still up to this day racks up citations in top journals and books across the social sciences.

Perhaps the key contribution of Healey and Barrett's paper is how it brings the structuration theory of Anthony Giddens (Giddens, 1984) to bear on questions of land and property development. They examine the approaches that were influential in analyses in this field at this point: institutional analysis, neo-classical theory, land economics and Marxist economics, but come to the conclusion that all these theoretical frameworks fail to address the crucial question of just how the actions of specific actors both reproduces and may alter the 'broader processes' that drive the strategies and interests of the actors involved in the land an property development spheres (Healey and Barrett, 1990:89). Based on Giddens' framework of analysis they suggest a 'relational approach' involving tracing the practical deployment (and simultaneous iterative reproduction) of *resources*, *rules* and *ideas* so as to understand the co-constitution of emergent structures and individualized agency. The crucial research task becomes to examine how '*external pressures are reflected in and affected by the way individual agents determine their strategies and conduct their relationships as they deal with specific projects and issues, and as they consider their future stream of activities*' (Healey and Barrett, 1990:90, emphasis in original).

Thus, Healey and Barrett propose an empirical research approach that recognizes the relative powers of both 'structure' and 'agency', as well as how these forces interrelate and are constituted, as a fruitful way towards an increased understanding of land and property development processes. They argue that such an approach enables the researcher to focus attention on how individual firms and agents interrelate in the negotiation of particular development projects and how, through this, land and property markets are constituted (Healey and Barrett, 1990:93). In this, their perspective has strong affinities with e.g. Fainstein's seminal study of urban property development (Fainstein, 1994). Healey and Barrett's important point of divergence from broader political economy-inclined studies of the subject was the crucial interest in just how wider societal forces are mediated, reproduced and/or altered in particular times and places, thus raising the question of how they may be more or less consciously 'tweaked' through active agency, a notion even more directly explored in Healey (1992a). In her contribution to this volume, Steele also succinctly teases out this divergence between the 'Healey' – structuration centred – perspective on urban development and the 'Harvey' perspective, which can be construed as more statically structure centred, at least when it comes to the circulation of capital, and diverging beliefs regarding the potential role of planning activity to contribute to a fairer and more democratic society, also within the broader structures of a capitalist economy (see also Huxley, 2009). Nevertheless, it cannot be denied that Harvey's thinking and works appears to have provided a crucial and also lasting influence on Healey's perspective on urban development generally, and property development specifically, as evinced by her recent description of his book *The limits to capital* (Harvey, 1982) as a 'very perceptive account' and 'landmark text' on the structural mechanics of capitalist circuits of accumulation (Healey, 2010, note 13).

Even though Healey partially (but only ever partially) turned away from political economy as a comprehensive framework of analysis already in the mid-1980s, one can in her works note a persistent interest in the structuring power of economic circuits of accumulation and value production which appears to have remained as a longstanding influence on her way of thinking. For instance, already in 1990 Healey and Barrett urge scholars who study urban development to pay closer attention to the effects of the ongoing 'globalization of real estate' (Healey and Barrett, 1990:94) and related shifts in investor and developer behaviour. They therefore call for 'attention to the wider context of the changing characteristics of those involved in the development process, and the factors which are driving these changes' as well as 'the way in which property development processes in the UK relate to the 'global processes of real estate investment'' (Healey and Barrett, 1990:97). In light of this appeal, McNamara's contribution in this part of the book, Chapter 11, written by someone who has not only had an insider-position in the British property investment market for almost three decades, who also possesses a solid academic grasp of the wider theoretical ramifications of shifts in economic structures and arrangements, appears as a crucial contribution to the contemporary scholarly understanding of the quite dramatic restructuring these sectors and related investment and development activities have undergone in Great Britain during the recent years.

Economic Shifts and the Fate of Places

Healey's main body of work on the subject of land and property development bears the double mark of the 1980s urban blight in Britain and the Thatcherite response of 'rolling back the state' and its associated vicious all-out attack on the British planning system which left planning as a public activity, as well as the planning profession, reeling. While holding on to various crucial reservations, Healey's perspective on the troubling developments in many of Britain's urban centres was deeply influenced by political economy thinking. It is thus important to remember that her emerging interest in processes of structuration and the potentials of agency in relation to property development dynamics by no means encompassed some form of disavowal of the role of emergent structures. Rather, in relating her earlier work on property development to the global financial crisis of 2008 and the associated property crisis, she has criticized how in the present 'climate of ideas' rules and regulations are often 'dismissed as an inheritance of outdated attempts to 'control' what happens or as an old-fashioned approach to territorial development' and relatedly stressed how 'legal and administrative procedures and practices have always played an important role in shaping the contexts in which innovative activity and experimentation take place. They have had a 'structuring' role' (Healey, 2010:11).

Relatedly she has cautioned that 'too much of an emphasis on uncovering the specificities of particular development pathways can neglect the way these specificities are shaped by powerful exogenous forces' (Healey, 2010:12). So just because she turned towards structuration theory to find better tools to conceptualize just how structures are generated, change and dissolve, has not meant that she has somehow denied their existence or influence on human action. As Engels (2015) states, her perspective by and large appears to resonate with Marx's claim that 'men [sic!] make history … but they do not make it just as they please; they do not make it under circumstances chosen by themselves, but under circumstances directly encountered, given and transmitted from the past' (Marx, quoted in Engels, 2015, current volume). Also in line with her early political economy influences, she has particularly displayed a keen awareness towards the importance of markets dynamics and how they are conditioned, as well as an interest in how they come to bear upon the unfolding fate of specific places. By way of her experience and learning she is not a believer in the planned economy, countering calls for strong governmental economic steering with the claim that 'the record of state control is not such a 'shiny' one, being prone to subversions by private interests, inefficiency and inertia' and laconically adding that, no matter what, it 'seems highly unlikely that market mechanisms will disappear from the complex urbanized societies of the 21st century' (Healey, 2010:12). From her view, what we need to learn more about is rather 'the way market activity is structured' and the role economic flows and transactions play in place development, and further: how they can consciously be altered or tweaked in a 'progressive' direction.

Some of the economic conditions that in this context take the form of an in-themselves difficult-to-affect 'powerful exogenous force' is the broad cyclicity of capitalist accumulation, even if local responses to these patterns can be highly variegated with radically different results (see also further below). Here Patsy subscribes to Harvey's (1982) analysis of the three circuits through which capital is accumulated in capitalist economies, which she highlights as a 'very perceptive account of the crisis-prone nature of each system and the potential for all to be in crisis at once' (Healey, 2010:18, n 13). She thus brings into sharp attention the influence of economic cycles of 'boom and bust' on different spatiotemporal scales and how they come to affect the unfolding fate of places. Engels (this volume) brings into specific focus the effects of long so-called 'Kondratieff-waves', but comes to the conclusion that crude mechanistic approaches do not do justice to the complex relationships through which such shifts are mediated into particular responses in specific times and places. This insight dovetails Healey's insistent claims that the consequences of economic conjunctures, as well as 'the booming and busting', are unevenly distributed socially and spatially. Further, responses are variegated, and dependent on a plethora of factors, including – but not limited to – cultural contexts, institutional settings and political leadership.

Sketching a research agenda for concrete avenues of inquiry into how processes of economic development, partially relayed through land and property development dynamics, affect particular places and locales, Healey and Barrett (1990:98) outline four potential major themes for such research:

1. A review of the changing forms of capital that flow in and out of the built environment which 'requires an understanding of the diverse sources of capital', for instance including not only profits of transnational companies, but also e.g. personal savings. Further, they call for an investigation into exactly how these flows are invested in property, implying that this makes a crucial difference in how development projects are selected and carried out, something which McNamara (2015) also highlights in his contribution. This demands attention to the nuances between the roles of different firms in the property development industry in different locales, and the particular interrelations between types of capital, firms and investment strategies and further how these have specific and variegated effects in particular instances of valuation and project selection.
2. Focus on the changing composition of firms involved in the development of the built environment, including e.g. the building and building materials industry. This implies investigations into which firms are involved in what subsectors, how the composition of these groupings are changing with time, how new firms and new business strategies develop, and an attention to the particular configurations of the industry in geographical and sectoral submarkets. Particularly through connecting this to an understanding of the constitution of 'the impact of such changes on the form of, and terms of access to, the built environment thus produced, and the consequences for local economies'.
3. Studying the role of the state in relation to these changing practices, including the tools of intervention employed and how they affect the changing dynamics in the two previously mentioned areas.
4. Finally, assessing the implications for local economies of all of the above in terms of capital flows, labour market demand and supply, building materials and land, the impact on land and property values, etc. Particularly so the resultant social, economic and environmental externality costs and benefits within local economies, as well as the relative distribution of these, concluding that '[i]t is this theme which presents the most challenges, both conceptually and empirically, yet it is clearly the

> most significant in assessing the contribution of property development to economic and social change' (Healey and Barrett, 1990:99).

What Healey and Barrett apparently try to grasp with their fourth outlined theme is what we perhaps with another wording could call the 'consequences not cared for' (or in economist-speak: 'negative externalities') of local economic development. These are perhaps particularly evident in the unstable conjunctures of economic upswing/downswing phases (cf. Engels, 2015) when calls to 'kickstart the economy' through loosened regulation are often at their most vociferous, which – if rapidly and strongly implemented – may lead towards capitalist dynamics in land and property development going almost completely unchecked and perhaps – presently – even 'feral', as Steele argues in her chapter. These situations generate forceful and, at least according to the Marxist model of capital accumulation which Healey draws upon, increasingly recurrent and intense cycles of boom and bust which become hard-felt in many places, for which the road to recovery can be long and painful; which further, if it even ever arrives, only does so patchily and unevenly distributed (see e.g. Harvey, 1982, 1989).

When Healey published her major contributions on land and property development in the late 1980s and early 1990s she depicted a British land and property development sector 'dominated' by increasingly mobile international finance, major economic restructuring, strong public opinions not only regarding economic growth and recovery but also increased demands for environmental quality. And finally: a UK government both pushing for less state intervention and higher degrees of deregulation in the development industry while at the same time actively directing support to 'lever' the private sector into areas adversely affected by economic change (Healey, 1992a; the situated concrete effects of the latter were later also studied by Healey and published in e.g. Healey, 1997). At the present, similar policies – albeit even more limited with regards to the level of active government intervention – are being implemented in the UK and large parts of Europe under the label of 'austerity' policies (for different understandings of these developments see e.g. Vigar, this volume; Lloyd and Peel, 2012; Lord and Tewdwr-Jones, 2012; Allmendinger and Haughton, 2013).

Seen in context of longer economic trends, there is nothing surprising about these developments, for just as Healey (2010) has noted, it has been historically recurrent that investors and development actors often demand 'release from 'bureaucracy' and 'regulatory constraint'' during 'boom periods', whereas in 'bust periods' the equally vociferous demands are instead for 'protection from instability and uncertainty' (Healey, 2010:14, cf. Adams, 1994; Adams and Tiesdell 2013). Reverberating her political economy inspirations and the argumentation by Engels in this volume, Healey notes that 'Planning systems and practices get buffeted by these demands and the economic winds that blow them', albeit in temporally, institutionally and spatially situated ways that are often not easy to predict (Healey, 2010:14).

Just like in the early 1980s, we presently live (and for a while now, have lived) in times when many experience that 'established arrangements for promoting and governing territories and localities have been destabilised' (Healey, 2010:11). Commenting on this, Healey has noted that some localities 'seem overwhelmed by this complexity and instability' while in others 'considerable governance capacity has evolved, through which responses to unstable conditions have been characterized by innovative experimentation, leading in some cases to transformations in governance cultures and practices' (Healey, 2010:11). As Engels illustrates, it is times of instability, historical broader swings and conjunctures of trends when lots of things are changing, that also present opportunities when the agency of such a practice as planning can actually have a potential to make a concrete difference in a specific time and place, with a possibility to affect the longer future trajectory of that specific context (cf. also Hillier, 2007). Or in the words of Healey: 'instability and crisis' should be used 'in a strategic way, as an opportunity to take stock, to re-think policies, projects and practices, and to build the intelligence and coalitions which could bring future benefits for the many not just the few in our localities' (Healey, 2010:21).

These are thus the conjunctures that offer an opportunity to build strategic frameworks through which development activity, when it revives, 'can be shaped in ways which promote wider objectives than just the maximisation of profits' (Healey, 2010:15). In such situations, 'the fortunes of urban areas ... are likely to depend not just on how national and global economic dynamics develop but on the local governance capacity to reduce the damaging impacts of a long period of economic difficulty ... In other words, it is the work of agency which generates new structuring possibilities when established structures weaken or collapse'

(Healey, 2010:18). But the crucial question then becomes, paraphrasing Marx, just *how* can localities develop the capacity to "act *for* themselves', transforming their processes and practices in doing so' (Healey, 2010:11, emphasis in original)?

The Agency of Planning in Land and Property Development

Healey does not write as a disinterested observer of land and property development processes, but rather as an engaged – and worried – public intellectual. At the time when she was penning her major contributions in this area of research she was witnessing first-hand a historical shift when planning as a practice and profession appeared to be pushed against the wall or even into the doldrums in the UK. A shift in which concerns for collective environmental and social qualities systematically were forced to stand aside to a more narrow focus on rates of return for the economic investments for capital providers and entrepreneurs, depicted as the only functioning motor for efficient urban renewal. Hence, being officially committed to the craft of planning, it is perhaps not surprising that Healey at this point in time is deeply engaged in questions concerning how planning practices might come to exert 'agency', i.e. make a (positive) difference in how local land and property development unfolds to the benefit of the 'many not the few' (cf. Healey, 2010). For, if broader structural shifts and trends can be taken up and responded to in very different ways in different places, an insight she has repeatedly come to in her empirical studies of urban development processes, shouldn't there be room to also *make* such a difference – for instance through planning interventions?

Relating to questions of this type, Healey (1992a) notes how from the 1970's and onwards British planners have been under pressure to become more 'market aware' in their work. This pressure constituted part of a broader shift in the relations between market and state towards more market-centred governance arrangements. The 'markets' in question that planners primarily deal with Healey defines as those in existing properties and in real estate development opportunities. She further argues that in a capitalist society where the private sector is the main provider of development, it is a correct assertion that planning needs to be market aware in regulating the development process, but that this by no means implies a smaller role for planning. Instead, '[o]n the contrary, development plans may bring positive benefits to different kinds of developers and to communities in seeking to benefit from development' (Healey, 1992a:13), as markets are only highly efficient 'in theory' while in practice while in practice being 'grossly imperfect'. Thus, the central argument presented by Healey is that 'in conditions where development activity is effectively privatised and communities care about their environment, development plans become more, not less, important' (Healey, 1992a:19). But for planning to fulfil its potential function as a corrective to the gross – and add to that, inescapable and fundamental – imperfections of the property development markets, Healey argues that planners must gain an understanding of the situated dynamics of land and property markets as the 'market-aware development plan' in her vision must be 'founded upon an understanding of these imperfections as they are manifest in particular areas' (Healey, 1992a:14).

Further, these insights raises the question of just *how* planning authorities should relate to these markets and make a difference in relation to them: should they have the ambition to 'follow', 'manage' or 'create' markets? In the article Healey traces then-recent shifts in British local plan-making practice, from the comprehensive urban structure plans of the 1950–70s, across the usage of 'plans' as mere records of the agreed principles resulting from pluralist bargaining processes in the 1970s and 1980s and then towards the emerging potential of the 'market-aware plan'. Discussing these different developments, Healey argues that one of the strengths of the early, comprehensive approach was that it 'internalised the treatment of externality costs and benefits within the calculus', driven by 'concerns for environmental quality and community needs' (Healey, 1992a:15). But due to the reality of property development investment being fundamentally market-led in the UK, and increasingly so during this period, plans were seldom realized as developers were simply not interested in slotting into the subordinated 'following' role allocated to them by local planning arrangements.

As a consequence, projects to increasing degrees began to be negotiated between individual developers and local councils on a 'project-by-project basis' in the 1970s and 1980s, leading to recurring accusations from local citizen groups that their interests, and the interests of local environmental concerns, were being sidestepped again and again, thus generating a very adversarial urban development climate. As Healey dryly notes, '[t]he main beneficiaries of a planning system in this form would appear to be consultants and lawyers, whose fee income is enlarged by the scale of negotiation and arbitration required' (Healey, 1992a:16). In this context, the relevance of plans suddenly appears to have again been actualized, as '[t]he value of the plan format is that it provides some semblance of accountability for the process', which might be sought to reassure local people that their interests are being seen to, or property developers and landowners that agreed upon projects will stand (Healey, 1992a:16). As Healey notes, in this form, the function of the plan becomes to record agreements about principles and directions arrived at between state and market on a case-by-case basis. However, 'the plan-as-record does little to structure the market and protect it from its own behaviour' (Healey 1992a:16).

At the time of writing, Healey identifies mounting voices for a more proactive local planning approach, to the surprise of some – primarily coming from the development and property industries, demanding better certainty (but nevertheless '*certainty-with-flexibility*') and clearer development visions from local authorities. Moving on from this insight, Healey analyses the different types of 'market aware' potential planning responses she sees as viable in this situation, focusing on the local development plan as either *following* markets, *managing* markets or *creating* markets (cf. also Brindley et al., 1989). Healey to a large extent appears to equate the 'following' approach with the formalization of the 'plan as record of negotiation' of the preceding period, dismissing this as a practice possessing 'little capacity to transform the qualities of areas, and thus contributes little to urban regeneration strategies' (Healey, 1992a:17). In contrast, the 'managing' approach to local development plan-making rather aims at macro-structural control of development through stipulating and enforcing the amounts and locations of development and infrastructure investment. While appearing as broadly sympathetic to this type of approach, Healey nevertheless cautions that it requires power to influence investment and the provision of local authority services, i.e. the plan must be 'expenditure-linked' and come with an implementation mandate and its own earmarked 'pouch of monies' (Healey, 1992a:18). Further, such a broad managing approach demands attention to the broad set of strategic relations that frame the dynamics of any specific place, and contribute to producing its qualities. The level of analysis demanded of this type of approach is thus very high and as Healey notes: '[a] 'market management' approach thus begins to have some aspects of the traditional comprehensive urban framework', signalling a wariness concerning the risks of producing plans that are considered 'unreasonable' by strong industry actors, but nevertheless communicating a cautious optimism regarding the more 'market-savvy' tangent of this new generation of plans to counter this risk.

Finally, considering the potential for planning to *create* markets in locales that appear to have fallen into disinterest or disrepute among contemporary property investors, Healey argues that this can be partially encouraged through 'development frameworks' that can contribute towards 'defining a new image for a place, and translating this into principles for the development of parts of an area' (Healey, 1992a:18). Such frameworks must be spatially selective but nevertheless quite 'comprehensive' regarding the areas they cover so as to 'define not only what may be appropriate and feasible in an area, and give some ideas about the order and 'packaging' of development'. They should also be tied to concrete resources, at least through outlining which subsidies and incentives might be available to realize the depicted potentials. She concludes that this 'creating markets' strategy is perhaps most efficient as pin-point initiatives within a broader 'managing' approach. Zooming out, Healey argues that there might be room for spatially selective and differentiated 'market-following' as well as 'market-creating' initiatives within a broader 'managing' approach, whereas the critical challenge for 'plan preparers' is to develop a broader macro-economic understanding so as to be able to 'deliberately *structure* the market' (Healey, 1992a:19, emphasis in original), further concluding that 'land and property development flourishes best when the state adopts a managerial mode' (Healey, 1992a:19, see further also Healey, 1992b).

Further developing this line of reasoning almost twenty years later, in her 2010 paper reproduced in this Part, Healey details how such a 'managerial mode' may be successfully exercised. Here she argues that 'spatial strategies and skilled co-ordination work' are the 'critical tools available not just to 'manage' land and property development back into activity, but to 'nurture' them', particularly in post-crisis situations (Healey, 2010:17). Healey stresses that this mode of operation demands sustained social and institutional work to build governance capacities. Particularly crucial is what Healey calls 'structural grasp', something that with other words perhaps could be called a 'relational sensibility' (see also Healey, 2004), i.e. 'the ability to see, in the fine details of localised phenomena, the wider social, environmental, economic and political forces which help to produce such details – to use a metaphor, the ability to 'see the world in a grain of sand'' and to recognize that '[n]o urban region is an isolated island and all depend in one way or another on their position in wider economic, social, political and environmental relations' (Healey, 2010:19). This, situated 'continual interaction, in thought and in wider discussion, between 'parts' and 'wholes'' is to Healey 'the essence of strategic thinking' that 'converts the sociological discussion of the relative importance of structure and agency, and the significance of focusing on the interactions between the two, into an active practice' (Healey, 2010:19).

Conclusions

Highlighting the dominance of the idea of 'trickle-down economics' in the field of UK property development policy in the 1980s, which meant that 'government policy towards urban regeneration is founded on the wealth-generating power of property investment – economic regeneration through property development', Healey and Barrett commented in 1990 (p. 96) that 'whether and how this equation works remains to be adequately investigated'. Twenty five years later we may perhaps venture to provide the tentative answer: in many places, not too well – thus re-actualizing Healey's 1992 request for more 'intelligent' actively guiding government policy and planning regarding land use and property development, based not only on aims for local economic growth, but also prioritizing social and environmental justice and development.

As is apparent from her writings about land and property markets, Healey believes in the power of planning to make a positive contribution to society, in sharp contrast to the Marxists which she otherwise share many analytical influences with. This belief is anchored in her faith in the potentials of agency, however situated, in making the world a little better through active endeavours in creatively 'translating' (cf. part 4) broad structural shifts. In certain periods this has implied the courting of market forces to actively 'create markets' with the aim of reconnecting neglected or marginalized places to wider flows of economic resources. In other periods and places it has instead meant a need for a managerial approach more focused on 'holding' so as to protect social and environmental qualities which the business models of global footloose capital tend to externalize if let run off leash.

Healey do not see it as a realistic prospect that the markets for land and property development will be socialized in Western economies in the near future, and therefore encourages planners to take the chance to 'assist market activity as long as this supports the values of social and environmental responsibility', particularly in the face of the post-2008 crisis prospect for many localities of a negative economic spiral leading towards a stagnated built environment 'scarred by poor maintenance, empty buildings and overgrown development sites' (Healey, 2010:17). In this quote we can see the haunting shadows of memory of the urban blight in 1980s northern England which still to this day echoes through many tarnished post-industrial landscapes. Still, it might be so that in our present era – the regulatory aspects of planning might be at least as important to develop so as to try to counterbalance the recurrent boom-and-bust cycle of 'letting loose' and 'reigning in' land and property development that has now recurred so many times in tandem with cyclical crises in capitalist accumulation. This is perhaps even particularly important at the present historical conjuncture when 'feral' urban capitalism completely sheds responsibility for all the negative externalities it produces, partially fuelled by increasing investor detachment from their investments by way of the 'trancheing' and 'packaging' of property assets, thus anonymizing the real locales of property investments to its formal owners, as described by McNamara in Chapter 11.

Of course, as so often, the crucial questions all come down to the difficult '*how*?' This question is particularly challenging within the context of the highly nuanced theoretical understanding of these processes that has been carved out by Healey, as any tentative answer is complicated by the related question of how 'inventive agency responses to changing conditions 'kick back' to alter those conditions' (Healey and Barrett, 1990:96). Through her recurring attention to such phenomena Healey, without using those precise words, has infused the political economy-oriented urban studies framework, which has functioned as a point of departure for so much of her research into land and property development, with a sense of complexity and uncertainty at the structure/agency structuration interface.

In her characteristic non-fussed anti-nostalgic fashion, she refuses to ignore the insight that comprehensive planning did simply not work in the 1970s UK, as a society then dominated by strong property rights and private capital, and that these fundamental institutional preconditions have not changed since. Further, now perhaps more than ever, uncertainty and relational complexity are ubiquitous not only in local property development markets, but also in the timing of the broader winds of fate of shifting economic fortunes on broader scales. But this also offers opportunities, since the highly variegated ways through which global processes of real estate investment touch ground evince the high degree of potential to locally negotiate and channel these shifting fortunes through local capacity building and strategic awareness in a manner that focuses not only on local economic development but also 'the social and environmental consequences of the particular forms which development processes may take, and to conduct evaluation in such a way that the distributive dimensions of outcomes are revealed' (Healey and Barrett, 1990: 98).

As the above quote so aptly illustrates, in all her studies of land and property development markets, the overarching concern of Healey – as it so often turns out to be – is with social justice and economic improvement for the 'many not the few' (Healey, 2010a). For while it might be so, as she pragmatically recognized already in 1990, that governments in the present historical epoch may be 'forced to intervene in a 'partnership' mode, whatever their ideological predispositions', this raises the need for a constant vigilance and a persistent raising of the question: 'what are the terms of these partnerships and what are the measures of their success?' (Healey and Barrett, 1990:96).

References

Adams, D. (1994). *Urban Planning and the Development Process*. London: Routledge.

Adams, D. and S. Tiesdell (2013). *Shaping Places: Urban planning, design and development*. London: Routledge.

Allmendinger, P., and Haughton, G. (2013). The evolution and trajectories of English spatial governance:'Neoliberal'episodes in planning. *Planning Practice & Research*, 28(1): 6–26.

Ball, M. (1983). *Housing Policy and Economic Power*. London: Methuen.

Brindley, T., Rydin, Y. and Stoker, G. (1989). *Remaking Planning: The politics of urban change in the Thatcher years*. London: Unwin Hyman.

Castells, M. (1998). *The Information Age: Economy, society and culture. Vol. 3, End of millennium*. Malden, MA: Blackwell.

Fainstein, S.S. (1994). *The City Builders: Property, politics, and planning in London and New York*. Oxford: Blackwell.

Giddens, A. (1984). *The Constitution of Society: Outline of the theory of structuration*. Cambridge: Polity Press.

Harvey, D. (1973). *Social Justice and the City*. London: Edward Arnold.

Harvey, D. (1982). *The limits to capital*. Oxford: Blackwell.

Harvey, D. (1990). *The Condition of Postmodernity: An enquiry into the origins of cultural change*. Oxford: Blackwell.

Healey, P. (1974). Planning and Change. *Progress in Planning*, 2(3): 143–237.

Healey, P. (1979). Networking as a Normative Principle with particular reference to local government and land use planning. *Local Government Studies*, 5(1).

Healey, P. (1982). Understanding land use planning: The contribution of recent developments in political economy and policy studies. In P. Healey, G. McDougall and M.J. Thomas (eds), *Planning Theory: Prospects for the 1980s*, Oxford: Pergamon Press, pp. 180–93.

Healey, P. (1983). *Local Plans in British Land Use Planning*. Oxford: Pergamon.

Healey, P. (1992a). Development plans and markets. *Planning Practice and Research*, 7(2): 13–20.

Healey, P. (1992b). Urban regeneration and the development industry. In Healey et al. (eds), *Rebuilding the City: Property-led urban regeneration*. London: E & FN Spon, pp. 15–38.

Healey, P. (1997). City fathers, mandarins and neighbours: Crossing old divides in new partnerships. In O. Källtorp and I. Elander (eds), *Cities in Transformation: Transformation in cities*. Aldershot: Avebury, pp 266–88.

Healey, P. (2004). The treatment of space and place in the new strategic spatial planning in Europe. *International Journal of Urban and Regional Studies*, 28(1): 45–67.

Healey, P. (2010). Re-thinking the relations between planning, state and market in unstable times. In P. Pinho and V. Oliveira (eds), *Planning in Times of Uncertainty*. Porto: FEUP edicoes, pp. 11–22.

Healey, P. and Barrett, S.M. (1990). Structure and agency in land and property development processes. *Urban Studies*, 27(1): 89–104.

Healey, P. and Nabarro, R. (eds) (1990). *Land and Property Development in a Changing Context*. Aldershot: Gower.

Healey, P., Davoudi, S., O'Toole, M., Tavsanogly, S. and Usher, D. (eds) (1992). *Rebuilding the City: Property-led urban regeneration*. London: E & FN Spon.

Hillier, J. (2007). *Stretching Beyond the Horizon: A multiplanar theory of spatial planning and governance*. Burlington, VT: Ashgate.

Huxley, M. (2009). Theoretical debates, planning practice and spatial processes, *Planning Theory & Practice*, 10(1): 139–42.

Lloyd, M.G., and Peel, D. (2012). UK enterprise policy–a twist on real estate, austerity and planning. *Journal of Surveying, Construction and Property*, 3(1): 1–7.

Lord, A., and Tewdwr-Jones, M. (2012). Is planning 'under attack'? chronicling the deregulation of urban and environmental planning in England. *European Planning Studies*, 21: 1–17.

Massey, D. and Catalano, A. (1978). *Capital and Land: Landownership by capital in Great Britain*. London: Edward Arnold.

Thornley, A. (1991). *Urban Planning under Thatcherism: The challenge of the market* (2nd edn). London: Routledge.

Chapter 7

Structure and Agency in Land and Property Development Processes: Some Ideas for Research*

Patsy Healey and Susan M. Barrett**

Introduction

All cities require the production of space in the form of both buildings and sites for various activities. Yet the processes through which this production has been accomplished have been given little attention in the social scientific literature on urbanisation and urban development. The role of landownership, the organisation of the construction industry, the nature of the finance invested in urban development and the significance of intermediaries, from developers to property consultants, lie hidden or are given little more than a passing reference in many historical accounts of urban development, with notable exceptions in the urban history research generated by Dyos's pioneering studies of Victorian cities (Dyos, 1961).

Until the mid-1970s, the only significant contributions to the field in Britain were a very few widely read case studies, notably Marriott (1969) and Ambrose and Colenutt (1973). This situation has been remedied more recently by work in the fields of Marxist economics,[1] public policy implementation,[2] urban geography,[3] and planning and estate management,[4] while there has been a considerable body of relevant research in the USA and Australia.[5] These empirical and theoretical contributions emphasise that land and property are not merely surfaces and spaces upon and within which the processes of production and consumption are played out. The way in which land and property are themselves 'produced' and consumed enters into the processes of economic production and consumption. Knowledge of the processes through which the built environment is produced and used, and in particular the processes of *land and property development*, is thus critical to our understanding of urban development and our attempts at managing urban development processes.

Understanding these processes is thus a necessary task, both to assist in the practical activities of urban management and property development and economic development, and to develop a critical capacity to evaluate such practices. A key issue in this task is the relationship between the strategies, interests and actions of the various agents involved in the development process – landowners, investors, developers, consultants, public agency planning officers, politicians and community groups – and the organisation both of economic and political activity and of values about land, property, buildings and environments which frames or *structures* their decision-making . Any analysis of development processes thus requires the development of an explicit approach to the relation between *structure*, in terms of what drives the development process

* Previously published in *Urban Studies* 27(1): 89–104 (1990).

** Susan M. Barrett is at the School for Advanced Urban Studies, University of Bristol, Bristol, UK. This paper draws on ideas developed with Christine Lambert and Paul McNamara. Paul McNamara's comments are particularly appreciated. A first draft was given at the School of Environmental Planning, Melbourne University, in March 1987, while Patsy Healey was on an ESRC Research Grant. Paper first received May 1988; in final form March 1989.

1 See Ball (1983) and the volumes of the Bartlett International Summer School 1–8, published by the Bartlett School, University College London ; also Smyth (1985).

2 See Healey et al. (1982, 1985); Rydin (1986); Short et al. (1986a, b); Witt (1987); Massey and Catalano (1978); Barrett et al. (1978a, b); Barrett and Whitting (1983); Adams (1985); Montgomery (1985); and Boddy (1982).

3 See Drewett (1973); Craven (1975); Goddard (1975); Daniels (1985); Dawson (1983); Davies (1984); Whitehand (1984); and Whitehand and Whitehand (1984).

4 See the new journal *Land Development Studies* for a range of material.

5 See Leitner (forthcoming), Leitner and Sheppard (forthcoming) and Feagin (1983) in the USA; Daly (1982, 1984), Sandercock (1976), Berry and Huxley (1985) and McLoughlin and Huxley (1985) in Australia; see also Boyle (1985) and Boyle and Rich (1984) for comparisons between USA and UK approaches to urban regeneration through property development.

and produces distinctive patterns in particular periods, and *agency*, in terms of the way individual agents develop and pursue their strategies. The approach adopted in this paper draws in particular on the work of Anthony Giddens, who argues for a relational approach between structure and agency in which 'structure' is established by the way agents operate: deploying, acknowledging, challenging and potentially transforming resources, rules and ideas as they frame and pursue their own strategies.[6] Structure, in terms of the framework within which individual agents make their choices, may be seen to inhere in the various *resources* to which agents may have access, the *rules* which they consider govern their behaviour, and the *ideas* which they draw upon in developing their strategies.

This kind of approach has been implicit in some of the more recent work arising from the study of the *implementation* of public policy initiatives in the land development field (see Barrett and Whitting, 1981, 1983). Allied to theory concerning the dynamics of economic and political organization, such an approach promotes an understanding of the 'development industry' as *actively involved* in its own reconstitution, while at the same time being driven to restructure by external pressures. Thus, the critical task for the analyst seeking to understand the processes of production of the built environment is an examination of how such *external pressures are reflected in and affected by the way individual agents determine their strategies and conduct their relationships as they deal with specific projects and issues, and as they consider their future stream of activities.*

It is proposed that, theoretically, and in terms of research strategies, a way to address this task is combine the insights derived from the traditions of institutional analysis, and in particular the studies of implementation processes within the field of policy analysis, with the neo-classical analyses of the operation of urban land markets and Marxist approaches to the way capital flows through the built environment. The present authors, in conjunction with Christine Lambert and Paul McNamara, have been working for some time on an outline of such a framework. This paper is both a record of this review process and a background for research now under way. It starts with a review of current research on the processes of the production of the built environment. The objective is to draw out the theories being used and the way in which the relations between structure and agency are established. The second section reviews contemporary tendencies in the processes discussed, as a basis upon which to identify some important research questions in the final section.

Existing Research

A major reason for the neglect of the study of land and property development processes by researchers in the UK until recently was that the subject was peripheral to the interests of the established academic disciplines (Healey and Barrett, 1985). The last decade has seen a significant change in this situation. In the UK, there has been a growing amount of work which focuses on the interaction between development interests and state agencies (Barrett et al., 1978b; Barrett and Whitting, 1981, 1983; Boddy, 1982; Elson, 1982; Healey et al., 1982; Barrett and Healey, 1985; McNamara, 1985; Rydin, 1986; Short et al., 1986a, b). Many of these studies are richly grounded empirically, but have focused largely on institutional forms of explanation. Several have been undertaken by policy analysts concerned with policy implementation, notably those by Barrett, Elson and Healey. The interests and strategies of actors and the nature of the relationships between them are identified, but the link to what generates these interests and strategies is often weakly developed, though usually noted. This opens such work to charges of empiricism and of a drift to pluralism in interpretation.

These studies have been complemented by research on particular sectors of development activity, notably Ball (1983), Dickens et al. (1985) and Smyth (1985) on the housebuilding industry; Barras (1984, 1985, 1987) on office development; Fothergill et al. (1987) and Adams (1986b, 1987) on industrial development; and Dawson (1983) and Davies (1984) on the retail development field. Ball's work in particular integrates the construction dimension with the property development aspects of the industry. In addition, there have been studies of specific agents, such as Goodchild and Munton's (1985) review of the role of landowners and McNamara's work on local estate agents (McNamara, 1984a) and surveyors (McNamara and Turner, 1987). Adams et al. (1985) have given particular emphasis to valuation practices in their analyses of inner city

6 See Giddens (1984).

property markets, while there has been a stream of comment on inner city land values,[7] and a major empirical study by Bruton, Gore and Nicholson of public sector vancant land holding (Bruton and Gore, 1980; Gore and Nicholson, 1985). Another line of research is directed at evaluating particular instruments of intervention and, among other things, their impact on development processes, and is represented notably by Elson (1986) on green belts, Healey et al. (1985) on development plans, Lambert and Underwood (1984) on local plans, and a rich array of case studies and consultants' reports evaluating some of the recent initiatives in UK land policy.

This growing body of work on the sectors of development and the relations between development processes and the state gives an idea of the complexity of the processes of built environment production. They involve the construction industry, the building materials industries, the financial sector, and the whole spectrum of social and economic activity in relation to demands and needs for land, buildings and location. To mediate the relationships between these, an array of development intermediaries has emerged in most developed countries. The state enters into these processes in diverse ways, through sectoral policies, as a development intermediary itself, and in order to safeguard particular interests and values. These studies also provide a range of hypotheses as to the relation between the interplay of agents in the negotiation of development projects and the wider economic context, and what the social, economic and environmental consequences of the activities analysed may be. Implicitly or explicitly, the studies referred to use some form of actor-based or institutional frame of analysis. But neither the individual studies nor the frameworks for analysis theorise the relations between individual actors and behaviours and the wider context in such a way that the dynamics of the relation between context and process, structure and agency, can be analysed.

Another line of inquiry, which has motivated some of the work summarised above, has started from a theoretical direction. Neo-classical economics, regional science and land economics have investigated the processes of spatial differentiation and the consequences of public goods provision, and have provided methods for evaluating public policy initiatives.[8] In the USA and Canada, there has been some attempt to move from micro-economic modelling to institutional analyses of land and property development processes using neo-classical concepts.[9] A particular interest has been the attempt to assess the economic consequences of land market regulation in terms of land and property prices (e.g. Dowall, 1984). In the UK, there has been an ambitious attempt to assess the economic costs of the British planning system (Cheshire et al., 1985).

This last study illustrates the inherent difficulties of adopting an idealised neoclassical model of market processes into which particular empirical measures are fed and modelled outcomes then compared either with each other, as in the Cheshire study, or with reality . Such models *presume* negotiation among agents as necessary but unproblematic in the processes of exchange, and assume certain structuring parameters, such as private property and a particular form of agent rationality. Thus, they do not explore negotiative processes empirically. They also tend to focus on selected distortions to ideal supply–demand relationships (such as the distortions of state intervention), rather than seeking to identify the range and relative significance of different distortions. The institutional research discussed above would suggest that forms of landownership, the organisation of agents and intermediaries, and the flows and forms of finance, were also significant distortions. The neo-classical approach has thus so far failed to address structure, agency and their relations directly, these being absorbed as assumptions into the theoretical parameters of the approach. Recent work by Evans (1988), however, has begun to explore some of the implications of incorporating institutional supply-side factors into a neo-classical model of the formation of land and property prices.

Marxist economics, by contrast, is explicitly concerned with the structural dynamics of the mode of production. However, there has been only limited interest in, and considerable controversy over, how land and property should be treated. Until recently, the focus of attention has been the problem of land rent, which has significance theoretically as an extraction by landowners of the surplus value generated by both capital and labour, an expression of the struggle between a 'pre-capitalist' landowning class and capitalist classes.[10] As

7 There is no comprehensive bibliography of this material, but a useful review which focuses on the development aspects of recent initiatives can be found in Cameron et al. (1987).

8 An extensive literature can be found in the journals *Land Economics* and *Urban Studies*; see also Harrison (1977), Evans (1985) and Hallett (1979).

9 See Brown et al. (1981), Dowall (1984) and Lin Leung (1987).

10 See Breughal (1975), Ball (1983), Massey and Catalano (1978) and Harvey (1982, 1985a).

such, it is both a drag on capitalist accumulation and a further drain on the ability of workers to enjoy the fruits of their labour. A number of interesting accounts exist of the way landowner power has conflicted with the demands of burgeoning urbanisation generated by economic development in one form or another, although it is clear in many of these studies that other factors are also at work.[11]

However, as Massey and Catalano (1978) have shown theoretically, and Goodchild and Munton (1985), empirically, landowners are by no means a homogeneous category, their interests and strategies being developed in response to a variety of orientations. The problem of identifying groups and interests in relation to land is a recurrent problem in much of the empirical work noted above (McNamara, 1988). Massey and Catalano (1978) identify three forms of landownership: former landed property, industrial landownership and financial landownership. Only landowners with a landed property interest can be seen as being in *opposition* to or parasitical upon capitalist accumulation. The others are integrated within the logic of the accumulation process. This point has been reinforced by the work of Ball (1983) and Ball et al. (1985), which shows clearly the way land and property are incorporated into the processes of production of goods, and the circulation of finance. The research pulled together under the aegis of the Bartlett Summer Schools on *The Production of the Built Environment*[12] has focused in particular on the construction industry and the complex interrelations of labour processes and land rent.

Harvey (1982, 1985a) provides a theoretical interpretation of the processes involved, through his framework of the circuits of capital and the role of the production of the built environment within them. His approach in effect provides a way of identifying how the dynamics of the mode of production *drive* the processes through which the built environment is produced, while at the same time recognising the spatial and temporal specificities of these processes. Harvey argues that the primary factor through which agents are bound into structural relations is through paths of capital flow, through *resources*.

However, the relation between this general conception of the structuring dynamics of development processes and the specific interests and strategies of individual agents remains tenuously established. Yet it is at the level of these interests and strategies that development processes are practically constituted, while it is through the way interventions 'bite' on these decisions and strategies that public policy is achieved or not. This suggests that much more attention needs to be given to the way individual firms and agents interrelate in the negotiation of particular development projects and how, through these transactions, land and property 'markets' are constituted and built environment investment decisions made. The 'institutional' research reviewed above also draws attention to the way politico juridical *rules*, and *ideas* and *ideology*, convey structural limitations and possibilities into the strategies and relationships of agents. Anecdotal evidence of contemporary examples suggests some complex interconnections here: for example, the House Builders' Federation's attempt during the 1980s to 'break' increasingly tough policies of environmental restraint (Rydin, 1986); the struggle over, and the consequences of, the significant changes to the Use Classes Order; and the way certain 'new' developers have been able to exploit the new rules and resources of the 'grant economy' to promote new kinds of development project. The analytical task is to link the institutional analysis of the development process with the dynamics of the economy as reflected in resource flows, and with political organisation and cultural values as reflected in rules and ideas.

This definition of the research task derives from a development of Giddens's approach to the relation of structure and agency. Giddens (1984) approaches 'structure' as

> the structuring properties [which allow] the 'binding' of time–space in social systems, the properties which make it possible for discernibly similar social practices to exist across varying spans of time and space, and which lend them 'systemic' form. (p. 17)

He further argues that systems (structures) are reproduced through the *rules* and *resources* which individuals draw upon in the production and reproduction of social life. Rules and resources, he proposes, are inherently

11 For studies of land owner power and economic development, see Fine (1985), Sutcliffe (1981) and Gilbert and Healey (1985).

12 See Note 1; see also Hooper (1985).

interrelated, and rules may be conceptually distinguished into those concerned with the constitution of *meaning* and those concerned with the *sanctioning* of modes of social conduct.

This approach provides little more than a way of focusing our ideas and empirical research onto the way in which 'structure' both affects and is changed by the way individuals act within the development process. It needs to be complemented with a development of the ideas within neo-classical and Marxist economics and within institutional analyses of policy processes. Such an enterprise will inevitably open up a series of further conceptual and methodological problems.[13] Nevertheless, Giddens's conceptualisation suggests it would be fruitful to research the way in which individual agents, as they operate in the development process, draw upon *resources*, *sanctioning rules* and *rules of meaning* to determine what they do. Within the context of development processes and the 'social systems' of which such processes are a part, this leads to a research emphasis on:

a. the *resources* for development, as channeled via the financial system and the interrelation of supply and demand;
b. the politico juridical *rules* which limit the construction of development opportunities; and
c. the *ideas and values* people hold about what they should build, what they would like to occupy and what kind of environment they seek.

To summarise, one result of the lack of academic attention to the processes of the production of the built environment has been that many analysts of spatial change have treated the conversion of economic and social processes into land use change and built form as essentially unproblematic. Demand, it was assumed, would be translated into supply, with the various agencies involved in development working collectively to provide development at 'the right time in the right place at the right price' (Lichfield and Darin-Drabkin, 1980). Implicitly, this formulation draws on the simple neo-classical model of the market. Such a treatment of urban development processes is also implicit in more recent work on contemporary economic restructuring (Massey, 1984; Cooke, 1986). Yet empirical research shows clearly that the relationship between demand and supply in the various processes which constitute the production of the built environment is problematic in many ways. Meanwhile, empirical reality is revealing these problems with stark clarity as urban areas face the complexities of recession and industrial restructuring, the internationalisation of finance, and the 'globalization of real estate'.

Contemporary Dilemmas

In the context of the discussion of structure and agency in the development process, it is particularly necessary to understand the characteristics of particular periods and the 'objective' possibilities of consumer demand for and investment returns from land and property development in particular circumstances, given the role of ideas, of 'talking up', or 'hype', in directing and realising the strategies of many of the agents involved. In this framework, the present period could be characterised as one dominated by:

1. international finance, with the barriers to financial mobility increasingly reduced as a result of information technology, allowing rapid transnational capital movements and the conversion of property into securities which can be traded in the international capital market;
2. major economic restructuring, with its significant spatial consequences, redistributing land and property values in complex ways;
3. a climate of articulate opinion strongly preoccupied with both economic regeneration and the quality of the environment;
4. a UK government caught in a paradoxical situation, pushing more of the effort for land and property development onto the private sector through strategies of deregulation and privatisation, while at the

13 Sayer (1984), for example, takes a different approach to the relations of structure and agency, while Hajer (1989) finds it necessary to evolve Giddens's framework into a more dynamic form.

same time providing the support needed to 'lever' the private sector into areas adversely affected by economic change.

Evidence of these forces at work is widespread in the anecdotal evidence provided in the property press. In response to new forms and flows of finance, new government policies, and the impact of economic restructuring on land and property markets, new actors have entered the field. Existing ones have changed their organisation, their strategies and their interests in land and property. In the UK, the financial institutions substantially increased their property holdings until the early 1980s. Since then, they have adopted more cautious strategies (Nabarro, 1990). Meanwhile, the 1973–74 slump and the bankruptcy of several development companies fostered larger, more diversified companies. At the same time, some larger conglomerates sought to diversify into property development, particularly housing, in the search for a balance of activities which would safeguard general levels of accumulation. Similar tendencies have occurred in other parts of the world, producing new international actors in the construction and development field.[14]

Overseas capital and development intermediaries have entered the UK market, partly as a result of problems at home (e .g. for Japanese investment) and partly responding to UK government policy which has fostered an open, deregulated financial market. In addition to foreign financial institutions, there is some evidence of American and Japanese property companies operating in the UK on an increasing scale at the present time, exploiting financial advantages (low interest rates in the USA and high rates of return relative to local ones in Japan). Some USA companies enter the UK market accustomed to a higher risk orientation than British firms adopt, challenging local assessments of property markets and local traditions of negotiation. Such firms may find it more effective to work in partnership with local firms to maximise their 'entry' possibilities into local economies and local politics, but such an infusion of new ideas could well change many traditions of British land and property development practice . A few parts of the country are affected by very specific international connections, such as the investment of Hong Kong capital in Manchester. There are also signs that new property developers are emerging locally, generated in part by the government's urban development initiatives: for example, John Hall at Gateshead, the Richardson brothers in Dudley, Paul Sykes in Sheffield and Eddie Healey in Rotherham. Competition for customers among the banks also encouraged support for property companies in the mid-1980s (Nabarro, 1990).

Meanwhile, as noted earlier, the established property consultancies are under pressure from financial investment managers and accountants who treat property as if it were like any other investment possibility (McNamara, 1990). This had led to some reorganisation.[15] There has also been a proliferation of new consultancies in the planning, design and development field. There are various reasons for this, but one of them is certainly the need for more research into market conditions, planning opportunities and how to negotiate constraints than has been the case among many development firms in the past.

These trends, widely recognised in the property field in the UK, suggest that the constitution of the 'development industry', in terms of the mix of firms involved in development and the objectives which direct their strategies at any place and time, is being significantly changed . The result is a reorientation of interests, strategies, working methods and relationships. On the one hand, these trends challenge established relationships through which the production of the built environment has been negotiated, which, both nationally and locally, have tended to display an oligopolistic form.[16] On the other, they introduce new perceptions of limitations and opportunities which can in turn change the parameters of land and property markets. Thus, John Hall in Gateshead is not only a new entrant to property development (a mining surveyor turned property developer), but he has also introduced a new product, the regional megastore, onto the market.

In terms of the structure–agency relation, the above example illustrates the way inventive agency responses to changing conditions 'kick back' to alter those conditions. What we so far know little about,

14 Interesting examples here are the Japanese construction firm, Kumagai Gumi, and the Italian property firm, Bellasconi.

15 The reconstitution is evident in the expansion of research teams. There have also been some mergers; for example, the absorption of several regional property consultancies into the international company, Chestertons.

16 Circumstantial evidence of these oligopolistic relationships is considerable; see McNamara (1984b), Wood (1982) and Nabarro (1989).

however, is how extensive these 'kick back' effects are. New actors, interests, strategies and relationships exploit new niches created by uneven development, the opportunities generated by the particular direction of capital flow or the 'grant economy'. Others seek to escape the consequences of economic decline in those areas adversely affected by uneven development. Do the changing composition and relationships of those involved in development produce different outcomes in terms of economic, social and environmental outputs and effects? Much current government policy towards urban regeneration is founded on the wealth-generating power of property investment–economic regeneration through property development. But whether and how this equation works remains to be adequately investigated.

One argument which could be used to support a positive connection between property development and economic growth can be derived from Harvey's discussion of the periodic need to transform the built environment (Harvey, 1985a). The spatial configurations suitable for capital in one period may act as an impediment, a drag, on capital's objectives in another, because of the need to reorganise space *in situ*, to develop formally 'greenfield' areas, or to extract capital investment from areas where values are falling. But to make a contribution to this effort of spatial reorganisation, the 'development industry' demands and probably requires government intervention. Thus, corporatist conceptions of partnerships between government and industry coexist, if uneasily, with ideological discourses of deregulation and letting the private sector 'take over' government functions. Harvey has argued that such spatial transformations are endemic in the way the built environment is produced in capitalist economies. In effect, new demands for space are being created in part through the reorganisation of the activities which create the demand for space, and in part through the entry of new forms of capital and new firms seeking to *construct* demand in different ways. This game is increasingly played out in the international sphere of product development, marketing and financial flows. Meanwhile, those whose economic and social horizons are rooted in localities, in particular cities, are ever more subject to the ebb and flow of the international game.[17] Following this economic line of argument, governments may be forced to intervene in a 'partnership' mode, whatever their ideological predispositions. This point is strongly confirmed by the experience of urban regeneration in the USA, where effective public–private partnerships are considered the key to success. But what are the terms of these partnerships and what are the measures of their success?[18]

The state is thus unavoidably a critical actor in the development process. This is so through the need not only to support the development process, but also to moderate its adverse externalities, safeguard social needs and conserve resources and environmental heritage. The way this state role is developed has a major *structuring* effect on the constitution of the development 'industry' in any country or region. In Britain, there has been a long tradition of urban and rural 'estate management', through which particular landscapes, buildings and urban designs have been maintained over time. Through the planning system, and supported by a powerful set of values about what the environment should be like, land and property development has been firmly regulated. By limiting the possibilities for speculation, and concentrating development opportunities, the state has in effect fostered the centralised and oligopolistic form which has characterised the 'development industry' in Britain. It has helped to sustain land value growth and to produce less risky and more protected markets than, for example, in the USA. The impact of policies of deregulation and privatization in this situation has a major destabilizing potential for the development industry in the UK.

At the same time, however, politicians are attracted to property development as a visible way of revitalising local economies. The phrase 'cranes on the skyline', used by politicians in recession-hit Australia to symbolise a political commitment to job-creating growth, is now deployed by John Hall to symbolise urban regeneration in the North East.[19] In effect, the state, locally and nationally, is caught within complex dilemmas in its relations within the processes of the production of the built environment. It is under pressure to deliver environmental quality, social facilities and economic opportunity, to reduce pollution and to safeguard built and natural environment. At the same time, in response to the 'fiscal crisis of the state' (O'Connor, 1973), the costs of collective urban management are kept down to satisfy vociferous ratepayers. In this context, partnership with

17 See Thrift (1987) on finance capital and localities.

18 See Boyle (1985), Boyle and Rich (1984) and also Moore and Pierre (1988).

19 See *The Newcastle Initiative*, a promotion document produced by the Newcastle Confederation of British Industry Action Team in June 1988, where the phrase 'cranes on the skyline' appears to symbolise economic development.

the private sector in various ways is an attractive option for governments forced into or committed to turning to the market for resources and development initiative. Many private firms welcome such arrangements as they search for investment outlets or fear devalorisation of their property assets. Government involvement thus offers a way of securing risk and supporting land and property markets. In doing so, it creates new forms of dependency in the development industry.

The 1980s have seen a stream of initiatives in Britain aimed at urban regeneration, or facilitating private sector property development. The list – Enterprise Zones, Urban Development Corporations, Urban Development Grants, Simplified Planning Zones, Public Land Registers, Urban Regeneration Grants, Inner City Task Forces, Garden Festivals, Housing Action Teams, City Grant – not only seems to be added to each year, but is complemented by a whole variety of changes to existing programmes and practices at both central and local level. So far, evaluation research has largely focused upon the impact of these innovations with respect to measures such as job generation, leverage ratios and property value effects.[20] Much less consideration has been given to the effects of these initiatives on the constitution of the development industry, the organisation of the development process, the economic, social and environmental effects of the development process itself, and the product of development in terms of cultural and environmental outcomes as well as economic ones. In particular, there has been little attention to the wider context of the changing characteristics of those involved in the development process, and the factors which are driving these changes. As a result, we are neither learning much about the replicability of such initiatives, nor about their consequences – for local economies, for the quality of the environment or for the way in which property development processes in the UK relate to the 'global processes of real estate investment'.

Research Questions

This paper has argued that the relation between context and specific events may fruitfully be approached by building on existing research work informed by an agency or institutional frame of reference, and linking this to conceptions of economic and political structure via Giddens's approach to the relation between structure and agency. This emphasises the interrelation between the two, rather than a deterministic relation between context and agency. It is suggested that, in research terms, the interrelation may be observed through the way in which agencies, individuals and organisations define and implement their strategies in relation to the *rules* they acknowledge, the *resources* they draw upon and seek to accumulate, and the *ideas and ideology* they assert in determining and justifying their strategies.

Given the current state of our knowledge in the field, four intersecting themes may be proposed, around which research is needed: the first would address the relation between the financial system as a whole and investment in land and development processes; the second would explore the way the *resources* (as above) and the *rules* of economic organisation constitute the types and strategies of firms involved in land and property development processes; the third would assess the way the state *structures* land and property development processes through its contribution to the constitution of *rules* and *resources*; and the fourth would evaluate the outcomes of these processes. The priority in policy terms is to examine outcomes in terms of local economic development, but it is important to explore other dimensions, in terms of the social and environmental consequences of the particular forms which development processes may take, and to conduct evaluation in such a way that the distributive dimensions of outcomes are revealed.

The first, the review of the changing forms of capital flow into and out of the built environment, requires an understanding of the diverse sources of capital, from the profits of transnational companies to personal savings; the different ways capital can be invested in property; and the place of property in the investment strategies of different kinds of firm. It should focus on the interrelationships between types of capital, types of firm and types of investment strategy, and on the implications for the way in which different firms come to

20 There have been several evaluation studies of government initiatives in the 1980s, but these mainly use variants of cost benefit analysis, or accounting audit techniques, and are dominated by economic criteria. See Roger Tym and Partners on *Enterprise Zones* in the early 1980s. Ecotec on *Urban Development Grant* (Pearce, 1988) and urban regeneration cases in 1988, and Roger Tym and Partners on *Derelict Land Grant* in 1987.

value types of land, property and location. Existing work which provides a useful foundation for such a review includes Plender (1982), Daly (1984), Oakeshott (1985), Scott (1986) and Shutt (1980).

The second focuses on the changing composition and strategies of the firms involved in the development and redevelopment of the built environment. This would have to include the building and building materials industries; the changing composition of the various firms involved in the different sectors of property development (commercial, industrial and residential) and in financial investment in property, and the property consultancies; the evolving strategies of different types of firm, and their consequences; the relations between firms and the extent of formation of local market management oligopolies; and the appearance of new firms and the nature of the opportunities they exploit. The emphasis here would be on the way changing strategies reconstitute the interests firms have in land, property and property redevelopment (see McNamara, 1988, on interests), and the way these interests find reflection in the negotiative practices through which action is undertaken. The objective would be to provide a basis for analysing the impact of such changes on the form of, and terms of access to, the built environment thus produced, and the consequences for local economies. Many of the research works cited in this paper, as well as the recent studies of economic restructuring and local economies, provide a valuable foundation for such a study (see especially Hausner, 1986, 1987; Boddy et al., 1986).

The third would examine the various ways in which the state impinges on these changing practices, in relation to the tools of intervention employed (forms of financial subsidy and taxation, forms of land dealing), the way in which these affect the demand for space, the rules within which individual firms develop their strategies, and the forms of development processes . Again, much of the research cited in this paper provides a valuable empirical foundation for such an analysis.

The final theme would involve an assessment of the implications of the above processes for local economies in terms of capital flows, labour market demand and supply, building materials and land, the impact on land and property values, and the implications of changes for these; the resultant social, economic and environmental externality costs and benefits within local economies and the distribution of these. It is this theme which presents the most challenges, both conceptually and empirically, yet it is clearly the most significant in assessing the contribution of property development to economic and social change.

Empirical research based on such an agenda raises challenging problems of research method and data sources. It involves arenas where many powerful actors operate, where secretive strategies are part of the battle for competitive success, where data is scarce and produced in ways which are often difficult to penetrate, and where publicly available documentation and public talk is often deliberately distorted for the purposes of competitive advantage. Nevertheless, there is a range of source materials which can be used.[21] The sensitised investigative researcher can penetrate development processes with considerable success (e.g. McNamara, 1985; Kent, 1986; Rydin, 1986). However, because of these difficulties, a methodological awareness of the problems involved is a prerequisite for an effective research contribution in this field.

Clearly, no one piece of research could address this substantial agenda. What is important, however, is the perspective from which any specific piece of research is undertaken. A particular perspective has been proposed in this paper which should not only increase our understanding of contemporary processes of the production and transformation of the built environment in the UK. It should also improve our awareness and theorisation of the way the practices through which development projects are negotiated are linked, via the interests and strategies of agents and the rules and resources upon which they draw, to the dynamics of the wider structure of economic and political organisation which these practices both constitute and reflect . In relation to policy, such an emphasis should help assess the nature of the contemporary 'period' and its broad consequences for the production of the built environment, as well as providing a basis for evaluating the possibilities for and consequences of specific interventions in development processes. More generally, such research should add to our understanding of the role of the processes of production of the built environment in local economic development, and of the role of the state in the development process.

---

21 Housing and construction statistics, National House Building Council data, development control records, commercial and industrial floorspace statistics, land registers, Investment Property Data Bank, reviews by the major property firms of local and national property markets, the property press, company reports.

References

Adams, C.D. (1985) Landowner involvement in the preparation of local plans. Paper to *Seminar on Research in Local Land Use Planning*, Oxford, June.

Adams, C.D. (1986) *The Accommodation Needs of Small Industrial Users in Inner Manchester*. Final Research Report, Department of Town and Country Planning, University of Manchester.

Adams, C.D. (1987) Industrial property: The nature of demand for small premises, *Estates Gazette*, 283: 533–5.

Adams, C.D., Baum, A.B. and Macgregor, B.D. (1985) The influence of valuation practices on the price of inner city vacant land, *Land Development Studies*, 2:157–73.

Adams, C.D., Baum, A.E. and Macgregor, B.D. (1987) Land values and land availability in inner city redevelopment, in: B. Robson (ed.) *Managing the City*. London: Croom Helm.

Ambrose, P. and Colenutt, B. (1973) *The Property Machine*. Harmondsworth: Penguin.

Ball, M., Bentevegna, V., Edwards, M. and Folin, M. (1985) *Land Rent, Housing and Urban Planning: A European perspective*. London: Croom Helm.

Ball, M. (1983) *Housing Policy and Economic Power*. London: Methuen.

Barras, R. (1984) The office development cycle in London, *Land Development Studies*, 1: 35–50.

Barras, R. (1985) Development profit and development control: The case of office development in London, in: S.M. Barrett and P. Healey (eds) *Land Policy: Problems and alternatives*. Aldershot: Gower.

Barras, R. (1987) Technical change and the urban development cycle, *Urban Studies*, 24: 5–30.

Barrett, S.M., Boddy, M. and Stewart, M. (1978a) *The Implementation of the Community Land Scheme*. SAUS Occasional Paper No. 3, School for Advanced Urban Studies, University of Bristol.

Barrett, S.M., Stewart, M. And Underwood, J. (1978b) *The Land Market and the Development Process*. SAUS Occasional Paper No. 2, School for Advanced Urban Studies, University of Bristol.

Barrett, S.M. and Whitting, G. (1981) *Local Authorities and the Supply of Development Land to the Private Sector*. Working Paper No. 19, School for Advanced Urban Studies, University of Bristol.

Barrett, S.M. and Whitting, G. (1983) *Local authorities and land supply*. Occasional Paper No. 10, School for Advanced Urban Studies, University of Bristol.

Barrett, S .M. and Healey, P. (eds) (1985) *Land Policy: Problems and alternatives*. Aldershot: Gower.

Berry, M. and Huxley, M. (1985) Property development, capital switching and state intervention: studies of Melbourne, Australia. Paper presented to the *International Sociological Association: Urban and Regional Research Committee Conference*, University of Hong Kong, 14–20 August.

Boddy, M. (1982) *Local Government and Industrial Development*. Occasional Paper No. 7, School for Advanced Urban Studies, University of Bristol.

Boddy, M., Lovering, J. and Bassett, K. (1986) *Sunbelt City: A study of economic change in Britain's M4 growth corridor*. Oxford: Clarendon Press.

Boyle, R. (1985) 'Leveraging' urban development: A comparison of urban policy directions and programme impact in the United States and Britain, *Policy and Politics*, 13: 175–210.

Boyle, R. and Rich, D. (1984) *In Pursuit of the Private City: A comparative assessment of urban policy orientation in Britain and the US*. Strathclyde Papers in Planning No. 3, Department of Urban and Regional Planning, University of Strathclyde.

Breughal, I. (1975) The Marxist theory of rent and the contemporary city, *Political Economy of Housing Workshop*, London.

Brown, H.J., Phillips, R.S. and Roberts, N.A. (1981) Land markets and the urban fringe: new insights for policy makers, *Journal of the American Planning Association*, 47: 131–44.

Bruton, M.J. and Gore, A. (1980) *Vacant Urban Land in South Wales*. Department of Town Planning, University of Wales Institute of Science and Technology.

Cameron, G., Monk, S. and Pearce, B. (1987) *Vacant Urban Land: A literature review*. Unpublished report for the Department of the Environment. Department of Land Economy, University of Cambridge.

Cheshire, P., Sheppard, S., Hooper, A. and Peterson, J. (1985) *The economic consequences of the British planning system*. Discussion Paper No. 29, Department of Economics, University of Reading.

Cooke, P. (1986) The changing urban and regional system in the United Kingdom, *Regional Studies*, 20: 243–51.
Craven, E. (1975) Residential expansion: The role of the private developer in the South East, *and* Private residential expansion in Kent, in: R. Pahl (ed.) *Whose City?*, 2nd edn, Harmondsworth: Penguin, pp. 105–44.
Daly, M.T. (1982) *Sydney Boom, Sydney Bust*. London: Unwin.
Daly, M.T. (1984) The revolution in international capital markets: urban growth and Australian cities, *Environment and Planning A*, 3, pp. 1003–20.
Daniels, P. (1985) *Service Industries: a Geographical appraisal*. London: Methuen.
Davies, R. (1984) *Retail and Commercial Planning*. London: Croom Helm.
Dawson, J.A. (1983) *Shopping Centre Development*. London: Longman.
Dickens, P., Duncan, S., Goodwin, M. and Gray, F. (1985) *Housing, States and Localities*. London: Methuen.
Dowall, D.E. (1984) *The Suburban Squeeze: Land conversion and regulation in the San Francisco Bay area*. Berkeley, CA: University of California Press.
Drewett, R. (1973) The developers: Decision processes and Land values and the suburban land market, in: P. Hall et al. (eds) *The Containment of Urban England*, Vol. 2. London: Allen and Unwin, pp. 163–245.
Dyos, H.J. (1961) *The Victorian Suburb: A study of the growth of Camberwell*. Leicester: Leicester University Press.
Elson, M.J. (1982) *Land Release and Development in Areas of Restraint: An investigation of local needs policies*. End-of-grant Report to SSRC, Department of Town Planning, Oxford Polytechnic.
Elson, M.J. (1986) *Green Belts*. London: Heinemann.
Evans, A.W. (1985) *Urban Economics: An introduction*. Oxford: Blackwell.
Evans, A. (1988) *The Theory of Land Values, a Reintroduction*; *On monopoly rent*; and *On absolute rent*. Discussion Papers in Urban and Regional Economics, Series C, Vol. 1, Nos. 34–36, Department of Economics, University of Reading.
Feagin, J. (1983) *The Urban Real Estate Game: Playing Monopoly with Real Money*. New York: Prentice-Hall.
Fine, B. (1985) Land, capital and the British coal industry prior to World War II, in: M. Ball, V. Bentivegna, M. Edwards and M. Folin (eds) *Land Rent, Housing and Urban Planning*. Beckenham: Croom Helm.
Fothergill, S., Monk, S. and Perry, M. (1987) *Property and Industrial Development*. London: Hutchinson.
Giddens, A. (1984) *The Constitution of Society*. London: Polity.
Gilbert, A. and Healey, P. (1985) *The Political Economy of Land*. Aldershot: Gower.
Goddard, J.B. (1975) *Office Location in Urban and Regional Development*. Oxford: Oxford University Press.
Goodchild, R. and Munton, R. (1985) *Development and the Landowner*. London: Allen and Unwin.
Gore, T. and Nicholson, D. (1985) The analysis of public sector land ownership and development, in: S.M. Barrett and P. Healey (eds) *Land Policy: Problems and alternatives*. Aldershot: Gower.
Gregory, D. and Urry, J. (eds) (1985) *Social Relations and Spatial Structure*. London: Macmillan.
Hajer, M. (1989) *City Politics*. Aldershot: Gower.
Hallett, G. (1979) *Urban Economics: Principles and policy*. London: Macmillan.
Harrison, A. (1977) *The Economics of Land Use Planning*. London: Croom Helm.
Harvey, D. (1982) *The Limits of Capital*. Oxford: Blackwell.
Harvey, D. (1985a) *The Urbanisation of Capital*. Oxford: Blackwell.
Harvey, D. (1985b) *Consciousness and the Urban Experience*. Oxford: Blackwell.
Hausner, V. (ed.) (1986) *Critical Issues in Urban Economic Development*, Vol. 1. Oxford: Oxford University Press.
Hausner, V. (ed.) (1987) *Critical Issues in Urban Economic Development*, Vol. 2. Oxford: Oxford University Press.
Healey, P. and Barrett, S.M. (1985) Land policy: towards a research agenda, in: S.M. Barrett and P. Healey (eds) *Land Problems: Problems and alternatives*. Aldershot: Gower, pp. 1–20.
Healey, P., Davis, J., Wood, M. and Elson, M.J. (1982) *The implementation of development plans*. Report for the Department of the Environment. Department of Town Planning, Oxford Polytechnic.

Healey, P., Doak, A., McNamara, P.F. and Elson, M. (1985) *The Implementation of Planning Policies and the Role of Development Plans. Volume I: Main findings, Volume II: Planning policy implementation in Greater Manchester and the West Midlands*. Report to the Department of the Environment. Department of Town Planning, Oxford Polytechnic.
Hooper, A. (1985) *The Role of Landed Property in the Production of the Built Environment*. BISS Papers No. 6, Bartlett School, University College London.
Kent, R. (1986) The contribution of landowners to the formulation of local planning policies. Unpublished MTP thesis, University of Manchester.
Lambert, C. and Underwood, J. (1984) *The Role of Local Plans in Land Policy and the Development Process*. Final Report to the ESRC, School for Advanced Urban Studies, University of Bristol.
Leitner, H. (forthcoming) Pro-growth coalitions, the local state and downtown development: The case of six cities, in: M.M. FISCHER and M. Sauberer (eds) *Society-Economy-Space*. Vienna.
Leitner, H. and Sheppard, E. (forthcoming) The city as a locus of production: The changing geography of commodity production within the capitalist metropolis, in: R. Peet and N. Thrift (eds) *The New Models in Geography*. London: Allen and Unwin.
Lichfield, N. and Darin-Drabkin, H. (1980) *Land Policy in Planning*. London: Allen and Unwin.
Lin Leung, H. (1987) Developer behaviour and development control, *Land Development Studies*, 4, pp. 17–34.
Marriott, O. (1969) *The Property Boom*. London: Pan.
Massey, D. (1984) *Spatial Divisions of Labour*. London: Edward Arnold.
Massey, D. and Catalano, A. (1973) *Capital and Land*. London: Edward Arnold.
McLoughlin, J.B. and Huxley, M. (1985) Australian urban studies: The state of the art, *Built Environment*, 11: 143–56.
McNamara, P.F. (1984a) The role of estate agents in the residential development process, *Land Development Studies*, 1: 101–12.
McNamara, P.F. (1984b) *Restraint Policy and Development Interests*. Oxford Working Paper No. 76, Department of Town Planning, Oxford Polytechnic.
McNamara, P.F. (1985) The control of office development in central Edinburgh. Unpublished PhD thesis, University of Edinburgh.
McNamara, P.F. (1990) The changing role of research in investment decision-making, in: P. Healey and R. Nabarro (eds) *Building the New Britain: Land and Property Development Processes in a Changing Context*. Aldershot: Gower.
McNamara, P.F. (1988) Interests in land and planning in action, in: P. Healey, P.F. McNamara, M.J. Eison and A.J. Doak (eds) *Land Use Planning and the Mediation of Urban Change*. Cambridge: Cambridge University Press, chapter 7.
McNamara, P.F. and Turner, G. (1987) *Chartered Surveyors and the Site Finding Process*. Occasional Paper Series, No.1, Department of Estate Management, Oxford Polytechnic.
Montgomery, J.R. (1985) The political economy of local authority land transactions. Unpublished CNAA thesis, Oxford Polytechnic.
Moore, C. and Pierre, J. (1988) Partnership or privatisation? The political economy of local economic restructuring, *Policy and Politics*, 10: 169–78.
Nabarro, R. (1990) The investment market in commercial and industrial development: some recent trends, in: P. Healey and R. Nabarro (eds) *Building the New Britain: Land and Property Development Processes in a Changing Context*. Aldershot: Gower.
Oakeshott, M. (1985) Institutional attitudes to property investment in the UK-last waltz of the dinosaurs around the 'prime property' temple, *Land Development Studies*, 2: 147–56.
O'Conner, J. (1973) *The Fiscal Crisis of the State*. New York: St Martin's Press.
Pearce, G. (1988) City grants-lessons from the Urban Development Grant programme, *The Planner*, 74: 15–19.
Plender, J. (1982) *That's the Way the Money Goes*. London: Andre Deutsch.
Rydtn, Y. (1986) *Housing Land Policy*. Aldershot: Gower.
Sandercock, L. (1976) *Cities for Sale: Property, politics and urban planning in Australia*. London: Heinemann.
Sayer, A. (1984) *Methods in Social Science*. London: Hutchinson.

Scott, J. (1986) *Capitalist Property and Financial Power: A comparative study of Britain, the United States and Japan*. Brighton: Wheatsheaf.

Short, J.R., Fleming, S. and Witt, S. (1986a) *Housebuilding, Planning and Community Action*. London: Routledge and Kegan Paul.

Short, J.R., Witt, S. and Fleming, S. (1986b) Conflict and compromise in the built environment: housebuilding in central Berkshire, *Transactions of the Institute of British Geographers*, N.S., 12, pp. 1–14.

Shutt, J. (1980) The new leviathans: a study of the pension funds in the UK and their investment in the land and property sector. Unpublished MPhil thesis, University of Birmingham.

Smyth, H. (1985) *Property Companies and the Construction Industry in Britain*. Cambridge: Cambridge University Press.

Sutcliffe, A. (1981) *Towards the Planned City*. Oxford: Blackwell.

Thrift, N. (1987) The big bang and its aftermath illuminating the City of London's role in British urban and regional development. Paper to the *Sixth Urban Change and Conflict Conference*, University of Kent, September.

Whitehand, J.W.R. (1984) *Rebuilding Town Centres: Developers, architects and styles*; Birmingham. Occasional Paper No. 19, Department of Geography, University of Birmingham.

Whitehand, J.W.R. and Whitehand, S.M. (1984) The physical fabric of towns: The agents of change, *Transactions of the Institute of British Geographers*, 9, pp. 231–47.

Witt, S.J.G. (1987) *District Councillors and House-builder Pressure in an Area of Growth: Central Berkshire*. Unpublished PhD thesis, University of Reading.

Wood, M. (1982) *High Wycombe: The implementation of strategic planning policy in a restraint area in the South East*. Oxford Working Paper No. 67, Department of Town Planning, Oxford Polytechnic.

Chapter 8
Development Plans and Markets*

Patsy Healey

State and Market in Land-use Planning

Since the late 1970s, planning authorities have been encouraged to consider market perceptions and market criteria when allocating land-use and development opportunities (Thomley, 1991). The objective was to change the relations between state and market in the development process, and to make local planning authorities more 'market-aware' in their work. What does this involve, and what are the implications for development?

The exhortation to be 'market-regarding' has presented planning thought with a considerable and ambiguous challenge. Planning has often been considered as opposed to the market. Planning focussed on *needs*, and on *users* of the city. Markets focussed on effective *demand*, and on the *exchange* value of land and buildings. An even stronger tradition portrayed planning strategies and decisions as 'above' the market, embodying a calculus of the 'public interest'. Within this conception, markets were restricted to operating within the niches created by the planning system. So does 'market-aware' planning mean abandoning planning and plans? Others have argued that planning in capitalist societies has always supported markets through the way the public interest has been identified (Scott, 1980; Pickvance, 1982). If so, what does the new emphasis mean? This raises questions about what markets land-use planning engages with and how planning authorities can use the levers of the planning system to support or restrain markets.

The planning system in Britain is inherently concerned with economic and social development. But its specific role has been focussed on the provision of land, buildings, locales, and environments, and the connections between them. In a society in which private landownership is predominant, land and property markets provide a key role in mediating the relations between the land and property demands of production and consumption activity and the provision of the space and physical structures needed to conduct such activities. The markets with which planning authorities engage are thus primarily *markets in existing properties* and in *real estate development opportunities*.

There has been much discussion of the impact of planning regulation and plans on land and property markets. Do constraints on land supply (through green belt policies) drive up prices? Evans (1987, 1991), and Cheshire and Sheppard (1989) argue that they do, and are therefore a cost to the economy. Do plans stabilise land and property markets, creating greater certainty, thereby reducing transaction costs and, by limiting the potential for 'over-building' in a property market, producing greater efficiency in the relation between supply and demand in land markets?

These debates raise important issues not only with regard to whether planning regulation, guided by plans, can effectively safeguard other criteria (e.g. environmental quality, community need) whilst 'managing' land and property markets. They also raise questions about what 'managing markets' means. Do development plans *create* market opportunities? Do they merely *follow* them, or do they *manage* markets to reduce adverse externality effects? These questions cannot be answered without considerable understanding of the nature of land and property markets and how these might vary, spatially, temporally, sectorally, and institutionally.

This paper argues that if the private sector is to be the main provider of development, then the planning system has to be 'market-aware' in regulating the development process. This does not undermine the role of development plans. On the contrary, development plans may bring positive benefits to different kinds of developers, and to communities in seeking to benefit from development. But to perform this function skilfully, a sound understanding of land and property markets is needed. These arguments are developed

* Previously published in *Planning Practice and Research*, 7(2): 13–20 (1992).

through a discussion of, first, some characteristics of land and property markets and how they vary; and, second, differing conceptions of planning, plans, and markets.

Land and Property Markets

Markets are mechanisms through which willing buyers and sellers determine the amount of a good which is produced and consumed through establishing a price related to other comparable transactions in a 'market-place'. In theory, markets are highly efficient mechanisms for optimising prices in relation to costs, and for fostering innovation to reduce costs and produce more goods for consumers. It is now widely recognised that in practice the perfect market is hard to find. Markets may be distorted through control by producers, or consumers, or middlemen, should any of these parties be in a position to establish oligopolistic or monopoly control. There may be constraints on supply, creating scarcity values. There may be time-lags between the initiation of production and eventual purchase by a consumer. Producers may make errors in predicting demand. If these errors are overestimates, over-supply of a good may result, driving down values. These imperfections may be treated in market analysis as 'short-term' problems in 'clearing', before supply and demand returns to 'balance'. More recently, considerable attention has been given to 'transaction costs', that is, the efficiency of the mechanisms through which values are arrived at, and products exchanged between producer and consumer.

Economists have long recognised that land and property transactions represent very peculiar markets (e.g. Hallet, 1979). Land is in fixed supply, so scarcity problems are inherent. Land and property are fixed in location. Because they are not 'consumed' through use, they are valued for investment purposes. They are unique in location, highly variable in physical form, and in the costs of production. Land and property markets are dominated by 'existing products', rather than additions to the stock of serviced sites and buildings. Land and property markets as a result trade in rights in existing sites and properties, as well as structuring the production of new ones. A planning application, apparently for development purposes, may in practice be merely a device to establish a value for trading purposes.

The problem of establishing a price is made more difficult because relatively few transactions take place. Landowners and land purchasers typically undertake very few transactions in their lifetime. This allows considerable power to accrue to 'middlemen'. Where development is involved, the time-lag problem is chronic, and escalates with the scale of development. Cycles of 'under-building' and 'over-building' in relation to demand are endemic as a result (Barras, 1985, 1987). Because land and property are typically 'lumpy and illiquid' products, with constraints on supply, the producers (landowners, developers, financiers) can readily dominate a market. Typically, land and property markets vary from place to place, depending not only on local demand, but on historical landownership patterns, and the institutional organisation of the property and development sector.

Faced with frequently gross imperfections in land and property markets, one reaction has been to take land out of the market. The conflicting interests between production 'capital' and 'land' (Massey and Catalano, 1987) help to explain why some countries have a tradition of municipal ownership of development land, notably Sweden and the Netherlands. In other West European countries, for example parts of the Federal Republic of Germany, production capital and community interests can dominate land and property markets, aided by landownership patterns characterised by small, fragmented holdings. In such contexts, land and property markets are likely to be 'user-driven'. State involvement is valued to assist in land assembly, to make suitable sites available for production activity and workers' housing.

In Britain, land markets are dominated by a 'rentier' conception of the value of land. Land and property markets are driven by land and property owners whose prime interest is in investment income as rent rather than in use values. A close relationship between land and property investment and the financial sector has been long-standing (Healey, 1988). Land and property perform an important function as a 'store' of value, with finance switching into and out of property according to conditions within the finance sector and, in today's global world, within the international economy (Harvey, 1982).

Land and property markets are structured by the difficult problem of matching demand and supply for a locationally fixed and 'lumpy' product. They are also driven by two different dynamics of demand, that for use and that for investment. The tension between these two dynamics leads to complex cycles in development behaviour and chronic problems in 'clearing'. The resultant cyclically means that any exercise in market management must attend to *temporal* variation in land and property market conditions.

There are also other major dimensions of variation. While it may be possible to conceive of a single urban *land* market, there are substantial variations in markets for different types of property. The residential sector in Britain has traditionally been largely distinct from commercial and industrial development, with its own cycles of demand and investment, and a distinct nexus of agencies and relationships (Ball, 1983). Within the commercial sector, office, retail, industrial, and leisure property perform, in terms of capital values and rental returns, on different time-scales.

While there are distinctive institutional relationships in each sector, landowners, developers, and investors may switch between sectors according to market conditions. This has been particularly evident with the large mixed-use projects, a new phenomenon in the development scene in Britain in the 1980s. As a result, planning authorities are often faced with renegotiating the balance of uses in a project in response to changing market conditions. Finally, not only do different market sectors behave in different ways, each sector contains a range of types of market, which may operate in relatively separate segments; for example, the market in managed work-spaces, business park premises, or golf courses. In each, the nexus of producers and consumers may be very different, in sources of capital, dynamics of demand, and spatial scale of the relationships.

All land and property markets are nevertheless *local*, linking demand for property in particular locations with specific sites and buildings. Because locations and building configurations are unique, the fine detail of *spatial variation* is critically important for land and property markets. At one level, only a few types of property in a few locations may be considered 'prime' by investors. These may be competing with prime sites in a quite different urban region. Those trading, refurbishing and developing smaller properties in an urban region are likely to deal primarily in 'secondary' property for local investors and their occupiers. Their advantage is a more sensitive understanding of property type and local differentiation. For some types of development, the fine detail of design and layout is critical. At a larger scale, the environmental qualities and status of neighbourhoods, as well as material attributes, such as access and infrastructure, play a major part in differentiating land and property markets within urban regions. It is for this reason that the property development industry has a keen interest in whether a neighbourhood or an industrial/commercial zone is improving or declining, and whether there is an opportunity to 'turn an area round' (and thus make significant land value gains for those in at the start).

Finally, land and property markets vary in their *institutional relationships*. If land market transactions are a difficult task accomplished through, typically, a sequence of negotiations between a nexus of landowners, occupiers, investors, developers, builders, and their agents, not to mention planning authorities and solicitors, then the 'power relations' between these various parties are likely to be critical. In the US, Logan and Molotch (1987) argue that 'rentier' growth coalitions dominated local politics until the rise of environmental politics. In Britain, the long-standing power of landowners, 'rentiers', and the increasing de-localisation of the ownership of industry and commerce left land and property interest as one of the few business interests seeking to play a powerful role in local politics in British urban regions (Saunders, 1979; McCrone and Elliott, 1989). But these local property networks have usually been closely linked to investors and development companies operating nationally through the property consultancy nexus (McNamara, 1983).

Land and property markets are thus grossly imperfect and vary in their behaviour in relation to patterns of user and investor demand, the timing of property cycles, types of use, location of building type, and their institutional relationships. The market-aware development plan is one which is founded upon an understanding of these imperfections as they are manifest in particular areas.

Development Plans and Land and Property Markets

Development plans are only one part of one mechanism (planning regulation) through which public policy affects land and property markets and the development process. National economic policy, taxation policy, the development activity of the public sector, the public sector's roles in land-use and environmental regulation, its role as provider of information, and its institutional relations all impinge on supply and demand conditions in land and property markets, on transaction costs within markets and for development, and on the institutional relations of the development industry.

A development plan as an 'intervention tool' is primarily an *informational* document (Keogh, 1981). It indicates how regulatory authorities may make (or have made) their decisions, provides information on the qualities and opportunities of an area, and about how other agencies may act. Plans are a political 'position statement', recording a planning authority's attitudes to the debates and conflicts over land use and development which may have taken place during the preparation of the plan. It may record proposals about who should pay for what. In contrast to continental European practice, neither local plans nor structure plans are legal documents defining property rights. The 'power' of the plan lies in the political-administrative sphere. So long as planning authorities and the Secretary of State uphold the plan, it plays a significant role in providing information on how property rights may be affected by the planning system. What role can such a tool have in land and property markets dominated, as in the UK, by investment considerations?

There have been major changes in the approach to development plans since the 1940s. Interestingly, the formal change from 'old-style' development plans to structure and local plans is less significant in retrospect than the changing conceptions of the relations between plans and markets. To explore these changing conceptions, plan-making practice since the war is conceptualised into three 'ideal types'.

1. the comprehensive urban structure plan (1950s–1970s),
2. the plan as a regulatory record of pluralist bargaining processes (1970s–1980s),
3. the market-aware plan (1980s–1990s).

In each case, the plan expresses a trade-off between market considerations, environmental quality, and community needs. But the approaches vary in how the trade-off is made and in the anticipated role of the plan in structuring land and property markets.

The Plan as a Comprehensive Urban Structure Framework

In this conception, plans contained no explicit treatment of land and property markets as such. The task was to 'balance' demand, arrived at through trend predictions of needs, and supply of land, property, and physical and social infrastructures. The plan thus represented a rational calculus of the combined land and development needs of the various activities which make up a community.

It internalised the treatment of externality costs and benefits within the calculus. Driven by concerns for environmental quality and community needs, it was assumed that land and property markets could take care of themselves. It was expected to direct a substantial public-sector effort in development, particularly with respect to infrastructure. It defined a framework of rules and opportunities within which land and property markets had to fit.

In this conception, the plan directed all investment in land and development. It structured patterns of land and property values, creating and inhibiting development opportunities. It provided certainty, reduced risks for developers and effectively structured land and property markets and private-sector development activity. But the certainty came without flexibility. This proved an increasing problem as the development initiative was tilted towards the private sector in the 1950s particularly in relation to housing and commercial development. Key problems arose over development timing (when were promised public-sector investments likely to come forward?) and who was to pay for what (the issue which grew into the short-lived Land Commission and Betterment Levy, and the much longer-lived practice of 'planning gain').

Plans in this form structured land and property markets, but as a secondary consequence of the power to control development, rather than as an explicit aim. The development industry, while using the plan inquiry and appeal systems to challenge plans and decisions, nevertheless benefitted from the certainty they provided, which stabilised development conditions. However, in addition to the problems noted above, tensions also arose as it became clear that the 'trends' and assumptions on which plans tended to be based did not reflect emerging trends in where firms wanted to locate and where people wanted to live. These trends were partly the result of other government initiatives, such as de-congestion policies and transport investment, which created the strategic opportunities within which such new demands could be met. Thus, the conceptions of urban structure built into the plans no longer matched the pressures for user and investor demand. Increasingly, development control decisions drifted away from the basis of plans, which were criticised as too rigid, outdated, and hence 'unreasonable' as a material consideration in development control.

This traditional model makes sense where the public sector plays a major role in the development process, as landowner and dealer, as builder and developer, and as infrastructure investor. It still underpins thinking about the planning system in Germany and the Netherlands (Davies et al., 1989). It never worked so well in Britain because of problems of public-sector coordination and because of sophisticated, oligopolistic, and increasingly investment-oriented land and property markets. Since the 1970s, its relevance has in any case diminished, as the development initiative has been steadily shifted towards the private sector. It lives on, however, in the language of plans. There are Unitary Development Plans which still refer to investments 'the Council will undertake', presenting in various ways the Council as the prime mover of development activity. Even in the most depressed of land and property markets, this is unlikely to be the case, and the availability of resources for projects is highly uncertain.

The Plan as a Record of Agreed Principles

An alternative concept was implicit in much of what happened in the 1970s and 1980s. Plans were side-lined if they did not reflect market interests, and development projects were negotiated on a project-by-project basis. In this case, the trade-off between market considerations and citizens' concerns with environmental quality and community needs is negotiated whenever an application is submitted for planning permission. The negotiations may be guided by development control criteria, but not by strategic considerations about amounts and locations for development. In this situation, public policy follows 'market demand', in the sense that it evolves as the market brings proposals forward. These market demands are then moderated by political considerations. While this creates opportunities for the 'politically aware' market entrepreneur, who sets out to 'second guess' and manipulate political reactions to specific cases, generally this approach leads to a high level of uncertainty in development markets, to high transaction costs with respect to negotiations at the level of project, and to uncertain parameters in the negotiations between developers and landowners over land prices. This is particularly the case where opposition to development is likely. In addition, adverse externality consequences may begin to appear, as environmental qualities sought by one developer are undermined by the actions of another, or as aggregate environmental costs of individual developments begin to undermine the viability of projects. The London Docklands provides merely a dramatic example of this problem. The politics of development may also become more hostile to developers, as local people criticise the amount and quality of new development and come to question the biases and policies of their political representatives. The main beneficiaries of a planning system in this form would appear to be consultants and lawyers, whose fee income is enlarged by the scale of negotiation and arbitration required.

In this context, the plan may begin to reappear, as a way of defining the rules which have evolved through bargaining over projects. These rules may take the form of development control policies, essentially an expanded building code. Or they may express more strategic matters in the form of performance criteria (for example, with respect to which range of business uses may be permitted, what infrastructure contributions and community benefits may be required, and what design standards will prevail in particularly sensitive areas). The value of the plan format is that it provides some semblance of accountability for the process. This may be sought to reassure local people that their interests have not been ignored. It may also reassure developers and landowners who fear that some get better treatment than others from the planning system.

In this form, the plan *records* agreements about rules and principles arrived at between state and market on a case-by-case basis. However, the plan-as-record does little to structure the market and protect it from its own behaviour. The plan gives no indication of how much development is appropriate in any period, or where it should be located. Whether the negotiation proceeds unfettered by rules, or with some form of development control code, what development occurs, where it occurs and when, is defined 'as and when' proposals come forward. Planning policy merely affects the terms in which the development takes place, through open negotiation or a rule-governed decision. It should be said that the planning system in Britain has rarely degenerated to this from, partly because of the continuing strategic role of structure plans in setting housing targets, defining green belts, and indicating other conservation zones. In practice, the plans produced out of this kind of practice were a kind of regulatory 'trend planning' as exemplified in Colchester by Brindley et al. (1989), a form of 'market following' to be discussed in the next section. But Tory Party advisers in the early 1980s certainly envisaged a regulatory system consisting merely of building codes and countryside conservation zones (Jones, 1982).

Table 8.1 summarises the roles of the plan in the definition of development opportunities in these first two conceptions of a development plan.

Table 8.1 The role of development plans in defining development opportunities: Two conceptions

	Comprehensive urban structure		**Bargaining**	
	Public sector as dominant developer	*Private sector as dominant developer*	*No plan*	*Plan as a record*
What is developed	define	define	'as and when'	'as and when'
Where	define	define	'as and when'	'as and when'
On what terms	define	define	negotiate at project level	define
When	define	define	'as and when'	'as and when'
By whom	ignore	ignore	reflect	reflect

Market-aware Development Plans

While welcoming the greater priority given to market considerations in the 1980s, the development industry has increasingly complained of the lack of certainty provided by the planning system. Initially, this demand was interpreted as a request to reduce political interference and bureaucracy in the operation of the planning system. Without firmly upheld frameworks provided in plans, this provided flexibility (it is always possible to influence political-administrative systems) but without strategic certainty and often high transaction costs.

Since the mid-1980s, the representatives of the development industry have argued for clear strategic frameworks, with market sensitivity over details, that is, *certainty-with-flexibility* (for example, the House Builders' Federation's defence of structure planning). Such a strategic framework should limit amounts of development and indicate locations for development, to help in coordinating development activity, and to 'clear' opportunities for development through environmental politics. Market sensitivity involves awareness of local conditions in different sectors of each main market.

But, as Brindley et al. (1989) illustrate, a market-aware development plan, and the treatment of particular markets within it, may take various forms. In all cases, it requires an explicit approach to the trade-off between market considerations, environmental quality, and community needs. This may be developed in three ways. The plan, or policies within it, may *follow* markets, as a formalised form of the practices described in the previous section. Or the objective may be to *manage* markets, through macro-structural controls such as the amounts and locations of development and infrastructure investment. Or policies may seek, in entrepreneurial fashion, to *create* markets, through promotional investment and marketing (see Table 8.2).

Following markets

If the approach is primarily to *follow* the market, the market leads the trade-off, taking precedence over environmental and community considerations. The plan provides a form of 'advanced clearing' through the political process. Based on careful discussion with the development industry and research on market behaviour, the plan takes the form of a strategic contract between state and market. It emphasises *what* development will be allowed to take place, in order to impose some discipline on the market for its own sake. This should offer some constraint on tendencies to 'overbuilding', and hence reduce the impact of

Table 8.2 Market-aware strategies for development plans

Strategy	Follow markets	Manage markets	Create markets
What to influence	record	define	define
WHAT – what sectors types qualities – how much	record	define	promote
WHERE	'as and when'	define	promote
on WHAT terms	indicate	define	negotiate at project level
WHEN and in what order	'as and when'	define	'as and when'
WHO	reflect	recognise	target

slumps in development activity. The plan should also specify the terms on which it will be allowed to take place, in order to reduce the time taken in negotiating individual developments. It is likely to be very explicit about developers' contributions. This should simplify negotiations between landowners and developers. The development plan then becomes a record of what is expected to happen, which has the advantage of providing information for all the developers, investors, and landowners involved. This approach essentially adds a strategic dimension to the plan-as-record-of-bargains. Many structure and local plans in the 1980s have this form (see Brindley et al. (1989) on Chelmsford). Stratford on Avon District Council's offer to developers and landowners to 'register' their interest in sites during the local plan preparation process is an interesting way of acquiring information to feed into the contract (O'Grady, 1990). Many development briefs are of this nature, and plans frequently refer to their preparation for strategic development sites.

The problems with this approach are that

a. It readily ignores community needs which are not well-represented in the market-place (e.g. social housing).
b. It tends to follow currently strong development sectors, and therefore may contribute to 'over-building' these and 'under-building' others.
c. It tends to incremental changes in the qualities of an area, which in retrospect even landowners and developers may find counter-productive to their activities.
d. It is unfair as between market interests, since in oligopolistic markets, some developers tend to have privileged access to the negotiations over plan content.

This approach thus focusses on 'investor' rather than 'user' interests. It takes a 'short-term' view, lacks any 'vision' of 'quality of space' beyond the project, and is likely to privilege those with good contacts with planning authorities. It has little capacity to *transform* the qualities of areas, and thus contributes little to urban regeneration strategies.

Managing markets

This approach involves taking a deliberate role in structuring the conditions in which land and property development markets are formed. The objective of such management is two-fold: to reduce the imperfections within land and property markets, and to give greater priority to environmental quality and community needs. However market, environmental, and community considerations are weighted and related, market management plans should express a strategic view of where development is to take place and why. Clear reasoning based on good information is essential, since landowners will demand to know why their site is not allocated for development while another is. It is critical, therefore, that the plan defines not only what will be allowed and on what terms, but where it will be allowed. While it may be prepared as a result of careful analysis rather than in consultation with developers, it should nevertheless reflect an awareness of the tendencies in different land and property markets in an area, in order to stand up to tests of reasonableness at appeal. The plan thus becomes a long-term strategic contract between state and market, providing firmness over strategy and flexibility over detail. The Solihull draft UDP (1990) provides a good example of what this involves. It is precise about how much development is allowed and where, but much more flexible on the terms in which development will take place, providing a baseline for negotiations.

In Solihull, the market is sufficiently buoyant to avoid the risk of serious 'over-building'. In many other areas, however, assessing how much land to allocate for development requires careful market appraisal work, based on understanding of the complex relations between user and investor demand with respect to the local economy in question. Considerations of development timing and location are critical.

A good example of a concern with *timing* may be found in the treatment of industrial land and office development in the Birmingham draft UDP (1990/1991). Its office development policy (4.33–4.35) seeks to monitor the 'office pipeline to avoid detrimental "over-supply" of office floorspace', thus smoothing the cyclicality in commercial property markets. Barras has argued (1985) in relation to office development, that development plans ought to adopt a 'counter-cyclical' stance, over-allocating in periods of slump, and under-allocating in periods of boom, to minimise the rise in values when property is in short supply at the start of a boom and to reduce the collapse in prices which can happen when projects started in boom conditions come on stream in depressed markets. A careful approach to the timing of development may also be necessary where areas are expected to change significantly, for example new settlement or major town expansion projects, or large-scale urban redevelopment projects. This of course implies that a development plan which is to play a market management role must obtain the 'power' to influence infrastructure investment as well as the provision of local authority services; in other words to be an 'expenditure-linked' plan. But this means careful consideration of investment strategies of both the public and the private sectors in an urban region, rather than the coordination of public investment which was the focus of the discussion of 'expenditure-linked' plans in the 1970s. The implementation sections of some of the new unitary development plans and district-wide plans are beginning to reflect this point.

The location of development is also a critical factor to consider in market management approaches. Developers will be concerned about two things in particular in assessing land allocation and phasing policies in a plan. The first is the locational concentration of the amount of development. It is not merely sufficient to assume that a particular urban region can accept so much development. As was recognised in many plans of the 1970s, the location of, for example, office development, is critical to the returns that can be expected. In a relatively weak market, it may be desirable to have firm locational policies, to reduce the uncertainty as to land and property values in that market. Manchester in its City Centre Local Plan of the early 1980s, deliberately sought to 'write down' land values on the periphery of the centre, in order to create more investment certainty in the remaining part of the centre (McNamara, 1986). The Strathclyde Structure Plan (1981) took a similar approach, given the over-allocation of peripheral housing land.

Secondly, given the increasing differentiation within markets in relation to environmental (and, regrettably, social) qualities, the particular qualities of places needs careful consideration this requires attention to the 'boundaries' between locales within an area. Are these to be open, permeable, or firm 'buffer' zones? An effective market management approach also requires attention to the strategic relations between city and countryside, between activities and communications, and between the elements of environmental systems such as energy flows and water and waste management.

A 'market management' approach thus begins to have some aspects of the traditional comprehensive urban framework. The difference is that, rather than expecting the market to fit into the niches created by the plan, the plan is based on a very sensitive appreciation of local land and property markets. This should help prevent the development industry collectively 'shooting itself in the foot', as was experienced in the late 1980s. In contrast to the 'market-following' approach, a close relationship with particular landowners, developers, and investors is not necessarily implied, except in the negotiation of objections to a plan, and those surrounding a particular project. There was, for example, no formal consultation with developers in the preparation of the Solihull UDP. It is nevertheless an excellent example of a market-aware management plan. Nor does the presence of such discussions indicate that the bias of the plan is more towards development than community objectives. In Siough's case, where close consultation has taken place with Slough Industrial Estates, the plan's objectives focus on using the benefits of development for environmental and community purposes. Sheffield's UDP expresses a similar consultative approach with the local development industry to achieve community benefits.

Creating markets

In some situations, there is little life in local land and property markets. This may apply to a whole conurbation, in an area with a weak local economy. More usually, it applies to parts of a conurbation. The preoccupation during the 1980s was to resurrect markets in areas of obsolete industrial land and property. The task here is to *create* markets. This may involve action at the level of the conurbation, as in the use of the plan to promote a vision of an area. The Birmingham UDP includes such a 'vision' statement in its introduction. The plan framework as a whole may also serve to locate and focus attention upon development opportunities in specific zones where it is hoped to foster market activity.

In current plans in the UK, this effort is masked by the activities of the Urban Development Corporations, which control many of these zones. Their remit has been specifically to create and release the potential for market activity in older industrial and commercial areas. The task is to transform a locale from one image and set of values to another in which substantially higher values can be realised. It is now widely recognised that a development framework is a critical tool for this purpose, defining a new image for a place, and translating this into principles for the development of parts of an area. A development framework 'adds value' to what would otherwise be the uncoordinated activities of individual landowners and developers. Such frameworks need to be located in a strategic context, so that investors can assure themselves about overall market conditions in an area. (The current tendency for competition between 'development opportunities' has inhibited development in local economies with weak land and property markets, as is being increasingly revealed as the property slump continues). Such frameworks must also reach down to the fine detail of place definition, from urban design principles, to infrastructure layout, and ways of managing 'public' (i.e. collectively used) spaces and landscapes. Such frameworks need to define not only what may be appropriate and feasible in an area, and give some ideas about the order and 'packaging' of development. They should also indicate what subsidies and other incentives might be available. Critically, therefore, they have to be expenditure-linked in some way.

There are other ways of creating markets, however. Where a location has opportunities through, for example, action nearby, or major infrastructure investment affecting an area, focussing policy attention on the zone may merely raise expectations about land and property values, and inhibit market activity. An alternative might be to 'ignore' the area from a policy point of view, but be ready to support initiatives (e.g. with flexible development control policies and compulsory purchase powers) 'as and when' these arise. It is arguable that this may be a more appropriate approach to Newcastle's West End, for example, than creating another 'policy zoning' through the City Challenge project. Other approaches (e.g. small factory units or managed workspace), and special exceptions to plan policies (e.g. social housing).

It is unlikely that a plan can 'create' strong markets throughout its area in a weak local economy. This approach requires selectivity, as to which markets and which segments of them to 'create', with intervention (or deliberate 'non-intervention') focussed on precisely what needs to be released to generate market activity. If this is not done sensitively, the approach merely creates new forms of subsidy-dependency. (In many weak markets, Enterprise Zones, Urban Development Corporations, and City Grant availability have merely shifted development activity around within a conurbation from one difficult area to another, see Davoudi and Usher, 1990). The 'creating markets' strategy is most likely to be effective *within* a market-managing approach. This suggests that the sophisticated market-aware development plan may very well include a market-following approach for one type of development or part of the city, a market-creating approach in another, within an overall market-managing approach. The critical challenge for plan preparers is to develop a macro-economic conception of the way land and property and development markets are constituted and operate at the level of the plan area, that is, to deliberately *structure* the market.

Conclusions

The central argument of this paper is that, in conditions where development activity is effectively privatised and communities care about their environment, development plans become more, not less, important. This is now recognised in the late amendment to the *Planning and Compensation Act 1991*, which prioritises plan policies over other material considerations. With respect to development activity, their role is to establish the regulatory framework within which a diversity of development projects can proceed. This regulatory framework must embody the principles of the calculus that has to be made between community concerns for environment and social needs, the demands of production and consumption activities for appropriate premises and locations, and the dynamics of land and property markets and development activity. Whatever the general arguments for a shift from what Harvey (1989) has called the managerial state to the entrepreneurial state, land and property development flourishes best when the state adopts a managerial mode (for this argument, see Ball, 1983; Stach, 1987; Weiss, 1987; Chintz, 1990).

This is of course not a new conclusion. What is new is the emphasis on an explicit understanding of the institutional relations of the private and public sectors in the development process, and a sophisticated understanding of both the relation between local economies and land and property markets, and of environmental and community politics and how these interlock with development objectives. It requires a shift of planning ideology and knowledge away from a 'user-led' approach to understanding development activity, and an assumption that land and property development markets will 'fit in' to whatever framework planning authorities arrive at. This does not mean a switch to an investor-oriented approach, concerned primarily with the maintenance and enhancement of property values. Rather, it requires the interrelations between user and investor markets to be carefully understood, both generally, and in terms of how they work out in particular times and places. Nor does such a change in orientation mean 'giving in' to markets. Instead the challenge is to develop a 'market-sensitive' style of planning, to understand precisely which trade-offs are being made as market considerations are being 'balanced' with those relating to environmental quality and community needs. A planning system informed by a clear idea of the dynamics of land and property markets is more likely to be in a position to strike good bargains for the community and the environment. Some plans are beginning to show how this may be done, for example Harlow Local Plan and Sheffield's UDP.

This is a challenging task. For it is not just a question of understanding land and property markets and the development process. As the fortunes of local economies adapt to changing international conditions, as new ideas about environmental quality pervade our thinking, and as the public sector's role changes from the 'providing' state to the 'enabling' state, planners have to help communities conceptualise what kind of place their areas can become, and what this means for the spatial structure of urban regions and the qualities of locales. This then has to be translated into what it means for local land and property markets and development activity. If we fail in this task, the prospect is not only of deteriorating environmental qualities and un-met social needs, but of a weakened development industry and less robust local economies.

Acknowledgements

This is a revised version of a paper originally given at a seminar on *Statutory Plans and Development Opportunities,* in the series *Planning and Markets* held at the Department of Planning and Housing, Heriot Watt University in October 1990. It will also appear in Ed. C Hague, (forthcoming) *Planning and Markets: British Practice and Prospects* (Gower/Avebury, Aldershot). The helpful comments of PPR reviewers are acknowledged.

References

Ball, M. 1983, *Housing Policy and Economic Power* (Methuen, London).

Barras, R. 1985, 'Development of profit and development control: The case of office development in London', in S.M. Barrett and P. Healey (eds), *Land Policy: Problems and alternatives* (Gower, Aldershot).

Barras, R. 1987, 'Technical change and the urban development cycle' *Urban Studies* 24(1): 5–30.
Brindley, T., Rydin, Y. and Stoker, G. 1989, *Remaking Planning* (Unwin Hyman, London).
Cheshire, P. and Sheppard, S. 1989, 'British planning policy and access to housing: Some empirical estimates' *Urban Studies* 26(5): 469–85.
Chinitz, B. 1990, 'Growth management: Good for the town, bad for the nation' *Journal of the American Planning Association* 56(1): 3–8.
Davies, H.W.E., Edwards, D., Hooper, A. and Punter, J.. 1989 *Development Control in Western Europe* (HMSO, London).
Davoudi, S. and Usher, D. 1990, 'Who is building what in Tyne and Wear', Urban Regeneration Project Paper 1, Department of Town and Country Planning, University of Newcastle.
Evans, A. 1987, *House Prices and Land Prices in the South East: A Review* (House Builders' Federation).
Evans, A. 1991, 'Investment diversion and equity release: The macro-economic consequences of increasing property values' *Urban Studies* 28(2): 173–82.
Hallett, G. 1979, *Urban Land Economics* (Macmillan, London).
Harvey, D. 1982, *The Limits of Capital* (Blackwell, Oxford).
Harvey, D. 1989, 'From managerialism to entrepreneurialism; The transformation of urban governance in late capitalism' *Geografiska Annaler* 71B: 3–17.
Healey, P. 1988, 'The British planning system and managing the urban environment' *Town Planning Review* 59(4): 397–417.
Jones, R. 1982, *Town and Country Chaos: A citical analysis of Britain's Planning system* (Adam Smith Institute, London).
Keogh, G. 1981, 'Land use planning and the production of market information', Discussion paper in Urban and Regional Economics, Department of Urban and Regional Economics, University of Reading.
Logan, J. and Molotch, H. 1987, *Urban Fortunes: The political economy of place* (University of California Press, Berkeley).
Massey, D. and Catalano, A. 1978, *Capital and Land* (Edward Arnold, London).
McCrone, D. and Elliott, B. 1989, *Property and Power in a City: The Sociological significance of landlordism* (Macmillan, London).
McNamara, P.F. 1983, 'The role of estate agents in the residential development process' *Land Development Studies* Vol. 1, pp. 101–112.
McNamara, P. F. 1986, 'Statutory plans for city centres' *The Planner* 72(6): 43–45.
O'Grady, S. 1990, in *Districtwide Local Plans – Emerging Issues* H Thomas et al., Working Paper 125, Department of Town Planning, Oxford Polyechnic.
Pickvance, C. 1982, 'Physical planning and market forces in urban development', in C. Paris (ed.). *Critical Readings in Planning Theory* (Pergamon, Oxford) pp. 69–82.
Saunders, P. 1979, *Urban Politics: A Sociological Interpretation* (Hutchinson, London).
Scott, A.J. 1980, *The Urban Land Nexus* (Pion, London).
Stach, P. 1987, 'Zoning – to plan or to protect' *Journal of Planning Literature* 2(4): 371–83.
Thornley, A. 1991, *Urban Planning under Thatcherism* (Routledge, London).
Weiss, M. 1987, *The Rise of the Community Builders* (Columbia University Press, New York).

Chapter 9
Re-thinking the Relations between Planning, State and Market in Unstable Times*

Patsy Healey

Unstable Times

For the past decade, analysts of urban and regional development dynamics in Western Europe have emphasised how established arrangements for promoting and governing territories and localities have been destabilised. The literature refers to fragmented versus integrated governance activity; multi-level versus hierarchical government systems; disjunctions between administrative, functional and cultural conceptions of place; diffuse and conflicting conceptions of place and spatial connectivity versus cohesive, integrated conceptions.[1] The result has been interpreted as a governance landscape of institutional complexity, mirroring the complexity of the relations which transect, intersect and clash in urbanised areas. Research into governance processes shows that some localities seem overwhelmed by this complexity and instability, unable to generate sufficient energy and cohesion to create coalitions with the capacity to become pro-active in shaping place futures. In other localities, in contrast, considerable governance capacity has evolved, through which responses to unstable conditions have been characterised by innovative experimentation, leading in some cases to transformations in governance cultures and practices. Such research has begun to show how and why some localities have greater ability to become pro-active sites of governance activity, with the capacity to 'act *for* themselves', transforming their processes and practices in doing so.[2]

This work has drawn on concepts which emphasise the mobilisation of actors and networks, the use of arenas and discourses and the formation of coalitions and 'regimes'. With a keen eye for contingencies and local specificities, analysts have either been cautious about over-generalisation, or have found that generalisations become unstuck when they hit the ground of diverse experiences. This has encouraged a research emphasis in fields such as urban and regional research on the activities and power of agency – of particular organisations and actors.[3] Such an emphasis is paralleled in practical affairs through the emphasis on the 'innovative entrepreneur', and the 'enabling and facilitating' public actor, taking initiatives and promoting projects. Both assume that pro-active governance demands strategic capacity, which can mobilise and co-ordinate development activity. In this climate of ideas, rules and regulations are often dismissed as an inheritance of outdated attempts to 'control' what happens or as an old-fashioned approach to territorial development (Albrechts 2001).

Yet, just as much as strategic energy, these legal and administrative procedures and practices have always played an important role in shaping the contexts in which innovative activity and experimentation takes place. They have had a 'structuring' role. Structural explanations for the changing and variable forms of urban and regional dynamics more usually focus on economic forces, as in Harvey's arguments for a shift from 'managerialist' to 'entrepreneurial' governance regimes, or Jessop's claim of a shift from 'welfare' to 'workfare' modes of regulating social order (Harvey 1989, Jessop 1995). Interpreters who wish to highlight the pervasive shifts in governance rhetoric and practices these days often express this is as a hegemonic emphasis on the policy primacy of concerns with 'economic competitiveness', and an expression of a

* Talk for the CITTA Conference: Planning in Times of Uncertainty, Porto, 15 May 2009. Previously published in P. Pinho and V. Oliveira (eds), *Planning in Times of Uncertainty*, Porto: FEUP edicoes, pp. 11–22 (2010).

1 See Gualini 2006, Healey 2006, Lefevre 1998, Salet and Thornley 2003.

2 See Le Galès 2002, Cars et al. 2002, Goodwin et al. 2006.

3 This has been a key emphasis in my own work.

'neo-liberal' ideology.[4] For some time, however, social science inquiry has been critical of too strong an emphasis on structuring dynamics. In a 'post-modernist' intellectual climate, there has been a critical scepticism about 'grand narrative' explanations. These have seemed to crowd out the ability to recognise specific local contingencies (Painter and Goodwin 2000, Jessop and Sum 2006). Yet too much of an emphasis on uncovering the specificities of particular development pathways can neglect the way these specificities are shaped by powerful exogenous forces.

The collapse of the global capitalist financial system from 2007 and its consequences for economic activity and public sector finance have brought the importance of structuring systems back into the forefront of attention. In particular, they have highlighted the social institutions necessary for 'markets' of various kinds to be constructed. The crisis has underlined the role of government regulations and practices in reducing the risk and uncertainty which undermine market activity. Just as, in late 1980s Britain, the celebration of the innovative property entrepreneur helped to create a steep boom/bust phenomenon, so has the hyper-celebration of the clever financial innovator come to haunt the global banking system. This leads some to argue that financial markets should be kept under firm regulatory control, perhaps even taken into full public control and management, while innovative entrepreneurs should be squeezed out of economic activity altogether. But the record of state control is not such a 'shiny' one, being prone to subversion by private interests, inefficiency and inertia.

Markets, as a mechanism for allocating resources and for mobilising creative energy, have been valuable ways of undertaking economic activity for millennia. Although markets, and the activities of merchants, have at times been squeezed out of social systems, it seems highly unlikely that market mechanisms will disappear from the complex urbanised societies of the twenty-first century. What instead needs to focus our attention, as researchers and policy advisers in the field of urban and regional dynamics, is the way market activity is structured. Markets, our welfare economics textbooks tell us, involve 'willing buyers' and 'willing sellers' in an interaction regulated by the 'price mechanism'. But what institutions 'structure' the values of goods to buyers and the production processes which makes goods available for sale? In what arenas do market transactions take place? Over what time periods are prices established and realised? What governance mechanisms intrude open these processes?

Such questions certainly need careful attention in the financial sector at the present time. But the markets that need special attention from those concerned with the spatial dimensions of urban and regional development are those involved in land and property development processes. It is on these processes and in relation to the markets relevant to them that urban planning has a powerful actual and potential leverage. We should be asking, in these present 'destabilised times', how are such processes faring? Are there any generalisations we can make across Europe about the impact on land and property development processes of shifts and instabilities in governance processes and of the shocks to the financial system and its general knock-on economic effects? Or are the experiences of different places and countries so variable that we need to maintain our attention firmly on local specificities? Are there general lessons about how to 'manage' local land and property development markets which we can exchange from one place to another? Or should our research and policy focus remain centred on agency power? Is the policy message that we should devote our attention to 'inventing' appropriate responses to the particular conditions in our own cities and regions?

Market Shaping and Land and Property Development Dynamics

Land and buildings on land have always presented a difficult problem to address in terms of a market. As many have argued, land as such is not generally 'produced' by human effort,[5] although its qualities and locational characteristics are deeply affected by the way we think about, use and invest in land and property, and by the role of access and ownership of land and property in the societies of which we are a part. How we think about

4 Even the idea and practices of 'collaborative planning' have been criticised as a 'neo-liberal' idea, see Bengs 2005 and for a critique of this view, see Sager 2010.

5 The exceptions are the substantial efforts in reclaiming land from the sea, and other ways in which 'natural' landscapes have been re-modelled on a large scale to make them habitable.

these issues affects how land and property 'values' are established, and to whom that value accrues. This in turn shapes the institutions – formal and customary law, regulations to set the relation between 'public' and 'private' interests in land and property development, what is seen to be acceptable public intervention in development activity, etc. – within which specific market activity develops. Land and property development institutions, with their organisations and practices, then develop within this context. These include the kinds of firms which engage in development activity, the 'marketplaces' where transactions are negotiated, and the terms of these transactions. Ball (1983, 1998) calls these institutions the 'structures of provision' through which building types are produced. Specific instances of development activity, the focus of both many planning research studies and local policy conflicts, express these 'layers' of institutional shaping. In other words, local land and property development activity occurs in contexts which are structured in particular ways. They may also, through generating particular localised ways of operating within structuring parameters, create another institutional layer. Such localised processes may, from time to time, unsettle and change the wider context. So the field of land and property development activity has an intrinsic social scientific interest because of the complexity of the interrelation of structuring forces and agency activity. Neither 'grand narrative' nor ethnographic 'case story' can by itself penetrate easily the role of land and property development dynamics in urban and regional development.

The above underlines that a lot of social, institutional work is needed to turn land and property, and hence land and property development activity, into a market form. And the institutions created are prone 'failure'. Development timescales are often long. Both buyers and sellers may be large corporations or landowners, able to act as oligopolies. How land is owned may be linked historically to the role and fate of past landed aristocracies, to the extent to which property is considered as an investment rather than for its use value, and, in the present period, to the closeness of the nexus between investment in land and property and other forms of financial investment. In the UK, as in the US, the investment emphasis has grown particularly close in recent decades, which is one reason why a collapse in housing property values has had such a massive consequence, causing not just the de-valuing of the holdings of many financial institutions, but of many households and firms. For many builders and developers, the consequences have been a massive 'destruction' of value, a sudden bust after a long and exaggerated boom. As in previous property busts, the land and property development sector has turned to government and demanded action. Whereas in boom periods, the demand has been for release from 'bureaucracy' and 'regulatory constraint', so in bust periods, the demand is for protection from instability and uncertainty. The 'innovative entrepreneur' is replaced by the risk-averse asset manager (Adams 1994, Healey 1998).

Planning systems and practices get buffeted by these demands and the economic winds that blow them. In the 1970s, some argued that planning activity merely reflected market dynamics, or the interests of powerful players in land and property markets.[6] Subsequent research inquiry also showed that planning institutions, strategies and practices had a role in shaping such markets. Such institutions operate 'structurally' by shaping expectations. In England, for example, most planning strategies express the deeply-embedded societal view that development should not be allowed to sprawl across the 'countryside'. In the Netherlands, planning strategies indicate not just where development will be encouraged and discouraged, but what investment projects are likely to come on stream, allowing developers to assess land and property value consequences. In both countries, the institutions and practices related to formal 'planning systems' help to structure negotiations over how the value of a development should be distributed, as between developers, owners and end-users, infrastructure and service providers, local communities, wider interests, and the state generally.

Some argue that land and its development are so problematic to treat in a marketised way that they should owned as well as regulated by the public sector – nationally, regionally and/or locally.[7] In times of property bust, developers, landowners and residents often look to the state to take over failed projects and derelict sites. Those who argue that the central aim of planning activity is to promote social justice and/ or environmental sustainability also doubt whether land and property development should be treated in a market form. This leads to an emphasis on the role of planning systems as a mechanism to restrict market

6 See, for example, Castells 1977, Scott and Roweis 1977.

7 For several decades from the 1940s, development land in the Netherlands and Sweden was primarily in public ownership.

activity which threatens such values. These arguments suggest that the role of planning strategies and other instruments is not to protect market operators from their own behaviour but to force them, if still given some opportunity, to take on these values. This can lead to planning strategies which are strong on rhetoric and to demands for strong regulatory powers. Whether the result is compliance (as in the Netherlands) or evasion (as in Italy) depends then on the relative strength of 'planning' institutions and other societal institutions in structuring land and property development practices (UN-Habitat 2009). However, another way forward is to work with awareness that land and property development activity benefits, in the long-term, from an institutional framework which 'smooths' the tendencies to boom and bust. This suggests that planning systems and practices have a potentially significant role in managing markets, as well as in promoting a recognition of responsibilities for the wider social and environmental contribution of any land and property development (Healey 1992a,b). A period of a collapse in value may stall land and property activity for a long while. But it may also be a period when the public sector has an opportunity to accumulate land and property assets, as private owners 'default', and to build a strategic framework through which development activity, when it revives, can be shaped in ways which promote wider objectives than just the maximisation of profits. What mechanisms are available to achieve this?

Managing Land and Property Development Activity

Table 9.1 summarises the main instruments available to the public sector through which to manage and shape land and property development markets. A critical instrument for a land and property 'market' to operate are rules governing property units, tenure and rights to sale. In some situations, these can be so complex that land and property proves an unattractive investment. Or people bypass legal means of occupying land, producing the semi-legal and informal settlements so common these days around many rapidly expanding cities outside North America and Europe (see UN-Habitat 2009). Planning systems come into play by giving some spatial specification to limits on land and property development rights, using instruments for deploying regulatory powers and powers for land acquisition. Planning systems also promote the co-ordination of development activity, and express the values of the 'public realm' which individual owners and developers are encouraged to absorb into their development projects. It is here that plans and strategies play a crucial role. However, if formal legal instruments are disregarded or distorted in the way they operate, then such plans and strategies may have little leverage on the development which actually takes place.[8]

The weight given to such instruments, even when available through legislation, varies from place to place, as does the way they are interpreted and used in practice. In Western Europe, persuasive spatial strategy-making became neglected in the 1980s, with the focus instead on the project, and on land use regulation (Healey et al. 1997, Albrechts 2001). In the 1990s, the significance of spatial strategy-making was given greater emphasis, partly to improve co-ordination, and partly to regulate the external effects which the promotion of individual projects created across the wider urban and regional space. These effects were evident not just in rising pollution and congestion due to the strategic neglect of key infrastructures such as transport investment. They were also manifest in the growing spatial inequalities across urban areas due to the investment focus on releasing market potential rather than attending to areas where market attention was lacking. The competition between projects and the lack of clarity about public infrastructure investment also created market uncertainties.[9] Some of those involved in the resurgence of spatial strategy-making very clearly saw the major role of spatial strategies in managing urban land and property development markets (Mazza 2004).

To illustrate the changing balance of planning instruments for managing land and property development markets, I now look at the case of England (see Table 9.2). This shows a steady shift from a strong public sector role as a provider, developer and owner of land and property, to an increasing role as a 'market manager',

8 Land allocation processes through formal plans may also be 'distorted' by practices which focus on political or private gains from development, rather than for the formal development purposes to which the plans are supposedly directed. See UN-Habitat 2009.

9 This has been especially evident in the UK.

Table 9.1 Mechanisms for managing land and property development activity

Mechanism	For example ...
Defining rules for the operation of land and property development markets.	Through rules governing land ownership, tenure and land/property transactions Through regulations governing how land can be used and developed. Through regulations governing the contributions private developers must make to the 'public realm' (as in development charges and developer contributions to public facilities and services)
Direct development: Acting as landowner and developer	Providing public housing, developing business parks, building new towns, and initiating and managing major area redevelopment projects.
Facilitating land assembly	Through powers of compulsory purchase (expropriation) for public purposes, such as a major development project.
Providing financial incentives	Grants and tax exemptions to reduce the costs of a private sector development project
Facilitating co-ordination and hence reducing transaction costs	Through negotiation and mediation processes
Providing spatial organising concepts, to focus development attention	Through spatial strategies and area master plans.

although this latter role is rarely explicitly acknowledged and continually contested. Instead, the public sector – and 'the planners' – have been caricatured as bureaucratic meddlers, following arcane procedures, inhibiting exciting designs and preventing the provision of needed buildings. Developers, in turn, are caricatured as money-grubbing capitalists, out to displace beautiful views and hard-working small businesses which appear in the path of their overblown ambitions.

Table 9.2 Shifts in the role of different mechanisms in England

	1950s–1960s	1970s–1980s	1990s–2000s	2010s
Defining rules	XX	XXX<	XXXX	????
Direct development	XXXX	XX>	X<	<<
Land assembly	XXX	X>	X<	??<<
Financial incentives		X<	XXX<	>>
Co-ordination	(via strategy)		XXX	<<<???
Spatial concepts	XXX	X>	X<	<<<???

Note: < = expanding; > = diminishing

What then is the next stage in this story? It is widely acknowledged that the British economy has, in Europe, been one of the most severely affected by the financial and economic crisis of 2007–2008. Land and property development activity in England has been battered by the financial collapse, especially in the buoyant south, where growth driven by the powerhouse of the financial sector. From the late 1990s, as the financial sector boomed, national government vigorously promoted the expansion of housing development to accommodate the rapidly rising demand and to increase the amount of low cost housing. Pressures were put on the planning system, which has long been a site of intense struggle between conservation and environmental values demands for new development locations, to release land for development. Planners and many development industry actors then struggled to contain new development initiatives within the framework of approved plans. Locating land released in planning frameworks was necessary to provide legitimacy to land allocation decisions.[10]

As a result of the negotiative efforts of many planners and developers, a good many sites were allocated for development during the 2000s, with substantial contributions negotiated for public infrastructure, justified through formal strategic plans. National government also claimed to demand that development sites were allocated through approved planning frameworks, but had difficulty in grasping that this meant accepting the validity of many environmental concerns and the importance of co-ordinating the allocation of development land with infrastructure investment. There has been a lot of experimentation and slow learning going on, but so far the ability of the public sector to provide clarity about development directions, regulatory requirements and infrastructure projects and their funding has been compromised by failures to understand the dynamics of different land and property development markets and struggles over where intense conflicts are to be resolved. Governance capacity in this area has thus been full of instability and uncertainty.[11]

The property development sector in the UK is dominated by large firms and investment companies. Since the 1970s, it has become increasingly intertwined with the financial sector. As a result, property boom turned to bust very quickly from 2007. Severely affected by financial collapse and consequent economic recession, firms went bankrupt or withdrew from development projects, leaving incomplete schemes and vacant sites across our cities. Developers have sought relief from contracts with municipalities which demanded significant contributions to major infrastructures and other public realm requirements. The output of development, instead of growing with the allocation of sites and the negotiation of infrastructure payments, has slumped. Some large scale residential developers ceased building on all development sites for a while, as they could not sell their stock.

Meanwhile, the public finances are overwhelmed by a debt mountain which will take many years to reduce. It is widely recognised the funds for capital investment, whatever government is in place, will be very limited. Yet planning issues, and especially the relation between development, conservation and infrastructure provision, are a key plank of the opposition agenda, which is also committed to giving more powers to local authorities and 'sweeping away' the agencies at national and regional level which have been struggling to get some inter-departmental and inter-municipal co-ordination under the present government.[12] The prospect for many localities is a relapse into a stagnation which in past periods of sluggish development activity leaves the built environment scarred by poor maintenance, empty buildings and overgrown development sites. That leaves spatial strategies and skilled co-ordination work as the critical tools available not just to 'manage' land and property development markets back into activity, but to 'nurture' them.

It is this which creates a significant opportunity for a market shaping role. Both local communities and private development interests would welcome more co-ordination between the allocation of land for development and the provision of infrastructure. But it is possible to do more. There is a chance to assist market activity so long as this supports the values of social and environmental responsibility. The struggles of the 2000s over the location and form of new development show such an 'attitude' developing both in redevelopment work and in developments on greenfield sites. If these transformations in attitudes and expectations can be carried forward into the 2010s, then the contribution of the combined efforts of public shaping and private development activity could prevent our cities turning bleak again, provide more

10 For accounts of this complex interaction between planning, housing and economic policy, see Dixon and Adams 2008, Watkins 2008.

11 For an example of this, see the recent experiences in the East of England (see Healey 2007, Chapter 5).

12 Labour administrations have tended to favour regional development agencies, and the 1997 incoming Labour government was committed to regional devolution in some form. This occurred in Scotland and Wales, but was more weakly developed in England. Since then, in England, opposition parties have emphasised devolution to the municipal level. Meanwhile, local authorities in some areas have taken various initiatives to strengthen local coalitions within and between municipalities. Both Tories and Liberals, the current opposition parties, are committed in a general way to such a municipal 'devolution' agenda.

of the much needed low-cost housing which has been lacking for so long, and significantly improve the environmental quality of the country's built stock. But achieving such transformations demands sustained social and institutional work to build governance capacities to act in this strategic, market management way.

Market Shaping and Governance Capacity

In the perspective of critical urban political economy, it is claimed that moments of severe crisis lead to substantial political and economic transformations. In the present moment, all three 'circuits' of David Harvey's analysis of capitalism are in crisis, with political systems struggling to patch up and stabilise fractured systems.[13] In such a context, the fortunes of urban areas across Europe are likely to depend not just on how national and global economic dynamics develop but on the local governance capacity to reduce the damaging impacts of a long period of economic difficulty. How effectively those involved in such activity are able to do this will depend on their capacity to create a co-ordinated agency capacity which can act strategically across an urban area. In other words, it is the work of agency which generates new structuring possibilities when established structures weaken or collapse. In England, with its over-centralised and heavy state, one of the most interesting developments in the 2000s has been the evolution of such capacity in a number of places, both in areas experiencing substantial growth, and in some of the older industrial urban agglomerations.[14] A significant motivation behind the alliances, development companies and partnerships which have emerged in these areas has been to create local capacity to make use of opportunities provided by national government programmes while limiting the tensions and confusions which are carried in the shifting sands of national investment and regulatory initiatives. Similar stories of the building up of local capacity are to be found across Europe (Salet and Thornley 2003), and also, in the US, in Portland, Oregon, and in the Salt Lake City area in Utah.[15] In such situations, local governance processes not only shape local development conditions and development opportunities. They may also provide beacons which shape how both private developers and the public sector agents work elsewhere, a kind of bottom-up transformation of the 'structures of provision' of property development.

What, then, seem to be key elements of such a governance capacity? One is a capacity that might be called 'structural grasp'. By this I mean the ability to see, in the fine details of localised phenomena, the wider social, environmental, economic and political forces which help to produce such details – to use a metaphor, the ability to 'see the world in a grain of sand'. No urban region is an isolated island and all depend in one way or another on their position within wider economic, social, political and environmental relations. Here David Harvey's analysis of the opportunities for urban regions in a new global economic order may still have value. Rather than merely all competing in a similar 'league', in which most were bound to fail, he suggested that a region could develop its position within the international division of labour or within the spatial division of consumption. Or it could compete for global command functions or to attract funds targeted at redistributing surpluses (Harvey 1985, Chapter 8). The economic story of Britain in the past 20 years can be read as the result of a strategy of competing to become Europe's global financial centre. Portuguese cities have competed to capture European Union structural (redistribution) funds. Many cities in Europe in the 2000s sought to locate themselves on a map of cultural activities and tourist destinations, Harvey's 'spatial division of consumption'. But at the same time, a more localised focus may provide a way forward, centred on promoting the livability and sustainability of urban areas as experienced in the daily lives of people across an urban area. Such a focus could be a way of creating an environment which both limits adverse impacts and generates values which attract wider attention. The instances of Vancouver, in Canada, and Portland, in the

13 See Harvey 1982. In this landmark text, Harvey describes the three circuits through which capital is accumulated in capitalist economies – through exploiting primary resources, through industrial production and through financial systems. He provides a very perceptive account of the crisis-prone nature of each system and the potential for all systems to be in crisis at once.

14 These are to be found both in areas experiencing growth pressures and in cities where urban regeneration has long been a priority.

15 For Portland, see Abbott 2001; for Utah, see Briggs 2008.

US, provide examples where such attention, sustained over 30 years, has created the localities which are now widely appreciated as 'good cities' in which to live, work and visit.[16]

Such 'structural grasp', linked to a capacity to see the relation of a specific situation to wider forces, and vice versa (how an evolving world might touch the many and diverse 'grains of sand'), is the essence of strategic thinking. It involves a continual interaction, in thought and in wider discussion, between 'parts' and 'wholes' (see Churchman 1979, Healey 2009). This converts the sociological discussion of the relative importance of structure and agency, and the significance of focusing on the interactions between the two, into an active practice. But this is no abstract conception. Such thinking needs to be fed with a good, detailed understanding of local dynamics, and the way different relations connect all kinds of different 'parts' to different 'wholes'. Just as we need careful work on ecological systems, and on the relations within various parts of the economy – the financial services, the wine industry, the logistics of the retail sector, etc. – so too do we need careful work on the relations and practices of the land and property development sector.[17]

But there is another aspect of governance capacity which has an important role in supporting strategic efforts to shape land and property development activity so that it contributes to improving urban and regional futures. Strategic power lies not just in control of specific organisations and formal instruments of government. It is sustained also by what is lodged in the culture of a political community. In England, too much sprawling development across our valued 'countryside' is always going to be resisted. The idea that urban development should be 'contained' within defined areas is not just some planning idea imposed on the country at large. The idea reflects and gets its power from its resonance with deep cultural values. In England, attitudes to the rural landscape are tied into conceptions of national identity and cultural value. This also provides a strong basis for support for a wider environmental agenda.

But such values do not just shape the principles expressed in planning systems and practices. The relations between cultural values and those embedded in planning activity are not just a one-way street. Struggles in and around the practices of a planning system can also help to develop and focus the values of a political community. The stories of the development of Vancouver and Portland demonstrate this well. They show how, over time, discussion and debate about how to shape neighbourhood change and how to manage urban growth slowly changed urban political cultures so that new understandings about the important qualities of both cities were generated. These in turn shaped how development industry actors thought about what projects to promote and the parameters within which they should be negotiated. The 'culture' as well as the laws and the work of municipal planners and other agencies framed the institutional context within which local land and property development 'markets' worked. This 'culture' generated expectations not only about what development should look like and how its impacts should be managed. It also shaped the development process itself, and in particular, the way that residents, other specific stakeholders, and citizens at large were expected to have a voice in shaping what happened in 'their' city.

The Academic Contribution

In the last quarter of the twentieth century, the dominant interpretation of the relation between state and market has been that the inertia and inefficiency of overly bureaucratic states has held back market innovation, and hence impeded economic growth. This encouraged all kinds of initiatives which contributed to the fragmentation and destabilisation of urban and regional governance institutions across Europe. The crisis of the late 2000s has brought back into a sharp focus the need for effective governance capacity, in order to help stabilise markets, creating sufficient certainty and direction to breathe life back into them. This is particularly important for land and property development markets, which are so prone to 'market failure'. What can urban

16 I use these examples in my own forthcoming book, which draws on experiences where planning ideas and practices are judged to have brought significant benefits to the quality of urban life, see Healey 2010.

17 Although there has been a steady stream of 'institutionalist' work on land and property development activity, particularly centred around David Adams in the UK, Barrie Needham, in the Netherlands, and at the Politecnico di Torino, in Italy, there has been surprising little academic work on the land and property development sector, particularly from a more economic and sociological point of view.

analysts and planning academics contribute to this re-thinking of the relation of state/market relations at the present time?

One contribution lies in our role as scholars and researchers. Through research, we can help to expand the knowledge base available to our political communities and to enhance the 'intelligence' of democratic debate.[18] We do this as part of our job description, but it would be particularly helpful to have more work on land and property development dynamics, focusing on the institutional relations of the sector. This demands careful, well-grounded empirical research which gets underneath the persuasive rhetoric and the aggressive language of dispute which often colours the reporting of development issues. We also need to develop skills in conveying this understanding to the different members of our political communities. Second, we often act as consultants to various actors involved in local land and property development processes. As we engage in such work, it would be valuable at the present time if we also thought about the contribution we inevitably make both to generating the 'intelligence' of our political community and to building governance capacity to manage markets into the future. Finally, we also play an important role as activists and 'public intellectuals', promoting some issues into the public realm. Here we need to pay careful attention to how our interventions affect the way issues are understood and discussed.

In other words, as academics we are not on the sidelines of these difficult and unstable times in our work as researchers, consultants, activists and public intellectuals. We are deeply involved in shaping governance capacity in our localities, and need to think carefully about the social, political and moral dilemmas such involvements generate for us. In conclusion, perhaps the key message we can offer our polities at the present period is to encourage other actors to use the perceptions of instability and crisis in a strategic way, as an opportunity to take stock, to re-think policies, projects and practices, and to build the intelligence and coalitions which could bring future benefits for the many not just the few in our localities.

References

Abbott, C. (2001). *Greater Portland: Urban Life and Landscape in the Pacific Northwest.* Philadelphia: University of Pennsylvania Press.

Adams, D. (1994). *Urban Planning and the Development Process*. London: Routledge.

Adams, D. and C. Watkins (2003). *Greenfields, Brownfields and Housing Development.* Oxford: Blackwell.

Albrechts, L. (2001). From traditional Land Use Planning to Strategic Spatial Planning: The case of Flanders. In L. Albrechts, J. Alden and A. da Rosa Pires (eds) *The Changing Institutional Landscape of Planning.* Aldershot: Ashgate, pp. 83–108.

Ball, M. (1983). *Housing Policy and Economic Power*. London: Methuen.

Ball, M. (1998).Institutions in British property research: A review. *Urban Studies* 35(9): 1501–18.

Ball, M., C. Lizieri, and B.D.MacGregor. (1998). *The Economics of Commercial Property Markets*. London: Routledge.

Bengs, C. (2005a). Time for a critique of planning theory. *European Journal of Spatial Development,* http://www.nordregio.se/EJSD/editorials.html (Editorial no. 3, June).

Briggs, X. da S. (2008). *Democracy as Problem-solving*. Boston, MA: MIT Press.

Cars, G., P. Healey, A. Madaniour, A and C. de Magalhães (eds) (2002). *Urban Governance, Institutional Capacity and Social Milieux*. Aldershot: Ashgate.

Castells, M. (1977). *The Urban Question*. London: Edward Arnold.

Churchman, C.W. (1979). *The Systems Approach* (2nd edition). New York: Dell Publishing.

Dixon, T. and D. Adams (2008). Housing supply and brownfield regeneration in a post-Barker review world: A comparison of policy and practice in England and Scotland. *Urban Studies* 45: 115–39.

Dewey, J. (1927/1991). *The Public and its Problems*. Athens: Swallow Press/Ohio University Press.

Goodwin, M., M. Jones and R. Jones. (2006). Devolution and economic governance in the UK: Rescaling territories and organisations. *European Planning Studies* 14(7): 979–95.

18 I make reference here to the pragmatist argument for cultivating the capacity for 'intelligent debate', as a basis for a richly democratic polity, Dewey 1927.

Gualini, E. (2006). The Rescaling of governance in Europe: New spatial and institutional rationales. *European Planning Studies* 14(7): 881–904.

Harvey, D. (1989). From managerialism to entrepreneurialism: The formation of urban governance in late capitalism. *Geografisker Annaler* 71B: 3–17.

Healey, P. (1992a). Development plans and markets. *Planning Practice and Research* 7(2): 13–20.

Healey, P. (1992b). The reorganization of state and market in planning. *Urban Studies* 29(3/4): 411–34.

Healey, P. (1998). Regulating property development and the capacity of the development industry. *Journal of Property Research* 15(3): 211–28.

Healey, P. (2006). Transforming governance: Challenges of institutional adaptation and a new politics of space. *European Planning Studies* 14(3): 299–319.

Healey, P. (2007). *Urban Complexity and Spatial Strategies: Towards a relational planning for our times*. London: Routledge.

Healey, P. (2009). In search of the 'strategic' in strategic spatial planning. *Planning Theory and Practice* 10(4): 439–57.

Healey, P. (2010). *Making Better Places: The planning project in the twenty-first century*. London: Palgrave Macmillan.

Jessop, B. (1995). Towards a schumpeterian workfare regime in Britain? Reflections on regulation, governance and the welfare state. *Environment and Planning A* 27: 1613–26.

Jessop, B. and N.-L. Sum (2006). *Beyond the Regulation Approach: Putting capitalist economies in their place*. Cheltenham: Edward Elgar.

Le Galès, P. (2002). *European Cities: Social Conflicts and Governance*. Oxford: Oxford University Press.

Lefevre, C. (1998). Metropolitan government and governance in Western countries: A critical review. *International Journal of Urban and Regional Research* 22(1): 9–25.

Mazza, L. (2004). *Piano, progetti, strategie*. Milano: FrancoAngeli.

Painter, J. and M. Goodwin (2000). Local Government after Fordism. In, G. Stoker (ed). *The New Politics of British Local Governance*. Basingstoke: Macmillan, pp. 33–53.

Sager, T. (2010). Role conflict: planners torn between dialogical ideals and neo-liberal realities. In: J. Hillier and P. Healey (eds) *The Ashgate Research Companion to Planning Theory: Conceptual Challenges for Spatial Planning*. Aldershot: Ashgate, pp. 183–214

Salet, W., A. Thornley, et al. (eds) (2003). *Metropolitan Governance and Spatial Planning: Comparative studies of European city-regions*. London: E&FN Spon.

Scott, A.J. and S.T. Roweis (1977). Urban planning in theory and practice: A reappraisal. *Environment and Planning A* 9: 1097–1119.

UN-Habitat (2009). *Planning Sustainable Cities: Global Report on Human Settlements 2009*. London: Earthscan.

Watkins, C. (2008). Microeconomic perspectives on the structure and operation of local housing markets. *Housing Studies* 23(2): 163–77.

Chapter 10
Planning Wild Cities

Wendy Steele

> The task of the planning enterprise is to critically interrogate the governance practices that currently exist and to help … develop different approaches where these are seen to be failing.
>
> (Healey, 2003, p.116)

Introduction

What kind of urban futures are we planning? Our cities are now capable of being seen from space: the super-structures of the super-species in the Age of the Anthropocene (Suzuki, 2010). As Okome (2002: 316) observes: 'we live in the age of the city. The city is everything to us – it consumes us, and for that reason we glorify it'. For economists such as Glaeser (2010) the triumphant city has conquered as 'humanity's greatest invention'. It is within this context that this chapter focuses – following Healey (1990, 1992, 2003, 2010) – on the need to better understand the planning and development nexus as a lens into issues around land, how places are produced and changed and the relational role of markets, planning instruments, and social/environmental movements.

Globally, cities are home to more than half the world's population, consume nearly three-quarters of its natural resources, and are highly vulnerable to the impacts of human-induced climate change. They act as the discursive site for society's contradictions and collective anxieties. As Beauregard (2003: ix) notes they are 'places of decay, moral turpitude and social disorganization'. Abani (2004: 7) describes the city as 'half slum, half paradise' and wonders how they can be so violent, ugly and beautiful at the same time. Whilst Davis' (2006) vision of a 'planet of slums' offers a dystopic image of a twenty-first-century urban world squatting in its own excrement surrounded by pollution and decay of its own making.

There is a restless churning of cities at both the global and local scale. Important questions around urban food, energy, water, housing and transport jostle for public and policy attention alongside ecological imperatives and issues of infrastructure provision and property development. Growth ideologies form a key defining driver within the context of late capitalism. For Castells (1977: 426) what could emerge if urban movements fail to undertake their collective political task is perhaps a new and more sinister urban form: 'the Wild City'.

The planning project is implicit here, as both outcome and response to the urban condition. From its earliest conception, planning has sought to correct harms arising from living in cities and to advance a better quality of life for urban residents (Hall, 2002). Since the late 1800s burgeoning metropolises have been the sites of national economy of global dimensions, but also collective anxieties, the object to debate injustice and inequalities, and the scapegoat for social ills. The purpose of urban planning was considered to be (at least in part) an instrument of social reform, closely associated with strong reformist ideals around improving the human misery and squalor (Eisenschitz, 2008).

For some urban planning is a remedial project 'generated as a reactive response to urban land use and development pathologies' (Dear and Scott, 1981: 15). To others the central question for planning should be 'what kind of role planning can play in developing the city and region within the constraints of a capitalist economy and a democratic system' (Campbell and Fainstein, 2005: 1). As Leonie Sandercock (1990) highlights in her seminal mediation on property, politics, power and urban planning in *Cities for Sale*, this constitutes a series of complex dilemmas for capitalist societies around questions such as 'whether popular immediate solutions to planning problems are to be preferred to long term more expert ones, whether a combination of popular movements and representative institutions can outweigh entrenched power groups, and whether more participatory and redistributive policies will in the end enhance overall human welfare' (p.46).

Finding ways to develop collective projects through political will and a sense of political necessity is the planning response to the capitalist project envisaged by Harvey (2014); one that seeks 'to think of how to organise a whole city, politically, around an anti-capitalist project ... to build a different kind of city and a different kind of urbanisation' (p.1). For Patsy Healey (2007: xi), however, the planning field, whilst full of frustrations and contradictions, remains a 'fine endeavour' focused on 'exploring what it takes to act in the world to pursue *collective purposes* which are seen to be inclusively beneficial in a particular social context, in ways which attend to the significance of the spatial dimension of all relations and to the particularities of place qualities as they evolve' (italics added). To this end the development of planning ideas and struggles with the logic of market, land and property development in practice is centrally part of the planning project – the governance of urban place and space.

The logic and structure of this chapter are as follows. The first section focuses on the notion of 'feral capitalism' and the implications of this for the urban planning project at the development/property interface. The second section turns to the temporal and spatial dimensions of the land question as a key consideration for understanding land development processes, spatial dynamics and their speed of impact on place qualities. Finally the chapter looks to the seeds of urban change and the role of the institutional lens into how partnerships between actors form, and how spatial strategies connect to wider policy cultures within the context of urban growth and development, particularly at the local scale.

Wild Cities, Feral Capitalism

Issues of land and property development, the relations between planning and markets and the implementation impacts of planning instruments all take place within what Harvey (2012: 157) describes as 'feral capitalism'. The problem, he argues is that we live in a society where capitalism has run amok amidst 'the animal instincts of the entrepreneur'. This affects both the social and environmental realms as Davis and Monk draw attention to in their book *Evil Paradises – Dreamworlds of Neoliberalism* whereby 'the earth is being held by the muscular arms of it entrepreneur-plurocrats, like it, or not' (Citigroup Research, 2006: 3).

The economy, as many have observed, offers both mode of seeing and means of organising cities but has no ethical platform, singular logic or essential form (Harvey 1996, 2012; Merrifield, 2002). Capitalism exists to a large extent parasitically, drawing momentum and sustenance from the earth and others as opportunity arises and provides (Mitchell, 2007). The promise, and indeed for those privileged within the system the reality, is one of endless possibilities. Nowhere is this expressed more powerfully and viscerally than in the global/local city. And for Harvey (1996), it is the shift[y]ing logic of capitalism that in turn plans the role of planning.

Marxist economics would suggest that the central mechanisms within capitalism fail in two key ways: [i] in the provision of public infrastructure; and [ii] in the management of urban land-use development and outcomes (Dear and Scott, 1981). Both of these failings in turn call for planning intervention which often exacerbates the very issues it sets out to resolve. As Kenny and Meadowcroft (1999: 1) observe 'the history of planning is littered with unintended consequences and undesired outcomes'. For Marx (1867: 812) planning equals:

> Improvements of towns, accompanying the increase of wealth, by the demolition of badly built quarters, the erections of palaces for banks, warehouses etc. the widening of streets for business traffic, for the carriages of luxury, and for the introduction of tramways etc., drive away the poor into even worse and more crowded hiding places.

Healey and Barrett (1990) argue that whilst Marxist economics provide useful directions for understanding the development process, it lacks the capability to address fundamental dimension of our understanding of development processes. In particular they point to the need to better understand the complex relations between the way actors behave in deploying resources to realise specific investments, and the broader processes that drive the strategies and interests of the various actors involved. This involves a renewed emphasis on the institutional role of resources, rules and ideology in land-use and property development as a constructive way forward for planning within an urban era dominated by:

- international finance and rapid transnational capital movement;
- major economic restructuring redistributing land and property values; and
- a stronger role for the private sector through deregulation and privatisation policies (p.94).

The state as a critical actor in development processes is highlighted. State-led planning system and associated mechanisms, whilst contextually specific, hold in common a number of broad characteristics including: a legal framework; a plan; a regulatory system; political decision-makers; and an appeals process (Carmona and Sieh, 2004). The planning literature has oscillated between the relative merits (or not) of 'conforming' (regulative certainty through zoning) *or* 'performing' (strategic flexibility through discretionary planning) planning systems (Rivolin, 2008; Steele 2011). The regulatory zoning approach is connected with greater levels of certainty in decision-making processes by offering greater levels of consistency and objectivity through an emphasis on planning prescription. The discretionary approach is linked more closely with flexible decision-making that is able to be responsive to individual circumstances as and provide scope for negotiation and creative decision-making processes. In planning practice however these differences and distinctions are often less clear (Steele and Ruming, 2012). The result is a series of planning systems that emphasise '*con*formance (of evidence and policy, policy and outcome) rather than *per*formance (or strategic "doing")' (Hillier, 2007: 21).

State-led regulatory zoning sought to shape the built/spatial environment by: a) dividing communities into distinct zones; and b) determining what 'could' or 'could not' be built within those discrete zones (Stair et al., 2008). The structure of zoning provided an economic, social and environmental tool for directing order and change through land-use controls designed to outline the parameters and methods of compliance: an economic tool to protecting property and land values based upon zoning qualifications; a social tool protecting public space and amenity by excluding unacceptable activities, for maintaining social control through the creation of exclusive areas of privilege, and facilitating the supply of additional housing by changing land-use type or density; and an environmental tool seeking to regulate the impacts of pollution and by segregating potentially noxious industrial uses.

A key purpose was to create certainty by removing and separating conflicting land uses and dysfunctional districts that might impede solid investments in land (Boyer, 1986). Although as Ferriss (1922: 8–9) noted 'early entrepreneurs, architects, and developers were quick to euphemistically interpret regulatory zoning not as a restriction but rather as 'a form-giving principle for new, modern skyscrapers'. Indeed planning-related laws and legislation were often only supported when they promised certainty for investment decisions, such as systems of zoning (Fischler, 1995, 1998; Freestone and Grubb, 1998; Hiller, 2013). Fischler (1995: 22) observes that zoning was often 'created and sustained under the political pressure of property owners and developers fighting to maintain or increase their assets'.

Development and community aspirations and planning reform ambitions around sustainability and economic growth have resulted in a combination of strategic, regulatory and discretionary characteristics – hybrid planning systems that attempt to embrace both flexibility and certainty such as 'flexible zoning' in the United States (Elliott, 2008) and 'plan-led discretionary systems' in the United Kingdom (Carmona et al. 2003). As John Sulman (1921: 173) observed nearly 100 years ago 'regulation and its evasion is as old as the buildings themselves, particularly if authorities are weak and corrupt'.

Rivolin (2008) suggests that the critical question for planning is not whether conforming or performing systems or plans are preferable but rather '*how the strategic and regulative functions of planning should be differently correlated in a planning system addressed to performing rather than conforming aims*' (p.171). The focus is on the performance of planning systems as a whole (incorporating both top-down and bottom-up processes) rather than a dichotomy between strategic (discretionary) and/or regulatory (zoning) planning systems. Rivolin (2008: 178) argues that *in principle* performing planning systems are 'more suitable than conforming ones if planning is to fulfil its social role'.

For Healey (1992) a 'performing' planning framework must embody 'the principles of the calculus that has to be made between community concerns for environment and social needs, the demands of production and consumption activities for appropriate premises and locations, and the dynamics of land and property markets and development activity' (p.19). This in itself requires a sound understanding of development and property markets as the basis for strategic planning and activity, and how these might vary 'spatially, temporally, sectorally and institutionally' (p.13).

Such a focus entails: better understanding the institutional relations particularly between private and public interests in the development process; the dynamic between local economies and land and property markets; and the processes of environmental and community politics and how these interlock with development objectives. As Healey (1992: 19) acknowledges:

> This is a challenging task. For it is not just a question of understanding land and property markets and the development process. As the fortunes of local economies adapt to changing international conditions, as new ideas about environmental quality pervade our thinking, and as the public sector's role changes from the 'providing' state to the 'enabling' state, planners have to help communities conceptualise what kind of place their areas can become, and what this means for the spatial structure of urban regions and the qualities of locales.

Within feral capitalism this process of understanding and translation leading to action is further complicated by predatory development funding practices (think Fannie Mae and Freddie Mac[1]) and the associated vast terrain of land speculation and accumulation (often by dispossession) that impact on those most vulnerable. For Harvey (2012) there is an inevitable trajectory of creative destruction through urban growth and development that results in social and ecological destruction. In theory a planning system 'informed by a clear idea of the dynamics of land and property markets is more likely to be in a position to strike good bargains for the community and the environment ' (Healey, 1992: 19). In practice how can these predatory growth and capital tensions be tamed by planning the wild city?

The Land Question Revisited

At the heart of these urban planning and sustainability tensions within wild cities is the land question. The processes of land and property development are central to urbanisation and the production of space, community and place. Within an urban context this includes serviced land rather than just raw land including the goods, infrastructure, human labour and market use value and exchange this type of land embodies (see Roweis and Scott, 1981). As Healey and Barrett (1990) highlight, 'the role of landownership, the organisation of the construction industry and the nature of funding mechanisms and acting intermediaries such as developers and planning consultants are critical to understanding urban development and urban development processes'.

If the land question is an expression of capitalism (alongside rights and social and environmental justice) then it is one that is made manifest in cities through housing development and infrastructure. Land, property and development processes reflect the relations between state and market, the nature and role of the planning system as well as shifting community perceptions, spatial dynamics and place qualities. For Castells (1977: 73) 'urban society is never merely spatial form but urban culture – values, norms and social relations – that possess a historical specificity and logic of organization and transformation'.

Feral capitalism has particular emphasis on speed which pays due homage to the three E's of economy, efficiency and effectiveness (Spigelman, 2001). The focus on urban land supply at low cost for development amidst calls for greater economic competiveness through reductions in planning controls have ignited public concern about the quality of space and place, liveability, equity and environment. Land reform through planning is central to what Lash and Urry (1994) describe as 'fast capitalism'; the 'speeding up and stretching

1 See for example V. Acharya, M., Rchardson, S.V. Nieuwerburgh and L.White (2011) *Guaranteed to Fail: Fannie, Freddie, and the debacle of mortgage finance*, New York: New York University; and D. Harvey (2012) *Rebel Cities: From the right to the city to the urban revolution*, London: Verso.

out' of the processes of production and consumption through the use of technology that forms part of the growth cycle of a hyper-competitive market. For Marx (1973: 538) as noted in the *Grundrisse*, 'the important thing is not the markets distance in space but the speed – the amount of time – by which it can be reached … the velocity of circulation the time by which it is accomplished is a determinate of how often capital can be realised in a given time'.

The combined quest for flexibility and speed leads to calls for reductions to regulatory guidelines and standards that might unnecessarily restrict development and cause delays. The urban land use appetite elevates the use of strategic management techniques and performance indicators in the pursuit of performance excellence, effective outcomes and value for money. These are mechanisms associated with 'corporatist governance that is associated with advancing private sector interests and keeping bureaucratic regulation to a minimum' (England 2004: 25).

Land regulation becomes about promoting rather than restricting entrepreneurial activity and development and seeking 'outcomes that aid market-led prosperity' (Booth, 2003). The significance of speed in this planning reform context is not so much about achieving social and environmental aims (i.e. the necessary adaptation and mitigation to human induced climate change), but rather the accelerated spirit of capitalism. Is planning as Healey and Barrett (1990) ask, opposed, above or about the dynamics of property markets?

Planning systems and reform contexts differ but the neoliberal land development narratives that exist in the Global North are similar and can be crudely schematised as follows:

(over)- regulation / planning = delay = time = money = market (in) efficiencies
(micro-economic) reform = speed = efficiency = supply

The changes in Australian land provision for housing for example have arisen from ongoing state-market planning reform and economic restructuring processes. The twin crisis of chronic housing shortage and poor housing affordability are further stressed by fluctuating national ambitions for a 'Big Australia' (Steele, 2012). The pressure to provide homes for the anticipated (albeit publicly contested) population growth by 2050 has an environmental and social cast as Australian housing systems struggle to respond to the existing challenges of urban change including climate change and the Global Financial Crisis (GFC).

The dominant land reform narrative within the Australian context is one that links planning approval speed with supply as a means of achieving housing affordability and equity. Over the last four decades this has led to calls for greater levels of micro-economic reform in order to streamline housing processes and reduce supply delay. Key technologies deployed for this purpose include an emphasis on 'one-stop-shop' processes designed to speed up and standardise planning and development assessment through increased web-based development applications and accelerated code assessment. The Planning Institute of Australia's website for example highlights these reforms as being 'important in promoting private sector activity in the context of the global recession' (PIA, 2011: 1).

According to the Australian National Housing Supply Council (NHSC) *State of Supply Report* (2010: 4) 'as the population grows supply shortages will put pressure on prices, and low income renters will be particularly adversely affected'. The report goes on to advance the case for significant supply side reform through streamlined development assessment processes (including more as-of-right development and code-based assessment) and removing regulatory barriers to new housing supply. The national ambition is to 'speed up the time it takes to bring new homes on to the market' (Plibersek, 2010: 1).

The political response to these issues has been to emphasise the need to 'close the gap' between housing demand and land release and supply through pressure on planning system reform. A powerful coalition of independent government advisory councils and property and development lobby groups work together to reinforce the meta-narrative that increased speed – achieved through micro-economic reform – is needed to redress 'an undersupply of housing right across the national housing market that is putting substantial pressure on housing affordability' (UDIA, 2011: 1). Key stakeholders include: the Commonwealth of Australia; the Council of Australian Governments (COAG); the Development Assessment Forum (DAF); the Urban Development Institute of Australia (UDIA); and the Property Council of Australia (PCA).

The focus on land and the concomitant need for a land reform agenda was reflected in the 2010 communiqué of the Council of Australian Governments (COAG) meeting. Three key areas for performance measurement highlighted included: system performance, outcome performance and process performance. The poor performance of housing supply in response to demand-side pressures was cited as the key reason for the endorsement of the housing reform agenda.

> COAG today endorsed a housing supply and affordability reform agenda and timeline. The agenda focuses on the primary areas where gains may be achieved in decreasing the time it takes to bring housing to the market, reforming government policies that act as barriers to supply or that artificially stimulate demand and ensuring efficient use of existing housing stock … The cohesive microeconomic reform agenda should substantially improve the responsiveness of supply in the housing market. (COAG 2010a: 15)

Further to this the *Australian National Housing Affordability Agreement* specifically outlined the need for 'planning reform for greater efficiency in the supply of housing' (COAG, 2010b: 7). This is consistent with the COAG (2010c) Competition and Regulation report *A National Partnership Agreement to Deliver a Seamless National Economy* which focuses on a range of institutional measures to improve the speed and efficiency of housing and development approvals through planning and development assessment (DA) reform. The development assessment reforms outlined by COAG mirror those including: the roll out of electronic development assessment (EDA) processing nationally; a system of national performance monitoring; the accelerated use of 'code assessment'; establishment of a set of supporting national planning system principles; and the assessment of benefits accruing from DA reforms.

Of these initiatives two are worth mentioning in more detail: code assessment acceleration and national performance monitoring. The first was the introduction of code assessable development for 'low risk single unit residential developments while noting the significant differences in definitions, processes and timeframes for planning approvals of low risk residential development'. This included a model code assessable template for residential development, with a view to this being expanded to include multi-unit dwellings, commercial and industrial development, and a potential web-based application within an online approval system. The reform intention was streamlining of process as reflected in the national performance measures for development assessment in the *First National Report on Development Assessment Performance: 2008/09* (Australian Government 2010: 45).

This effects the not-for profit sector as well. As outlined in the *Regulation and Growth of the Not-for-Profit Housing Sector* discussion paper (2010) the Australian government calls for reform processes that promote greater 'efficiency in the social housing sector through fostering greater competition between housing providers' (p.18). The pressure is for 'market savvy' not-for-profit operations envisaged in Australia to operate as 'a small number of larger, commercially sophisticated, not-for-profit housing providers operating alongside the existing State and Territory-run housing authorities' (Plibersek, 2010: 2). The construction of housing affordability is inextricably linked to the need to improve the speed and sophistication of land and market supply. Achieving housing affordability is the coupling of growth and supply speed with micro-economic reform. Yet there is little evidence within the Australian context that the last four decades of growth-led reform have improved either housing affordability or equity (Forster, 2010; Stilwell, 2000).

The need for speed grips the capitalist imagination where 'time is money' and increased market growth and efficiency is progress. Tomlinson (2007) argues that acceleration rather than deceleration is the driving leitmotif of late western modernity. In general he criticises the lack of any plausible models for transformative societal change (p.147). To this end he points to the need for 'slow values' (i.e. an emphasis on balance, effort, measure, frugality and proportion) as crucial underpinnings for governance *modus operandi* at all scales in response to the 'aimless trajectory of fast capitalism and the (possibly disastrous) failure of governments locked into high-growth economic policies to confront environmental threats' (Tomlinson, 2007: 154).

The urban land question involves the polarities and tensions of urbanisation and deliberate urban intervention through planning systems and mechanisms. Roweis and Scott (1981) argue that within the context of capitalism the logic linking land ownership and use can never achieve economic efficiencies is the way the market desires: 'that is that the allocation of urban land to competing uses undermine whatever general tendencies may otherwise exist in the direction of efficiency' (p.145). Efficiency in this sense is when it is 'not possible to increase the output of desired goods and services except by decreasing the output of some other desired goods and services thereby creating a spatial pattern of differential locational advantage (e.g. through the modification of existing infrastructure or private redevelopment)' (p.144). As Friedmann (2010: 1) observes in globalised land and property development processes 'speed, movement, and power have been valued more than the fragile social infrastructure of place-based communities'.

Harvey (2012: 87) argues that 'what matters is not the particular mix of institutional urban arrangements, but rather the unified effect of collective political action' that is needed to address the excesses of capitalism and generate much needed urban change. By contrast Healey (2010) draws attention to the importance of the institutional ways in which 'we think about, use and invest in land and property in the societies of which we are a part' (p.4). Her argument is that the dynamic at the heart of the urban land question is best addressed through a fine-grained focus on the layers, contours and contexts of institutional shaping – including the structures of planning provision (see Ball, 1998). For Patsy Healey it is these collective processes that unsettle and change the wider urban context. These activities in turn lay down the necessary seeds for urban change.

Seeds for Urban Change

What then are the viable alternatives for planning that are both sensitive and responsive to the interdependencies between economic development, environmental protection, and social equity? Patsy Healey (2010) observes we live in unstable urban times – fragmented, diffuse and conflicting. On the one hand the rate and scale of urban change is for many like a windy day that never ends (Gleeson, 2006). On the other hand the frustratingly slow transition to low carbon cities in the face of a climate of socio-economic and environmental change creates its own political and practical tensions.

Reflections in the planning literature emphasise the role of planning as intimately linked with the discourses of sustainability and the 'challenges of co-existence in shared spaces' (Healey, 2008: 431). Central to this endeavour are deeply existential questions that go to the core of professional identity and purpose: 'who are the planners?' and 'what do planners do? This in turn has led to more critical questions related to 'planning for what and whom?' and how planning education and practice might be better focused to achieve progressive democratic, environmental and economic aims (Friedmann, 2008).

The idea of 'seeds for change' is about creatively confronting the urban/planning crisis (see White, Sutton, Pears, Mardon, Dick and Crow, 1978). For some the future of western cities is bleak where the 'rich will simply hide out in their castles desperately trying to consume all the good things in their lifetime ... that ruse of reason by which the neoliberal order both acknowledges and dismisses the fact that the current trajectory of human existence is unsustainable' (Davis and Monk, 2007). For Harvey and Wachsmuth (2012: 264) it all comes down to questions around: 'what is to be done and who the hell is going to do it?'

One answer is to envisage urban revolution: reclaiming the city for anti-capitalist struggle in which the left actively engage with but also create urban alternatives through the creation of urban commons. A social practice of *communing* – places and spaces off-limits to market exchange and valuation logics (Harvey, 2012). This builds on the 'right to the city' as outlined by Lefebvre, which focuses on 'the right to participate in urbanity, the right to appropriate the city not merely as an economic unit, but as a home and an expression of lived experience' (Merrifield, 2002: 156).

Lefebvre (1974: 18) refers to this as the 'social (spatial) practices inherent to the forms under consideration'. Our collective political task he suggests is to seek to 'imagine and reconstitute a totally different kind of city out of the disgusting mess of globalizing, urbanizing capital run amok' (see Harvey 2013: xvi). Whilst Harvey (2000: 17) focuses on redefining, 'in a subtle way the terms and space of political struggle in extraordinary times'. Feminist collective Gibson-Graham (2005) emphasise the need to foster 'spaces of hope' in order to unsettle the seemingly hegemonic neoliberal logic through community. As Harvey (2012: 1) observes, 'if you're interested in urban or environmental questions, you're bound to depart a little from fundamentalist Marxists and find yourself in some sort of peculiar space in the middle'.

For Healey the seeds of urban change are there, particularly at the local level, but the complexity of unpacking the interrelation of structuring forces and agency activities surrounding the planning and development nexus within wild cities can be overwhelming. A key element of transformative urban capacity Healey (2006) argues, lies in the quality of local policy cultures: the actors, networks, resources and arenas; how partnerships form; and spatial strategies connect to wider political and policy cultures. In particular she emphasises the power collaborative approaches and the importance of institutional capacity-building including (but not limited to):

- generating new policy discourses about the qualities of places;
- developing collaboration among stakeholders in policy development delivery,;
- widening stakeholder involvement beyond traditional power elites,
- recognising different forms of local knowledge, and
- building rich social networks as a resource of institutional capital through which new initiatives can be taken rapidly and legitimately.

This emphasis on collaborative planning through an institutional approach she argues shifts the task of urban planning from simply being about 'building places', to fostering the capacity in territorial political communities for ongoing 'place-making' activities at the local scale. This is about better understanding how places are produced and change and critically rethinking the complex inter-relationship between planning, land, growth and development.

Conclusion

We live in wild cities. If we are to believe Patsy Healey (2010) part of our role as planning scholars and researchers within the conditions of feral capitalism is threefold: to expand the knowledge base available to our political communities; to enhance the 'intelligence' of democratic debate; and to develop skills in conveying this understanding to the different members of our political communities. (pp.15–16). In her words:

> … perhaps the key message we can offer our polities at the present period is to encourage other actors to use the perceptions of instability and crisis in a strategic way, as an opportunity to take stock, to re-think policies, projects and practices, and to build the intelligence and coalitions which could bring future benefits for the many not just the few in our localities. (Healey, 2010: 15)

Planning wild cities is a transformative planning agenda. Yet whilst informed and infused by revolutionary calls to arms (see for example Harvey), Healey's approach is not the anti-capitalist struggle but one that sees planners playing an important role as activists and public intellectuals by promoting key issues and transforming governance processes. Her socio-political location, grounded in her belief in the progressive potential of the planning project, is driven by the need to better understand for whom, for what purpose, and driven by what reform values planning systems, plans and practices purport to 'perform'.

For urban planners seeking to support sustainability-led change charting the task ahead remains complex, turbulent and unclear. Calmer, cooler in her passion for planning, the Healey legacy offers an almost therapeutic approach to navigating the planning and development nexus of the wild city. Through her lens the planning project becomes an opportunity to focus on how real places and spaces are produced and change, and the role of planning and community within it. For Healey herein lie the collective seeds of change.

References

Abani, C (2004) *Graceland*, New York: Grove Press.

Acharya, V., Richardson, M., Nieuwerburgh S.V. and White, L (2011) *Guaranteed to Fail: Fannie, Freddie, and the debacle of mortgage finance*, New York: New York University.

Australian Government (2010) *Regulation and Growth of the Not-for-profit Housing Sector: Discussion paper*, Canberra: Commonwealth of Australia.

Beauregard, R. (2003) *Voices of Decline: Post-war fate of U.S cities*, New York: Routledge.

Booth, P. (2003) *Planning by Consent: The origins and nature of British development control*, London: Routledge.

Boyer, C. (1986) *Dreaming the Rational City: The myth of American city planning*, Cambridge, MA: MIT.

Campbell, S. and Fainstein, S. (2005) *Readings in Planning Theory*, Oxford: Blackwell Publishers.

Carmona, M., Carmona, S. and Gallant, N. (2003) *Delivering New Homes: Processes, planners and providers*, London: Routledge.

Carmona, M and Sieh, L, (2004) *Measuring Quality in Planning: Managing the performance process*, Oxford: Spon Press.

Castells, M (1975) The wild city: an interpretation of research and analysis on the US urban crisis, *Kapitalistate*, 4(5): 1–30.

Castells, M. (1996) *The Rise of the Networked Society*, Oxford: Blackwell Publishing.

Citigroup research (2006) *Revisiting Plutonomy: The rich getting richer*, 5 March, 3 [???].

COAG (2010a) *Council of Australian Governments Meeting Communique*, 19–20 April, Canberra.

COAG (2010b) *National Affordable Housing Agreement*, Council of Australian Governments, Canberra.

COAG (2010c) *National Partnership Agreement to Deliver a Seamless National Economy*, Report to the Council of Australian Governments, Canberra.

Davis, M (2006) *Planet of Slums*, London: Verso.

Davis, M and Monk, D (2007) *Evil Paradises: Dream worlds of neoliberalism*, London: The New Press.

Dear, M and Scott, A (1981) *Urbanization and Urban Planning in Capitalist Society*, New York: Methuen Inc.

Eisenschitz, A. (2008) Town planning, planning theory and social reform, *International Planning Studies*, 13(2): 133–49.

Elliot, D. (2008) *A Better Way to Zone: Ten principles to create more liveable cities*, Washington: Island Press.

England, P. (2004) *Integrated Planning in Queensland*, Annandale: The Federation Press.

Ferriss, H. (1922) The new architecture, *The New York Times Magazine*, 19 March, pp. 8–9.

Fischler, R. (1995) Strategy and history in professional practice. In H. Liggett and D. Perry (eds) *Spatial Practices*, Thousand Oaks: Sage, pp.13–58.

Fischler R. (1998) Towards a genealogy of planning: Zoning and the welfare state, *Planning Perspectives*, 13: 389–410.

Forster, C (2010) *Australian Cities: Continuity and change*, Melbourne: Oxford University Press.

Freestone, R. and Grubb, M. (1998) The Melbourne Metropolitan Town Planning Commission 1922–1930, *Journal of Australian Studies*, 22(57): 128–44.

Friedmann, J. (2008) The uses of planning theory: A bibliographic essay, *Journal of Planning Education and Research*, 28: 47–257.

Friedmann, J (2010) Place and place-making in cities: A global perspective, *Planning theory and Practice*, 11(2): 149–65.

Gibson-Graham, J.K. (2006) *The End of Capitalism (As We Knew It): A feminist critique of political economy with new introduction*, Minneapolis: University of Minnesota Press.

Glaeser, E. (2011) *Triumph of the City*, New York: Penguin Books.

Gleeson, B (2006) *Australian Heartlands: Making space for hope in the suburbs*, Sydney: Allen Unwin.

Hall, P. (2002) *Cities of Tomorrow: An intellectual history of urban planning and design in the twentieth century*, Malden, MA: Blackwell.

Harvey, D. (1996) *Justice, Nature and the Geography of Difference*, Oxford: Blackwell.

Harvey, D (2000) *Spaces of Hope*, Los Angeles: University of California Press.

Harvey, D (2012) *Rebel Cities: From the right to the city to the urban revolution*, London: Verso.

Harvey, D and Wachsmuth, D (2012) What is to be done? And, who the hell is going to do it? In Brenner, N., Marcuse, P and Mayer, M (eds) *Cities for People, not for Profit*, London: Routledge. pp. 264–74.

Harvey, D (2014) Interview with David Harvey with Mahon, M, *The White Review*, accessed online 7/5/2014 at http://www.thewhitereview.org/interviews/interview-with-david-harvey/

Healey, P. and Barrett, S.M. (1990) Structure and agency in land and property development processes, *Urban Studies* 27(1): 89–104.

Healey, P. (1992) Development plans and markets, *Planning Practice and Research*, 7(2): 13–20.

Healey, P. (2003) Collaborative planning in perspective, *Planning Theory*, 2(2): 101–23.

Healey, P (2007) *Urban Complexity and Spatial Strategies: Towards a relational planning for our times*, New York: Routledge.

Healey, P. (2008) 'Editorial', *Planning Theory and Practice*, 9(4): 431–4.
Healey, P. (2010) Re-thinking the relations between planning, state and market in unstable times. In P. Pinho and V. Oliveira (eds) *Planning in Times of Uncertainty*, Porto: FEUP edicoes, pp. 11–22.
Hillier, J. (2007) *Stretching beyond the Horizon: A multiplanar theory of spatial planning and governance*, London: Ashgate.
Kenny, M. and Meadowcroft, J. (1997) *Planning Sustainability*, London: Routledge.
Lash, S. and Urry, J. (1994) *Economies of Signs and Space*, London: Sage Publications.
Marx, K. (1973) *Grundisse*, London: Penguin.
Merrifield, A (2002) *Dialectical Urbanism*, New York: Monthly Review Press.
Mitchell, T (2007) Dreamland. In M. Davis and D. Monk (eds) *Evil Paradises: Dreamworlds of neoliberalism*, London: The New Press, pp. 1–34.
NHSC (2010) *National Housing Supply Council: 2nd State of Supply Report*, Canberra: Commonwealth of Australia.
Okome, O (2002), Writing the Anxious City. In Okwui Enwezor et al. (eds), *Under Siege: Four African Cities – Freetown Johannesburg*, Kinshasa: Ostfildern-Ruit.
Plibersek, T. (2010) Forward, in Australian Government (2010) *Regulation and Growth of the Not-for-profit Housing Sector: Discussion paper*, Canberra: Commonwealth of Australia, pp. 2.
Rivolin, U. (2008) Conforming and performing planning systems in Europe: An unbearable cohabitation, *Planning Practice and Research*, 23: 167–86.
Roweis, S and Scot, A (1981) The land question. In M. Dear and A. Scott (eds) *Urbanization and Urban Planning in Capitalist Society*, New York: Methuen, pp. 123–59.
Sandercock, L. (1990) *Property, Politics and Urban Planning: A history of Australian city planning 1890–1990*, New Jersey: Transaction Publishers.
Spigelman, J. (2001) Economic rationalism and the law, *University of New South Wales Law Journal*, 30 June, p. 19.
Stair, P., Wooten, H. and Raimi, M. (2008) *How to Create and Implement Healthy General Plans: A toolkit for building healthy, vibrant communities through land use policy change*, Los Angeles: Public Health Institute.
Steele, W (2011) Strategy-making for sustainability: An institutional learning approach to transformative planning practice, *Planning theory and Practice*, 12(2): 205–23.
Steele, W (2012) Do we need a slow housing movement?, *Housing, Theory and Society*, 29(2): 172–89.
Steele, W and Ruming, K. (2012), Flexibility versus certainty: Unsettling the planning shibboleth in Australia, *Planning Practice and Research*, 27(2): 155–76.
Stilwell, F. (2000) *Changing Track: A new political economic direction for Australia*, Annandale: Pluto Press.
Sulman, J. (1921) *An Introduction to Town Planning in Australia*, Observatory Hill: National Trust of Australia.
Suzuki, D (2010) *The Legacy: An elders vision for a sustainable future*, Vancouver: Greystone Books.
Tomlinson, J. (2007) *The Culture of Speed: The coming of immediacy*, London: Sage Publications.
UDIA (2011) *National Land Supply Report Issues Damning Verdict on Housing Supply*, media release, 9 February.
White, D., Sutton, P., Pears, A., Mardon, C., Dick, J., and Crow, M. (1978) *Seeds for Change*, Melbourne: Patchwork Press.

Chapter 11

The Changing Nature of Property Investment: Implications for Urban Planning

Paul McNamara

Introduction

In her groundbreaking paper with Susan Barrett, Patsy Healey outlined the value of using Giddensian structure-agency theory to better understand how the built environment is produced, consumed and managed (Healey and Barrett, 1990, Giddens 1984). Incorporating aspects of phenomenology, she later used structuration theory to great effect to explore how urban governance structures and institutions evolve (Healey 2007).

Rejecting the primacy of either deterministic ('top-down') or voluntaristic ('bottom-up') societal drivers, Healey and Barrett (1990) recommended analysts to link the economic, political, legal and ideological structures that frame the decisions of property market actors with their individualistic interests, strategies and actions, to reveal how structures both constrain and empower actors and how actors, through their actions, evolve those structures. Healey and Barratt further promoted the application of this approach to understanding better how the 'development industry' evolves, the 'ideas and ideologies' its actors assert when designing and justifying strategies and the 'rules' and 'resources' they operate with when defining and implementing their strategies. They concluded by outlining four specific areas for further analysis, namely:

- the relationship between the financial system and the property market, including what sources of capital are involved and the various ways in which it is invested in property;
- the changing composition and strategies of firms involved in the development and redevelopment of the built environment;
- the way in which the state impinges upon these changing practices and the tools it uses to intervene in and structure land and property development processes; and
- the implications of the above processes for local economies, including capital flows, land and property values, labour and land markets.

This chapter addresses the first two of these challenges in particular and comments about their implications for the practice of urban planning going forward. In particular, it explores why different investors invest in property, how the property industry[1] has collectively changed the nature, theory and practice of 'property' investment and how all of the above has changed the constitution of the property investment community with which urban planners interact. The chapter concludes by assessing whether these changes have implications for those setting urban strategies today.

However, before commencing our review of how the UK commercial property market[2] has evolved since Healey and Barrett first posed their challenges, it is helpful to remind ourselves of how it has performed over recent decades.

1 Development is simply one aspect of property investment. As such, I prefer to use 'property industry' as a more inclusive term than 'development industry'.

2 For reasons of space and the limits of the author's experience, the focus here is perforce restricted to a review of UK commercial real estate.

Changes in the Nature and Practice of Property Investment since Healey and Barrett (1990)

Fourt and Matysiak (2013) divide the period since 1990 into four stages, namely,

1. 1990–1992 Adverse/bear market conditions;
2. 1993–2003 Steady market conditions;
3. 2004–2006 Favourable/bull market conditions; and
4. 2007–2009 Adverse/bear market conditions.

Since 2009, there has been a modest recovery in property values followed in late 2013 by a reacceleration of investor interest.

From the above we can see that Healey and Barrett published their 1990 paper just as the late 1980s UK property development bubble was bursting. Limited new property supply through the early Thatcher years, a consumer-led economic expansion driven by low interest rates and reduced taxation and a loosening of planning and financial regulations (especially in the City of London) had together laid the foundations for a surge in development. Inefficient 'endogenous transmission mechanisms' within the market (Barras, 2009) then conspired to exaggerate the true levels of demand for property and developers, using easy credit from eager banks, materially oversupplied the market. By the end of the cycle, 25 lending banks had stopped trading (Benford and Burrows, 2012).

Following the subsequent price correction and related reduction in development, property prices slowly recovered with investors slowly returning to the market through the remainder of the 1990s.

In the early 2000s, conceived alternatively as chronic overproduction in developing economies or chronic overconsumption in western economies, a global savings imbalance burgeoned (Davies, 2010). With its focus on western sovereign bonds, this drove real interest rates down leading investors to seek higher yielding, bond-like, assets to better maintain prior levels of achieved return. At the same time, responding to the economic fall-out of the 9/11 attack in New York and the bursting of the 'dot.com bubble', western central banks sought to underpin economic growth and, in the USA at least, increase homeownership through maintaining aggressively low interest-rates.

Perceived as a relatively 'safe' asset providing a decent income return, capital poured into commercial property around the globe and prices rose sharply. With its long leases and trusted property market infrastructure, the UK was a favoured market. With borrowing rates below prevailing income yields, UK property investment became self-financing and, by gearing up with the support of a new generation of eager bankers, investors stretched what equity they had across multiple deals in pursuit of even greater profits. As prices spiralled, property appeared (irrationally) as a high returning but low risk investment, attracting in yet more capital. Meanwhile, in the US but also increasingly in Europe, banks were using new financial instruments[3] to package mortgage loans into portfolios for onward sale to others. In so doing, they profited while passing risk onto others.

Without material economic or rental growth, this pricing bubble finally burst in the UK in 2007. The capital value of the UK commercial market, having grown by almost 50 per cent to £867bn in the previous three years, peaked in the second half of 2006 (see Mitchell, 2014: 6). In 2007, property prices plunged and, as investors scrambled to liquidate and offload assets, market confidence evaporated and the banks stopped lending. Swathes of existing bank loans became 'non-performing' once again requiring major governmental intervention to rescue the banks. By mid 2009, the market had lost nearly one-third of its 2006 peak value. Mercifully, the 'bust' occurred just as investors, feeling increasingly starved of prime investments, had begun to develop their own. Other than in central London, new development had proved limited.

As in the early 1990s, UK commercial property prices then fell too far, triggering a small bounce-back in 2010. However, with tenant demand for property undermined by the impact of the 'credit crunch' on the real economy, capital value growth proved anaemic until mid 2013 when investor interest began to pick up again.

A cynic might say that this seemingly endless cyclical pulse of property markets, which was just as observable in the UK in the early 1970s as in the period described above, implies that nothing much changes

3 For example Commercial and Residential Mortgage Backed Securities, (CMBS and RMBS).

in the UK commercial property market and planners should develop mechanisms to deal with the excesses and vacuums it creates. However, though clearly powerless to dampen the most recent cycle, significant changes have been occurring in the UK commercial property market since Healey and Barrett (1990) that are of relevance to those managing urban spatial strategies today. Not least of these are changes in the structure of property ownership. As we shall see, the constitution of the property investment community, the way investors think about property investment and the motives and drivers of property investors are all markedly different now to those in 1990, all of which could have relevance to the way urban strategists think about their work.

The Significance of Property to Investors

Surveys of institutional investors (see PMI and PRUPIM 2008) regularly identify the main attractions of property as its relatively high and steady level of income return and, by providing total returns (capital changes plus income returns) unsynchronised with those from other investment types, its ability to smooth and de-risk the overall flow of returns received by investors over time.

However, to understand how and why the nature and practice of property investment and the structure of the property industry changes over time, we need to go further than this and set property investment in its socio-economic context.

Under a capitalist mode of production, property rights are mostly commoditised and transferrable, with the vast majority necessarily being held by some person or entity. The reasons for holding such rights vary and include enjoying the utility that a property provides for production, consumption or exchange or earning an income from leasing a property to a third party to do so. With respect to the latter, the pattern of cash flows delivered to landlords depends crucially on the nature of the leases negotiated. These leases are socially-determined, hammered out both through individual negotiations and through factional struggle between landlords and tenants, with the relative strength of the factions varying over the economic and property cycle.

Different types of investors have different investment aims and the nature of leases and their related cash flows largely determine how attractive property investment is to them. A long-term income growing with both the real economy and inflation, secured on a long lease to a creditworthy tenant has obvious attractions for insurance companies and pension funds who take monies regularly from their clients in exchange for promising payouts long term. However, should the nature of leases or the aims of investors change, then property's relevance might also change and, as a result, the universe of investors might change.

Crudely, a property's value is equivalent to the income received divided by a 'capitalisation rate' expressed as a percentage. So, a property generating £10 of income, with a capitalisation rate of 10 per cent is valued at £(10/0.10) or £100. The capitalisation rate employed to do this reflects

- the rate of return for a risk-free investment (usually a government bond which forms a basic reward to investors for locking capital up for a future period);
- an additional rate of return to compensate for the risks of investing in properties where unforeseen things can happen, and
- the perception of how investment income will grow.[4]

Over and above the effects of generalised economic growth, most long-term landlords seek to increase value by reducing a property's risk by renegotiating a longer lease with a tenant, finding a better tenant or increasing rental income through changing its use, refurbishing or redeveloping it.

Capitalisation rates should reflect medium term rather than current expectations. However, short-term perceptions of risk premium and rental growth fluctuate markedly through property cycles. Investors get more confident in an upswing, perceiving greater rental growth and fewer risks. This drives capitalisation rates lower and 'value' estimates higher. This can attract in short-term investors seeking to 'buy low, sell high'. Naturally, the opposite is true in a downswing; risk premia rise, rental growth expectations fall, capitalisation rates rise and values fall. As such, the aim for short-term investors is timing a profitable exit from the market.

4 At a gross level, rental growth is driven by real economic growth and inflation. However, investors receive 'net' rental growth after expenditures to maintain the physical and functional attractiveness of their properties to tenants.

Hence, amongst other things, the nature of property as an investment, how its returns meet investor aims and the stage in the market cycle all affect the universe of property investors at any given time. Such changes are capable of description using concepts of structure and agency. As we shall see, wider societal forces of the type described by Giddens are as much at work in the property investment industry as any other walk of life.

The Changing Universe of Property Investors

Probably the biggest change that Healey and Barrett would notice in the structural composition of investors in UK commercial property since 1990 would be the diminished importance of domestic long-term financial institutions. Alongside UK-based property companies, these institutions dominated the investable UK commercial property market at that time.

Following Massey and Catalano's (1978) analysis of UK landownership patterns, McNamara (1985) revealed how these investors had come to own 40 per cent of central business district frontages in Edinburgh in twenty year period to 1978.

Key and Law (2005) estimated the value of the UK commercial property market in 2003 at £611bn. The 'investment market' constituted 40 per cent of this total[5] and included 60 per cent of the UK's office and retail stock and a quarter of UK industrial stock. At that time a third was owned directly[6] by UK insurance companies and pension funds with overseas investors, listed property companies and private property companies each owning a further seventh.

Through a combination of, amongst other things, significantly increased overseas investment opportunities and the slow demise of 'with-profits' insurance policies for which property assets were a useful liability match, the proportion of long-term insurance company capital invested in direct property has slowly dwindled from 20 per cent in the early 1980s to just 3 per cent now.[7] Similarly, as pension funds have 'matured' and increasingly focused on the need to pay out to pensioners, they have edged away from 'risk-bearing assets' like property and shares towards more predictable investments like corporate and sovereign bonds. Their allocations to direct property have fallen from 12 per cent to 2 per cent.[8]

A focus on the last 10 years shows more rapid and dramatic changes in the ownership structure of UK commercial property, in particular the growth in overseas ownership and collective property investment vehicles of varying types.

Mitchell (2014) estimates the 2013 value of the UK commercial property market at £647bn, having peaked in 2006 at £867bn and plunged to £587bn in mid-2009. Through the absorption into the investment market of historically owner-occupied property types, such as supermarkets and hotels, he records the percentage of overall stock now considered 'invested' as having grown to 56 per cent.

With respect to the evolving structure of ownership, Mitchell (2014) shows in Table 11.1 that overseas investors are now the largest single property investor type in the UK by value. However, we should note that this is largely a central London based phenomenon where, as Mitchell notes, they now own over half the total, high-value, stock of the City of London (see also, Lizieri et al. 2011).

The other major new investor type is the collective investment scheme. Frodsham (2012) notes that, fuelled by debt, capital inflow and price inflation, unlisted funds grew from £6bn to £48bn in the six years to 2007 then falling back through devaluations and sales. However, as Table 11.1 shows, they are now the second largest investor type in the UK.

Mitchell shows long-term insurance company holdings continuing to fall as asset allocations to property halved over the period. However, we should note that insurance company based fund managers, capitalising

5 Following a definitional review and new information, these figures have been revised by Mitchell (2014) to £580bn and 50 per cent invested.

6 'Direct' property refers to the actual ownership of buildings. As we shall see, many pension funds and insurers have been replacing their direct investments by investing indirectly through units in collective investment vehicles.

7 http://www.ons.gov.uk/ons/publications/re-reference-tables.html?edition=tcm%3A77-327001

8 These figures fail to capture that, given recent perceptions of shares as risky and bonds as poorly returning, some pension funds have been indirectly reinvesting in property through collective investment vehicles. (See NAPF 2013 for Defined Benefit schemes and Harrison et al. 2013 for Defined Contribution schemes.)

Table 11.1 Changes in the UK commercial property investment universe, by investor type

Investor type	Per cent of investment market owned in 2003 (%)*	Per cent of investment market owned in 2013(%)	Implied net change in investment*
UK insurance companies	20	11	–£17bn
UK segregated pension funds	11	8	0
UK and Channel island collective investment schemes	10	16	+£32bn
UK REITs and listed property companies	14	14	+12bn
UK private property companies	18	14	0
UK traditional estates and charities	5	4	+£2bn
UK private investors	3	3	+£2bn
UK other	5	5	+£3bn
UK SUB TOTAL	*85*	*76*	
Overseas investors**	15	24	+£47bn
TOTAL	*100*	*100*	

* These figures are estimated from Mitchell (2014) Table 4.1. Changes in investment include both investment and disinvestment but also include valuation changes.
** All those domiciled outside the UK, excluding foreign owned fund managers, insurance companies and pension funds investing UK sourced capital
Source (Mitchell, 2014)

on their in-house investment expertise, have been central to the growth in collective property investment vehicles (Mitchell 2014: 12).

Pension funds have continued to switch how they invest in property, moving away from direct property to indirect investment through collective vehicles. As such, their direct property holdings have fallen slightly in absolute terms over the past decade but, as Mitchell points out, if their indirect exposures are properly accounted for their beneficial interest in UK property has almost doubled over the last decade.

The above shows that the UK commercial property market is now significantly more internationalised and unitised than in 1990 leaving many planners with a new cast of actors with potentially different aims and objectives to work with. However, while market level statistics can capture the changing composition of the property investment community they say nothing about how these various agencies operate and think, and how they interact with the wider investment community and government bodies.

In the following sections we review how investment thinking in the property industry has evolved since 1990 and how its relationships with government and the wider institutional investment industry have changed.

Property within the Investment Industry: Pressures for Change[9]

When Healey and Barrett were writing, property investment was both part and apart from mainstream institutional investment. Even before the early 1990s property crash, pressure was building inside and outside the property industry to make property a more transparent, 'modern', investment class. One outcome of this pressure was the establishment of a new and, subsequently, influential property industry organisation.

9 What follows is necessarily a personalised account from a senior researcher working in the same major investment institution for 25 years.

At a meeting in 1986 to discuss the novel securitisation of a major property investment by a bank rather than any chartered surveying practice, an influential and respected Financial Times journalist made clear to senior property professionals present that 'investment surveyors were not only relatively ignorant of property's place in the wider capital markets but were also held in "less than high regard" in the City' (Wyatt and White 2013: 45). The frankness of this significant message ultimately led those present to establish an off-shoot from the Royal Institution of Chartered Surveyors (RICS). Initially called the 'Investment Surveyors Forum' (ISF),[10] this group had the explicit aim of improving the understanding of property as an investment type, both inside and outside the property industry.

Simultaneously, and with much the same drivers, the property industry saw the launch of an transformative new property investment performance measurement service (Investment Property Databank), major new property research capabilities established by financial institutions,[11] two new property research journals, and the founding of a Society of Property Researchers (SPR).

By 1990, many major financial institutions had already converted their internal investment departments into semi-autonomous fund managers to manage both internal and, for a fee, external capital. This greater external focus increased the need to achieve observably good property investment performance, both in absolute terms and relative to those provided by competitor organisations. Thus, in April 1988, the highly influential Chief Executive of fund manager Prudential Portfolio Managers, Mick Newmarch, lectured the newly formed ISF on the need for property to become more like other investment classes, insisting its professionals 'spend more … time thinking about investment in property, and less … about buildings' and 'approach a property as an investment analyst approaches a security'. He went on further to demand 'to be spoken to in (his) language … the language used in other markets', concluding that property investment needed to modernise to be confident of further (institutional) support.

Newmarch's Chief Surveyor at Prudential and first president of the ISF, Michael Mallinson, followed this with a speech to the ISF insisting that '(Investment surveyors) must be able to express property in an ordered and disciplined manner within an investment framework … and apply this knowledge in a rigorous, intellectually valid and comprehensible manner to the holding of real estate as a financial investment' (Mallinson, 1989). Mallinson (2013:122) subsequently wrote that, when the ISF was formed 'the (property investment) profession was thought to be in the Dark Ages'.

However, the discomfort senior multi-asset investment managers felt about property in the late 1980s was not confined to the process of property investment. It also related to the very nature of property investments themselves which were seen as 'lumpy', heterogeneous and illiquid, making property investment unduly risky even for the biggest of investors. The risk reducing benefits of adding property to multi-asset investment portfolios was well appreciated but there was a growing belief that property would cement itself better into the broader investment universe if it adopted the investment language and techniques of other asset classes and found ways to facilitate smaller lot size, standardised and liquid means of investing. In the following section, I review how the property industry subsequently responded to these challenges.

Innovation within the Property Market: Developing Investment Concepts and Techniques

The drive towards more 'disciplined' property investment is epitomised by the IPF, a major educational and research funding organisation with a mission 'to enhance the understanding and efficiency of property as an investment'.[12] Through extensive education and research programmes, IPF has been central in changing how property investors think since Healey and Barrett were writing. With SPR, IPF has transformed the language and technique of property investment and how property investors think about what they do (see Table 11.2). This has occurred primarily through importing and customising investment approaches used in other asset classes, underpinned by the establishment of important new market databases (McNamara 2013).

10 Later renamed the Investment Property Forum (IPF)

11 For example, Prudential Assurance Company established a team of 10 property researchers in 1987.

12 See the 'About the IPF' page on the IPF website at www.ipf.org.uk

Table 11.2 Ideas development in UK commercial real estate since 1990

Driver	Intellectual development
Improved investment decision-making at asset level	Better evaluations of asset 'worth' through better rental forecasting, improved understanding of asset depreciation and better identification of the sources and impacts of risks on asset value
Improved investment decision-making at portfolio level	Improving overall risk-adjusted returns by assembling portfolios of assets more 'scientifically'. Importing and customising portfolio risk management techniques from other asset classes
Improved understanding of property as an investment	Examples include understanding how property returns correlate with those from other asset classes under different economic circumstances, the causes of property illiquidity, and what proportion of multi-asset portfolios should be allocated to property. Assessing potentially new property markets and new forms of property investing
Improved understanding of the changing context for property investment	Examples include understanding changes in the sources of capital for the market, changing regulatory regimes, and the significance of green issues

However, the new found sophistication articulated in Table 11.2 ultimately proved powerless to prevent the most recent property investment cycle. Many long-term investors may have sold profitably as the bubble inflated but, in the main, they were left floundering in the wake of the £270bn debt-capital unleashed into the market by lending banks (Benfield and Burrows, 2013). This raises the question about whether there were parallel theoretical and practical developments occurring in property banking over this same period?

In contrast to the 'equity' side of the property investment industry, the practices of property banking remain opaque to outsiders (including many on the 'equity' side of the business). Data on property lending is very limited[13] and there are very few textbooks, no dedicated academic journals and only a few vaguely relevant university courses on the topic. Property banking seems 'learned on the job'. This makes it hard to establish how thinking in property banking has evolved since 1990.

There has been an Association of Property Lenders (APL)[14] since 1991. Its current aims are: 'to promote the interests of property lenders, foster a better understanding of property finance within the banking and property sectors, provide a recognised body as a forum for discussions between lenders and other property industry institutions, keep … members up to date with events in the market and provide relevant networking opportunities'.[15]

Primarily a networking organisation before 2009, it is noteworthy that the APL has become noticeably more active since the global financial crisis organising seminars, newsletters and working with other trade bodies on a variety of bank-related industry fora (such as the Bank of England Commercial Property Forum and the Property Industry Alliance Debt Group). In late 2010, with IPF's help, it created a Real Estate Lending Forum (RELF) to 'carry out and disseminate research and inform those involved in real estate finance and investment about issues affecting property finance'.[16] RELF's first research project in 2011 estimates the likely volume of property loans for the year ahead. This has now become an annual survey.

13 Bill Maxted, of De Montford University produces a UK Commercial Property Lending Report which, by virtue of its uniqueness, has become an established feature of the market since modest beginnings in 1997.

14 Until recently, called the Association of Property Bankers (APB), the APL currently has 300+ individual members from 60 UK and overseas organisations, http://www.the-apl.com/about-us.html

15 See the Association of Property Lenders' website at www.the-apl.com

16 See https://www.ipf.org.uk/home/vision_for_real_estate_finance/special_interest_groups/real_estate_lending_forum/default.aspx_

A more recent industry development in property banking is the establishment of a 'Real Estate Finance Group' (REFG) comprising leading property industry figures and senior property bankers. In October 2013, it published '*A Vision for Real Estate Finance in the UK*' (REFG and IPF, 2013) which, amongst other things, called for a mandatory database of all UK property loans, real-time risk analysis of loan books, compulsory accreditation for lending team leaders, pro-cyclical mechanisms to damp credit growth as property markets rise, and a greater diversification of lenders beyond the traditionally dominant banks. With its aim to develop a market and regulatory regime to help smooth future market cycles, REFG's recommendations throw a harsh spotlight on what the property banking sector appears to have been missing since Healey and Barrett were writing, and before.

Somewhat akin to soul searching amongst institutional investors following the 1990s crash, one senses a new, post-crisis, response amongst property bankers to the political, social and investor opprobrium expressed towards it following the most recent property crash. As with the institutions in the early 1990s, the current response seems focused on materially improving investment education and market data. We wait to see whether intellectual and practical progress is made but, without it, there is no guarantee that the amplitude of future property cycles and their related catastrophic effects on the UK economy and cities can be dampened.

Evolving Property as an Investment Asset

Given the criticisms that commercial property was 'lumpy', heterogeneous, illiquid and, therefore, potentially unattractive to many investors, the property industry has devoted substantial energy in recent decades to innovating (not always successfully) ways to make property more widely investable in smaller, more homogeneous, tradable chunks. Figure 11.1 presents an 'evolutionary tree' of property investment types, much of which has occurred since Healey and Barrett were writing.

The second intellectual challenge pursued by property investors has been to find ways to 'optimise' their investment portfolios to ensure overall returns are being maximised given the risks taken. While a portfolio's return is the weighted average of its constituent asset returns, portfolio risk is not similarly determined and depends largely on the correlations over time between individual asset returns. Despite regional and metropolitan property market returns being strongly correlated in the UK and individual properties moderately so, property investors have learned to reduce portfolio risk by spreading their investments across markets and by acquiring more assets.[17] This encourages investors to gain exposure to a greater range of less than perfectly positively correlated local markets. The challenge for investors, especially small scale investors, was then how to operationalise this when individual property investments are so 'lumpy'?

The securitised and unitised investment routes shown in Figure 11.1 clearly ease this problem. While investing in such vehicles has marginal impact on returns, it increases exposure to more markets and properties and, thereby, improves risk-adjusted returns. Investing in 'funds of funds' extends this effect even further.

As we have noted, property fund managers moved quickly to service this newly realised requirement and the numbers of collective investment vehicles burgeoned with sector- and location-based funds as well as funds allowing investors to preselect preferred levels of risk and return. Small investors can now assemble more optimised portfolios appropriate to their specific needs. Clearly, the above problem is even more difficult for individual savers and, after much campaigning by the property industry for regulators to allow them, fund managers have moved swiftly to open investment funds to so-called 'retail investors'.

Since international property markets tend to be less correlated than domestic regional markets, international property investment has also grown significantly since 1990 with British investors increasingly venturing overseas and a wide range of international investors attracted to British property. Generally deemed 'safer' by investors venturing into unfamiliar markets, such investment initially favours 'world cities'. Thus, as noted

17 While the underlying returns volatility (a definition of risk) in property markets cannot be counteracted by investors, judicious selection of assets can reduce risk specific to the assets in a portfolio. A simple portfolio of two equally-sized assets with equal but perfectly negatively correlated returns will remove all asset-specific risk other than that pertaining to the market. Two similar but perfectly positively correlated assets will not reduce risk. Because most assets are not perfectly positively correlated, every additional asset added to a portfolios marginally reduces risk until, eventually, the returns volatility in the portfolio approaches that of the market.

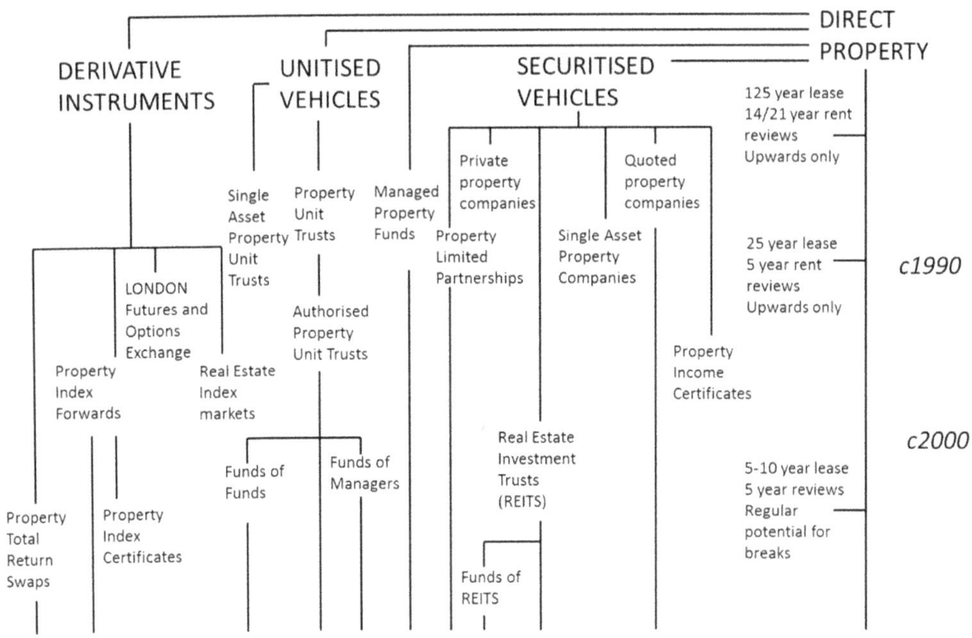

[1] Note that these various means to access property returns relate mainly to superstructures of rights sitting above actual properties. The properties underpinning these funds are, for the most part, still owned and managed by a specific entity. The exception is 'property derivatives' where no properties are involved.

Figure 11.1 The evolutionary tree of UK property investment

already, London remains the overwhelming focus for foreign investment into the UK. However, as experience develops, international investors increasingly push into regional markets.[18]

Finally, we should note that important changes have also been wrought in direct property investments since 1990. If occupiers and owners mutually enjoy the economic benefits bestowed by a property's location, their interests are at odds with respect to what they want from the leases they negotiate.

When Healey and Barrett were writing, UK leases were typically for 25 years, with rents reviewed every five years on an upwards-only basis. These arrangements developed in post-war Britain as investors fought hard to protect their real returns from the damaging impacts of inflation on their rental income, by reducing historically very long lease and review periods, allowing them to more quickly capture any rental uplifts occurring. However, tenant business dynamics have also changed and such leases have increasingly been criticised as unfair and inappropriate for modern business practice. Powerful retail and office tenants have been resisting such terms and their representative groups have lobbied government in recent decades for action to force greater lease flexibility. This has especially been so in market downturns when tenants are at their strongest, and has resulted in marked changes. BPF/IPD (2012) record that lease lengths have now stabilised at around 5 years on existing buildings and, of interest to UK planners given it affects the medium term economics of development, average lease lengths on new properties have now fallen to 10 years.

18 The arrival of substantial levels of non-domestic investment into 'safe haven' London property markets inflated prices. This led many UK investors to sell down and focus on other locations and land use sectors.

Links between Property Investors and Government

Having reviewed how theory and practice in UK commercial property have evolved, I now turn to review how the property industry's relationship with government has evolved in recent decades.

As we shall see, property investors exhibit a variety of interests and will pursue them actively at both central and local government level. Perceived threats and opportunities result in the investors engaging government, both individually and collectively through their various trade bodies. This is especially so at central government level which is where I begin my review.

Property Investors' Involvement with Central Government Level

Many issues unite the property investment industry. These do include land use planning but have more typically related to taxation, lease terms and, more recently, investment management regulation and sustainability. Recent notable property industry campaigns have included removing 'double-taxation' from property company share ownership, combating threats to upward-only rent reviews and other leasing provisions, and engaging UK and European governments about the level of capital reserves financial institutions must hold to cover the effect of pricing volatility in property markets.

The UK has multiple property industry trade bodies representing professional, functional, sectoral and economically-based interests. Each interacts separately with government in pursuit of its interests which sometimes include aspects of planning policy. However, following reproach in 2001 from a senior government Minister[19] they have also established a supra-trade body, the Property Industry Alliance (PIA, founded 2006) to coordinate their views and speak to government with one voice.[20]

These various trade bodies are not defined by interaction with government. Rather, they operate around broad commonalities of interest such as professional and business standards, education and research.[21] For example, the IPF focuses on the 'efficient operation' of the property market, whilst the Association of Real Estate Funds (AREF, founded 2002) focuses on standards of behaviour and transparency amongst investors, fund managers and investment advisers.

However, there is one trade body that is clearly defined economically and which consistently engages with central government on land-use planning issues, the British Property Federation (BPF, founded 1963). With its stated mission to represent the interests of real estate owners and investors and create conditions in which the real estate industry can thrive, BPF's website[22] lists three objectives, namely,

- to raise the profile of the real estate industry with political stakeholders, the media, and the public;
- to improve legislative, fiscal and regulatory conditions that affect (the real estate) industry and so enhance the benefits (it) can bring to the UK; and
- to encourage best practice within the industry as a means of increasing long-term value and improving stakeholder perception.

It also believes that, since its membership includes some of the biggest real estate investors in the UK, it can provide needed expertise to legislators and regulators in a range of policy areas, including land use planning.

19 At an industry conference in 2001, Peter Mandelson (McKee, 2001) told the property industry that to be heard by government it must demonstrate its economic value and speak with one voice.

20 For many years the Department of the Environment had a Property Advisory Group advising it on trends in commercial property and the implications of planned policy developments on the property market. This group of senior property professionals operating in a personal capacity was superseded by a Property Consultative Group where officials and ministers met a roundtable of various property trade bodies. Nominally at least, there is still a group of senior property industry figures who meet with civil servants to discuss trends in the commercial property market but the meetings are occasional, informal and ministers do not currently attend. The Bank of England has had a 'Commercial Property Forum' since 1993. (See http://www.bankofengland.co.uk/publications/Pages/speeches/2013/701.aspx)

21 Examples include setting industry standards for reporting investment performance and providing information on client charges, etc.

22 http://www.bpf.org.uk/en/about_us/who_we_are.php

BPF's most comprehensive statement on investors' interests in urban planning is laid out in its *Planning Manifesto: Making Planning Work* (BPF, 2008). In it, BPF recognises freely that planning safeguards the quality of life for individuals and communities and stands as one of government's most important post-war achievements. However, it also expresses dismay that planning has become a bureaucratic and peripheral process rather than the creative force it should be. While expressing support for planning and the certainties it brings, BPF is unequivocal in its desire for major planning applications to be determined quickly and expertly. It argues for more planners to be freed from dealing with minor planning issues to work on strategic matters, mandatory training for councillors and the incentivisation of pre-application discussions with planning authorities.

Elsewhere, BPF advises businesses on how best to engage in neighbourhood planning. It also campaigns to keep development-related infrastructure costs simple and transparent and collects data on how fast major planning applications are dealt with in London.

However, despite the major changes I have described in the daily operation and thinking of the UK commercial property industry, the focus of its interaction with government with respect to urban planning, in general, remains unchanged. As when Healey and Barrett were writing, BPF remains very clearly the voice of commercial property owners and investors on planning matters and the needs it expresses on behalf of the property industry (namely, vision, speed and fairness) have remained reasonably constant over the intervening period.

The presence of nearly 2,500 words of property industry testimonials on its website,[23] includes the following, suggests its members believe it to be a potent representative of their interests with respect to land use planning.

> The BPF is at all levels is immensely well connected and well informed. … .One of the BPF's greatest strengths is its ability credibly to engage government. … Ministers and civil servants listen and positive results follow. *Timothy Pugh, Berwin Leighton Paisner*

> The BPF provides an important property industry voice and influence on planning matters … . *Graeme Steer, SEGRO*

> The BPF is continuously at the leading edge of policy development: plugged into government, engaged with its members and, increasingly, acting as a bridge between the public and private sectors. *Marnix Elsenaar, Addleshaw Goddard*

Investors' Involvement with Land-Use Planning at Local Government Level

Given it is a contentious area of urban management that reveals environmental, community and business interests in high relief, it is hardly surprising that most of the academic attention given to the interaction between investors and planners at local government level relates to the development process. However, we should not forget that urban planning also affects the prospective returns from 'standing' property investments held predominantly for income and capital return in a locality, which are collectively much more significant in value terms.

As elsewhere, since Healey and Barrett were writing the property industry has seen a substantial increase in the amount of data relating to the occurrence of, and returns from, UK development activity (IPD, 2013). These confirm development as a high risk activity that, interestingly, often fails to provide commensurately high returns with less 'speculative' industrial developments providing the best development returns since 1983 (median total return, 8.4 per cent) and more speculative office developments, the worst (4.5 per cent).

Statistics like these and the investment experiences that underlie them have, over time, made the investment community generally more wary of development as an investment process, dampening enthusiasm for it amongst some. Changes in property ownership structures are also relevant in this respect. International investors often see non-domestic investment as risky enough in its own right and collective

23 http://www.bpf.org.uk/en/members/testimonials.php

investment vehicles needing to provide consistently good short-term performance on a rolling basis, may well not want their capital tied up in the lengthy, uncertain and risky pursuit of development returns. Both might reasonably exhibit a diminished appetite for development compared to longer-term financial institutions and property companies. Given the generally reduced involvement of the latter in the market, this might provide opportunities for new generations of developers in the future.

However, with respect to development, the industry trend mentioned above that is of most significance for UK urban planners is probably the shortening of lease lengths recorded for new properties (BPF/IPD, 2012) which must affect the economics and viability of development. With only a 10 year guarantee of lease income (rather than the 25 years available in 1990), undertaking developments has become even more risky than when Healey and Barrett were writing. This might lead to land values being driven lower or developments being delayed medium-term until rents rise sufficiently to restore profitability. Both of these issues should be of concern to planners.

The value of planning in providing consistency and stability to property investors in a local area was recently reported by Jackson and Watkins (2011:2329) who noted '(it) is widely held [amongst investors] that regulation of a market provides a form of restriction that, by virtue of its historic embeddedness, has become necessary and attractive for investment and investors'

However, they also noted that, with respect to the appeal of any given locality, local planning policies and the 'attitudes' of local planning authorities were likely to be subsidiary considerations for property investors compared to market conditions and local economic prospects. (Jackson and Watson 2011: 2330) They go on to report that the two aspects of local planning most relevant to proprty investors considering investing (or, presumably, maintaining holdings) in a locality are the short-term impacts of already permitted developments on future rental growth in the area and the implications on potential future returns from asset management of the planners' longer term vision for the locality.

Reflections on the Evolution of the UK Property Industry and its Relationships with Government

At the highest level of analysis, one could argue little has changed in UK commercial property since Healey and Barrett (1990). The commodified property market existing today would have been immediately recognisable to them then. However, beneath that high-level structural bound of the nature of land rights in a capitalist mode of production, almost everything else has changed and some of those changes are relevant to urban planners.

From a political economy perspective, perhaps the greatest change that has occurred over recent decades (and requires far greater treatment than a brief essay like this can offer) is the continued emergence of 'popular property investment'. Academic discourse should not stay uncritically at the level of 'the investor'. It needs to look through that simple descriptor to identify and understand the ultimate beneficiaries of 'investment'. We have seen that shareholders sharing the profits of property companies or companies holding property on their balance sheet have been replaced relentlessly in the post-war UK commercial property market, first by insurance policyholders, pensioners and personal savers, both domestic and foreign. While these institutionalised owners and their property fund managers might be driven by fiduciary responsibility and the corporate and personal pursuit of fees and bonuses to behave much the same as their capitalist forebears, even if few recognise or exercise their interests relating to it, it appears that through our insurance policies, pensions and savings, we are nearly all property investors now. As such, whatever governments do, at central or municipal level, when managing property investment and development activities, will affect individuals indirectly in a general financial sense as well as very specifically in the places they inhabit.

We have seen Giddensian structuration theory at work within the property industry itself with critical parent institutions deploying their 'authoritative' and 'allocative' rights to impose ideas about investment on property subsidiaries. This, along with the agency of individual property investors and the collective action of their trade bodies, has transformed how property investment is conceived and executed. New ideas and techniques have been developed in pursuit of improved risk-adjusted returns, increased capital allocations to property and higher management fees, underpinned by a brand new sub-industry servicing the concomitant need for data, research and advice.

Collectively, this has transformed the UK commercial property industry which now offers many more ways to access property market returns and draws capital from a far wider array of sources than Healey and Barrett would have witnessed in 1990. To protect the swathes of new 'non-professional' retail investors, property investment is now noticeably more regulated by government. Within the UK, the market is also significantly more international in nature and has extended beyond traditional retail, office and industrial sectors. The 'equity' part of the market at least has developed better tools to evaluate the worth of property investments and has generally embraced the need to consciously diversify portfolios across sectoral, regional and, increasingly, national property markets.

No such development of ideas and techniques is, as yet, discernible in the all-important property banking sector. However, paralleling developments in the institutional property market in the early 1990s, widespread public anger and a clear political concern to avoid the economic havoc and social hardships wrought by any future property cycles has greatly increased the pressure on the property debt sector to improve its data, education and business standards.

Land use planning issues have attracted investor and trade body attention over recent times. However, it seems clear that, even though irked intermittently at individual investor level, property investors collectively appreciate the benefits of urban planning and support the need for it. This said, of late the wider property industry and government have been much more concerned with developing more professionalised and transparent investment business models (Forbes, 2012) and as such the property industry's main interactions with government lately have centred around issues of new controls on fund management and capital reserve requirements. Before the financial crisis, it was lease and tax-related issues that dominated. Going forward, green issues will also be a major focus of attention and policy dialogue and, as with planning, the industry will seek to work with government to find solutions that achieve government aims without undue disruption for the operation of the market (See Deloitte Real Estate, 2014).

In all these policy areas, we see individual property investment actors ready to fight their own corner but also willing to form alliances to pursue their collective interests with government. These days, it seems the multiplicity of UK property trade bodies is matched by a similar multiplicity of advisory groups across government departments for them to connect with. However, much of this interaction is driven by ensuring industry regulation is appropriate (rather than weak).

So, if major changes have been wrought in what the UK commercial property industry looks like and how it thinks and acts, does this suggest any changes in how it engages with urban places and the spatial strategies to govern them?

If anything, through an invigorated BPF and the newly formed PIA, the commercial property industry seems better organised and connected to central government now than when Healey and Barrett were writing. Although planning is just one of a number of areas of interest, the property industry appears keen to ensure urban planning remains 'market aware'.

Looking at the potential impacts of the changes I have described on how property investors interact with planners at local level, with the exception of the marked changes in leases, most of the modernisation of property investment has mainly affected the process of investment and the superstructure of rights above real properties rather affecting the nature of physical properties. As such, there should be little direct impact on local planning. However, some indirect impacts may be relevant.

The replacement of longer-term institutional investors by unlisted collective investment vehicles judged more frequently by clients on their rolling short-term performance could result in investors trading assets more frequently. This could result in two opposing effects. It could lead investors to put pressure sooner on local planners to capture any available asset management gains through changes of use, refurbishment and possible redevelopment, prior to sale. Alternatively, it could lead other investors to engage less with local planning as they do not expect to hold assets long enough to realise the benefits of complex negotiated asset management initiatives.

Healey (2007) rightly describes urban areas as interpenetrated political arenas where actors operating at very different spatial scales negotiate and determine the evolution of places. In this respect, growing internationalisation could result in local planners having to deal with less emotionally connected, more passive and remote investors. More positively, planners seeking to 'create' markets (Healey 1992) may well now find themselves able to cast their nets far wider in search of capital. Paradoxically, the arrival of international capital in such profusion into central London could push major UK investors more noticeably into regional and non-traditional markets. Planners should be prepared for this.

The conscious spreading of investments across more locations and land use sectors to reduce portfolio risk has probably resulted in local planners dealing with a wider range of investors now, each with less obvious local interest. This should not concern local planners overly. However, in an 'information rich' market where informed investors continuously check they are not unconsciously out of step with competitors, planners need to remain informed about the latest ideas and trends sweeping the property market. Views can become widespread and affect perceptions of local markets quickly and notions such as 'business parks have had their day' and 'retailers are withdrawing to the 50 top centres' could quickly impact investor perceptions of local property markets.

Conclusion

In her recent work, Healey (2007) has argued convincingly that urban areas and their governance structures are complex and ever evolving through the interplay of structure, agency and phenomenology. So too, it would appear, is property investment and the industry and market that engage in it. Notwithstanding that the nature, form and practice of both urban governance and property investment both evolve, in truth the requirements made of the former by the latter vary little; property investors will continue to press planners at all government levels for clarity, constancy of vision, speedy responses and accessibility.

However, within the overarching structural bounds of property rights under a capitalist mode of production, we have witnessed powerful agency at work in recent decades changing almost everything about how the UK commercial property market operates. Healey and Barrett's promotion of Giddensian structure–agency seems well founded and the four research areas they highlighted in 1990 remain as pertinent for the next thirty years as for the last. For example, we might well ask how current government regulations on the need for capital reserves will affect remaining institutional investor appetites for property investment and lending and, if they continue to withdraw from the market who will replace them. We might ask if new forms of investment like property index derivatives (which pay property returns without the need to actually own rights in property) could result in some investors leaving the direct market altogether? These and a hundred other questions about the future evolution of the commercial property market remain outstanding and their resolution will very much determine the identity of who it is that urban planners will deal with in the future and what their strategies might be.

References

Barras, R. (2009) *Building Cycles: Growth and instability*, Chichester: Wiley–Blackwell.

Benford, J. and Burrows, O. (2013) Commercial property and financial stability, *Bank of England Quarterly Bulletin*, Q1, pp. 48–58.

British Property Federation (2008) *Planning Manifesto: Making planning work*, British Property Federation, London.

British Property Federation and Investment Property Databank (2012) *The BPF / IPD Annual Lease Review 2012*, British Property Federation, London.

Davies, H. (2010) *The Financial Crisis: Who is to Blame?* Cambridge: Polity Press.

Deloitte Real Estate (2014) *Carbon Penalties & Incentives: A review of policy effectiveness for carbon reduction and energy efficiency in the commercial buildings sector*, Commissioned by the Green Construction Board and Green Property Alliance, London.

Forbes J. (2012) *Unlisted Funds – Lessons from the crisis*, Report for The Association of Real Estate Funds, pwc, London.

Fourt, R. and Matysiak, G. (2013) Analysing holding periods, *FOCUS, The Journal of the Investment Property Forum*, 23, April, pp. 13–16.

Frodsham, M. (2012) *A Decade of Fund Returns*, IPF Research Programme Short Papers Series, No.6, Investment Property Forum, London.
Giddens, A. (1984) *The Constitution of Society*, London: Polity Press.
Giddens, A. (2001) *Sociology* 4th edn, Cambridge: Polity Press.
Harrison, D., Blake D. and Key, T (2013) *Returning to the Core: Rediscovering a role for real estate in defined contribution pension schemes*, The Pensions Institute, Cass Business School, London.
Healey, P. (1992) Development plans and markets, *Planning Practice and Research* 7(2): 13–20.
Healey, P. (2007) *Urban Complexity and Spatial Strategies: Towards a relational planning for our times*, Royal Town Planning Institute Library Series. New York: Routledge.
Healey, P. (2010) Re-thinking the relations between planning, state and market in unstable times. In P. Pinho and V. Oliveira (eds) *Planning in Times of Uncertainty*, Porto: FEUP edicoes, pp. 11–22.
Healey, P. and Barrett, S.M. (1990) Structure and agency in land and property development processes: some ideas for research, *Urban Studies*, 27(1): 89–104.
Investment Property Databank (2013) *IPD UK Development Report 2012*, IPD, London.
Jackson C. and Watkins C. (2011) Planning policy and retail property investment in the UK, *Urban Studies*, 48(11): 2321–38.
Key, T. and Law, V. (2005) *The Size and Structure of the UK Property Market*, IPF Research Paper, jointly funded by the IPF Educational Trust and the IPF Research Programme.
Lizieri, C., Reinert J. and Baum A.E.(2011) *Who Owns the City 2011, Change and Global Ownership of City of London Offices*, Research sponsored by Development Securities plc, Real Estate Finance Group, University of Cambridge.
McKee W. (2001) Property for the people: let's lobby the Mandelson way, *Property Week*, 14 December.
McNamara, P.F. (1985) The control of office development in central Edinburgh, PhD thesis, University of Edinburgh.
McNamara, P.F. (2013) Evolution of property development, in *Developing Property as an Asset Class: 25 years, 1988–2013*, Investment Property Forum, London, p. xii.
Mallinson, M. (1989) What is an investment surveyor? Summary of the 1989 address to the Investment Surveyor's Forum, republished in Investment Property Forum (2014) *Developing Property as an Asset Class: 25 years, 1988–2013*, Investment Property Forum, London, pp. 48–53.
Mallinson, M. (2013) IPF: 25 years on, *Developing Property as an Asset Class: 25 years, 1988–2013*, Investment Property Forum, London, p. 122.
Massey, D. and Catalano, A. (1978) *Capital and Land: Landownership by capital in Great Britain*, London: Edward Arnold.
Mitchell, P. (2014) *The Size and Structure of the UK Property Market 2013: A decade of change*, Investment Property Forum Research Paper, Investment Property Forum, London.
NAPF (2013) *Trends in Defined Benefit Asset Allocation: The changing shape of UK pension investment*, National Association of Pension Funds, London.
Office of National Statistics, Time Series Data: *MQ5: Investment by Insurance Companies, Pension Funds and Trusts*.
PMI and PRUPIM (2008) *Real Estate Investment: What's on the Horizon*, Pensions Management Institute, London.
Wyatt, A. and White A (2013) Founding the Investment Property Forum, *Developing Property as an Asset Class: 25 years, 1988–2013*, Investment Property Forum, London, pp. 45–7.

Chapter 12
Re-thinking the Relations between Capitalism, Urban Development and Historic Shifts in Planning Practice

Benno Engels

Introduction

For students of economic history, the global financial crisis of 2008–2009 and the global economic recession it subsequently triggered would not have come as a total surprise. Capitalism has experienced throughout its history similar economic 'shocks' on an almost regular basis every 30 years (Barras, 2009). For others, it was a sober reminder that we still live in an economic system that is prone to boom and bust cycles, despite advances in the diagnostic skills of economists. As in the past, the signs were there of what would happen, but we chose to disregard them bar the so-called doomsday warnings of a few (Schiller, 2005; Pettifor, 2006). The end of every upswing in a capitalist economic cycle is marked by an investment boom into the property market, only to be followed by an inevitable downswing. While the particularities of this crisis – a build-up of housing and consumer debt fuelled by the reckless lending practices of large parts of the global banking sector that were exposed by the subprime mortgage loan crisis in the USA – may be different to previous crises, the end result has, however, been the same: an economic recession (Martin, 2011; Gamble, 2009; Gowan, 2009).

The subsequent aftermath of the global financial crisis offers, an opportunity for us to pause and re-think the relationship that is said to exist between the urban development process, the cyclical nature of capitalism and urban planning. While there is much to be done on documenting how urban planning practice undoubtedly will be required to respond to the changes engendered by the global financial crisis this chapter will instead contend that just as much can be learnt by looking back in time to find out how urban planners and planning theory had responded to similar cyclical tendencies in capitalism. A focus on the broader cyclical tendencies and their influence can in part be found in the research which Patsy Healey has conducted on the changing nature of land use planning in England during the 1980s and 1990s.

Structure and Agency in Urban Land Use Planning

Even the most cursory examination of Patsy Healey's work will reveal the significant contribution she has made to several key areas of land use planning and how it has changed in England during the 1980s and 1990s (Healey, 1992a, 1992b; Healey and Barrett, 1990; Healey et al., 1988). Probably one of the most discernible themes in Healey's work is a recognition that the land development process consists of competing stakeholders one of whom is the urban planner. Central to Healey's understanding of this process is the notion of 'agency' and how different types of market agents interact to produce specific local outcomes, within a regulated market that is partly defined by the local development 'plan'. Agents and the market strategies they pursue must therefore be understood within their particular locality and historically-specific context. But specific locales do not exist in isolation from each other or from 'powerful exogenous forces' (Healey, 2009:3). In fact, 'the nature of the economy, the mode of production into which localities are linked, the political and administrative institutions which frame local possibilities, and the cultural traditions to which people in a locality variously relate, all affect the material resources and values extant in a locality' (Healey et al., 1988:6). The role that the structural context plays in urban land use development and planning agency is therefore another theme in Healey's work.

Recognition of the role that structural forces play, naturally led Healey to consider the issue of 'historicity' or the temporal dimension of what she was investigating, for the land development process was literally changing before her eyes in England during the 1980s and 1990s. The trigger for this change was an end to the post war economy long boom in the mid-1970s which ushered in a Thatcher-led Government that sought to resuscitate the British economy through a neo-liberal reform agenda. Urban land development was to be encouraged, but the country was perceived to be shackled with a cumbersome restraint-oriented urban planning system. Healey documented, through a series of case studies, how local development plan-making was now required to be more 'market aware' in order to facilitate market-led development (Healey, 1992a, Healey et al., 1988).

The final theme emerging in Healey's work is the issue of why these temporal specific changes to land use development and planning practice occurred. While reference is made to a series of broader structural economic factors they do not appear fully integrated into her work. This, therefore, represents the jumping-off point for the remainder of this chapter in which I seek to explore the relationships that can be found to exist between urbanisation processes, cyclical tendencies in capitalism and shifts in urban planning practice.

Long Waves, Social Structures of Accumulation and Capitalist Urbanisation

Economic historians of varying persuasions agree that cyclical tendencies 'have been, and continue to be, an intrinsic characteristic of capitalism' (Barras, 2009:11). Where the disagreement begins is with regard to the regularity, length and possible causes of such cyclical tendencies. The approach adopted here draws upon long wave theory associated with the work of Nikolai Kondratieff at the start of the twentieth century. He analysed economic data which suggested that capitalism had since the late eighteenth century experienced a series of long waves lasting between 50 or 60 years (Kondratieff, 1935). Subsequent research done by others has revealed that the world economy has progressed through four such long waves and that we are presently experiencing a fifth at the beginning of the twenty-first century (Mager, 1987, O'Hara, 2006). A common misconception found in the literature about long waves is that they are a kind of economic cycle that must conform to the behavioural properties associated with other types of cycles. Long waves, however, are not like other orthodox shorter economic cycles.

First, long waves can and do vary in length, some lasting 40, 50 or 60 years. The term cycle infers that economic activity occurs in a predictable and identical manner. This is one reason why some theorists have resisted labelling them as 'cycles' as opposed to 'waves or swings' (Gordon et al., 1982). Second, they possess upswing and downswing phases, but they are not symmetrical in duration or amplitude. An upswing phase, for example, will possess periods of downswing, but the overall tendency will be one of rising economic prosperity. Similarly, a downswing phase will possess periods of economic upswing, however these will be of shorter duration within a generally downward tendency. Third, long waves are also characterised by corresponding social, cultural and political changes, for they represent distinct stages in the historical development of capitalism. This will be further explored below. Fourth, while disagreement surrounds the possible cause of long waves – exogenous versus endogenous factors to the capital accumulation process – their historical occurrence is not seen as deterministic. Instead, the structural properties of long waves are thought to merely define the parameters within which economic, social and political activity of individual countries may occur, always being contingent on local and historically-specific conditions. People and nation-states, are presumed to be capable of shaping their own history, including when and how they respond to economic fluctuations associated with long waves. Not all countries and their economies respond in an identical manner. Fifth, differences will therefore exist between countries with regard to their progression through different phases of a specific long wave, due to the uneven temporal and spatial development of capitalism across the world. It is the very 'complexity and uneven development' of long waves, argues O'Hara (2006:3), that is such an 'essential part of the[ir] motion'. One theoretical framework that tries to take account of the differing capacity of countries to respond differently to long wave fluctuations is the neo-Marxist notion of 'social structures of accumulation'.

Originally proposed in the late 1970s by David Gordon (1978a, 1980), the social structures of accumulation approach – hereafter SSA – was further developed in the 1980s and 1990s by his colleagues Bowles, Edwards and Reich (1982, 1994). It can be regarded as a 'middle range or intermediate' theoretical approach that seeks in a non-reductionist manner to build a linkage between macro level structural forces and a micro level focus on the 'prevailing political, economic and social institution[al agency] ... to be found in specific historical periods' of respective countries (Mavroudeas, 2012:316). Three core assumptions underpin the SSA approach. First, capitalism cannot exist in 'a vacuum or in chaos', it is reliant upon an external supportive institutional environment where the actions and decisions of agents are paramount (Gordon et al., 1982:23). A SSA is therefore said to consist of those institutions 'that directly and demonstrably condition capital accumulation and excludes those that affect it only tangentially' in individual countries (Gordon et al., 1994:15). Institutions are conceived here broadly and include all those 'economic, political, cultural, social and ideological in character' that are related to profit generation and reinvestment activity (McDonough et al., 2010:2). Second, internal contradictions in one, such as the economy, 'can threaten the stability of a SSA and correspondingly, instability in the SSA can undermine the economy' (Gordon, 1980:20). In turn, the restoration of stability in one is not possible unless stability is restored in the other. Each new long wave must be accompanied by a new or restructured SSA otherwise the conditions needed for the restoration of economic prosperity will be stalled. Third, all SSA and their institutions are historically and geographically contingent, similar to the historically-specific long waves of capitalistic economic development.

Corresponding changes in an economy and its supporting SSA gives rise to distinct stages in the historical evolution of capitalism at both global and nation-state scales (Gordon, 1984; Mandel, 1980; O'Hara, 2006; Ross and Trachte, 1990). The periodisation of capitalism into a number of successive historical stages corresponding to specific long waves has also been etched into the urbanisation processes evident in many capitalist countries. There is now a substantial body of empirical research from a number of theoretical perspectives which has documented the synchronisation that exists between the cyclical activity evident in national economies and that of cities (Abramovitz, 1964; Lewis, 1965; Whitehand, 1987). Some urban theorists have subsequently contended that capitalism has not only gone through a series of distinct historical stages of development, but that this has been mirrored in correspondingly distinct historically-specific urban forms (Berry, 1983; Gordon, 1977, 1978, 1984; Ross and Trachte, 1990; Walker, 1981). Moreover, transitions between the different stages of capitalist urban development at a locality-specific scale had been facilitated through the physical planning activities of urban SSA's, also known as 'city growth coalitions' and 'urban regimes' (Imbroscio, 1997; Lauria, 1997; Logan and Molotch, 1987).

The impact of capitalist structural forces on both urbanisation and the regulatory capacity of urban planning during different phases of a long wave has attracted the attention of a few urban planners. Roweis (1981), for instance, has argued that during periods of economic and territorial instability, such as the downswing phase, significant pieces of urban reform legislation are often passed and urban planning is allowed to be more interventionist. In contrast, it is during periods of relative economic and territorial stability, that the progressive reform momentum declines and the capacity for urban planning intervention is said to decline. If urban planning theory and planning practice can indeed be regarded as being 'embedded in the historical dynamic of capitalist urbanization' (Scott and Roweis, 1977:1097), then it readily follows that their development must also have been conditioned by and synchronised with the fluctuating fortunes of capitalism and the cities it has produced. It now remains to be seen if these assertions are borne out by the historical evidence.

Historical Stages in Urbanisation and Development of Planning Practice

A number of caveats are needed before it is possible to begin the historical analysis. First, urban planning is to be understood as any 'concerted effort by [a] public authority to guide the development of land in the interests of economic efficiency and the common welfare' (Sutcliffe, 1981a:1). Second, no attempt will be made here to suggest that the emergence of a new approach to planning immediately leads to its universal adoption even within a single country. This was not the case. Third, when a new planning approach was eventually adopted, rarely was it a smooth or un-opposed process. Abandonment of old ways is often fraught with doubt and it takes time for a new approach to be comprehensively adopted. Fourth and finally, due to space constraints, the following analysis is largely restricted to England, Germany and the United States of America – but this should not be interpreted as suggesting that changes in planning practice were not also evident elsewhere.

The First Long Wave, 1770s to 1840s

The antecedents of modern urban planning are to be found in the emergence of industrial capitalism in Britain and the first economic long wave that lasted from the early 1770s to the late 1840s. It is now well documented that the industrial revolution not only transformed how things were made but also the settlement pattern of nineteenth-century England as people moved to its expanding towns and cities in search of work (Sutcliffe, 1980). Much of this industrial driven form of urbanisation was both unregulated and unplanned. Many observers of this period including Friedrich Engels (1845) vividly captured in their writing the squalor and misery that was to be found in England's early industrial towns. Not only was much of the housing for the 'new' industrial working classes poorly built and overcrowded, but there was no clean water, problems existed with sewerage and waste disposal, there was poor ventilation, as well as no access to open space (Cherry, 1988; Hall, 1974).

A failure to plan the growth of these rapidly expanding industrial towns and cities can be attributed to two factors. First, urban or town planning as we know it today, had as yet not emerged as a distinct activity of the state. Second, urban planning – which has been defined as a deliberate act of intervention into the operation of the capitalist market – ran counter to the prevailing ideology being espoused within the then national SSA. The shift from mercantile to industrial capitalism was underpinned by classical or laissez-faire liberalism which subscribed to notions of individualism and a business environment free of state interference. For much of this first economic long wave, the British SSA was in turmoil, as a new economic class of industrial and commercial entrepreneurs sought to wrestle political power away from the landed aristocracy. In this political climate, industrial expansion was therefore allowed to go unchecked, despite the rapid deterioration that it caused in the urban environment.

Remedial action had to be taken in existing towns especially as political unrest emerges as the 1st long wave of capitalist development came to an end in the 1820s. Periodic boom and bust conditions – 1825–1826, 1836–1837, 1839–1842 and 1846–1848 – punctuate a two and a half decade downswing in British capitalism that trigger a range of political movements and anti-capitalist protests (Hobsbawm, 1962). Surprisingly, England did not become engulfed in urban revolution which broke out in 1846–1848 in many European countries, but the threat of worker-led uprising was real and must have acted as a warning to the British Parliament. No less significant a reason was the regular occurrence of public health epidemics including cholera in 1831–1833, influenza in 1837–1838 and cholera again in 1848–1849 (Melosi, 2000). The first tentative steps would subsequently be taken by state authorities that can be considered the origins of modern urban planning (see Table 12.1).

Following the introduction of the Municipal Corporations Act of 1835, newly incorporated towns and cities were aided in their efforts to tackle their public health problems by the 1848 Public Health Act, championed by Edwin Chadwick (Cherry, 1988). The Act sought to address the issue of poor sanitation but these measures were not made mandatory, as anti-regulatory sentiments persisted in the national English SSA. Some towns and cities such as Liverpool and Leicester embraced this legislation and attempted to reform their urban environments, despite the cost of remedial work falling onto local ratepayers. Implementation occurred in some towns and cities because of the progressive leadership within the local SSA (Garrard, 1983); level of working class agitation for reform (Fosters, 1974); the nature of a towns industrial base and its capacity to generate fiscal revenue to undertake improvements (Hennock, 1963), plus the extent of urban problems (Ashworth, 1954). Not until the 1866 and 1875 Sanitation Acts which made improved sanitation and building standards compulsory did all English towns and cities actively go down this urban planning reform pathway (Cherry, 1988).

Exogenous factors could and did work in tandem with endogenous factors to bring about reform. For example, the last great cholera outbreak of 1866 in England is often credited with the introduction of the 1866 Sanitation Act, which made it compulsory for 'all local authorities to inspect their districts and to suppress nuisances' (Cherry, 1988:28). It has also been argued that this Act had its origins in the then Liberal Government trying to appease London's influential urban property owning interests. The preceding 15 years had witnessed a building boom on the outskirts of the city but sizeable amounts of this housing stock stood empty by the mid 1860s. The legislation made it compulsory to sewer all existing housing stock and limit the number of inhabitants per house. Overcrowded and unsewered inner city housing was now ripe for clearance and its inhabitants could relocate to the suburbs where another government scheme provided them with

Table 12.1 Long waves and selected urban planning developments since the start of the nineteenth century in England, Germany and the USA

Long Wave	Upswing Phase	Downswing Phase	English planning reforms	German planning reforms	American planning reforms
1st	1770–1816	1817–1848	1835 Municipal Corporations Act; 1848 Public Health Act;	1808 Cities Organizational Reform Act; 1831 Local Government & Franchise Reform Act; 1845 Berlin Street Planning Law;	
2nd	1849–1873	1874–1896	1866 Sanitation Act; 1872 Public Health Act; 1875 Public Health Act 1888 Local Government Act.	1874 Expropriation Law; 1875 Street Lines Law; 1880s Sanitation reforms; 1891 Frankfurt Zoning Ordinance;	1855 Report & Plan of Sewerage for the City of Chicago; 1864 Sanitation Survey of New York; 1866 New York City Metropolitan Health Law; 1870s US Sanitary Commission; 1879 US National Board of Health; 1890 City Beautiful Movement;
3rd	1897–1919	1920–1945	1898 Housing of the Working Classes Act; 1898 Garden City Movement; 1906 Housing Law; 1909 Housing, Town Planning Act; 1919 Housing Act; 1932 Town and Country Planning Act; 1945 Distribution of Industry Act;	1900 Saxon Urban Development Law;	1901 New York Tenement House Law; 1909 Plan of Chicago City; 1916 New York Zoning Ordinance; 1924 US Dept of Commerce: State Zoning Enabling Act; 1926 Euclid vs Ambler US Supreme Court case; 1928 US Dept of Commerce: City Planning Enabling Act; 1933 New Deal program; 1937 US Housing Act;
4th	1946–1974	1975–2008	1947 Town and Country Planning Act; 1952 Town Development Act 1953 & 1954 Town and Country Planning Acts; 1959 Town and Country Planning Act 1963 Town and Country Planning Act 1968 Town and Country Planning Act 1969 Skeffington Report 1980 Local Government, Planning and Land Act		1949 US Housing Act; 1954 US Housing Act; 1956 Federal Aid Highway Act; 1959 US Housing Act; 1962 US Highways Act;

Sources: Dates for the long waves are adapted from Mager (1987) and Went (2000). Information regarding the historical developments in English urban planning came from Ashworth (1954), Cherry (1988), Greed (1993), Melosi (2000) and Ward (2002); German urban planning developments came from Sutcliffe (1981b) and Ward (2002); American urban planning developments came from Feagin (1989), Melosi (2000) and Scott (1969).

cheap housing loans (Foster, 1979). A cholera outbreak may have been the public pretext for introducing new sanitation legislation but it might not have been the only reason. Whatever the real reason may have been the English sanitation movement of the 1840s-1860s bequeathed modern urban planning two things. First, it reinforced the belief that changing the built environment could solve social problems, thereby avoiding the need to pursue more fundamental societal reforms. Second, spatial reform could also be undertaken without compromising market capitalism, but it had to remain a constrained form of state liberalism.

The Second Long Wave, 1850s to 1890s

The British economy recovered in the 1850s and 1860s, marking the economic upswing of a 2nd long wave, where industrial output and urban growth accelerated beyond previously recorded levels. Industrial capitalism was also exported to northern Europe and the USA which triggered population growth and rapid urbanisation in turn. It would be the Germans who took the next leap forward in planning practice prompted by a need to better deal with urban sprawl caused by rapid industrialisation (see Table 12.1). German municipal authorities were ill-equipped to deal with such challenges. A property investment boom that began in the later 1860s came to an abrupt end in 1873 as the rapidly expanding world capitalist economy crashed. It triggered the start of a turbulent economic period in international capitalism that lasted from 1873 to 1897 and which is sometimes referred to as the first Great Depression. This was another long wave downswing.

In a newly unified Germany under the guidance of Bismarck, the 1870s depression forced the central government to demand larger annual subsidies from the states, a fiscal decision that had flow-on effects for local government and their urban planning capacities (Rosenberg, 1943). With more state revenue being directed to central government fewer funds reached local government. Two important urban planning laws were passed in 1874 and 1875 that helped shape the legislative basis of urban planning. Municipal authorities were granted the legal powers to compulsorily purchase land and buildings for all projects of public utility and their powers were also extended to regulate town extensions and transfer the cost of new infrastructure onto property owners (Sutcliffe, 1981b). All the German states adopted these new provisions by the early 1880s.

A need for additional urban planning intervention was fuelled by another property investment boom in the late 1880s and early 1890s, punctuated by the infamous Panic of 1893 and a global recession that marked the end of the second long wave. Amid these fluctuating periods of economic and urban growth, land use zoning would be conceived, as a solution to better regulating urban sprawl (see Table 12.1). Originally proposed by Reinhart Baumeister in 1876, it took a reform-minded mayor of Frankfurt, Franz Adickes, concerned about growing public unrest with tenement housing in his city, to adapt Baumeister's idea and introduce the world's first differentiated building regulations in 1891. The so-called Frankfurt ordinance consisted of three land use zones or districts – residential, mixed and industrial – across the whole city, where building height and density were specified (Sutcliffe, 1981b). Planning control had been extended to an entire city. This was another step forward in comprehensive plan making, yet again, linked to a unique combination of local situation and innovative leadership of a specific SSA.

Fluctuating economic fortunes during the 1880s and 1890s in England also shaped urban planning thought. Structural unemployment contributed to a further deterioration of inner city housing which prompted municipal authorities to undertake slum clearance (Cherry, 1988). It also prompted some urban reformers to devise new techniques such as the 'social survey'. The most famous exponent of the social survey was Charles Booth whose work on poverty in London was in part triggered by rioting amongst the working classes of the East End of London in February 1886 (Stedman-Jones, 1971). Through the use of maps, Booth had made his study of urban poverty 'apprehensible in scientific terms' (Hewitt, 2012:249), thereby bequeathing a valuable tool to urban planning. The other significant development in English urban planning thought at this time was the 'garden city' idea (see Table 12.1). Ebenezer Howard's (1898) solution to inner city poverty and squalor was a series of decentralised new towns that were separated from each other and existing urban centres by an agricultural green belt. What differentiated the proposal advanced by Howard was that his townships were not to be owned by benevolent industrialists but via community ownership. Each garden township was to be comprehensively planned. The utopian ideals underpinning Howard's garden city ran counter to private property ownership and capitalism. Initially popular, its appeal quickly waned as the British economy recovered and entered a new economic upswing phase in the pre-World War I period.

The Third Long Wave, 1890s to 1940s

The late 1890s marked the start of a 3rd long wave whose economic upswing phase was distorted by World War I. A return of economic growth triggers house construction activity and urban expansion in many capitalist countries (Ward, 2002). The return of suburban sprawl to English cities lead to a call for the adoption of German-style statutory planning reforms resulting in the Housing and Town Planning Act of 1909. This first formal piece of English planning legislation was devised by a reformist oriented Liberal government intent on redressing the cumbersome land ownership and taxation systems that were asserted to be causing a shortage in working class housing. Its policy intent was modified in the House of Lords whose members regarded it as an attack on landed property (McDougall, 1979; Offer, 1981). As a result, the 1909 Act only empowered local authorities to prepare planning schemes for new suburban development, but it did allow for land use segregation based on modified German zoning principles (Cherry, 1988). By the outbreak of World War I, a hundred schemes had been authorised (Burke, 1971; Sutcliffe, 1981). But the next step forward in urban planning thought and practice would not come from England, but rather from a rapidly urbanising USA.

In the USA, there were equally strong anti-planning sentiments to be found within the property owning classes, who either controlled or lobbied government to resist land reform. If a city authority attempted to regulate any aspect of the property market that had not been sanctioned by laws of the state or federal legislature then it often resulted in legal action in the courts. This significantly stymied urban planning activity even though the need for it was growing in the period before and during World War I when an economic upswing resulted in over-crowded, congested and inefficiently organised industrial cities. The so-called 'City Beautiful Movement' which started in the early 1890s and lasted until the 1920s did not have any answers to these problems (see Table 12.1). Instead, American planners turned to Germany, where land use standards had first been devised. The initial application occurred in New York in 1916 with the introduction of a comprehensive city-wide zoning ordinance – land use type, location and height – designed to pacify a powerful property lobby who wanted to stop the further incursion of garment manufacturers into an exclusive retail and housing area (Feagin, 1993). This innovative application occurred in New York city due to the efforts of key individuals like Edward Bassett and Robert Whitten, who successfully convinced the state legislature to set-up a city planning commission, to investigate and recommend how land use conflict could be resolved (Scott, 1969). Planning reforms in the USA would be introduced in a piecemeal manner because of the complicated three-tier system of governance – local, state and national – that existed here, together with differences in the political configuration of SSA's found at these various governance levels.

The end of World War I ushered in a short economic downturn that lasted from 1920 to 1921 before American and international capitalism experienced a decade of fluctuating economic activity that promoted urban growth until the late 1920s. By 1920, 51 per cent of the United States population was already living in urban areas, as suburbanisation accelerated, fuelled by rising living standards and private car ownership (Scott, 1969). In response, 82 of the 93 largest cities followed the example of New York and introduced zoning laws to better manage the urban development process (Feagin, 1989). However, an anti-planning sentiment still persisted in many local and state SSAs, reluctant to commit fiscal resources to city planning. It was not until the landmark ruling in support of zoning by the US Supreme Court in the *Village of Euclid v Ambler Realty Company* case in 1926 that the legal challenges over zoning declined (Scott, 1969). By the end of 1927, 525 cities had adopted zoning ordinances, thereby extending comprehensive plan making across this rapidly urbanising nation (Scott, 1969). A side-effect of the widespread adoption of land use zoning was outward urban sprawl. This led, in turn, to the introduction of the national Standard City Planning Enabling Act in 1928 which further reinforced city-wide comprehensive plan making (see Table 12.1). But the start of the 1929 Great Depression would slow this process.

Once again, international and American capitalism had entered another long wave downswing phase, whose length and severity was on this occasion without parallel in the history of capitalism. It would last in some countries until the late 1930s. Millions were made unemployed as businesses went bankrupt. Political tensions mounted and protest sporadically broke out across the USA in the form of rent riots, relief protest and hunger marches by the unemployed (Piven and Cloward, 1972). In late 1932, the American people elected a new president who was given a mandate to deal with the economic crisis. Roosevelt's reform agenda, commonly known as the New Deal, would not only help revive a depressed economy during the 1930s but also institutionalise government intervention into the economy, constituting a modification of the national SSA. America was being shifted from a classical liberal to a state liberal market system. However, Roosevelt's effort to reform the administrative and economic operation of US society along Keynesian principles was strongly resisted both inside and outside the national legislature, including within his own party.

From a planning perspective, the New Deal encouraged all US states to engage in 'planning' which included undertaking regional land use audits, flood mitigation proposals, forecasting population growth and to a lesser extent re-organising cities (Scott, 1969). Public works funding was only supplied if the states agreed to engage in certain kinds of planning and reporting. To comply, state planning boards had to be established in each state to collect and interpret data, formulate and evaluate project goals, assign priorities, plus suggest how they might be implemented using performance measures and costing standards (Scott, 1969). These innovations were partly based on Luther Gulick's notion of rational public sector decision-making (Stone, 1990). At one level, this represented a frugal approach to the spending of public monies by a federal administration that only had limited fiscal resources, but a large number of political opponents. On another level, this form of fiscal administration using a cost-benefit mode of analysis can also be said to represent the rudimentary beginnings of rational planning because the actual 'processes of planning' were now being considered to be just as important as the final project proposal.

The Fourth Long Wave, 1950s to 2010

Few had anticipated that with the end of World War II, world capitalism might experience a period of sustained economic growth that would last for more than two decades, an upswing phase of a 4th long wave that for some countries persisted until the early 1970s. During this post-war economic long boom, the US economy experienced almost uninterrupted economic growth, punctuated only by a few short-lived slumps in 1948–1949, 1953 and 1958–1961. Federal housing acts were passed in 1949, 1954 and 1959, plus acts to build highways in 1956 and 1962 (Scott, 1969). Billions of dollars were injected into the American economy and its cities to maintain economic growth (see Table 12.1). If a city wanted to tap into these federal funds then they had to prepare a comprehensive plan for both the metropolitan area and targeted neighbourhoods (Scott, 1969). Underpinning the preparation of such plan making was an increasing reliance upon analytical social science methods to both collect and interpret data that would also be used by federal administrators to assess competing proposals within a cost-benefit analysis framework (Wildavsky, 1966).

The enormity of the US federal government's urban programs during the 1950s and 1960s led some observers to re-think the theoretical notions that underpinned the practice of rational comprehensive planning. Banfield (1959) argued that planning had to be seen as a sequential, rational, problem-solving activity involving the formulation and comparative evaluation of different options. But the sheer size of some projects which required an extended period of time for completion, led to calls for 'middle-range' planning (Meyerson, 1956). Others questioned whether rational comprehensive planning was even at all possible. Lindblom (1959) suggested that an 'incremental' rather than a rational form of decision-making was more realistic and achievable. As urban planners collected more data upon which to base their decisions, computers and computer modelling was absorbed into rational comprehensive planning. By the mid 1960s, a systems theory of the city and urban planning is proposed (Taylor, 2003), a new conceptual understanding that further brought into question long-range, end-state, master plans. Instead, flexible middle-range plans were considered more practical, but this would require a further development in rational planning. This next development would manifest itself as the 'concept or structure plan'. It would not occur in the USA but in England at the very end of the post-war economic long boom of the late 1960s.

In the mid-1960s, as British and world capitalism was entering the last phase of the post-war economic long boom, the then Conservative Government decided to reform planning legislation (see Table 12.1). This resulted in the introduction of the Town and Country Planning Act of 1968 which represented the next significant development in the practice and re-theorisation of urban planning. The Act 'stressed the policy basis of plans and sought to avoid the excessive elaboration of physical and land use details that dominated the blueprint approach' of comprehensive plan making (Ward, 2011:122). This was achieved by distinguishing between a structure and local plan. The structure plan was to be unshackled from the minutiae of old style plan making and would now focus on broad strategic goals that could be more responsive to the needs of the market (Greed, 1993). In turn, local plans were prepared, but they had to both reflect the goals of the broader structure plan and be responsive to changes in local market conditions (Ambrose and Colenutt, 1975).

Although the 1960s was a period of general economic prosperity it was also a period of growing public unrest on both sides of the Atlantic. Public objections were raised against the wholesale bulldozing of inner city neighbourhoods for housing renewal projects and highways (Jacobs, 1961; Goodman, 1971). Some urban planning academics now turned away from rational comprehensive planning and began to propose an array of alternate approaches including 'advocacy planning' (Davidoff, 1965). Many of these alternate approaches called for greater public participation in the plan making process. Planning legislation would be re-written during the early 1970s to permit more public participation and thereby pacify public unrest, but it often brought criticisms of tokenism, especially in the USA (Arnstein, 1969). More substantial structural challenges would, however, confront rational comprehensive planning practice with the onset of the 1970s global economic recession.

As in the past, the prelude to the recession accelerated the switching of investment into the built environment. This had already started in the late 1960s, extending the boom in commercial and residential property that was evident in many of the major world cities (Ambrose and Colenutt, 1975; Harvey, 1978). Both the economy and the property market crashed together in early 1974 following a sudden rise in world oil prices and short-term interest rates. It was the worst global recession since 1929 and it would last until 1976 but the general economic downturn would continue for decades. Those national governments that had been in power when the global recession started, lost office at their next electoral cycle, because economies failed to respond to the counter-cyclical measures recommended by Keynesianism. The continuation of the global recession led to the election of what we now call neoliberal governments in England, USA, Australia, New Zealand and Canada.

In this new ideological regime planning was to be both contained and reduced. As a result, the 'post-war consensus of a regulated mixed economy and its welfare state ended on which planning had rested for its legitimacy' (Ward, 2011). All newly elected neoliberal national governments of the 1980s and 1990s, on either side of the Atlantic, set about overhauling their planning systems in order to make them more responsive to the evermore unpredictable economic times. Making the planning system, and therefore the practice of state planners, meet the needs of the market was in itself not new. Rather, this more recent phase would attempt to re-assert the view that urban planning existed to facilitate the market, not stifle it. This often involved restructuring local government, amending planning acts, streamlining planning procedures, simplifying land use zones, plus cutting budgets and programs. Those academic planners who were critical of these changes and the new style of neoliberal planning began to rethink the purpose and nature of planning practice from a variety of theoretical perspectives including neo-marxism, critical theory and post-modernism (see summaries in Allmendinger, 2001; Hillier and Healey, 2008; Sandercock, 1998). It is too early to establish whether these latest theoretical developments are having any significant impact as the last three decades of neoliberalism have limited the opportunities for experimentation with government-led strategic and statutory planning practice.

Conclusion

Patsy Healey has sought to understand the nature of urban development and planning practice within capitalist societies from a non-reductionist perspective, where due regard is given to structure and agency, as well as to how they interact to produce locality-specific outcomes at different points in time and space. My reinterpretation of the historical spatial development of capitalism and urban planning practice in this chapter has been guided by these principles. Too often, the history of urban planning practice has either been written as the achievements of significant individuals and a benevolent middle class, or as a consequence of structural forces, where only token regard is given to human agency and contingent local factors. Such accounts are overly one-sided and reductionist in nature. In contrast, Pasty Healey has sought to understand the roles and interactions between market agents and societal structures rather than opt for one or the other.

In this chapter, I have suggested that the notion of SSA operating at different spatial scales offers a useful, non-reductionist way of re-examining the historical development of planning practice, which I have then analysed with regard to three different capitalist societies. The analysis revealed that developments in urban planning practice did not happen in a uniform manner across either time or space. In fact, developments in urban planning practice emerged in different cities, regions and countries, at different points in time due to the unique combination of locality-specific agency and institutional factors.

However, 'men make history', as Karl Marx famously claimed, 'but they do not make it just as they please; they do not make it under circumstances chosen by themselves, but under circumstances directly encountered, given and transmitted from the past' (Marx, 1852:10). Patsy Healey too has realised this claim. In turn, my analysis in this chapter has asserted that economic long waves have conditioned, but not determined human and institutional agency. For instance, land use zoning was introduced in Germany during the 1890s in response to urbanisation issues that were also evident in England and the USA at this time, but it took several more decades before zoning was adopted in the latter two countries. Furthermore, certain shifts in planning practice were closely associated with periods of economic downswing such as the 1840s, 1870s, 1930s and 1970s, when urban planning was used to help address problems that were being accentuated by the difficult economic times. This is consistent with claims made by Roweis (1981) that new planning interventions are more likely to occur in periods of economic and territorial instability. There were, however, also instances where urban planning was used to help address problems caused by an economic upswing such as the 1860s, 1910s, 1920s, 1950s and 1960s. The SSA approach provides the conceptual space for locality-specific differences to be taken into account, even though long waves of capitalist economic development have acted as structural constraints within which state intervention and urban planning has clearly occurred. To that extent, this chapter has observed and added to the intellectual tradition that Patsy Healey has bequeathed to the field of urban planning studies.

References

Abramovitz, M. (1964) *Evidences of Long Swings in Aggregate Construction Since the Civil War*. New York: National Bureau of Economic Research.

Allmendiner, P. (2001) *Planning in Postmodern Times*. London: Routledge.

Ambrose, P. and Colenutt, B. (1975) *The Property Machine*. London: Penguin Books.

Arnstein, S. (1969) 'A ladder of citizen participation', *Journal of the American Institute of Planners*, 35: 216–24.

Ashworth, W. (1954) *The Genesis of Modern British Town Planning*. London: Routledge and Kegan Paul.

Banfield, E. (1959) 'Ends and means of planning', *International Social Science Journal*, 11(3): 1–8.

Barras, R. (2009) *Building Cycles: Growth and instability*. London: Wiley-Blackwell.

Berry, M. (1983) 'The Australian city in history: critique and renewal'. In: L. Sandercockand M. Berry (eds) *Urban Political Economy: The Australian case*. Sydney: George Allen & Unwin.

Burke, G. (1971) *Towns in the Making*. London: Edward Arnold.

Cherry, G. (1988) *Cities and Plans: The shaping of urban Britain in the nineteenth and twentieth centuries*. London: Edward Arnold.

Davidoff, P. (1965) 'Advocacy and pluralism in planning', *Journal of American Institute of Planners*, 31: 103–15.

Engels, F. (1845) *The Condition of the Working-Class in England* (2nd English edition, 1973). Moscow: Progress Publishers.

Feagin, J. (1989) 'Arenas of conflict: zoning and land use reform in critical political-economic perspective', In: C. Haar and J. Kayden (eds) *Zoning and the American Dream*. Washington: Planners Press.

Foster, J. (1974) *Class Struggle and the Industrial Revolution*. London: Methuen & Co.

Foster, J. (1979) 'How imperial London preserved its slums: Book review essays', *International Journal of Urban and Regional Research*, 3(2): 93–114.

Gamble, A. (2009) *The Spectre at the Feast: Capitalist crisis and the politics of recession*. London: Palgrave Macmillan.

Garrard, J. (1983) *Leadership and Power in Victorian Industrial Towns, 1830–80*. Manchester: Manchester University Press.

Goodman, R. (1971) *After the Planners*. New York: Simon and Schuster.

Gordon, D (1977) 'Class struggle and the stages of American urban development'. In: D. Perry and A. Watkins (eds) *The Rise of the Sunbelt Cities*. Beverly Hills: Sage Publications.

Gordon, D. (1978a) 'Up and down the long roller coaster'. In: Union of Radical Political Economics (eds) *US Capitalism in Crisis*. New York: Union of Radical Political Economics Collective.

Gordon, D. (1978b) 'Capitalist development and the history of American cities'. In: W. Tabb and L. Sawers (eds) *Marxism and the Metropolis: New perspectives in urban political economy*. New York: Oxford University Press.

Gordon, D. (1980) 'Stages of accumulation and long economic cycles'. In: T. Hopkins and I. Wallerstein (eds) *Processes of the World-System*. Beverly Hills: Sage Publications.

Gordon, D. (1984) 'capitalist development and the history of American cities'. In: W. Tabb and L. Sawers (eds) *Marxism and the Metropolis: New perspectives in urban political economy*, 2nd edition. New York: Oxford University Press.

Gordon, D., Edwards, R. and Reich, M. (1982) *Segmented work, divided workers: The historical transformation of labor in the United States*. London: Cambridge University Press.

Gordon, D., Edwards, R. and Reich, M. (1994) 'Long swings and stages of capitalism'. In: D. Kotz, T. McDonough and M. Teich (eds) *Social Structures of Accumulation: The political economy of growth and crisis*. London: Cambridge University Press.

Gowan, P. (2009) The crisis in the heartland: Consequences of the new Wall Street system, *New Left Review*, 55 (Jan-Feb): 1–15.

Greed, C. (1993) *Introducing Town Planning*. London: Longman.

Hall, P. (1974) *Urban and Regional Planning*. London: Penguin Books.

Harvey, D. (1978) 'The urban process under capitalism: a framework for analysis', *International Journal of Urban and Regional Research*, 2(1): 101–31.

Healey, P. (1992a) 'Development plans and markets', *Planning Practice and Research*, 7(2): 13–20.

Healey, P. (1992b) 'The reorganisation of state and market in planning', *Urban Studies*, 29(3): 411–34.

Healey, P. and Barrett, S. (1990) 'Structure and agency in land and property development processes: some ideas for research', *Urban Studies*, 27(1): 89–104.

Healey, P., McNamara, P., Elson, M. and Doak, J. (1988) *Land Use Planning and the Mediation of Urban Change*. Cambridge: Cambridge University Press.

Hennock, E. (1963) 'Finance and politics in urban local government in England, 1835–1900', *The Historical Journal*, 2: 212–25.

Hewitt, L. (2012) 'The Civic Survey of Greater London: Social mapping, planners and urban space in the early twentieth century', *Journal of Historical Geography*, 38: 247–62.

Hillier, J. and Healy, P. (eds) (2008) *Critical Essays in Planning Theory*, Volume 1, *Foundations of the Planning Enterprise*. London: Ashgate.

Hobsbawm, E. (1962) *The Age of Revolution, 1789–1848*. New York: Mentor Books.

Imbroscio, D. (1997) *Reconstructing City Politics*. Thousand Oaks: Sage Publications.

Jacobs, J. (1961) *The Death and Life of Great American Cities*. New York: Random House.

Kondratieff, N. (1935) 'The long waves in economic life', *Review of Economic Statistics*, 17 (6): 105–15.

Laria, M. (ed. 1997) *Reconstructing Urban Regime Theory*. Thousand Oaks: Sage Publications.

Lewis, J. (1965) *Building Cycles and Britain's Growth*. London: Methuen Press.

Lindblom, C. (1959) 'The science of muddling through', *Public Administration Review*, 19(2): 79–99.

Logan, J. and Molotch, H. (1987) *Urban Fortunes: The political economy of place*. Berkeley: University of California Press.

Logan, T. (1976) 'The Americanization of German Zoning', *Journal of the American Institute of Planners*, 42(4): 377–85.

Mager, N. (1987) *The Kondratieff Waves*. New York: Praeger Publishers.

Mandel, E. (1980) *Long Waves of Capitalist Development: A Marxist interpretation*. Cambridge: Cambridge University Press.

Martin, R. (2011) 'The local geographies of the financial crisis: From housing bubble to economic recession and beyond', *Journal of Economic Geography*, 11: 587–618.

Marx, K. (1852) *The Eighteenth Brumaire of Louis Bonaparte*, 7th printing in 1977. Moscow: Progress Publishers.

Mavroudeas, S. (2012) 'The social structures of accumulation approach'. In: B. Fine and A. Saad-Filho (eds) *The Elgar Companion to Marxist Economics*. London: Edward Elgar.

McDonough, T., Reich, M. and Kotz, D. (eds) (2010) *Contemporary Capitalism and its Crises: Social structure of accumulation theory for the 21st century*. Cambridge: Cambridge University Press.

McDougall, G. (1979) 'The state, capital and land: The history of town planning revisited', *International Journal of Urban and Regional Research*, 3(3): 361–80.

Melosi, M. (2000) *The Sanitary City: Urban infrastructure in America from colonial times to the present*. Baltimore: Johns Hopkins University Press.

Meyerson, M. (1956) 'Building the middle-range bridge to comprehensive planning', *Journal of the American Institute of Planners*, 22(2): 58–64.

Offer, A. (1981) *Property and Politics, 1870–1914: Landownership, law, ideology and urban development in England*. Cambridge: Cambridge University Press.

O'Hara, P. (2006) *Growth and Development in the Global Political Economy*. London: Routledge.

Pettifor, A. (2006) *The Coming First World Debt Crisis*. London: Palgrave.

Piven, F. & Cloward, R. (1971) *Regulating the Poor: The function of public welfare*. New York: Vintage Books.

Rosenberg, H. (1943) 'Political and social consequences of the Great Depression of 1873–1896 in Central Europe', *The Economic History Review*, 13(1): 58–73.

Ross, R. and Trachte, K. (1990) *Global Capitalism: The New Leviathan*. New York: State University of New York Press.

Roweis, S. (1981) 'Urban planning in early and late capitalism'. In: M. Dean and A. Scott (eds) *Urbanization and Urban Planning in Capitalist Society*. New York: Methuen.

Sandercock, L. (1998) *Towards Cosmopolis*. New York: John Wiley & Sons.

Schiller, R. (2005) *Irrational Exuberance*, 2nd edition. New York: Princeton University Press.

Scott, A. and Roweis, S. (1977) 'Urban planning in theory and practice: a reappraisal', *Environment and Planning Series A*, 9: 1097–1119.

Scott, M. (1969) *American City Planning Since 1890*. Michigan: University of California Press.

Sutcliffe, A. (ed.) (1980) *The Rise of Modern Urban Planning, 1800–1914*. London: Mansell Publishing.

Sutcliffe, A. (1981a) *The History of Urban and Regional Planning: An annotated bibliography*. London: Mansell Publishing.

Sutcliffe, A. (1981b) *Towards the Planned City: Germany, Britain, the United States and France, 1780–1914*. Oxford: Basil Blackwell.

Stedman-Jones, G. (1971) *Outcast London*. London: Penguin Books.

Stone, D. (1990) 'Administrative management: Reflections on origins and accomplishments', *Public Administration Review*, Jan/Feb, pp. 3–20.

Taylor, N. (2003) *Urban Planning Theory since 1945*. London: Sage Publications.

Walker, R. (1981) 'A theory of suburbanization: Capitalism and the construction of urban space in the United States', In: M. Dean and A. Scott (eds) *Urbanization and Urban Planning in Capitalist Society*. New York: Methuen.

Ward, S. (2002) *Planning the Twentieth-Century City: The advanced capitalist world*. New York: John Wiley & Sons Ltd.

Ward, S. (2011) *Planning and Urban Change*, 2nd edition. London: Sage.

Went, R. (2000) *Globalization: Neoliberal challenge, radical responses*. London: Pluto Press.

Whitehand, J. (1987) *The Changing Face of Cities: A study of development cycles and urban form*. Oxford: Basil Blackwell.

Wildavsky, A. (1966) 'The political economy of efficiency: Cost-benefit analysis, systems analysis and program budgeting', *Public Administration*, 26: 292–316.

Part III.2
Doing Planning Work

Introduction to Part III.2
Doing Planning Work

Jonathan Metzger

Patsy Healey joined the emerging research field of planning studies in the late 1960's from a background that was different from that of many of her colleagues. In contrast to the traditions of architecture, land surveying or engineering that dominated planning education at the time, Healey had her initial academic schooling in the social sciences, with a Bachelor's degree in geography. In her subsidiary studies in anthropology the tutoring of Mary Douglas introduced her to the meticulous fieldwork-traditions of the discipline. Further, and perhaps just as important, she had worked as a practising municipal planner in a London Borough in the 1960s, so thus held first-hand experience of the turns and frustrations of planning in practice. The empirical study of 'really existing' planning practices and an appreciation of their situated specificities, complexities and nuances have, since the very beginning, also become something of a hallmark of Patsy's scholarly endeavour. As she herself notes, when she commenced her research career, planning research of such an inclination was not very much in fashion (Healey, 2015). Healey became part of a group of scholars who would revolutionise planning theory in the early-1970s by introducing social scientific thinking into the field, including, for instance, the comprehensive systems analysis proposed by McLoughlin (1969), or the somewhat abstract prescriptive models for rational-procedural planning introduced by Faludi (1973). In contrast to these scholars, Healey, perhaps influenced by her early introduction to empirical fieldwork in the anthropological tradition, was instead inclined towards building her research on concrete studies of the complex interactions of *in situ* planning practice.

Notwithstanding her thoroughgoing interest in studying existing practice, Patsy Healey has never settled on simply providing analytical descriptions of such activities. Rather, she has utilised the study of planning practice as it *is* empirically observed in the present to also discuss, in a normative register, what planning should be able to *become*, i.e.: visions of what planning *could* (and *should*) *be*. This mode of scholarship is present in some of Healey's earliest writings on planning (see e.g. Healey, 1974a), but also becomes increasingly prominent – perhaps particularly coming into focus in her important role in the development of communicative planning theory, for instance in *Collaborative Planning* (1997a), which is among the most cited publications in contemporary planning scholarship.

Part III.2, Doing Planning Work, focuses on Healey's persistent interest in the practice of planning, its context and potential, and highlights her concern with theory and action both with regard to local planning authorities' regulatory development plans and also to broader practices of strategy formation and policy processes. The papers included in this part thus reflect the importance of fine-grained empirical analysis attending to the shifting situational contexts in which planning work is performed, the potential function and impact of different types of planning at different places in different times, and the specific challenges and possibilities associated with these projects or ambitions. They highlight how there is not one all-encompassing definition to be found for planning practice, but that there may nevertheless be 'contingent universals' (Healey, 2012a) that can function as beacons pointing the way towards 'better' or more 'progressive' planning practices. It could definitely be argued that almost any paper by Patsy Healey contains a careful analysis or discussion of planning practice. The papers by Healey reproduced in this section have been chosen to illustrate the breadth and persistence of this totality of work with a particular focus on contributions that showcase the development of her thinking on the potential of planning practice: what it *could* be and do.

Already in 1979, more than twenty years before this became a common perspective in the field of planning, Healey not only discusses the possibility of understanding planning practice as a networking practice – utilising concepts such as 'action space', 'decision networks' and 'connective planning' – but also presents a deep-reaching critical analysis of networking as a proposed normative model for planning practice. Her critique is based on an analysis of research on the role of 'reticulist' skills in governmental policy work emanating from the field of operational research at that time (particularly in Friend, Power and Yewlett, 1974). Healey is interested, not only in posing the empirical question concerning whether this is the way planning 'really works', but also in discussing the more far-ranging implication concerning whether this is a *good* model for how planning work *should* be undertaken. Already at this time, Healey thus traverses the empirical and normative registers in what is to become a somewhat distinct fashion for her, through at the same time posing the question concerning 'how things are done' together – but not conflated – with the question concerning how they should be done, envisaged as two different but inevitably interrelated issues.

Healey has never been afraid to take to task the prevailing dominant patterns of planning practice in any given time or place when she has seen them to be severely lacking in living up to expectable standards of democratic, inclusive and just governance. A persistent point of aggravation appears to have been planning approaches that either ignore (at their own peril) to engage a broader public on reasonable terms as 'stakeholders' deserving a voice in these processes, or which fail to leverage successful 'episodes' of planning into broader institutional change. Both these issues come together and interlock in her critique of public-private partnership projects in urban development (1997b), where she makes the argument that such projects often set the cart before the horse in jumping into concrete 'delivery' of outcomes before taking the care to build up the local institutional capacity necessary to really make a significant structural shift in the fate of disadvantaged places. Arguing against such projects that are often framed as handing a 'lifeline' to the mainstream economy for disadvantaged groups and places, Healey – with a somewhat utopian fervour – argues in this paper that planning work towards institutional transformation should not be concerned with throwing out lifelines to the old imperial iron-side of the capitalist economy, but rather with constructing new 'rafts', i.e. economical practices based on more democratic and just principles than those we have today.[1]

This utopian streak (see Part I), conceptualising planning as a transformatory practice (also Part IV) which may fundamentally help us shape how we look upon the world and each other (and through this also transform the material realities around us), is perhaps at its most evident in her seminal writings on communicative planning theory. She (Healey, 1992) describes how the postmodern philosophical movement coupled with the vicious assault on the planning profession mounted by Margaret Thatcher's conservative government in the 1980s rocked the foundations of British planning as a practice and profession. Asking herself whether planning is passé as a form of public activity, Healey draws upon the writings of philosopher Jürgen Habermas to argue that this is not the case. Rather, she argues, planning can be reinvented as an arena of the public sphere in which different groups and interests can find a voice and come to build the mutual understandings and trust which are necessary to generate the common ground and partially shared ideas of desirable futures for a specific place. The way to achieve this is to generate parameters for entering into a respectful debate, built upon reciprocity and a mutual respect for divergences in ways of thinking, ways of expression, and ways of getting on in the world. In the wake of the heated debate that broke out around the idea of 'communicative rationality' (which to some extent underpinned the communicative theories of planning), Healey moved on to other intellectual inspirations – but without losing sight of the deep engagement with the question of how approaches for more democratic, just and open planning processes can be developed. This continuous interest is also evident in her writings on strategic spatial planning from the 1990s and 2000s, for instance in Healey (2004). This paper is important since it draws together influences from the new relational geography with Healey's thoughts on institutional change and transformative governance practice to reason around how spatial concepts may come to exert situated agency in a way that may contribute to (or conversely, hinder) the 'opening up' of planning and governance practice.

This interest in the potential agency of 'ideas' in planning processes, the forms they take, and how they may travel between contexts, has always been a central interest of Healey's ever since the beginning of her academic career (see also Part IV as well as e.g. Healey 1974b). But since this agency is always situated there is no telling that a concept or model that does 'good' or 'progressive' work 'here', however this is defined, will do the same job in a different context. Combining this intuition with an interest in the role played by planning *research* and *researchers* in enacting specific realities (see also Healey, 2008), Janice Barry (this volume) pursues a suggestion from Patsy Healey to critically examine her own previous application of Healey and colleagues' Institutional Capacity Development framework [ICD] (see e.g. Healey et al., 2003) to analyse

1 For a specification of the meaning of 'utopian' in this context, please refer to Part I.

planning processes involving both state and indigenous actors in the Canadian province of British Columbia. Paralleling Healey's (1997b) interest in 'the terms of collaboration', Barry thoughtfully shows how what from one perspective came across as productive and appropriate intellectual tooling, from another perspective appeared to obscure some of the most crucial and fundamental issues of contention. These issues related to the question of Indigenous sovereignty – questions that 'fell out the frame' of analysis, thus generating an 'unintended risk of obscuring, underplaying or misrepresenting Indigenous peoples' unique and sovereign authority' (Barry, this volume; cf. also Healey, 2007 on framing and Healey, 1974b on 'reference frameworks').

Also taking an interest in the often-excluded 'others' of planning, in Chapter 17 Ananya Roy discusses what planning theory and practice could potentially learn from the mobilising practices of 'poor people's movements' in the Global South through understanding these situated practices as 'templates and grammars of participation and learning'. These political communities are incomplete and emergent, only taking their form through struggle and mobilisation, constructing the subjectivity of their movement as they go along (cf. also Metzger, 2013a). Drawing on postcolonial theory and theorists of planetarity, Roy further argues that we must recognise the radical alterity through which global humanity is constituted – which means that even if we recognise 'planning' as a general concept contingently evolving through experiences and debates, the 'community of inquiry' of planning cannot be conceptualised as a 'geocommon'. Instead, building on Fraser's philosophy of scales of justice, and inspired by poor people's movements, she argues that communities of inquiry today must be seen as coinciding with communities of risk of different reach and scales that come together '*for* themselves' to recognise their mutual implications and shared interests (see also Introduction to Part III.1).

Roy argues that participation is crucial to democratic place governance. Through the mutual engagement of democratic participation we may come to work across divides and also begin to form new subjectivities and identities across previous divides, to care more for each other and take more and better care of the future of the places where we live our lives (cf. also Metzger, 2014). In his contribution to the present volume, in Chapter 18 Geoff Vigar discusses a specific device purported to generate such participation, engagement and care – the British 'neighbourhood plans' enabled by the 2011 Localism Act. Highlighting and building upon Healey's sociological-institutional approach (see also Part IV) Vigar examines this 'institutional design for very local plan-making' to tease out some of the implications and outcomes that processes for producing neighbourhood plans have had so far in the British context. He comes to the conclusion that they do sometimes, in some places, lead to 'something better and rather unexpected', but that this depends on the circumstances. What is troubling about these circumstances is that it appears as if there is a clear socioeconomic divide in which places successfully pursue this type of planning endeavours, centre-staging questions of socio-spatial in/equality, social justice and the risk of generating planning mechanisms through which the already well-off get even better, but leaving the less fortunate even further behind. Vigar concludes by arguing that even if planning frameworks and contexts to some degree are shaped by broad institutional shifts in, for example, national legislation, '[a] focus on institutions, particularly on the social relations, the micro-politics, of what is going on in planning work is vital to the understanding these broad shifts'. This call for sensitivity to context – the insight that 'what works here might also work there', or again, it might not, or it may do the complete opposite of what was intended – is also a thorough-reaching theme in Healey's whole oeuvre. And this is also why we can never afford to stop attending to the specifics of concrete, situated planning work and pay careful attention to the 'how' as well as the 'where' of planning.

Following the above presentation of the papers in Part III.2, the remainder of this introduction will focus on impressionistically describing how Healey's perspective and thinking on planning practice has developed through her career. It will particularly centre-stage her conceptualisation of the situated agency of planning practice, its potential to 'make a difference' and the various strands of social scientific theory that appear to have been crucial inspirations for her in the development of her ideas.

Agency in Context

Patsy Healey brought with her into the field of planning studies an academic background and personal experiences which, from the very beginning of her scholarship, inspired her to pursue an interest in the 'thick description' (cf. Geertz, 1973) of the micro-politics of on-going planning practice (see e.g. Healey, 1974b; Healey and Underwood, 1978). Long before the concept or approach came into vogue in the social sciences, Healey's research approach showed signs of a truly 'abductive' scientific intuition (see further e.g. Alvesson and Sköldberg, 2009), building upon the co-constitutive and creative interplay of theories and empirical studies, which, in her case, has primarily consisted of ethnographic fieldwork of contemporary planning practice. Healey and Underwood (1978) here stands out almost as a text-book example of how to creatively engender this type of theory-empirics interplay where a particular issue of interest guides the assembly of a theoretical discussion, which, in turn, is allowed to reshape the formatting of the research task and research questions, and where the empirical results are then utilised to reshape the theoretical framework.

It might appear a strange thing to claim about someone who is generally considered as one of the greatest planning theorists of all time that Healey, in actuality, never appears to have been very interested in 'doing theory' per se. Rather, she is interested in the specific difference certain ideas can make in a particular time and place (see also Healey, this volume). Theory seems to always have been a tool for Healey, to be used to further our understanding of the conundrums raised by her empirical ethnographic inquiries. Concerning 'theory' as an abstract project, Healey – carrying first-hand experience of much ostensibly 'progressive' but in practice often destructively dogmatic and self-referential Marxist theory of the 1970s – instead cautions us that there exists 'a danger' that the 'richness of research and theoretical development in our field' will lead to planning scholars reading 'the world emerging around us too much through the weight of our own scholarship'. In line with this research ethos, Healey has always been wary of abstract, context-independent and purportedly universal models of the world. This scepticism is already evident in some of her earliest work (e.g. Healey, 1974) and reappears throughout her publications. It could easily be speculated that Healey's pronounced penchant for ethnographic studies of the micro-politics of planning practice is what first led her to take an interest in the work of a scholar who, to some degree, intellectually, might appear as her trans-Atlantic scholarly kindred spirit, John Forester (see also further below).

Through her contextually situated understanding of planning practice, Healey has not only been interested in the difference context makes to planning, but also in how planning can come to affect and to some extent shape this 'context', i.e. the potential situated agency of planning to affect the unfolding fate of places and how this agency can be generated and harnessed for 'progressive' inclusionary and democratising ends. This aspect of her work appears to have become increasingly prominent and elaborate over time. In her earliest work (e.g. Healey, 1974; Healey and Underwood, 1978) Healey relays a somewhat vague idea that planning should be able to contribute to 'societal betterment' but also describes the observed broad disappointment among the practitioners she studied when actual planning practice repeatedly failed to live up to this promise. In particular, together with Jacky Underwood (1978), she argues that planning practice appears unable to assume a leading role in urban development processes and that the agency of planning practice is severely limited and generally restricted to marginal adjustments. This is why they also admonish planners to try to come to grips with the insight that their work is a governmental practice similar to all such – only focused on a specific area of intervention. A year later, on a similarly critical note regarding the potential agency of planning practice to make a positive difference in society, Healey asks herself whether the conscious application of networking skills (see further below) can '*deflect* power from existing positions and hence *create* new patterns of resource flows' (Healey, 1979:61, emphasis in original). But she arrives at a generally negative conclusion to this question based on the reasoning that 'if interest conflict is recognised as fundamental to social organisation, then situations must be anticipated where neither better communication nor better bargaining will resolve an interest conflict' (Healey, 1979:65).

Two decades later, Healey (1997b) still displays thorough-going scepticism regarding existing planning practice, at least concerning the efficacy of individual, context-insensitive attempts at urban governance interventions. Writing in opposition to the projectified order of the day of urban development in Britain at that time, her account is a nuanced critique of the then new-found political infatuation with public-private partnership arrangements in urban renewal projects. Reminiscent of her earlier critical remarks concerning 'connective planning' (1979) Healey here also homes in on the fallacies and dangers of being insensitive to the 'where of things', the structural context of all-too-well-intended initiatives. Concerning the specific initiative discussed in the text, she notes that no matter whether this project undertaken in Cruddas Park, an underprivileged neighbourhood in inner city Newcastle-upon-Tyne, was imagined as a 'partnership of equals'

– in practice 'people were not equals' for '[t]hey had different experiences, different priorities and wielded differential power resources in the wider world' (Healey, 1997b, 270). Therefore, 'building up partnership relations was shot through with tensions and inequalities in power relations', which meant that '[w]hat to some looked like a partnership of equals because they sat round a table together, for others seems like patriarchal patronage or just plain terrifying' (Healey, 1997b:270).

A major fault of the Cruddas Park Initiative, as analysed by Healey, was that it became enacted as a showcase for business intervention in social service provision, with the intent to put on display the productivity of business approaches to social issues and how 'trickle down'-effects of business stimulating policies supposedly would also come to the benefit of socioeconomically weaker neighbourhoods. In the spirit of 'Enterprise', all façade and little substance (cf. Law, 1994), the project catapulted itself into 'doing' without any proper prior thinking ahead, sensing or reckoning. Therefore, there was a 'low appreciation of the systemic nature of some of the issues addressed. Problems were more likely to be identified as existing due to specific people in specific places than as relating to questions of institutional structure, or the "way things go on"' (Healey, 1997b:275). The lack of focus on building thorough-going institutional capacity turned the Cruddas Park Initiative into a smoke-and-mirrors 'flashy' one-off run by self-ordained 'city fathers' and civil servant 'mandarins', imposed from above, approaching local residents with the attitude that they as outsiders to the neighbourhood were there 'to lead not to learn' (Simey, 1995:269, cited in Healey, 1997b:269).

The 1997 article evinces a strong impression that it should be possible to do planning *better* than this, and not only in relation to the minutiae, but also concerning the overarching mission or purpose of planning work – a better planning for a better society. To illustrate her thinking on this point, Healey utilises a maritime metaphor, depicting the 'mainstream' capitalist economy not as a 'stable flat-bottomed raft, with nooks and crannies for all ways of living', but instead a 'high-sided imperial machine, if somewhat battered by global competition … dominated by principles of hierarchy, class and rational order'. Displaced from this mainstream 'imperial machine', Healey notes that residents of underprivileged neighbourhoods in 'the harsh seas of life "on the edge"' have often constructed their own alternative 'lifeboats' (Healey, 1997b:285). Healey describes how the overarching purpose of the Cruddas Park Initiative was to extend 'lifelines' from the mainstream economy-ship to the residents of the alternative economy lifeboats. But this is an ambition of which Healey fundamentally questions the validity, for why should the residents of the lifeboats make an effort to 'climb up' into the mainstream ship, 'merely to be thrown off again', or perhaps only to find that their small lifeboats, often crafted 'with a great deal of social and psychological effort', might have disappeared (Healey, 1997b:285)? Should planning then keep on twining lifelines to the imperial machine? Or should the whole ship, the mainstream capitalist economy, be 'demolished'?

In contrast to many of those who cite her work, but similar to thinkers such as Gibson-Graham (1996; also see Steele, this volume), Healey radically opts for the second option, arguing that public-private partnership projects of the kind seen in Cruddas Park 'do not change systems' but merely 'serve to highlight the institutional barriers to systemic 'learning' and 'transforming'' (Healey, 1997b:285; see further Duggan and Ronayne, 1991; Taylor, 1995; Thake, 1995). Therefore, for the necessary 'coherent long-term project of institutional transformation' to ever be realisable, we should now work to 'demolish the ship and strengthen the raft', a raft which 'would not be like the welfare state of old, in which [the state] delivered services but without much sensitivity and respect for people's circumstances' (Healey, 1997b:286). Instead, it would be a 'project of institutional transformation' which would 'emphasise the relational work of building links, rather than conducting experiments' (Healey, 1997b:286). It would promote the collective rather than the individual, and would foster collaboration rather than 'promoting competition between areas and among initiatives in capturing the 'scraps' of resources available for community development' (Healey, 1997b:286).

Healey stops just short of conceptualising this as a transformation towards a fully-fledged socialist society, envisaging a role for both 'business' and 'citizens' on the raft, but built on a higher degree of mutual respect and balance of resources and power. According to her vision, it would be a raft characterised by its 'nooks and crannies' which would enable it to 'operate in respectful ways towards all citizens, rather than expecting everyone to fit into the categories of official systems' and guided by a 'moral requirement' towards acting respectfully towards all those who are affected by an issue (Healey, 1997b:286). As Healey notes based on her analysis of the Cruddas Park Initiative, if one is serious about building such 'institutional capacity' towards more thorough-reaching structural change, this implies quite substantial investments of time, commitment and resources to work on a longer time-scale with building up mutual trust and understanding, even if this can be achieved in a single initiative, stabilising and extending such developments more widely is difficult if not consciously and programmatically pursued. This type of insight then came to feed into Healey's influential thinking on institutional capacity building (see e.g. Healey et al., 2003), which was a prime inspiration for Janice Barry and André Sorensen in the work they discuss in this volume (see Barry Chapter 19, Sorensen Chapter 25).

As is evident from the above, even if still sometime scathing in her critique of existing planning practices, Healey's idea of the potential agency of planning to contribute to a better society does appear to develop from a thorough scepticism in the 1970s towards a more hopeful vision, particularly in the 1990s. To some extent this shift can perhaps be attributed to important influences from US planning scholars such as John Forester (e.g. 1989) and also Judith Innes (e.g. De Neufville, 1987) with whom Healey became closely acquainted in the late 1980s. Following these new influences, the point of gravity of Healey's work during the 1990s appears to have begun to substantially shift from primarily a (critical) description and diagnosis of the present state of existing planning, towards more directly also confronting questions concerning how planning could be performed in more 'progressive' and normatively 'better' ways from a pluralistic-democratic viewpoint.

The Communicative Turn

Patsy Healey has always been particularly interested in the ideas and practices that can make an actual difference: that – if we may use such presumptuous wording – could potentially function to make the world a little bit better (see also the Introductions to Parts II, IV and Healey, this volume). In a critical review of the potential to draw upon Friend, Power and Yewlett's (1974) idea of 'connective planning' based on informal 'reticulist judgements' as a normative ideal for planning, Healey argues that even if the authors' analysis probably resonates with the experiences of many planning practitioners, one should nevertheless be cautious about accepting this at face value as a desirable ideal or organising principle for planning practice. She draws upon theoretical resources from both political economy and social anthropology to make the point that 'connective planning' is a deficient ideal for planning practice since it is 'insufficiently concerned with issues of macro-structure' (Healey, 1979:59). This is due to the lack of understanding that network relations are asymmetric which, in turn, obscures the fact that it is 'the command of resources which determines who is a member of a decision network' and that the purpose of networking to a large extent is 'the manipulation of those with command over resources' (Healey, 1979:58). Instead she raises as the central issue of content the question regarding 'in whose interests are these creative, innovatory operational styles to be deployed?' (Healey, 1979:59). Through highlighting the problem of democratic accountability in relation to this type of informal or semi-formal governmental practices, Healey's line of critique thus not only prefigures the development of thinking concerning 'democratic network governance' in the 2000s (see for instance, Torfing and Sörensen, 2007) but also the strong contemporary critique of the proliferation of amorphous and semi-formal governance arrangements in contemporary planning practice (see e.g. Allmendinger and Haughton, 2011; Metzger, 2011; Metzger et al., 2014).

Even if 'connective planning' clearly was not the way forward for planning practice in the 1980s according to Healey's vision, she nevertheless, in the years to come, found rekindled hope in Giddens' structuration theory that planning might contribute to positive societal developments (see Part III.1). As mentioned above, the influence from American planning scholars, such as Forester and Innes, also appears to have been critical for the further development of her thinking on how better planning might contribute to better societies. As a consequence of this close alignment, from the early 1990s onwards, Healey came to be identified, and to some degree identify herself, as part of the theoretical movement of 'communicative planning theory'.

Healey's critique of connective planning in 1979 can be seen as part of a broader search for ways to navigate an increasingly fractured landscape of British urban governance which, nevertheless, still had planning as a self-evidently central institution. Compared to this, the early 1990s evinced a very different situation for British planning. At this point planning in the UK was an institution and profession scorned and publicly derided by a central Government which programmatically did all in its power to diminish and circumscribe any potential societal agency of planning practice. Thus, at this point in history, the challenge that confronted British planning academics who held a belief in the potential merits of planning as a societal institution not only concerned figuring out how to make planning better, but more fundamentally, how to salvage any remnants of a profession in the face of directly hostile attacks from a Government which questioned the fundamental raison-de-être of planning practice. Further, this political attack on the institutions of British planning also coincided with the broader break-through of 'postmodern' thinking in academia, which was perceived to challenge many of the fundamental assumptions of planning practice; a shot below the waterline of mainstream planning as an expert profession. Taken together, these developments appeared to generate something of a 'perfect storm' which fuelled fundamental uncertainty concerning the legitimacy, relevance and efficacy of any planning ambition.

In the face of the dual challenge of, on the one hand academic post-modernism challenging the 'grand narrative of progress' and, on the other hand, political calls to 'cut red tape' and 'streamline' government, Healey's innovative response was to boldly suggest that planning needed to be reinvented and make a place for itself in this new governance landscape through becoming not less, but more. Planning practice should, according to her vision, function as a form of coordinating practice for place-focused strategic policy development and delivery in governmental decision making: 'a process for collectively, and interactively, addressing and working out how to act, in respect of shared concerns about how far and how to 'manage' environmental change' (Healey, 1992:152). A main source of inspiration behind this thinking appears to have been her engagement with the American planning scholars previously mentioned, further strengthening her conviction that the planning endeavour can, and should, be about much more than just 'plan-making', and also rekindling and reinfusing with energy her belief in the potentials of planning practice in the aftermath of the (at least from a planning perspective) largely depressing 1980s.

The seminal paper 'Planning Through Debate' (1992) is part of a wider project of Healey's at this time (including also e.g. Healey and Gilroy, 1990; Healey, 1997b; Healey, 1998) to attempt to resurrect what she considers the 'spirit of planning' – which she has since then continued to pursue in different forms (see e.g. Healey, 2010, 2012b). This project is conceptualised by Healey as 'a democratic enterprise, aimed to promote social justice and environmental sustainability' (Healey, 1992:143). She asks, in the face of the postmodernist critique of grand narratives and monolithic logo-centrism and the increasingly obvious problem of framing one singular 'public interest'; 'what kind of planning can be compatible with our contemporary understandings of a democratic attitude' (Healey, 1992:144)? Healey then goes on to highlight 'the promise a communicative form of planning in promoting and realising a progressive democratic attitude' as a potentially productive approach which may offer 'the hope that "progress", a "project of becoming" is still possible' (Healey, 1992:146, 152). The new vision for planning which Healey promoted at this time was to some extent based on a Habermasian idea of communicative rationality. It was underpinned by a pluralistic ethos which nevertheless strives towards finding or developing a common ground and collective trajectory towards the future based on, and with respect for, deep differences, where

> 'right' and 'good' actions are those we can come to agree on, in particular times and places, across our diverse differences in material conditions and wants, moral perspectives and expressive cultures and inclinations. We do not need recourse to common fundamental ideals, or principles of 'the good social organisation' to guide us. Planning, and its contents, in this conception, is a way of acting which we can *choose*, after *debate*. (Healey, 1992:151, emphasis in original)

She further adds that: '[i]t is this direction which, in the present author's view, holds an important promise and challenge for planning, and, more generally, for democracy' (Healey, 1992:152).

Healey broadly suggests that this general direction for planning can be pursued through a focus on developing practices for shifting 'systems of meaning'; an intuition which will later come to mature into her 'sociological-institutionalist' approach to planning [cf. also Sorensen, Chapter 25 in Part IV]. Healey's creative reconceptualization of the purpose of planning practice in her work from this period thus reimagines planning as an arena to enact democracy 'in practice'. In these collective ventures '[w]e may be able to agree on what to do next, on how to "start out", and "travel along" for a while', and even if we cannot know 'where this will take us' such planning exercise can help us 'act with hope and ambition to achieve future possibilities' (Healey, 1992:156). Healey makes the suggestion that environmental development issues are particularly suited for this purpose in contemporary fragmented societies through enacting an ethos of living 'together but differently … in shared environments' (Healey, 1992:156).

In a fashion also reminiscent of the thinking of scholars such as Leonie Sandercock (1998), Healey further argues that the living enactment of this pluralistic, democratic ethos demands that we open up planning processes to new voices and ways of expression by way of 'discussion of moral dilemmas and aesthetic experience, using a range of presentational forms, from 'telling stories' to aesthetic illustrations of experiences', which, as a consequence, then results in planning processes where '[s]tatistical analysis coexists in such processes with poems and moral fables' (Healey, 1992:154). If 'opened up' in this way, Healey reasons, this would lead to a situation in which the progressive potential of planning practice is realised in a fashion in which '[u]ltimately, the transformative potential of communicative action lies in the power embodied in "the better argument"' (Healey, 1992:155). This conviction can, of course, be accused of being naïve, and this reaction is something that she appears to have expected when she wrote that, in the eyes of many, her proposition probably would come across as 'too leisurely … too unfocused and diffuse … idealistic and innocent' (Healey, 1992:159). And accordingly, just as she predicted, the paper generated some substantial criticism, particularly coming from political economy-inspired scholars and by those inspired by the thinking of Michel Foucault (see e.g. Allmendinger and Tewdwr-Jones, 1998; Yiftachel and Huxley, 2000).

The ensuing debate, which pitched ideas about the relative merits of conflict or consensus as the proper goal of planning practice against each other, has fundamentally shaped the following twenty years of planning theory, and shows no signs of subsiding yet. The writings of Healey often occupy a central position and point of contention in these debates, but for the record it is worth noting that in contrast to other communicative planning theorists, Healey was never particularly focused on consensus-formation per se as the focal point of planning work. Rather, she repeatedly highlighted how agreements in planning processes are always partial and temporary and are more about '"taking a breath" and "sorting things out"' than some final settlement of issues (Healey, 1992; cf. also e.g. Pløger, 2014). Instead, Healey puts a much stronger emphasis on the importance of mechanisms for ongoing critique and also stresses the problematique related to the fluidity and complexity of individual and group identities. This insight also opens up complicated issues for the pluralistic model of democracy by arguing that '[i]f collective concerns are ambivalent and ambiguous … a communicative process should allow acknowledgement that this is so, perhaps unavoidably so', which means that 'the dilemmas and creative potentials of ambiguity [should] enrich the inter-discursive effort, rather than being washed out in the attempt to construct a one-dimensional language' (Healey, 1992:156). Thus, upon closer scrutiny, Healey's 'take' on communicative planning appears as a somewhat more politically conscious alternative to the American version. Whether it was still too naïve will likely remain a moot point. As for Healey's own work after the big 'communicative planning debate' it appears as if she held on to the democratising ethos of, for instance, *Planning Through Debate*, but by-and-large moved on to new inspirations which could function as 'hooks' or underpinnings for her argument, turning more towards interpretive policy analysis and classical pragmatism than the philosophy of communicative rationality.

Strategic Spatial Planning and Relational Geography

Parallel with the development of her work on communicative planning Healey also built stronger links with the continental European spatial planning community during the 1990s, perhaps foremost through her deep engagement in the formation and development of AESOP, the Association of European Schools of Planning. Through these engagements Healey came into contact with the resurgence of thinking on 'strategic spatial planning' that was occurring in Europe at this time through the combined contributions of scholars such as Louis Albrechts in Belgium, Barrie Needham, Willem Salet and Andreas Faludi in the Netherlands, Abdul Khakee in Sweden, Claus Kunzmann in Germany, and Alessandro Balducci and Luigi Mazza in Italy, among others. In a number of texts Healey developed the idea of a new strategic spatial planning together with other key academics in this endeavour (see e.g. Healey et al., 1997, Albrechts et al., 2003).

One of her most important individually authored texts in this part of her broader oeuvre is Healey (2004). In this article Healey specifically highlights the spatial imagery of strategic spatial planning 'episodes' asking herself the question: what difference does it make which the images and concepts are with which we present space and place in planning endeavours? As so often, the focus of Healey's interest is specifically in the ideas and practices that may potentially help articulate common concerns and further function to mobilise collective action around them. Hence, Healey writes that she understands 'strategic spatial planning' as 'self-conscious collective efforts to re-imagine a city, urban region or wider territory and to translate the result into priorities for area investment, conservation measures, strategic infrastructure investments and principles of land use regulation' where '[t]he term "spatial" brings into focus "the where of things", whether static or in movement' (Healey, 2004:46). Further, in the context of the composite term 'strategic spatial planning', *strategic* should be taken to mean a framework, implying selectivity, 'a focus on that which really makes a difference to the fortunes of an area over time', and *planning* should be understood as 'a developmental movement from past to future' fuelled by an imagination 'able to project a transgenerational temporal scale, especially in relation to infrastructure investment, environmental management and quality of life' (Healey, 2004:46).

Admitting that none of these things are easily achievable in practice, Healey argues that rather than therefore abandoning the whole idea, a lot of effort instead has to be put into the design of processes so as to find out 'how significant stakeholders can be mobilised to develop strategic agendas in a 'diffused power' context, and become cohesive enough to develop 'collective actor' power' (Healey, 2004:46). Based on this interest Healey then suggests that we generally must pay closer attention to the concepts in play in episodes of strategic spatial planning, and specifically to the spatial concepts that are articulated and evoked in such contexts. The reason for this is that these concepts make a difference – or in Healey's words, do 'institutional work' – as 'the vocabulary takes on a meaning in relation, potentially, to understanding place qualities and the spatiality of relations, to mobilising activity (in coordination, investment, new regulatory approaches, etc.), to legitimising action, and to dreaming about possible futures (a better quality environment and place to live and do business, a different kind of politics, etc.)' (Healey, 2004:47).

In her analysis of these geographical concepts Healey draws upon then-recent developments in human geography, i.e. the 'new relational geography' spearheaded by scholars such as John Allen, Doreen Massey, Ash Amin and Nigel Thrift (e.g. Allen et al., 1998, Amin and Thrift, 2002), which she had begun to explore from the early 1990s onwards. As Healey notes, in this emerging tradition, 'distance' is not understood as the quantity of physical space between two objects in Euclidean space. Rather, space is understood as topological and relational, where things seemingly physically proximate to each other may be worlds apart topologically but nevertheless closely related to other places and things on the other side of the globe. Thus, 'the social relations which transect a specific piece of territory may each have a different spatial reach, just as they may have different temporalities. They may or may not intersect as they pass 'over' or 'under' each other' (Healey, 2004:47). These spatial networks may thus coexist in a specific site without in any way infusing it with meaning as a specific 'place', 'still less mobilising that meaning to do political work' (Healey, 2004:47). Spatial concepts and vocabularies 'grasp' these relations in different ways and are therefore 'significant in mobilising attention, an imaginative effort which builds from and contributes to shaping conceptions of identity', that 'this place here' has some integrity as a specific place with specific boundaries and particular qualities (Healey, 2004:49; cf. Metzger, 2013b). Such spatial conceptions thus influence how we act in the world and, therefore, 'act back' on the spatial materialities of the relations they schematise, '[t]hus, the creation of material objects and the construction of conceptions of objects are co-generative processes'. Hence the formation of the spatial patterning of the materialities of social relations and place qualities is co-emergent with the 'naming of these spatialities and qualities' (Healey, 2004:49).

In a similar 'co-generative' way, Healey argues, a spatial imaginary may co-relate with particular political ideas pointing towards specific solutions and types of interventions. Thus ambitions to 'mould a new kind of planning "politics"' must go hand in hand with a transformation of 'the spatial vocabulary' (Healey, 2004: 51). The struggles over spatial concepts are therefore

> part of wider political struggles over the approach to area and territory development and over modes and arenas of policy development and delivery. These struggles and their outcomes are important because, if the strategic spatial organizing concepts influence government investment (especially in infrastructure and urban development) and help to shape land use regulation practices, then there will be significant material effects on emergent socio-spatial dynamics. (Healey, 2004:61)

As a consequence, concepts are interesting because they are potentially performative and reality-shaping rather than mere reflections of an external world, and can carry strong agency. They not only 'affect the structuring of political debate and struggle over the impacts of projects, the distributive justice of investments and regulatory principles, and over imagined futures', by extension they also quite materially influence the unfolding fate of places through influencing the behaviours of government in developing infrastructure and market actors in acquiring sites and developing project proposals (Healey, 2004:64).

Nevertheless Healey, as so often, cautions the reader not to assume that the tangent of the agency produced around a concept is absolute and stable. Rather, we must understand it as evolving, situated and contextual for '[i]n their usage, the metaphors may act transitively, to carry meanings from one institutional site to another. But metaphors may also be co-opted to serve very specific local purposes', again highlighting the importance of being attentive to the local institutional contexts of any planning endeavour (Healey, 2004:63). Shifts in spatial vocabularies can only be meaningful within broader strategies to shift the institutionalised way of 'going about things' in a particular place. What is important is thus not the particular words used to describe spatial phenomena, but rather the encouragement of 'a relational geographic imagination' which can function to 'explore the many material relations and mental images of place and place quality which are locally-important, and the interaction between these and the wider relations of which they are a part'. 'It is the locally-specific realization of a relational geographical imagination that could results in spatial strategies which have the potential to deliver real progressive effects for the quality of life in specific places' (Healey, 2004:65).

Conclusions

In her scholarly endeavour, Patsy Healey has persistently focused on attending to and reasoning around 'really existing' planning practices, evincing a deep on-going interest in questions concerning how planning work is performed in specific, spatio-temporally, culturally and institutionally situated episodes. Healey's thinking around the practices of planning have gone through a number of metamorphoses, from her early social anthropology-inspired writings, across more strongly argued political economy-perspectives in the later 1970s, towards the more sanguinely formulated prospects of the 'communicative turn' in the 1990s, and further ahead into more cautious discussions of the potential merits and risks of various inclinations of new urban governance models and strategic spatial planning.

Throughout this body of work, Healey repeatedly raises questions concerning the preconditions of planning work and the difference these make in formatting the field of the possible for planning practice. But she also turns the question around, asking what difference planning can make in influencing or changing this context, which must never be taken as immutable and wholly given. Increasingly obvious over time is also her concern with not only that which *is*, but also with what could potentially *become*, relating to a concern about how planning practice may come to function in a better way from a broadly 'progressive' perspective. That is, how planning practice can become more democratic, open, allowing and empowering to the weak by giving those groups a say in the unfolding fate of society on terms with which they may themselves be comfortable. The way Healey has gone about this has at times attracted an amount of criticism. To borrow an expression from Brand and Gaffikin (2007), the experience of trying to introduce collaborative planning 'in an uncollaborative world' resulted in many great disappointments and some grave misinterpretations – probably both intentional and not – of the ideas and ideals invested in this concept. This is not the place to attempt to assess the fairness of these critiques in various aspects, but we can however conclude that Healey's ideas on the subject have become a central pivot of discussions, perhaps to some degree *the* centre, for much of the scholarly debate in planning theory for the past two decades.

References

Albrechts, L., Healey, P., and Kunzmann, K.R. (2003). Strategic spatial planning and regional governance in Europe. *Journal of the American Planning Association*, 69(2): 113–29.

Allen, J., Massey, D. and Cochrane, A. (1998). *Rethinking the Region*. London: Routledge.

Allmendinger, P., and Haughton, G. (2010). Spatial planning, devolution, and new planning spaces. *Environment and planning. C, Government & policy*, 28(5): 803.

Alvesson, M. and Sköldberg, K. (2009). *Reflexive Methodology: New vistas for qualitative research.* (2nd edn) Los Angeles: SAGE.

Amin, A. and Thrift, N. (2002). *Cities: Reimagining the urban.* Malden, MA: Polity.

de Neufville, J.I. (1987). Knowledge and action: Making the link. *Journal of Planning Education and Research*, 6(2): 86–92.

Duggan, C. and Roynane, T. (1991). *Working Partners? The state and the community sector.* Dublin: Work Research Centre.

Faludi, A. (1973). *Planning Theory*. Oxford: Pergamon.

Forester, J. (1989). *Planning in the Face of Power*. Berkeley: University of California Press.

Friend, J.K., Power, J.M. and Yewlett, C.J.L. (1974). *Public Planning: The inter-corporate dimension.* London: Tavistock.

Geertz, C. (1973). *The Interpretation of Cultures: Selected essays*. New York: Basic Books.

Gibson-Graham, J.K. (1996). *The End of Capitalism (as we knew it): A feminist critique of political economy.* Cambridge, MA: Blackwell Publishers.

Healey, P. (1974a) The problem of ideology. *The Planner*, 60(3): 602–4.

Healey, P. (1974b) Planning and change. *Progress in Planning*, 2(3): 143–237.

Healey, P. (1992) Planning through debate. *Town Planning Review*, 63(2): 143–62.

Healey, P. (1997a[/2006]). *Collaborative Planning: Shaping places in fragmented societies.* London: Macmillan.

Healey, P. (1997b) City fathers, mandarins and neighbours: Crossing old divides in new partnerships. In O. Kalltorp, I. Elander, O. Ericsson and E. Franzen (eds) *Cities in Transformation: Transformation in cities* Aldershot: Avebury, pp. 266–88.

Healey, P. (1998). Collaborative planning in a stakeholder society. *Town planning review*, 69(1): 1.

Healey, P. (2004). The treatment of space and place in the new strategic spatial planning in Europe. *International Journal of Urban and Regional Research*, 28(1): 45–67.

Healey, P. (2007) *Urban Complexity and Spatial Strategies: Towards a relational planning for our times.* London: Routledge.

Healey, P. (2008). Knowledge flows, spatial strategy making, and the roles of academics. *Environment and planning. C, Government & policy*, 26(5): 861.

Healey, P. (2010) *Making Better Places: The planning project in the 21st Century*. London: Palgrave Macmillan.

Healey, P. (2012a) The universal and the contingent: Some reflections on the transnational flow of planning ideas and practices, *Planning Theory*. 11(2): 188–207.

Healey, P. (2012b). Re-enchanting democracy as a mode of governance. *Critical Policy Studies*, 6(1): 19–39.

Healey, P. (2015) Finding my way: a life of intellectual exploration and institution-building. In: Haselberger, B., Saija, L. and Knight, J. (eds) *The Future of Planning: Personal stories in the evolution of planning thought*. London: Routledge (forthcoming).

Healey, P., and Gilroy, R. (1990). Towards a people-sensitive planning. *Planning Practice and Research*, 5(2): 21–9.

Healey, P., and Underwood, J. (1978). Professional ideals and planning practice: A report on research into planners' ideas in practice in London Borough planning departments. *Progress in Planning*, 9, 73–127.

Healey, P., Khakee, A., Motte, A. and Needham, B. (eds) (1997). *Making Strategic Spatial Plans: Innovation in Europe*. London: UCL Press.

Healey, P., Magalhães, C. de, Madanipour, A. and Pendlebury, J. (2003) Place, identity and local politics: Analysing initiatives in local governance. In: M.A. Hajer and H. Wagenaar (eds) *Deliberative Policy Analysis: Understanding governance in the network society*, Cambridge: Cambridge University Press, pp. 60–87.

Law, J. (1994). *Organizing Modernity*. Oxford: Blackwell.

McLoughlin, J.B. (1969). *Urban and Regional Planning: A systems approach*. London: Faber and Faber.

Metzger, J. (2011). Neither revolution, nor resignation: (re) democratizing contemporary planning praxis: a commentary on Allmendinger and Haughton's 'Spatial planning, devolution, and new planning spaces'. *Environment and Planning. C, Government and Policy*, 29(2): 191–6.

Metzger, J. (2013a). Placing the stakes: The enactment of territorial stakeholders in planning processes. *Environment and planning A*, 45(4): 781–96.

Metzger, J. (2013b). Raising the Regional Leviathan: A Relational-Materialist Conceptualization of Regions-in-Becoming as Publics-in-Stabilization. *International Journal of Urban and Regional Research*, 37(4): 1368–95.

Metzger, J. (2014). Spatial planning and/as caring for more-than-human place. *Environment and Planning A*, 46(5): 1001–11.

Metzger, J., Allmendinger, P. and Oosterlynck, S. (2014). The contested terrain of European territorial governance: new perspectives on democratic deficits and political displacements. In J. Metzger, P. Allmendinger and S. Oosterlynck (eds) *Planning Against the Political: Democratic deficits in European territorial governance*. New York: Routledge.

Pløger, J. (2014). Impossible common ground: planning and reconciliation. In J. Metzger, P. Allmendinger and S. Oosterlynck (eds) *Planning Against the Political: Democratic deficits in European territorial governance*. New York: Routledge.

Sandercock, L. (1998). *Towards Cosmopolis: Planning for multicultural cities*. Chichester: John Wiley & Sons.

Simey, M. (1995) The politics of social change, *Social Work*, 66(4): 1–5.

Sørensen, E. and Torfing, J. (eds) (2007). *Theories of Democratic Network Governance*. Basingstoke: Palgrave Macmillan.

Taylor, M. (1995). *Unleashing the Potential: Bringing residents to the centre of regeneration.* York: Joseph Rowntree Foundation.

Tewdwr-Jones, M., and Allmendinger, P. (1998). Deconstructing communicative rationality: A critique of Habermasian collaborative planning. *Environment and planning A*, 30(11): 1975–89.

Thake, S. (1995). *Staying the Course: The role and structure of community regeneration organisations.* York: Joseph Rowntree Foundation, York Publishing Services.

Yiftachel, O., and Huxley, M. (2000). Debating dominance and relevance: Notes on the 'communicative turn' in planning theory. *International Journal of Urban and Regional Research*, 24(4): 907–3.

Chapter 13
Networking as a Normative Principle with Particular Reference to Local Government and Land Use Planning*

Patsy Healey**

Network-Building as an Approach to Organisational and Inter-Organisational Co-Ordination

The problem of co-ordination is a major theme in both justifications for more planning and in the practice of planning. This reflects one of the central concerns of any manifestation of planning with inter-relating phenomena and integrating action around a specified direction.

Co-ordination has been discussed in a number of ways in public administration and planning theory, different theoretical perspectives leading to different normative models for co-ordination. Where centralised direction is advocated, the mechanism for co-ordination often proposed is a centrally agreed plan which specifies, goals, programmes and targets for the various organisational parts or sectors involved in implementing the defined strategy.[1] Examples here are many National Economic Plans or certain styles of corporate planning in British local government.

Such an approach has been extensively criticised as being over-specific and inappropriate for situations in which organisations are constantly faced with new and unpredictable demands, and unpredictable resources with which to act. In the post-war United States, and in the United Kingdom from the mid-1960s onwards, an alternative model of rational decision-making processes was proposed.[2] This replaced centrally adopted action-specific plans with a goal-orientated decision process supplemented by rational methods for arriving at appropriate means to achieve the ends specified in the goals.[3] The common goals were the mechanism for coordination and the rational methods were to ensure the design of appropriate actions.[4]

This approach has itself been attacked both for its political model of a consensus situation in which common goals are accepted and maintained, and for the intellectual demands involved in devising rational means of achieving goals in any situation.[5] For some critics, notably Lindblom (1965), any notion of co-ordination via common frameworks (such as centrally-adopted plans) or common goals and values (as in rational decision-making processes) is abandoned. A pluralist model of society is advocated, operating on *laissez-faire* principles. Co-ordination occurs, not by formal co-operation within and between organisations, but through the process of *partisan mutual adjustment*, by which each agency acts independently, anticipating and adjusting as far as possible to the actions of others when designing its own actions, just as individual firms operate within the market.

Others have turned to the behavioural model of organisation as systems of actors, all involved in responding to pressures from the organisational environment through agency learning behaviour.[6] This approach assumes a more or less pluralist organisation of society. The emphasis in more recent work has

---

* Previously published in *Local Government Studies*, 5(1): 55–68 (1979).

** The author wishes to thank Ian Bentley and the students of Urban Design at Oxford Polytechnic for prompting her to write this paper.

1 Essentially, this involves co-ordination by rules defining what action to take.

2 See Friedmann, 1973, for the introduction of these ideas into planning theory discussions.

3 See Simon, 1961, and several articles in Ed. Faludi, 1973.

4 Van Gunsteren, 1976, would classify both approaches as examples of rational-central rules, although in the second case the rules define the methods, not the actions.

5 See Banfield, 1959, and Braybrooke and Lindblom, 1963.

6 See Lawrence and Lorsch, 1967, for this model in organisation theory, and Friend and Jessop, 1969, and Faludi, 1973, for its application to planning organisations.

been on organisational responses in a 'turbulent' environment.[7] In the field of local government studies in the United Kingdom, one of the most significant contributions from this approach has been the work of John Friend and others. Initially their interest was in the pressures which might lead to inter-agency co-operation in the making of 'strategic choices' (Friend and Jessop, 1969). A subsequent study looked at the way a jointly agreed strategy was implemented by the agencies concerned (Friend, Power and Yewlett, 1974). It was in this study that the notion of a *decision network* linking relevant organisational representatives was proposed. The concept of inter- and intra-organisational networks connecting relevant actors as a mechanism for co-ordination has received considerable attention in recent years, especially in the United States. The relevance to British local government has also been emphasised (Harris, 1976). How far does this notion of network-building, or 'reticulism' as Friend, Power and Yewlett (1974) have called it, provide a justifiable and viable alternative mechanism for organisational co-ordination?

In this paper, I explore the concept of networks by looking in more detail at Friend, Power and Yewlett's ideas, and expanding on them on the one hand by emphasising the differential distribution of control over resources within a network and on the other by referring to the well-developed discussion of social networks in social anthropology (which Friend, Power and Yewlett acknowledge as the source of their ideas on networks). I relate the discussion from time to time, as Friend, Power and Yewlett do, to the practical side of British local government and in particular to land use planning, in which co-ordination problems are especially acute.[8]

Friend, Power and Yewlett's Concept of Decision Networks[9]

As outlined above, Friend, Power and Yewlett's approach derives from a view of organisations as systems within a pluralist model of political organisations. Organisations interact responsively with their environment, their central concern being the design and operation of responsive ('learning') behaviour. Each organisation is an independent entity, co-operating with others where mutual benefit can be gained. In the earlier study by Friend and Jessop (1969), inter-agency co-operation was described in relation to the making of 'strategic choices', desirable to reduce mutual uncertainty in respect of each agency's actions. Friend, Power and Yewlett, however, put more emphasis on how actors in various agencies develop contacts with each other in order to implement agreed programmes, both the definitions of common strategies and the co-ordination of implementation being part of a general process of 'connective planning'.

Friend, Power and Yewlett postulate the existence of a 'policy system' (a set of arrangements to deal with a class of decision problems). This policy system covers an *action space* or field of decisions. Members of the policy system are linked to each other, or to those in a related policy system, through a *decision network*. This is 'an open network of communications among people acting either within policy systems or across the interfaces between them, which may influence the commitments reached on any specified class of decision problem' (p. 43).

The members of the network are seen as exploring more effective ways of operating. In other words, they are involved in innovating new ways of doing things in order to ensure effective organisational responsiveness. Some members of the network are more strategically placed than others, at nodal positions. Such people have a key role to play in using and developing the network. 'It is likely to be at such nodal points that judgements will be faced most frequently as to which further links in the network to activate in the course of exploring particular decision problems'. These judgements are seen as reticulist or networking judgements. 'The use of reticulist judgements in activating, fashioning, re-fashioning and – occasionally – creating decision networks provides one of the most important dynamic elements of the decision-making process' (p. 45). Consequently they advocate the development of skills in making such judgements. 'The making of reticulist judgements – in other words the mobilisation of decision networks in an intelligently, selective way – depends on a capacity to appreciate both the structure of the problem situations and the structure of organisational and political relations that surround them' (p. 364).

7 See for example Schon, 1971, and Argyris and Schon, 1974.

8 This is because land use planning involves co-ordinating public and private sector development activity, as well as intra-public sector co-ordination.

9 See especially chapters 2 and 13.

Thus the decision network is seen itself as dynamic, changing in form over time as the actors involved learn more about each other's actions and the dimensions of their action space in relation to the policy system as a whole. The network may include formally structured connections, but is in principle non-formal, to allow dynamic, creative and hence in theory responsive behaviour.[10]

Friend, Power and Yewlett then go on to explore in detail a range of policy systems involved in the process of town expansion in Droitwich, each network being illustrated by a complex interpersonal/interagency diagram (focussed on a key respondent – see the two examples illustrated in Figure 13.1). Central to their conclusions is the need for organisations to provide for the development of reticulist activities. They hypothesise that 'the practical influence of public planning depends on the disposition among agencies of skills and resources relevant to the selective activation of interagency networks' (p. 348).

Networks and Interests

For many working in the public sector on tasks that involve organisational co-ordination, the concept of interpersonal networks formalises their existing perception of their work situation as a web of social relationships. How reasonable though is Friend, Power and Yewlett's assumption that network interactions are centrally about the communication of information, and how supportable is their hypothesis that a major influence on network structure is the reticulist skills of network members?

Networks, in their view, are communication channels. Information flows across the web of relationships. Yet they do not distinguish information in any way which might allow an examination of the impact of a particular flow of information. They do, however, acknowledge that inter-organisational relationships are often about bargaining and negotiating, and that the kind of co-operation which they seek to achieve via the development of decision networks ('connective planning') is not likely when problems are complex and inter-related (pp. 40–41). In other words, they recognise that where there is insufficient basis for agreement, Lindblom's 'partisan mutual adjustment' is the more likely style of operation. There is an implication here that what is being explored within network relationships is related to what is being offered and exchanged between network members and what interests network members may be seeking to maximise.

This suggests that the structure of the network, the pattern of links and the locations of 'nodes', may be related to interests and to what is being exchanged. As Friend, Power and Yewlett describe it, a decision network is structured partly by who is involved in the network, which would appear to be those with functional responsibilities to participate in a particular policy system, and partly by the reticulist skills of network members. Yet it could be argued that it is as much the command over resources which determines who is a member of the decision network, and it is the manipulation of those with command over resources which the reticulist is out to achieve.[11]

By focusing on reticulist skills rather than bargaining over resources as the key to successful inter-organisational co-ordination, Friend, Power and Yewlett are emphasising the qualities of the salesman rather than his product. Yet the qualities of the salesman are in the long run less important than the value of his product.

In effect, Friend, Power and Yewlett's approach does not allow them to deal explicitly with the distribution of power and influence within networks. Such political questions are treated as external to the network, although the connection between decision networks, policy systems and the wider structures of power and interest in society are sketchily made.[12] Once questions are raised about the distributions of power and interest within a network, about the connection of this to wider political structures and about the impact of network-building (reticulism or networking) on both, then the problem of the accountability and legitimacy of the

10 Friend, Power and Yewlett refer to networks as 'aformal' (p. 45).

11 Friend, Power and Yewlett suggest that uncertainty about another agency's activities is the driving force leading to inter-organisational co-ordination via network-building. In the land use planning field, such uncertainty is likely to be about the resources that agency has to allocate and when/where and to what it will allocate them.

12 Friend, Power and Yewlett use a representative democracy model for each agency, officials being accountable to politicians and their constituents. Their concept of multiple agencies in overlapping environments is essentially pluralist, however. There is no discussion of the possible inter-relationships of criteria within the environment, and the curious position is reached whereby central government ministries are treated as of equivalent status to local government.

Network relationships of Group Housing Architect for provision of town development housing

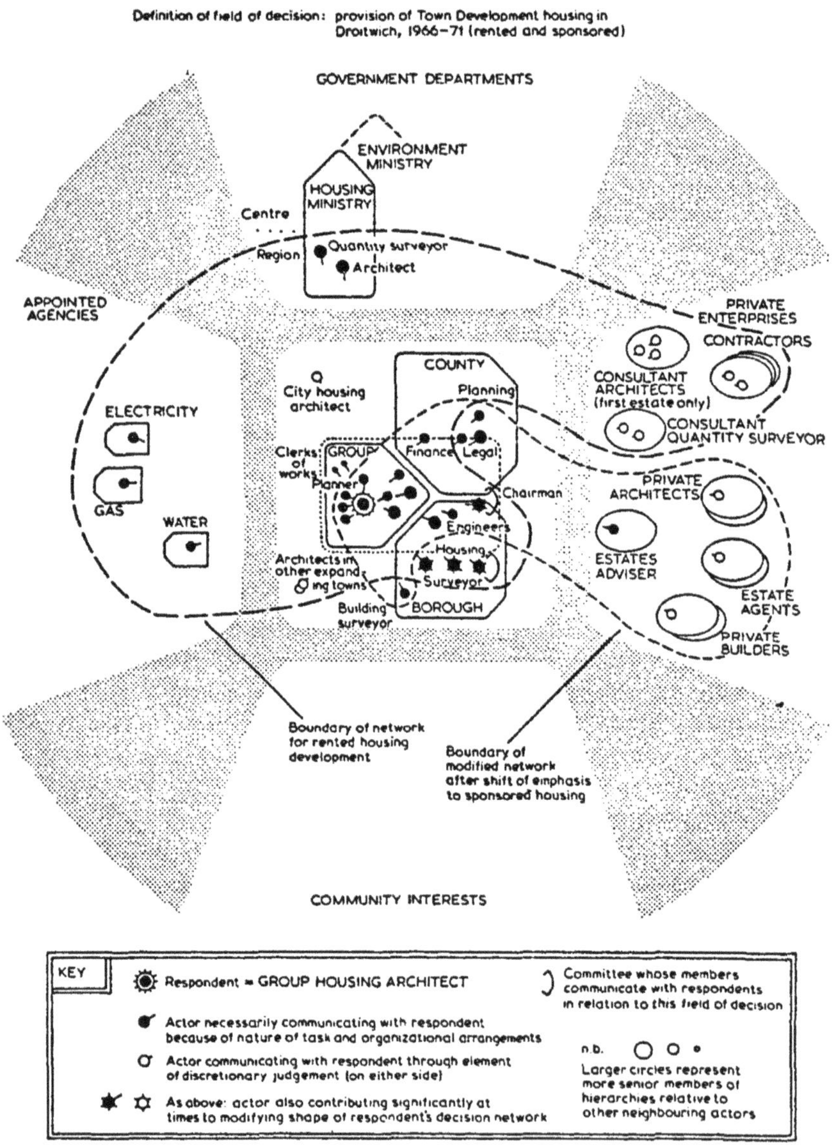

Figure 13.1 Example of a Friend, Power and Yewlett network diagram

actions of organisational agents arises. Put another way, one is forced to ask – in whose interest are these creative, innovatory operational styles to be deployed?

A Resource-based Approach to Organisational Networks

It has been suggested that there is a missing dimension to Friend, Power and Yewlett's work. An American author, Benson, interested in the concept of inter-organisational networks, has attempted to provide this (Benson, 1975), although from a perspective he refers to as 'political economy'. This is neither pluralist nor systems-focussed, but stresses the forces which structure the distribution of resources between interests in society as a whole. Benson comments that 'inter-organisational research and theory have been insufficiently concerned with issues of macro-structure. The larger patterns of post-industrial society have, for the most part escaped the attention of inter-organisational analysts' (p. 230).

He argues that 'a link can be made between the inter-organisational (and inter-personal) dimension and the realm of the wider social structure via a focus on resources, bearing in mind that these are differently distributed between groups in society'. Organisation members are seen as engaged in processes of resource acquisition, resources being of two types, money and authority.[13] Some organisations are more powerful than others because they directly command more resources and/or because of their connection to dominant interest groups in the wider social structure. 'An agency may be able to mobilise forces external to the network as a means of controlling resource flow within it' (p. 233).

Two examples provide possible illustrations of this. In the first, a Labour councillor used his good Labour Party connections to attempt to influence (via the relevant minister) central government allocations to the local authority for which he worked as a corporate planner. In the second, the availability of central government finance for inner city programmes is reported to be leading to changing patterns of power and influence within relevant local authority administrations.

Benson then goes on to argue that the agencies with most command over resources are the most strategically located within a network, and have an 'enhanced' bargaining power vis-à-vis peripheral organisations (p. 233).

They can acquire more resources though having more to start with. In contrast to Friend, Power and Yewlett's approach, then, Benson's network structure is more than just the web of interpersonal relations. It must include weightings which reflect the power relations between members, and which itself can be linked to the wider distribution of command over resources in society. Thus while Benson might agree that the form of the network does have some influence over 'the commitments reached in any specified class of decision problem',[14] the form of the network itself is not an independent variable. What becomes of Friend, Power and Yewlett's concept of the reticulist when viewed in this light?

For those in strategic positions, commanding considerable resources, networking activities may be seen as a way of manipulating and increasing these resources. The reticulist assembles more power to himself and his agency.[15] What is the situation of the would-be reticulist with limited command over resources? Can he, by the exercise of 'reticulist' skills, *deflect* power from existing positions and hence *create* new patterns of resource flow. For Friend, Power and Yewlett, reticulism is a creative, innovatory operational style and hence should have this potential. Certainly many structure and local planners are hoping that the development of mutual awareness among the various agencies and individuals with interests in the course of development in an area will lead to some change in the form, location and timing of development. However, Benson would seem to suggest that the planners here would have to offer some gain to the more powerful interests in return for such changes.

Benson's approach, by focusing on the location and flow of resources within networks, provides a more precise way of structuring the description of organisational networks. It should assist in identifying more clearly the opportunities for influencing events by network-building, and, perhaps more significantly, the limitations of networking. For more advice on the 'mapping' of networks, I now turn to the work of social anthropologists.

13 'the right or responsibility to carry out programmes of a certain kind' (p. 232).

14 Friend, Power and Yewlett's hypothesis (p. 43), as quoted earlier.

15 A few structure planning teams, closely linked to County resource allocation, may use their control over the resources which Districts need to influence locational decisions within the Districts.

The Discussion of Networks in Social Anthropology[16]

The concept of social network, which by the late sixties was a major research interest among many anthropologists, was developed from the pioneering work of Bott (on family and neighbouring relationships), and Barnes (on Norwegian fishing communities). The concept has been helpful in describing social relationships which are either outside or cut across institutional structures and it allows consideration of the indirect influences of one person on another (i.e., via an intermediary). The premise underlying the concept is that the morphology of social networks can be used to interpret social action.[17] The anthropological discussion stresses two theoretical problems about the network concept that are worth highlighting here. The first is the intellectual validity of linking inter-personal or behavioural analysis with that based on institutional structure. Benson attempts this, seeking an objective description of network structure and relationships. The second questions Benson's approach further by suggesting that network description is an essentially subjective exercise. This relates to the empirical problem of defining the limits of any network.

Despite these theoretical limitations on the network concept, anthropological work provides useful guidance on how to describe networks. Mitchell (1969) in a summary article divides the characteristics of social networks into *morphological* criteria and *interactional* criteria.[18] Under morphological criteria, he lists the following:

1. *anchorage*: this relates to the point made earlier that network description is egocentric, working outwards from a particular individual (e.g. the 'key respondent' in Friend, Power and Yewlett's diagrams).

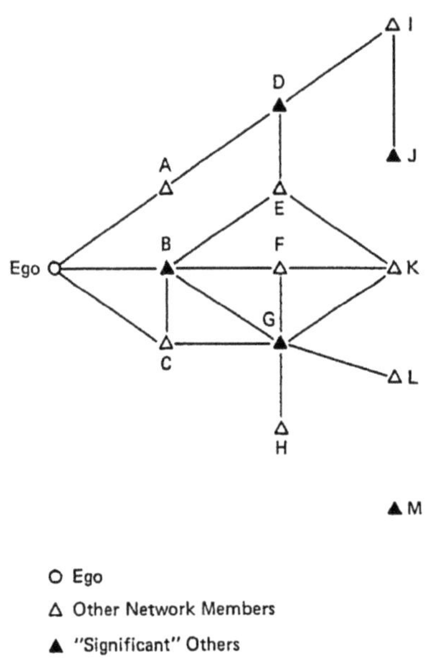

Figure 13.2 Example of social network

16 The main sources I have used here are Ed. Mitchell, 1969, and Ed. Boissevain and Mitchell, 1973, concentrating especially on Mitchell's own work.

17 Mitchell, 1969, pp. 5–6. Compare Friend, Power and Yewlett's statement 'an open network of communications among people ... may influence the commitments reached' (p. 43).

18 The characteristics of networks described below can of course be treated more precisely in a formal mathematical way.

2. *reachability*: the extent to which significant people (who are important to ego) can be reached via network relations. In the above diagram, B, D, G, J and M are significant people but M cannot be reached from ego's network. 'Significance' here could be related to Benson's concept of command over resources. (B and G would no doubt be seen by Friend, Power and Yewlett as occupying nodal positions).
3. *density*: the degree to which the people that ego is connected with are connected to each other. The diagram above represents a fairly open network.
4. *range*: the number of people with whom ego is in direct contact. In the diagram above, there are only three. If that network were at maximum density, there would be thirteen. (This does not necessarily imply that direct contacts are important contacts (see on). In the diagram, only B is an important direct contact.)

The following are Mitchell's interactional criteria:

1. *content*: for analytical purposes Mitchell distinguishes three types of content, although in practice these may co-exist.[19] The three types of content are:
   a. *communication*: the diffusion of information (as in Friend, Power and Yewlett's use of the concept).
   b. *exchange*: 'individuals are related to each other by sets of transactions which have implications for the actors other than the simple act of exchange itself. Participants are bound together in a 'series of expectations and obligations' (Mitchell, 1969, p. 25). Benson characterises network transactions in this way, which also aptly describes the kind of bargaining relations in which a developer-applicant agrees to reduce residential density in a proposed scheme in order to maintain a favourable relationship with a planning authority to assist negotiations on future schemes.
   c. *normative content*: here relationships are affected by the 'expectations each may have of the other because of some social characteristic or social attribute the other may possess' (Mitchell, 1969, p. 26). Typically this refers to the kind of obligations built up within family relationships, or in special groups. Within the public sector, it is sometimes suggested that a 'public service ideology' may perform a similar function. 'Dense' networks, especially if they have a high degree of 'durability' (see below) might be expected to have a considerable normative content.
2. *directedness*: an interaction initiated by ego to A may be quite different in effect from the reverse, due to the different attributes of A and B. Sociometric studies describe how strong personalities attract acolytes. Benson would no doubt approach this characteristic in terms of the resources controlled by the relevant parties.
3. *durability*: some network links are only developed around a particular issue (for example, motorway action groups). Others are regularly used over many years.
4. *intensity*: this refers to the strength of a network link, and may modify the morphological quality of range. Ego's strongest links are not necessarily with those he is in direct contact with, nor with those he sees most frequently.[20]

The above criteria would seem to provide a valuable 'organisational mapping device' for planners operating in complex organisational relationships as well as for analysts studying those relationships. Such a network map could help the former to identify relevant links to activate when searching for more resources in relation to some project or policy. It might also serve to locate gaps in the network which could be remedied by exercising 'reticulist' skills (perhaps forging a link to M, in the earlier diagram). Combined with Benson's approach it might help to identify situations where developing inter-personal relationships is unlikely to solve

19 Where a network is limited to one type of content, it is often referred to as single-stranded, 'uniplex'. Alternatively it may be many-stranded or 'multiplex'.

20 Mitchell's example here is of the family member living overseas.

coordination problems. For the analyst, network description as a mapping device might help to identify some of the influences on resource allocation.

Mitchell's discussion of the content of networks reinforces the earlier conclusion that Friend, Power and Yewlett's limitation to communication is unnecessarily narrow. Although communication obviously occurs, it seems more reasonable to describe decision networks as essentially about exchange, the various parties being engaged in an ongoing series of trade-offs. Bearing in mind Mitchell's point that types of content may co-exist, however, what happens when an organisational network takes on a normative element? This might be seen as a desirable reinforcement in the drive for co-operative 'connective planning'. But what if the partners involved represent organisations with different interests, as when local authority planners negotiate with developers. Is there a danger that the former may be unduly influenced by the values of the latter? The central question at issue here is the accountability of public officials, entering negotiating relationships with other agencies and the private sector on behalf of their organisation. What mechanisms are there to prevent such an official being 'taken over' by the values of those he is bargaining with? That the networks of local authority planners engaged in 'local planning work' often have such a norm-reinforcing element is illustrated in a recently completed empirical study of planners in the London Boroughs.[21]

Concluding Comments

As suggested above, the concept of social network applied to inter- and intra-organisational relations provides a useful mapping device for anyone interested in access to and influence over resource allocations in the public sector. Thus the local authority planner may find it useful to map out an inter-personal network as an adjunct to the list of agencies and groups he may have to consult in relation to, for example, local plan preparation. Similarly, the analyst of local government behaviour may use the tool as part of the study of the influences on local plan preparation or the allocation of resources in general. However, both practitioner and analyst need to be aware of the bias implicit in the network concept towards interpreting social action in terms of individual interactions. Those working in organisations often interpret the interest conflicts they encounter at work in terms of 'personalities', and some theoretical approaches to organisational behaviour seek an individualistic or socio-psychological level of explanation. Most approaches to the study of organisations, however, require some kind of generalisation from actions and values to broader categories such as institutional structures or interests. Possibly the network concept as applied to public sector organisations belongs most appropriately to the kind of theoretical framework associated with the notion of 'urban managerialism'.[22] However, it may well be that network description could be a valuable research tool within several types of theoretical investigation.

Friend, Power and Yewlett, though, put forward the notion of reticulism, or networking, as a normative principle, as an operational style. They claim it could lead to a more creative and responsive style of inter- and intra-organisational behaviour, in contrast to the formal bureaucratic procedures common in local government. In particular it would improve co-ordination between and within public sector organisation. Whether one accepts this prescription depends on one's view of organisations and their activities. Friend, Power and Yewlett's assumption is that lack of co-ordination between organisations is the result of lack of knowledge about each other's problems and interests. Hence, developing better methods of communication, as through inter-personal contacts, will overcome the block to effective co-ordination. They might also accept, as Benson does, that even where there are conflicts of interest between organisations, these may be resolved in particular instances by negotiating and bargaining.

However, if interest conflict is recognised as fundamental to social organisation, then situations must be anticipated where neither better communication nor better bargaining will resolve an interest conflict. In this case, one might expect to see organisational participants either abandoning co-ordination attempts and resorting to 'partisan mutual adjustment' or being subjected to imposed solutions by the most powerful authority. Examples of both situations come to mind fairly readily from the field of land use planning.

21 See Healey and Underwood, 1978, especially chapters 3 and 5.

22 See Pahl, 1975, especially Part 3, and Pahl, 1977. Pahl himself suggests a link with social network analysis as used by social anthropologists (Pahl, 1975, p. 207).

To some extent Friend, Power and Yewlett recognise this, although they suggest that the determining factor is the complexity of the problem situation (p. 41). They give little consideration to the role of central government, which may step in to resolve a county-district conflict (over a structure plan policy, for example), or impose a particular policy (as in the housing and transport fields).

I have already made reference to a further difficulty with the networking idea as an operational style, the problem of accountability. As applied to organisations, the concept of network building involves both relationships formalised within bureaucratic procedures for accountability (such as answerability to chief officers, and to councillors), and informal contacts developed through personal initiative. The danger of such a non-formal structure is that, in the course of these inter-personal relationships, organisation members may lose sight of the interests and values which legitimate their involvement in these relationships in the first place. Examples of such situations can be illustrated by the following anti-ideal types:

1. *The unprincipled bargainer* who promotes his private interest rather than a legitimate public sector interest. This is what in extreme form we understand as corruption.
2. *The naive bargainer* who thinks he has negotiated a gain to his legitimate interests but has been effectively short-changed (as in Ambrose and Colenutt's description of planners and property developers, Ambrose and Colenutt, 1975).
3. *The innocent bargainer* who is unwittingly taken over by his opposite number's values (which we felt to be the case with some local authority planning groups we observed; Healey and Underwood, 1978).

From certain viewpoints a fourth type might be identified:

4. *The value-neutral managerialist bargainer*, who actively promotes inter- and intra-organisational co-ordination to improve the efficiency of decision-making without an awareness of the different values and interests which may have a stake in the issue in hand. (Some recent critiques of corporate planning in local government would take this line; Benington, 1976, Cockburn, 1977).

To summarise, just as Jowell is concerned about the officer-developer bargaining relationships evolving in development control, which bypass some of the restrictions on the sphere of planning imposed by the courts (Jowell, 1977), so many might view Friend, Power and Yewlett's reticulist as a very dangerous animal indeed. However, as Hague et al. (1975) have shown, the problem of the accountability of public sector officials and of those spending public money is not specific to inter-personal networks or particular public sector organisations. It is a general problem reflecting the increasing complexity of the public sector role in society, and the increasing 'interpenetration' of the public and private sectors.[23]

Maintaining control over the way the public sector operates while releasing its creative energies to act appropriately in any situation is one of the central problems of contemporary public administration. 'Networking' as a normative principle may offer a creative and responsible operational style but it implies considerable exercise of administrative discretion. How can society, or any groups within it, be satisfied that this discretion is not being abused?

References

Ambrose P. and B. Colenutt, 1975, *The Property Machine*, Harmondsworth: Penguin.

Argyris, C. and D.A. Schon, 1974, *Theory in Practice: Increasing professional effectiveness*, San Francisco: Jossey-Bass.

Banfield, E.C. 1959, 'Ends and means in planning', *International Social Science Journal*, XI(3): 361–8.

Benington, J. 1976, *Local Government Becomes Big Business*, CDP Information and Intelligence Unit, London.

23 Hebbert, 1975, points out that accountability problems are particularly difficult in local government as there are several 'overlapping interests' to which local government officials may be held accountable.

Benson, J.K. 1975, 'The inter-organisational network as political economy', *Administration Science Quarterly*, 20(2): 229–49.
Boissevain, J. and J.E. Mitchell (eds), 1973, *Network Analysis in Human Interaction*, The Hague: Mouton.
Braybrooke, D. and C.E. Lindblom, 1963, *A Strategy for Decision*, New York: The Free Press.
Cockburn, C. 1977, *The Local Slate*, London: Pluto Press.
Faludi, A. 1973, *Planning Theory*, Oxford: Pergamon Press.
Faludi, A. (ed.), 1973, *A Reader In Planning Theory*, Oxford: Pergamon Press.
Friedmann, J. 1973, *Retracking America*, New York: Anchor Press.
Friend, J.K. and W.N. Jessop, 1969, *Local Government and Strategic Choice*, London:Tavistock.
Friend, J.K., J.M. Power and C.J.L. Yewlett, 1974, *Public Planning: The intercorporate dimension*, London: Tavistock.
Hague, D.C., W.J.H. Mackenzie and A. Barker, 1975, *Public Policy and Private Interests: The institutions of compromise*, London: Macmillan.
Harris, R.J.P. 1976, 'Inter-authority decision-making: Some implications for local government', *Local Government Studies*, 2(3): 17–26.
Healey, P. and J. Underwood, 1978, 'Professional ideals and planning practice', *Progress in Planning*, 9: 73–127.
Hebbert, M. 1975, 'Notes on public accountability', *Reading Geographer*, 4: 24–32.
Jowell, J. 1977, 'Bargaining in development control', *Journal of Planning and Environment Law*, July, pp. 414–33.
Lawrence, P.R. and J.W. Lorsch, 1967, *Organisation and Environment: Managing differentiation and integration*, Boston, MA: Harvard.
Lindblom, C.E. 1965, *The Intelligence of Democracy*, New York: Free Press.
Mitchell, J.C. 1969, 'The concept and use of social networks', in J.C. Mitchell (ed.), *Social Networks in Urban Situations*, Manchester: Manchester University Press.
Mitchell, J.C. (ed.), 1969, *Social Networks in Urban Situations*, Manchester: Manchester University Press.
Pahl, R.E. 1975, *Whose City?* (revised edition), Harmondsworth: Penguin.
Pahl, R.E. 1977, 'Managers, technical experts and the state: Forms of mediation, manipulation and dominance in urban and regional development', in M. Harloe (ed.), *Captive Cities*, London: Wiley.
Schon, D.A. 1971, *Beyond the Stable State*, New York: Random House.
Simon, H.A. 1961, *Administrative Behaviour*, 2nd edition, New York: Macmillan.
Van Gunsteren, H. 1976, *The Quest for Control*, London: Wiley, London.

Chapter 14
Planning through Debate[1]: The Communicative Turn in Planning Theory[2]*

Patsy Healey

This article is about what 'planning' can be taken to mean in contemporary democratic societies. Its context is the dilemma faced by all those committed to planning as a democratic enterprise, aimed to promote social justice and environmental sustainability. The dilemma is that the technical and administrative machineries advocated and created to pursue these goals in the past have been based on what we now see as a narrow scientific rationalism. These machineries have further compromised the development of a democratic attitude, and have failed to achieve the goals promoted. So how can we now support a renewal of the enterprise of planning? If we can, what are its forms and principles?

The article is written specifically for those planners in Britain, in planning schools and in planning practice, who have shared a particular experience of the 1970s and 1980s. The 1970s provided us with a soft social and environmental commitment and a hard political critique of the enterprise of planning. In this critique, planning was a site of struggle between class forces for control of the management of the urban environment. By the 1980s, this critique had itself dissolved into a search for a less one-dimensional view of conflict and cleavage in society and a more nuanced appreciation of the diversity of the experience of urban life and environment. This search for a democratic pluralism[3] took place, however, against the harsh backcloth of the Thatcherite hegemonic agenda. This set out to destroy not merely democratic socialist thought and practices, but the very enterprise of urban management and planning which was the object of the democratic socialist critique.

The Thatcherite project has now been brought to a remarkably sudden halt as a political idea, though many of the practices it instituted remain.[4] Citizen responsiveness environmental sustainability, as vague political principles, are now widely asserted, as in the general idea of environmental planning and the specific principle of a plan-led regulatory land-use planning system. But what kind of a planning can be compatible with our contemporary understandings of a democratic attitude? And how can the concept of planning survive the contemporary philosophical challenges to materialism, modernism and rationalism, these central pillars of the traditions of 'modernity' which dominated Western thought from the middle of last century until late into the present one?[5] How can there be a 'planning' without 'unifying' conceptions of systems and structures, based on scientific knowledge, from which to articulate hypotheses as to key relationships and appropriate

* Previously published in *Town Planning Review*, 63(2): 143–62 (1992).

1 Debate is used here in preference to 'argumentation', as a more collaborative and positive word. Others see 'debate' as involving opposition between two sides. It will become clear that this is not what I associate with the word.

2 This is a very substantial development of ideas initially sketched in Healey, P., 'Planning through Debate' (Paper given to *Planning Theory Conference*, Oxford, April 1990). A shorter version will appear in F. Fischer and J. Forester (eds), *The Argumentative Turn in Policy Analysis and Planning*, Durham, NC, Duke University Press. My thanks to my sister Bridget who allowed me to write this and read Habermas while on holiday. My thanks to Huw Thomas, John Forester, Seymour Mandelbaum, Jean Hillier, Jack Ellerby, Michael Benfield, Beth Moore Milroy, Gavin Kitching, Judith Allen, Michael Synnott and Nilton Torres, for their critical attention to an earlier draft.

3 See, for example, Rustin, M., *For a Pluralist Socialism*, London, Verso, 1985.

4 See Thornley, A., *Urban Planning under Thatcherism*, London, Routledge, 1991 for a discussion of its impact in the planning held.

5 For discussions of the meaning of 'modernity' and 'post-modernity' in relation to planning, see Friedmann, J., *Planning in the Public Domain*, Princeton, NJ, Princeton University Press, 1987; Moore Milroy, B., 'Into Postmodern Weightlessness', *Journal of Planning Education and Research*, 10(3) 1991, pp. 181–7, and other articles in this issue of the *Journal*, and Goodchild, B., 'Planning and the Modern/Postmodern Debate', *Town Planning Review*, 61(2) 1990, pp. 119–37.

interventions? How can decisions be arrived at without systematic 'rational' procedures for knowledgeable and collective 'deciding and acting'?

Throughout the past decade, signs of alternative conceptions of planning purposes and practices have been increasingly identified and debated in planning theory. One route to imagining alternatives has focused on substantive issues, moving from material analyses of options for local economies exposed to global capitalism, to concerns with culture, consciousness, community and 'placeness'.[6] Another has taken a 'process' route, exploring the communicative dimensions of collectively debating and deciding on matters of collective concern.[7]

The problem with the substantive route is its a priori assumptions of what is 'good/bad', 'right/wrong'. Local economic development is presented as often 'good', national economic intervention as oppressive, 'bad'. By what knowledge and reasoning has this been arrived at? If such principles are embodied in our plans, will we not have fallen yet again into the trap of imposing the reasoning of one group of people on another? Does the process route offer a way out of this dilemma of relativism which treats every position as merely someone's opinion, and hence the dominance of a position pursued through planning strategies and their implementation as nothing more than the outcome of a power game?

This article argues that it can. The argument is explored firstly by a brief review of the idea of planning and its challenges. The article then identifies five directions for the management of the urban environment which seem to be prefigured in present discussions. Of these, it is argued that the conception of planning as a communicative enterprise holds most promise for a democratic form of planning in the contemporary context. The article concludes with some implications for the systems and practices of environmental planning.

Throughout, the contextual 'locus' of the article is environmental planning in Britain, although this merely allows the purposes and practices for planning to be developed in a specific context. More generally, the challenge for planning in the contemporary era lies at the heart of our efforts to reinterpret a progressive meaning for democracy in Western societies.[8]

The Idea of Planning and its Challenges

As with so much of Western culture, the contemporary idea of planning is rooted in the enlightenment tradition of 'modernity'.[9] This freed individuals from the intellectual tyranny of religious faith and from the political tyranny of despots. Such free individuals, in democratic association, could, it was believed then and since, combine in one way or another, to manage their collective affairs. By the application of scientific knowledge and reason to human affairs, it would be possible to build a better world, in which the sum of human happiness and welfare would be increased. For all our consciousness of the errors of democratic management in the past two centuries, it is difficult not to recognise the vast achievements that this intellectual and political enlightenment has brought.

---

6 This is evident particularly in discussions on locality, place and local economic development. See, for example, Cooke, P.N., *Back to the Future*, London, Unwin Hyman, 1990; Massey, D., 'The Political Place of Locality Studies', *Environment and Planning A*, 23 (2) 1991, pp. 267–81.

7 See Forester, J., *Planning in the Face of Power*, Berkeley, CA, University of California Press, 1989; Throgmorton, J., 'Planning and Analysis as Persuasive Storytelling: The Case of Electric Power Rate-making in the Chicago Area' (paper presented to the ACSP Congress, Austin, Texas, November 1990).

8 'Democracy' is, of course, used in contemporary debate in a wide and confused range of meanings (Williams, R., *Keywords* [2nd edn], London, Fontana, 1988). By a 'progressive meaning', I align myself with the position adopted by British authors such as Held, D., *Models of Democracy*, Oxford, Polity Press, 1987, who argue for a notion of democracy based on the principle of autonomy in both political and economic spheres, in a system which promotes 'discussion, debate and competition among many divergent views' (p. 280). Within this conception, open debate, access to power centres, and general political participation are key requirements for democratic public life (p. 284). It is principles such as these which have helped to fuel the *Charter* 88 constitutional movement in Britain.

9 See Note 6, and also Bernstein, R.J. (ed.), *Habermas and Modernity*, Cambridge, Polity Press, 1981; Berman, M., *All That's Solid Melts to Air*, London, Verso, 1983; Harvey, D., *The Condition of Postmodernity*, Oxford, Blackwell, 1989.

This modern idea of planning, as Friedmann has described in his authoritative account of its intellectual origins,[10] is centrally linked to concepts of democracy and progress. It centres on the challenge of finding ways in which citizens, through acting together, can manage their collective concerns, with respect to the sharing of space and time.

In this century, Mannheim's advocacy of a form of planning which harnessed systematised social scientific knowledge and techniques to the management of collective affairs in a democratic society proved inspirational for the influential Chicago school of rational decision making.[11] A procedural view evolved which presented planning as a progressive force for economic and social development in a world where democracy and capitalism were seen to co-exist in comfortable consensus.[12] It challenged populist 'clientelism' (as in Chicago in the 1950s)[13] as much as idealist totalitarianism.

But as with any progressive force, procedures developed with a progressive democratic intention may be subverted for other purposes. In the early 1970s, this subversion was identified with the power of capitalist forces to dominate everyone's life opportunities. Environmental planning, it was argued, put the needs of capital (through regional economic development and the implicit opportunities for land and property markets created by planning regimes) before citizens and the environment.[14]

However, a more fundamental challenge to the Mannheimian notion of planning was gathering force, through the critique of scientific reason itself. German critical theorists and French 'deconstructionists' elaborated ideas which challenged reason's dominance of human affairs. Reason, understood as logic coupled with scientifically-constructed empirical knowledge, was unveiled as having achieved hegemonic power over other ways of being and knowing, crowding out moral and aesthetic discourses. Further, rationalising power dominated the very institutions set up in the name of democratic action, the bureaucratic agencies of the state. Following Foucault's analysis, planning could be associated with the dominatory power of systematic reason pursued through state bureaucracies.[15] Evidence for this seemed to be everywhere, from the disaster of high-rise tower blocks for the poor to the dominance of economic criteria justifying road projects and the functional categorisation of activity zones which worked for large industrial companies and those working in them, but not for women (with their necessarily complex lifestyle), the elderly and disabled, and many ethnic groups forced to discover ways of existing on the edge of established economic practices.

This 'challenge to systematised reason', and with it, to the planning enterprise, strikes at the heart of the enlightenment project, or, as we now understand it, the project of 'modernity'. The challenge is now labelled as 'post-modernist', drawing on a terminology first developed in art and architectural critique.[16] But whereas post-modernism in architecture is primarily a critique of a particular paradigm and style *within* Western art and architecture, philosophical post-modernism challenges the foundations or two hundred years of Western thought.

The post-modern challenge to Western thought is both progressive and regressive in its potential, as was the idea of systematised reason. It is also highly diverse, with different lines of development. Only some of these claim to *replace* the 'project of modernity' with that of post-modernity. Others, following the position of the economic geographer Harvey, and the critical theorist Habermas, seek new ways of reconstituting the 'incomplete' project of modernity. Some of the strands of post-modernist debate leave space for a form of planning, i.e. for collective activity. Others dismiss planning as, variously, impossible, irrelevant or oppressive.

10 See Friedmann, J., op. cit., Note 5.

11 Mannheim, K., *Man and Society in an Age of Reason*, London, Routledge, 1960; Friedmann, J., *Retracing America*, New York, Anchor, 1973; Friedman, J., op. cit., Note 5; and Faludi, A., *Critical Rationalism and Planning Methodology*, London, Pion, 1986.

12 See discussion in Friedmann, J., op. cit., Note 11.

13 See Meyerson, M. and Banfield, E., Politics, *Planning and the Public Interest*, New York, Free Press, 1955.

14 This position was most forcefully articulated in Castells, M., *The Urban Question*, London, Edward Arnold, 1977. See also Ambrose, P. and Colenutt, B., *The Property Machine*, Harmondsworth, Penguin, 1973; and Scott, A. and Roweis, S.T., 'Urban Planning in Theory and Practice', *Environment and Planning A*, 9, 1977, pp. 1097–1119.

15 See Habermas, J., *The Philosophical Discourse of Modernity*, Cambridge, Polity Press, 1987 for a helpful debate on the work of Adorno, Marcuse, Foucault and Derrida.

16 See Note 9 above.

Moore Milroy, reviewing the development of the post-modernist debate in planning thought, identifies four 'broad characteristics' to the challenge post-modernism presents to modernism.

> It is *deconstructive* in the sense of questioning and establishing a sceptical distance from conventional beliefs and, more actively, trying both to ascertain who derives value from upholding their authority and to displace them; *antifoundationalist* in the sense of dispensing with universals as bases of truth; *nondualistic* in the sense of refusing the separation between subjectivity and objectivity along with the array of dualisms it engenders including the splits between truth and opinion, fact and value; and *encouraging of plurality and differences*.[17]

This double challenge, to the tendency for progressive values to be destroyed by the very systems created to promote them, and to the systems of technocratic rationalist thought which have underpinned so much of Western and Eastern bloc thinking about planning, seems so powerful as to be fatal to the idea of planning. Is there any way out? It is argued in this article not only that there is, via the development of communicative forms of planning, but also, following Harvey and Habermas, that some directions of the 'post-modern challenge' to planning need to be actively resisted as in their turn regressive and undemocratic. Current debate suggests several 'routes for invention' of a new planning. Five such 'routes' or directions are outlined here. These directions are not necessarily exclusive. Their presentation in the planning literature varies in its coherence. They are offered as a sketch of possibilities, through which to foreground the promise of a communicative form of planning in promoting and realising a progressive democratic attitude.

Directions for a New Planning

The five directions discussed are:

1. A retreat to the bastions of scientific rationalism as expressed through neo-classical economics. Planning is reformulated to provide a framework of rules to ensure collectively experienced impacts are addressed through the price mechanism.
2. An idealism based on fundamental moral or aesthetic principle. Planning purposes and practices would be directed to realising this principle.
3. A relativism in which self-conscious individuals assert their own principles and 'mutually' adjust when they get in each other's way. Planning has little purpose in this route except as deconstructive technique, to reveal 'dominatory' systems in order to remove them.
4. Enlarged conceptions of democratic socialism beyond economic struggles over material conditions, to incorporate other loci of 'cleavage', such as gender and race, and allow more space for cultural issues (moral and aesthetic). This refocuses the purposes and practices of planning around a reformulated substantive agenda.
5. A communicative conception of rationality, to replace that of the self-conscious autonomous subject using principles of logic and scientifically-formulated empirical knowledge to guide actions. This new conception of reasoning is arrived at by an inter-subjective effort at mutual understanding. This refocuses the practices of planning, to enable purposes to be communicatively discovered.

The Principle of Price

This conception ignores most of the debates and challenges just discussed, and continues the rationalist project. In conformity with the post-enlightenment tradition, individuals are constituted as autonomous subjects, confronting the object-world. They allocate their resources according to their subjectively perceived wants, and their material opportunities. Public policy facilitates this allocatory process by authoritative structures (rules) based on 'market information' about supply, demand, and the blockages to market exchange. Environmental planning comes into play to conserve assets which are not readily traded in the market place

17 Moore Milroy, B., op. cit., Note 5.

(national parks, wildlife reserves, historic buildings, agricultural land) and to ensure that the actions of individuals do not impose excessive costs on neighbours, communities and environments. As far as possible, such a planning should proceed by pricing strategies which require everyone to internalise these external costs. Only when this is difficult to enforce or where positive conservation is required, should regulatory intervention be used. David Pearce's approach to environmental economics provides a clear example of how 'environmental sustainability' objectives could be achieved in this paradigm.[18]

Though hesitantly and inconsistently, it is this route which was followed by the British government in the 1980s.[19] Some planning theorists have been developing its dimensions in the planning field.[20] It has been a dominant tradition for some years in the United States, and is now being vigorously developed in Britain in ideas for traffic management and an 'impact fee' approach to planning gain. But, as environmental debates clearly illustrate, it proceeds ignorant of any doubts about the supreme power of scientific rationalism, and assumes that most aesthetic and moral issues can be converted into priceable preferences.[21]

It is not hard to see the dominatory force of scientific rationalism at work here. This is not to argue that such an approach is *never* appropriate; merely that it is but one possibility among many. We may criticise its practitioners for their failure to grasp that it 'lives together' with other ways of making policy issues manageable, and we may criticise governments and knowledge production institutions for prioritising it above others. If post-modernism has any progressive meaning, it must mean that this direction for planning turns away from, rather than towards, the challenge of 'making sense together while living differently'.

Idealist Fundamentalism

Several strands of contemporary post-modern critique focus on unmasking the corrupting power of scientific rationalism at the heart of our thought, to reveal deeper unifying principles which hold our world together.[22] From this perspective individuals are constituted not as autonomous subjects responsible for their own actions, but as bearers and interpreters of a metaphysical principle. This principle becomes the locus of moral and aesthetic order, and its contemplation fosters a reflective 'interiorisation' of experience rather than 'acting in the world'.[23] The preoccupation is with existence, with *being*, rather than a collective 'exteriorised' enterprise of *becoming* something different and better in the world. While such a fundamentalism has progressive force in releasing and legitimating people's search for a moral basis for their lives, it also contains within it a dominatory potential. Examples of this can be seen in the adoption of religious codes in public spheres (for example, the British 1988 Education Act's requirement for a 'daily corporate act of worship of an essentially Christian nature') or proposals for environmental actions irrespective of their economic and social consequences justified in terms of scientifically and aesthetically constructed notions of ecological apocalypse. Planning in this context becomes either an irrelevance to the contemplative interior life or an expression of the metaphysical principle (as in, for example, Chinese *fung schiu* or Islamic principles for environmental planning derived from the Koran). In essence, this direction merely replaces one uni-dimensional hegemony (rationalism) with another (a particular moral principle). It is hard to see how such an approach can advance the project of progressive democratic pluralism.

18 Pearce, D., Markandya, A. and Barbier, E.B., *Blueprint for a Green Economy*, London, Earthscan, 1989.

19 See Thornley, op. cit., Note 4.

20 See Sorensen, A., Towards a Market Theory of Planning', *The Planner*, 69 (3) 1983; pp. 78–80.

21 For good critiques of this 'direction' with respect to environmental issues, see Hajer, M., 'Bias in Environmental Discourse: an Analysis of the Acid Raid Controversy in Great Britain' in Fischer, F. and Forester, J. (eds), *The Argumentative Turn in Policy and Planning*, Durham, NC, Duke University Press (forthcoming); and Grove White, R., 'Land, the Law and Environment', *Journal of Law and Society* (forthcoming).

22 See the discussion of Nietzsche's Dionysian Search and Heidegger's justification of Nazism in Habermas, J., op. cit., Note 15, and Moore Milroy's discussion of the fundamentalism in some 'post-modern' thought (Moore Milroy, op. cit., Note 5).

23 See discussion in Sennett, R., *The Conscience of the Eye: The Design and Social Life of Cities*, London, Faber & Faber, 1990; and Habermas, J., op. cit., Note 15.

Aesthetic Relativism

Others elevate experience and the aesthetic mode to the central dimension of human life. This focuses on the self-conscious autonomous individual, existing, being, to be extricated from the oppression of functional systems based on scientific rationalism. Within this conception, there is no unifying metaphysical source, to be contemplatively revealed once the reasoning dominance has been unmasked. Instead, all interpretations are valid. The unifier of humanity is merely the experiential capacity. This leads to a celebration and enjoyment of differences,[24] but experienced individually rather than collectively. No criteria seem to be available for distinguishing one person's interpretations and actions from another's, since to distinguish would involve recourse to reason or idealist beliefs. All have equivalent standing; all have validity through interior reflection. The much-criticised outcome is a potentially regressive idealist nihilism. The dominatory potential within this strand of post-modernism is of enraged anarchistic violence between individuals and groups struggling to stake out the territory within which their purposes and practices can prevail. The Western media's portrayal of inter-ethnic and factional strife in the former USSR and Yugoslavia provides examples of what this could mean. If planning has any role at all in this direction, it is to stake out and defend boundaries and at the same time to foster the celebration of difference. But without a discursive reasoning capacity, it is difficult to see what practices could constitute such a planning. To argue this is not to reject the importance of aesthetic and emotional experience in forming our understandings and values. It is the *prioritising* of a particular dimension of experience and understanding (in this case the aesthetic) above all others which compromises the project of progressive democratic pluralism. The progressive challenge is instead to find ways of acknowledging different ways of experiencing and understanding, while seeking to 'make sense together'.

Extending Modernity's Tolerance

Another route has been to develop the socialist project beyond a preoccupation with material conditions and economic classes. This project, whether in Marxian or democratic humanist forms, aims to develop a society in which the conditions of material existence are adequate for all, and in which everyone has the opportunity to work in conditions where we are justly rewarded and respected for the work we do and in which we have real control over the economic and political conditions of the societies in which we live. Marxist analysis conceived of individuals as self-conscious and reflective. But people's perceptions and worth were seen as constituted by the material conditions of the societies in which we live, and specifically by the mode of production. Through scientific historical analysis, people could become aware of this, and through collective action, change the conditions of our existence. Planning thus became the means for redesigning less oppressive societies than those dominated by feudal, capitalist and colonial power.

But scientific materialism as the basis for the socialist project in retrospect engendered a domination by state bureaucrats pursuing scientific management principles in the name of working class power (in Eastern European countries, often in highly corrupt forms). Moral principles and aesthetic consensus were interpreted within a set of scientific 'laws' about economic class interests. By the 1970s in Britain and elsewhere, many socialist thinkers were identifying similar tendencies within the welfare state machinery of Western capitalist economies.[25] Critiques developed which firstly highlighted power-distributing cleavages other than economic class, notably those of gender and race, and secondly sought to break out of a 'totalising' scientific rationalism. The new socialism of the 1980s in Britain has been concerned with developing a pluralist understanding of people's needs, values and ways of experiencing oppression. Appreciating diversity and recognising differences are key elements in this conception, requiring collective action to be informed by principles of tolerance and respect. There is not one route to progress, but many, not one form of reasoning but many. The socialist project thus comes to focus on both restructuring the control of economies and the flow of the fruits of material effort, while at the same time discovering ways of 'living together differently

24 See Habermas' discussion of Derrida and difference in Habermas, J., op. cit., Note 15.
25 See Cockburn, C., *The Local State*, London, Pluto, 1977.

but respectfully'.[26] Planning retains its traditional importance in socialist thought, but the planning enterprise is refocused to recognise diverse forms of disadvantage.

The frame of reference of these efforts remains a struggle for opportunities for the disadvantaged against a systematically understood capitalist world order. This provides a frame of reasoning which interprets and selects among the various claims for attention which a pluralist socialism can generate.[27] But where does this frame of reasoning come from and what gives it its authority? Is it merely providing a slightly more sensitive development of the notion of class interests? And does this really accommodate the claims and arguments of the different ethnic communities in Britain, or the anger of those oppressed by racial and gender prejudice? These 'voices' may argue that the pluralist socialist project of 'living together but differently' dominates them by failing to *listen* to their different ways of *experiencing*. It requires acceptance of a belief in the analyses propounded, in a particular interpretation of what 'living together' and 'difference' mean. The planning frameworks developed within this route thus cannot escape the critique of scientific rationalism. In other words, the pluralist socialist project is still founded on systematised rationality and scientific understanding of social structure in its conception of 'living together' and of 'difference'.

Communicative Rationality

It is here that Habermas' search for a reformulation of modernity's concept of reason offers a way forward. Habermas argues that, far from giving up on reason as an informing principle for contemporary societies, we should shift perspective from an individualised, subject–object conception of reason, to reasoning formed within inter-subjective communication. Such reasoning is required where 'living together but differently' in shared space and time drives us to search for ways of finding agreement on how to 'act in the world' to address our collective concerns. Habermas' communicative rationality has parallels within conceptions of practical reasoning, implying an expansion from the notion of reason as pure logic and scientific empiricism to encompass all the ways we come to understand and know things and to use that knowledge in acting. Habermas argues that without some conception of reasoning, we have no way out of fundamentalism and nihilism. For him, the notion of the self-conscious autonomous individual, refining his or her knowledge against principles of logic and science, can be replaced by a notion of reason as inter-subjective mutual understanding arrived at by particular people in particular times and places, i.e. historically situated. Both subject and object are constituted through this process. Knowledge claims, upon which action possibilities are proposed, are validated in this conception of reasoning through discursively establishing principles of validity, rather than through appeal to logic or science, although both may well be considered as possibilities within the communicative context.[28]

In this way, knowledge for action, principles of action and ways of acting are actively constituted by the members of an inter-communicating community, situated in the particularities of time and place. Further, the reasoning employed can escape from confines of rational–scientific principles to include varying systems of morality, and culturally-specific traditions of expressive aesthetic experience. 'Right' and 'good' actions are those we can come to agree on, in particular times and places, across our diverse differences in material conditions and wants, moral perspectives and expressive cultures and inclinations. We do not need recourse to common fundamental ideals, or principles of 'the good social organisation' to guide us. Planning, and its contents, in this conception, is a way of acting which we can *choose*, after *debate*.

Habermas' conception of communicative action has been criticised in the context of the present discussion on two grounds. Firstly, by holding on to reason, it retains the very source of modernity's dominatory potential. Secondly, Habermas would like to believe that consensual positions can be arrived at, whereas contemporary

26 See Rustin, M., op. cit., Note 3 and Massey, D., op. cit., Note 6. Interestingly, this thinking parallels some ideas developed by Mel Webber on 'persuasive planning' for pluralist, democratic societies in the 1970s, which aimed to foster debate and encompass difference (Webber, M., 'A Difference Paradigm for Planning' in Burchell, R.W. and Sternleib, G., *Planning Theory in the 1980s*, Rutgers, NJ, Center for Urban Policy Research, 1978).

27 See Rustin, M., op. cit., Note 3.

28 See Habermas, J., op. cit., Note 15.

social relations reveal deep cleavages, of class, race, gender and culture, which can only be resolved through power struggle between conflicting forces.[29]

Habermas justifies his retention of reasoning as a legitimate guiding principle for collective affairs on the grounds that, where collective 'acting in the world' is our concern, we need to engage in argumentation and debate. We need a reasoning capacity for these purposes. We cannot just engage in aesthetic presentation or moral faith if at some point we are faced both with 'making sense together' and 'working out' how to act together. This does not mean that the language of morality or aesthetics is excluded from our reasoning. Habermas argues that our intersubjective practical reasoning draws on the store of knowledge and understandings of technique, morality and aesthetics. In this way, our collective reasoning is informed by, situated within, the various 'lifeworlds' from which we come to engage in our collective enterprises.[30] Our inter-subjective arguments may involve 'telling stories' as well as 'doing analyses'.[31] Thus the narrative mode should accompany and intersect with experiential expression and the analytical mode. But in the end, the purpose of our efforts is not these (doing analysis, telling stories, rhetoric), but *doing something*, i.e. 'acting in the world'. For this, we need to discuss what we could and should do, why and how. There is an interesting parallel here with Walzer's notion of principles of justice for different spheres of social activity.[32]

But does not this process of collective argumentation merely lead to a new and potentially dominatory consensus, as the agreement freely arrived at through argument in one period imposes itself on the different differentiations of the next? Habermas proposes to counteract this possibility through criteria to sustain a dynamic critique within the reasoning process. Claims should be assessed in terms of their comprehensibility, integrity, legitimacy and truth.[33] Forester has since developed these as heuristic questions for planners to use in critiquing themselves and others as they search for a progressive power-challenging planning.[34]

The mutual understandings and agreements reached for one purpose at one time are thus revisable as the flow of communicative action proceeds. Habermas himself would clearly like to see stable consensuses emerge, societies built around principles of mutual understanding. Several planning theorists have also proposed the development of a communicative 'metalanguage' or a 'meta-discourse' for planning discussion.[35] Such an enterprise parallels the search noted above by the 'new left' in Britain for forms of a democratically pluralist participation.

But a metalanguage, however full of internal principles of critique, unavoidably contains dominatory potential. It could all too easily settle into assumptions of understanding and agreement detached from those whose ways of being, knowing and valuing are supposed to be reflected in the agreement. To be liberating rather than dominating, inter-communicative reasoning for the purposes of 'acting in the world' must accept that the 'differences' between which we must communicate are not just differences in economic and social position, or in specific wants and needs, but in *systems of meaning*. We see things differently because words, phrases, expressions, objects, are interpreted differently according to our frame of reference. It is this point, long understood in anthropology[36] and emphasised in phenomenology, which underpins the strength of the relativist position. It is here that the present author would part company with Habermas, in order to recognise the inherent localised specificity and untranslatability of the systems of meaning. We may shift our ideas,

---

29 See Moore Milroy, B., 'Critical Capacity and Planning Theory', *Planning Theory Newsletter*, Winter 1990; pp. 12–18; and Sennett, R., op. cit., Note 23.

30 Habermas, J., op. cit., Note 15.

31 See Innes, J., *Knowledge and Public Policy: The Search for Meaningful Indicators*, New Brunswick, Transaction Publishers, 1990; Mandelbaum, S., 'Telling Stories', *Journal of Planning Education and Research*, 10 (3) 1991, pp. 209–14; and Forester, J., 'The Politics of Storytelling in Planning Practice' (Paper to the ACSP Congress, Austin, Texas, November 1990), for an appreciation of the role of 'story-telling' in policy analysis.

32 Walzer, M., *Spheres of Justice: A Defence of Pluralism and Equality*, Oxford, Blackwell, 1983.

33 Habermas, J., *The Theory of Communicative Action Vol. 1: Reason and the Rationalisation of Society*, London, Heinemann, Polity Press, 1984.

34 Forester, J., op. cit., Note 7.

35 See, for example, Hillier, J., 'Deconstructing the Discourse of Planning' and Throgmorton, J., 'Impeaching Research: Planning as a Persuasive and Constitutive Discourse' (both Papers presented to the ACSP/AESOP Congress, Oxford, July 1991).

36 See Geertz, C. *Local Knowledge: Further Essays in Interpretive Anthropology*, New York, Basic Books, 1983; and Bourdieu, P., *In Other Words: Essays Towards a Reflexive Sociology*, Oxford, Polity Press, 1990.

learn from each other, adapt to each other, 'act in the world' together. Systems of meaning or frames of reference shift and evolve in response to such encounters. But it can never be possible to construct a stable consensus around 'how we see things', merely a temporary accommodation of different, and differently adapting, perceptions.

The critics of modernity argue that the system of meaning proposed by scientific rationalism has dominated and crowded out all other systems of meaning. If communicative action is to transcend this dominatory threat, its concern should rather be to develop understandings and practices of inter-discursive communication, of translation rather than superimposition. For, as Geertz argues, no one system of meaning can ever fully understand enough.[37] It can merely search for ways of opening windows on what it means to see things differently.

Developed in this way, this direction is for a new form of planning through *inter-discursive* communication, a way of 'living together differently through struggling to make sense together'. Its openness, its exteriorising quality, its internal capacity for critique should counteract any potential to turn mutual understanding arrived at one historical moment into a repressive cultural regime at the next. It offers the hope that 'progress', a 'project of becoming', is still possible. It is this direction which, in the present author's view, holds an important promise and challenge for planning, and, more generally, for democracy, as Forester argues.[38]

Planning as a Communicative Enterprise

Environmental planning has been understood in this article as a process for collectively, and interactively, addressing and working out how to act, in respect of shared concerns about how far and how to 'manage' environmental change. Mannheim argued that scientific rationalism provided the central resource for this enterprise.[39] The collapse of the uni-dimensional domination of scientific rationalism has now demolished this route to invention for planning. Apart from the vestigial endeavours of a politically dominant economics, any recourse to scientific knowledge or rational procedures must now be contained within some other conception of what makes for democratic 'acting in the world'. Habermas offers an alternative which retains the notion of the liberating and democratic potential of reasoning, but broadened to encompass not merely rational–technical forms of reasoning, but moral appreciation and aesthetic experience. This wider understanding of what we know, and how we know it, rooted as much in 'practical sense'[40] as in formalised knowledge, is brought into collective 'deciding and acting' through inter-subjective communication rather than the self-reflective consciousness of autonomous individuals. The effort of constructing mutual understanding as the locus of reasoning activity replaces the subject-centred 'philosophy of consciousness', which, Habermas argues, has dominated Western conceptions of reason since the Enlightenment.[41] Through it, the specificities of time and place, of culture, society and personality, of 'habitus' as Bourdieu puts it,[42] are expressed, and constituted. For Habermas, a conscious inter-subjective understanding of collective communicative work is a force to sustain an internally critical democratic effort, resisting the potential domination of 'one-dimensional' principles, whether scientific, moral or aesthetic.

What can planning mean in this context of post-rationalist, inter-communicative, reasoned, many-dimensional, 'thinking about and acting' in the world? What purposes and practices should it have? A communicative approach to knowledge production – knowledge of conditions, of cause and effect, moral values and aesthetic worlds – maintains that knowledge is not pre-formulated but is specifically created anew in our communication through exchanging perceptions and understanding and through drawing on the stock of life experience and previously consolidated cultural and moral knowledge available to participants.

37 See Geertz, C, op. cit. Note 36.

38 Forester, J., 'Envisioning the Politics of Public Sector Dispute Resolution' in Silbey, S. and Sarat, A. (eds.), *Studies in Law, Politics and Society*, Vol. 12, Greenwich, CT, JAI Press, 1992, pp. 83–122.

39 See a recent reassessment of Mannheim's thinking by van Houten, D., 'Planning Rationality and Relativism', *Environment and Planning B: Planning and Design*, 16 (2) 1989, pp. 201–14.

40 See Bourdieu, P., op. cit., Note 36.

41 See Habermas, I, op. cit., Note 15, pp. 196–297.

42 Bourdieu, P., op. cit., Note 36.

We cannot, therefore, predefine a set of tasks which planning must address, since these must be specifically discovered, learnt about and understood through inter-communicative processes.

Nevertheless, ongoing processes of debate about environmental matters have created a thought-world, a contemporary 'common sense' within which, however fluid and in need of critique it may be, the elements of a substantive agenda are evident. The contemporary rediscovery of environmental planning is fuelled by a widespread and inter-discursive concern with managing economic development, enriching our cultural life, avoiding polarising and segregating tendencies in life styles and life opportunities, and undertaking all these within an attitude to the natural environment which is both respecting and sustaining of long-term ecological balances. The general purposes of environmental planning situated in this context are to balance these connecting but often contradictory aims. But what constitutes the 'balance' in particular times and places cannot be known in advance. 'Standardised' approaches to 'balancing', which have a long history in planning thought and practice, encapsulated in substantive 'blueprint' plans, merely 'dominate' the situations they land upon.

This shifts attention from the substantive purposes of environmental planning to the practices by which purposes are established, actions identified and followed through. What does a communicative rationality suggest as appropriate when addressing environmental management issues in contemporary Western democracies, and how could their conversion into a 'process' blueprint be avoided?

The outlines of appropriate practices for an inter-communicative planning are beginning to emerge through the work of a range of planning theorists during the 1980s. This work has been influenced not only by Habermas, but by other and often conflicting contributors to the post-modern and anti-rationalist debate, notably Foucault and Bourdieu, and by an increasing number of 'ethnographic' studies of planning practice.[43] An attempt is made to summarise this 'new' planning direction through ten propositions.

1. Planning is an interactive and interpretive process, focusing 'deciding and acting' within a range of specialised allocative and authoritative systems, but drawing on the multi-dimensionality of 'lifeworlds' or 'practical sense', rather than a single formalised dimension (for example, urban morphology or scientific rationalism).[44] Formal techniques of analysis and design in planning processes are but one form of discourse. Planning processes should be enriched by discussion of moral dilemmas and aesthetic experience, using a range of presentational forms, from 'telling stories' to aesthetic illustrations of experiences. Statistical analysis coexists in such processes with poems and moral fables.[45] A prototype example here might be some of the new initiatives in Britain in working to help tenants and residents improve the quality of their living environment.
2. Such interaction assumes the pre-existence of individuals engaged with others in diverse, fluid and overlapping 'discourse communities', each with its own meaning systems, and hence knowledge forms and ways of reasoning and valuing. Such communities may be nearer or further from each other in relation to access to each other's languages, but no common language or fully common understanding can be arrived at. Communicative action thus focuses on searching for achievable levels of mutual understanding for the purposes in hand, while retaining awareness of that which is not understood (i.e. we may not understand why someone says 'no', but we should recognise the negation as valid; that we know there is a reason but it cannot [yet| be understood by us).[46]
3. Such interaction involves respectful discussion within and between discursive communities, respect implying recognising, valuing, listening and searching for translative possibilities between different

43 See Forester, J., op. cit., Note 7, Throgmorton, J.(op. cit., Note 7, and also Hoch, C, 'Conflict at Large: A National Survey of Planners and Political Conflict', *Journal of Planning Education and Research*, 8(1) 1988, pp. 25–34; Hendler, S., 'Spending Time with Planners: Their Conflicts and their Stress' (Paper presented to the ACSP Congress, Austin, Texas, November 1990).

44 See Innes, J., op. cit., Note 31, and Healey, P., 'A Day's Work', *Journal of the American Planning Association*, 58(1) 1992, pp. 9–20.

45 See Mandelbaum, S., op. cit., Note 31, and Forester J., op. cit., 31.

46 This goes beyond Habermas' argument into the ideas offered by 'ethnographic' scholars such as Bourdieu and Geertz.

discourse communities.[47] A prototype example here might be the public participation exercise undertaken on Sheffield City Centre's Local Plan.[48]

4. It involves invention not only through programmes of action, but in the construction of the arenas within which these programmes are formulated and conflicts identified and mediated. Such a planning thus needs to be reflective about its own processes.[49] The Sheffield City Centre Local Plan exercise is one example among several in Britain which illustrate this sensitive attention to arenas within which planning work gets done.
5. Within the argumentation of these communicative processes, all dimensions of knowing, understanding, appreciating, experiencing and judging may be brought into play. The struggle of engaging in inter-discursive communicative action is to grasp these and find ways of reasoning among the competing claims for action they generate, without dismissing or devaluing any one until it has been explored. Nothing is 'inadmissible' except the claim that some things are 'off agenda' and cannot be discussed. All claims merit the reply: 'we acknowledge you feel this is of value. Can you help us understand why? Can we work out how it affects what we thought we were trying to do? Are there any reasons why the claim cannot receive collective support?'[50]
6. A reflexive and critical capacity should be kept alive in the processes of argumentation, using the Habermasian claims of comprehensibility, integrity, legitimacy and truth. But the critical intent should not be directed at the discourses of the different participative communities (not: 'we are right and you are wrong'; 'we are good and you are bad'), but at the discourse around specific actions being *invented* through the communicative process (e.g. 'watch out: this metaphor we are using blocks out the ideas our other colleagues are proposing', or 'this line of thinking will be dismissed as illegitimate by central government. Do we really think it is illegitimate? Are we really going to challenge their power? OK, so how?').[51] A sensitive illustration of this was the discussion around developing the women's agenda for the Greater London Development Plan as described by Allen.[52]
7. This inbuilt critique, a morality for interaction, serves the project of democratic pluralism by according 'voice', 'ear' and 'respect' to all those with an interest in the issues at stake. This is no easy matter, as interests overlap and conflict, with the conflicts experienced within each one of us magnified in the inter-discursive arena. The important point is that the morality and the dilemmas are addressed inter-discursively, forming thereby both the processes and arenas of debate.
8. The literature on negotiation counsels us that apparently fixed preferences may be altered when individuals and groups are encouraged to articulate their interests together.[53] Interaction is thus not simply a form of exchange, or bargaining around pre-defined interests. It involves mutually reconstructing what constitute the interests of the various participants – a process of mutual learning through mutually searching to understand.[54]
9. It is not only innovative, but has the potential to change, to transform material conditions and established power relations through the continuous effort to 'critique' and 'demystify'; through increasing understanding among participants and hence highlighting oppressions and 'dominatory'

47 The emphasis on respect is powerfully expressed in John Forester's work. Geertz (op. cit., Note 36) highlights the challenge of translation.

48 Alty, R. and Darke, R., 'A City Centre for People: Involving the Community in Planning for Sheffield's Central Area', *Planning Practice and Research*, 3, 1987, pp. 7–12.

49 See Forester, J., 'Anticipating Implementation: Normative Practices in Planning and Policy Analysis' in Fischer, F. and Forester, J. (eds), *Confronting Values in Policy Analysis: The Politics of Criteria*, California, Sage, 1987; and, at a more organisational level, Bryson, J. and Crosby, B., The Design and Use of Strategic Planning Arenas', *Planning Outlook*, 32 (1) 1989, pp. 5–13.

50 The importance of this listening and learning attitude is emphasised in Forester, J., op. cit., Note 7. See also Throgmorton, J., op. cit., Note 35, and Healey, P., op. cit., Note 44.

51 See Throgmorton, J., op. cit., Note 35, and Tait, A. and Wolfe, J., 'Discourse Analysis and City Plans', *Journal of Planning Education and Research*, 10 (3) 1991, pp. 195–200 for the critical 'deconstructive' analysis of planning texts.

52 Allen, J., 'Smoke Over the Winter Palace: The Politics of Resistance and London's Community Areas' (Paper presented to the Second International *Planning Theory in Practice* Conference, Torino, Italy, September 1986).

53 I am indebted to Seymour Mandelbaum for the phrasing of this sentence.

54 See Forester, J., op. cit., Note 38.

forces; and through creating well-grounded arguments for alternative analyses and perceptions, through actively *constructing* new understandings. Ultimately, the transformative potential of communicative action lies in the power embodied in 'the better argument',[55] in the power of ideas, metaphors, images, stories. This echoes Bourdieu's point that how we talk about things helps to bring them about.[56] In this way, diverse people, with experience of different societal conditions and cultural communities, are encouraged to recognise each other's presence and negotiate their shared concerns. Through such processes of argumentation, we may come to agree, or accept a process of agreeing, on what should be done, without necessarily arriving at a unified view of our respective lifeworlds. The critical criteria built into such a process of argument encourages openness and 'transparency', but without simplification. If collective concerns are ambivalent and ambiguous, such a communicative process should allow acknowledgement that this is so, perhaps unavoidably so. So the dilemmas and creative potentials of ambiguity enrich the inter-discursive effort, rather than being washed out in the attempt to construct a one-dimensional language.[57]

10. The purpose of such an inter-communicative planning is to help to 'start out' and 'go along' in mutually agreeable ways based on an effort at inter-discursive understanding, drawing on, critiquing and reconstructing the understandings we bring to discussion. The inbuilt criteria of critique, if kept alive, should prevent such 'starting agreements' and 'travelling pacts' consolidating into a unified code and language which could then limit our further capacities at invention. We may be able to agree on what to do next, on how to 'start out', and 'travel along' for a while. We cannot know where this will take us. But we can act with hope and ambition to achieve future possibilities. Neither the 'comprehensive plan' nor 'goal-directed' programmes have more than a temporary existence in such a conception of communicative and potentially transformative environmental planning.[58]

Systems and Practices for Environmental Planning

How can this conception of communicative practices for constructing and critiquing understanding among diverse discursive communities assist in the development of 'systems' for environmental planning, of local realisations of these, and of the specific contents of local planning systems? The very concept of a 'system' immediately conjures up notions of dominatory practices which impose themselves on our actions. Yet with respect to our mutual environmental concerns, a key purpose of communicative action is to work out what 'rules' or 'codes of conduct' we can agree we need, to allow us to 'live together but differently' in shared environments.

Planning systems consist of formal rules to guide the conduct, the resource allocation and management activities of individuals and businesses. But they are more than a set of rules. The rules derive from conceptions of situations (contexts), problems experienced in these situations, ways of addressing these problems and of changing situations. It is where planning effort is deliberately focused on *changing* situations that we can speak of a planning with transformative intent.[59]

'Urban design' or 'physical blueprint' approaches to environmental planning focused on 'transforming towns'. Ideas of urban existence were consolidated into principles of urban structure and form, and from these to rules to govern proposals for development projects. Debates were confined to principles of urban form, conducted primarily within a narrow expert group (architects, engineers) legitimated by paternalist notions of 'planning *for* people'. It was supported by a narrow architectural engineering discourse about the relative merits of different urban forms, drawing on aesthetic and moral principles. The 'dominatory' consequences of

55 As Habermas, J., op. cit., Note 15, claims.

56 Bourdieu, P., op. cit., Note 36 is referring to Marx's idea of class.

57 See Forester's discussion of Nussbaum's work (Forester, J., op. cit., Note 31).

58 I am indebted to correspondence about the qualities of a democratic plan with John Forester for my thinking here, as well as the ideas of Sennett, R., op. cit., Note 23, on designing with the grain of diversity.

59 Following the usage of Friedmann, J., op. cit., Note 15, 'transformative' here refers to changing the context deliberately as well as acting within the context.

this for our towns and cities are notorious. This was essentially a continuation of a pre-enlightenment tradition of city planning carried forward into the context of nineteenth- and twentieth-century industrialisation and urbanisation.

The Mannheimian conception of planning as the 'rational mastery of the irrational'[60] provided a more appropriate realisation of a 'modern' conception of planning. Translated through the Chicago school, this became the rational comprehensive process model of planning which has since been so influential in planning practice. This focused on the processes through which goals were formulated and strategies for achieving them devised. Here, rule generation operates on two levels – the methodological rules for arriving at a plan or programme, and the criteria necessary for realising that programme. Both were designed to be recursive, with feedback loops via monitoring procedures intended to sustain an internal critique of planning principles. Planning effort was focused on comprehensive understanding of urban and environmental systems and the 'invention' of sets of objectives and guidance principles for the comprehensive management of these systems. Rules to govern change in systems were expressed as performance criteria, linked back to objectives. The vocabulary of this approach is still influential in plan-making practice in Britain, in the way strategy is identified and expressed and rules for development control articulated. In this rationalist conception, citizens contribute to the process, but only by 'feeding in' their rationalised goals, rather than debating the understandings through which they come to have their goals. The concerns of politicians and citizens are in effect 'translated', 'converted' into the technical scientific language of policy analysts and urban and regional science. The metaphors of this language focused around images of process forms, of strategy and programmatic action. The dominatory potential of the rational procedural model lies in the claims to 'comprehensiveness' of what was primarily a narrow, economistic and functionalist conception of the dimensions of lifeworlds. The critical capacity of the monitoring feedback loops merely shifted priorities within the discourse. It did not provide a mechanism for critiquing the discourse itself.

Pluralist conceptions of interest mediation, of the kind first proposed by Davidoff,[61] but later widely developed, seemed to reflect more clearly the reality of environmental planning politics. The practice of environmental planning has been described by many in Britain, including the present author, as one within which environmental perceptions and interests were asserted and mediated[62] The strategies, rules and the way rules were used were the product of bargaining processes among conflicting interests. But, as Forester argues, this treats each interest as a source of power, bargaining with others to create a calculus which expresses the power relations among the participants. Its language is that of prevalent political power games. It is not underpinned by any effort at 'learning about' the interests and perceptions of the participants and with that knowledge, revising what each participant thinks about each other's and their own interests. Only if this could happen could a creative inventive form of environmental planning develop, rather than merely a power-broking planning.[63]

The focus of an inter-subjective communicative argumentation is exactly at this point. It starts by recognising the potential diversity of ways in which concerned citizens (citizens with an 'interest' in issues) come to be concerned. Citizens may share a concern, but arrive at these through different cultural, societal and personal experiences. 'Understanding each other' must therefore be accepted as a challenging task which is unlikely to be more than partially achieved. The language of inter-discursive communication, as already discussed, uses multiple modes, moving between analysis, moral fables and 'poems'.

The struggle within such inter-discursive communication is to maintain a capacity for critique. This requires the development of a critical, interactively reflexive habit. Of course, the dynamics of the ongoing flow of relations means that people cannot pause to reflect collectively at every instant. What it means is that 'taking breath' and 'sorting things out' should become a normal part of the practical endeavour of planning

60 Mannheim, K., op. cit., Note 11.

61 Davidoff, P., 'Advocacy and Pluralism in Planning', *Journal of the American Institute of Planning*, 31, November 1965, pp. 331–8.

62 Healey, P., McNamara, P.F., Elson, M.J. and Doak, A.J., *Land Use Planning and the Mediation of Urban Change*, Cambridge, Cambridge University Press, 1988; Brindley, T., Rydin, Y. and Stoker, G., *Remaking Planning*, London, Hutchinson, 1989; Blowers, A., *The Limits of Power: The Politics of Local Planning Policy*, Oxford, Pergamon, 1980.

63 See Forester, J., op. cit., Note 38.

work. The Habermasian criteria help here, but reflection is also required as to the arenas of the communicative effort itself. Are there other concerned people who should be involved? Are there other ways of understanding these issues, discursive practices, which we should include? How should the position we have reached be expressed to maximise its relevance to all of us, allowing us to 'move on' but yet minimise the potential that what we have agreed will live on beyond our need for it and come to dominate us?[64] Through these processes of active discursive critique, ideas for action may be invented, and necessary codes of conduct for the collective management of shared concerns be identified and agreed upon.

This conception of a planning 'invented' through reflective processes of inter-subjective communication within which are absorbed internal criteria of critique, is suggestive of ways in which existing processes of plan-making, conflict resolution and implementation programmes might be transformed. Specifically, the active presence of a planning in this form will be reflected in the language and metaphor used within the various arenas constituted for environmental planning work. It would reflect efforts at honesty and openness, without losing a recognition of the layers and range of meanings present among those concerned with the issue in hand. It would acknowledge with respect the limited scope for mutual understanding between diverse discourse communities, while struggling to enlarge that understanding. It would accept other limits – to power, to empirical knowledge, to the resolvability of moral dilemmas – but seek to enable the world-of-action to 'start out' or 'move on' towards something better, without having to specify precisely a goal. Rather than Lindblomian marginal adjustments to the present,[65] its language would be *future seeking*, but not, like its physical blueprint' and 'goal-directed' predecessors, *future defining*. Its images and metaphors would draw on both the experiential and abstract knowledge and understanding of those involved, recognising the interweaving of rational–technical, moral and aesthetic dimensions in our lives. It would seek to 'reason between' conflicting claims and conflicting ways of validating claims. It would not force one dimension of knowledge to dominate over another. It would be courageous, challenging power relations through critique and the presentation of alternative arguments. It would reflect the internal critical monitoring practices of participants. It is thus by the tone of its practices that it would be identified.

The Dialectics of a New Planning

To those seeking specific substantive solutions to particular problems, the planning outlined here may seem too leisurely. With environmental disasters near at hand, can we afford to take the time to invent answers? To those seeking knowledgeable actions, this planning may seem too unfocused and diffuse. What happens if mystical perceptions or aesthetic reification crowd out the useful empirical and theoretical knowledge we have about cause and effect? To those conscious of the scale of inequalities in power relations, it may seem idealistic and innocent. Does it not merely cocoon us into a naive belief in the power of democratic discussions, while the forces of global capitalism ever more cleverly conceal the ways they oppress us?

To these doubts there are two replies. One is that to engage in any other strategy is to generate once again forms of planning which have inherent within them an anti-democratic 'dominatory' potential. Each is one-dimensional, drawing on the power of design, of moral imperative, of scientific reasoning or, unmasked, as a direct power struggle, drawing on the possibility of replacing one dominant power source with another. The second is that the practices involved are not so far from our experience. Prefigurative examples can be found in Britain in some of the work of the 'new left' for example, in the Greater London Council (GLC), particularly in dealing with women's issues;[66] and recently in a few of the new efforts in plan-making in Britain resulting from requirements to prepare Urban Development Plans and District Development Plans.[67] More generally, some branches of the environmental and feminist movements have been moving in this direction. Further

64 I have developed ideas on the constitution of *process forms* in Healey, P., 'Policy Processes in Planning', *Policy and Politics* 18 (1) 1990, pp. 91–103.

65 Lindblom, C.E., *The Intelligence of Democracy*, New York, Free Press, 1965.

66 See Allen, J., op. cit., Note 52.

67 See Healey, P., 'The Communicative Work of Development Plans', *Environment and Planning B: Planning and Design*, forthcoming.

prefigurative potentials can even be recognised in contemporary management theory's emphasis on group culture formation and empowerment, rather than management through hierarchical authoritarian structures[68] At a broader level, the 'struggle for democracy' in Eastern Europe and China has highlighted awareness in Western societies as to what democracy might mean. It is in Britain perhaps that this awareness has most progressive potential, since a critical eye finds so few guarantees of democracy in our political and legal systems.

'Inventing democracy' is thus, for British people, an issue which is moving increasingly sharply into focus. It is a time, as noted at the start of this article, for the invention of democratic processes. The field of environmental concerns is one of the critical arenas within which such invention is being demanded and tested.[69]

However, there are many democracies which might be invented. Learning and listening, respectful argumentation, are not enough. We need to develop skills in translation, in constructive critique, in collective invention and respectful action to be able to realise the potential of a planning understood as collectively and inter-subjectively addressing and working out how to act in respect of common concerns about urban and regional environments. We need to re-work the store of techniques and practices evolved within the planning field to identify their potential *within* a new communicative, dialogue-based, form of planning. This article has drawn on the work of a number of planning academics searching within the 'lifeworld' of planning practice for a better understanding of these skills. What is being invented, in planning practice and planning theory, is a new form of planning, a respectful argumentative form, of *planning through debate*, appropriate to our recognition of the failure of modernity's conception of 'pure reason', yet searching, as Habermas does, for a continuation of the Enlightenment project of democratic progress through reasoned inter-subjective argument among free citizens.

Yet, as the planning community explores the hopefulness of this new approach, it is important to remember the experience of past efforts at 'democratic making'. Habermas offers the theory of communicative action as an inter-subjective project of emancipation from fundamentalism, totalitarianism and nihilism through deliberate efforts in mutual understanding through argument. But this can only succeed for more than a historical moment so long as the processes of internal critique are kept constantly alive; if what Habermas calls 'the lifeworld' is constantly brought into the collective thinking about 'acting in the world' in respect of common affairs; and if the communicative effort of mutual understanding is sustained as a critical as well as a creative process. Either we succeed in keeping a critical dialectics alive within communicative action, or we remain caught within the dialectics of totalising systems. As the opposition of 'capitalism' versus 'communism' collapses, perhaps there is a hope that, through dynamically critical communicative processes, the democratic project of 'making sense together while living differently' can develop as a progressive force.

68 See Handy, C, *Understanding Organisation* (3rd edn), Harmondsworth, Penguin, 1985.

69 Grove White, R., op. cit., Note 21 makes this point very cogently.

Chapter 15

City Fathers, Mandarins and Neighbours: Crossing Old Divides in New Partnerships*

Patsy Healey**

From 'Trickle Down' to 'Reconnecting'

This paper examines the experience of an experiment in partnership in urban regeneration in a disadvantaged inner city neighbourhood in Newcastle, UK. The experiment drew business, government and local people into a formalised 'three-way partnership', aimed at helping residents back into work in the formal economy, and involving local people more actively in shaping programmes which could make a difference to the quality of their lives. The experiment was hugely ambitious. It aimed to build links between the world of residents in neighbourhoods where daily life is often harsh and tense and that of business life in the formal economy. At the same time, it also sought to co-ordinate the various agencies involved in delivering training and welfare services to residents, as well as develop community cohesion among residents. This multi-dimensional relation-building work took place in a context of competing and fragmented urban governance. It illustrates a particular approach to alliance-building and partnership in cities and urban regions which has developed in recent years in Britain to address and guide the processes of socio-spatial transformation experienced in the context of a post-Fordist economy and a post-welfarist state.

In the 1980s, the policy pre-occupation in Britain was with economically-abandoned zones. Transforming these through property-led regeneration provided the imagery of an economic alternative which would sustain local economies and maintain social relations. In neo-liberal political regimes such as Britain, it was believed that supply-side injections of subsidy into physical transformation would 'trickle down' to disperse benefits throughout urban economies. In parallel, the labour force was put under pressure to change attitudes. The expectation of a secure lifetime job in a single factory was to be replaced by more flexible working practices, to enable adjustment to the dynamics of new companies with new skill demands and less interest, in Britain anyway, in a lifetime-committed labour force. In support of this shift, the welfare benefit system was transformed to reduce 'dependency' and encourage people back to work. In this way, the universalistic framework of the welfare state was transformed towards a 'workfare' state (Jessop 1991).

The reality which has emerged in the 1990s has revealed the fallacies in this economic model. Firstly, we now talk about the phenomenon of job-less growth. New economic activities are certainly developing and expanding. Old companies have been re-structured and many now compete well in the international market place. New 'inward investment' can be attracted. But this does not compensate for the jobs lost. Secondly, not only has economic growth been less assured than expected, the new economic activities demand a polarised workforce, some with highly developed skills, and unskilled service workers prepared to work for low pay for unsociable hours, often part-time. Getting access to these well-paid jobs may require a level of education that seems unattainable for many whose families lived off the old industrial jobs where education was of little relevance to factory work. One consequence has been a transformation of household relations in many working class homes, as women go out to work in hard manual service jobs for low pay, while the men stay at home, unemployed and often unprepared to take on housework (Campbell 1993, Wheelock 1990).

* Previously published in O. Källtorp, I. Elander, O. Ericsson and M. Franzen (eds) *Cities in Transformation: Transformation in cities*, Aldershot: Avebury (1997), pp. 266–88.

** This article is based on a joint research work by the author and Rose Gilroy, Suzanne Speak and James Wood (Wood et al., 1995).

So the distance between the new formal economy and the social worlds of the former workforces of the old economy drift apart. In such a context, benefits from the new do not 'trickle' easily to those suffering from the dislocation of the old.

The 'zones of disinvestment' which now confront policy makers in British cities are thus not the obsolete industrial and port zones which attracted attention in the 1980s, but the neighbourhoods where poverty and disadvantage are concentrated. There has been a history of attempts since the 1960s to understand and to improve conditions in such neighbourhoods (Gibson and Langstaff 1982). Meanwhile, the quality of life in many such neighbourhoods has steadily deteriorated. The formal economy provides decreasing levels of income, while changes to the benefit system reduce the state benefits flowing into such areas. Levels of educational achievement decline as children turn away from school, accepting that it will offer little of benefit. Health declines as people suffer stress and poverty. Into the vacuum flow alternatives, particularly the explosive mixture of organised crime and drugs (Simey 1995). These tendencies exacerbate socio-spatial segregation as those who can leave the areas with the most difficult living conditions. To add to this, housing policies in Britain in recent years have concentrated the unemployed, the jobless and those on very low incomes in the estates of the local authority rented sector. This in turn encourages the 'labelling' of neighbourhoods, which feeds into informal 'red-lining' of particular areas by employers and those who provide credit and insurance services. In many such neighbourhoods, it becomes impossible to get a washing machine on hire purchase, insure a car, or get sent a mail-order catalogue.

People in such situations live with a daily struggle over the means of existence, and daily tensions within the family between alternative value systems. Should they turn towards the 'formal economy' and get the qualifications for a skilled job? Should they turn towards government and organise to demand better support and services? Should they turn towards the informal economy, and even the high-risk but potentially high-return benefits of crime, and the informal power structures which hold this together? Should they organise themselves to build up community cohesion and co-operation to support each other and confront their difficulties?

As people living in such conditions struggle with their daily lives, the 'formal' systems – the formal economy and the firms which provided jobs, and the political system and the government agencies which delivered welfare services – have been challenged both morally and materially. Morally, the business community and politicians cannot shrug off all responsibility for some response to the increasing isolation of people from formal economic systems. Materially, in the concentrated arena of urban space, people fear the fact and threat of crime, violence and riots. While it is people in the neighbourhoods who suffer most from these phenomena, the threat of social unrest affects city centres, nearby residential neighbourhoods, local business and areas anywhere without expensive investment in security. Thus the coping strategies adopted by people in such neighbourhoods touch the worlds of those in positions of power.[1]

During the 1980s, a range of policy initiatives focused on neighbourhood improvement. These ranged from investment in physical refurbishment, to improve design and management of tenanted property and provide better local facilities, to developing economic opportunities, through community businesses and training opportunities (McArthur 1993, Haughton 1990, Atkinson and Moon 1994, Blackman 1995). Many of these initiatives encouraged the active involvement of residents in defining and managing programmes. These schemes often brought benefits, both in material terms, in the social status of the neighbourhood, and through the sense that government systems cared what happened to people. But many schemes did little to halt the continuing disintegration of both social life and the physical stock. Nor did neighbourhood improvement necessarily do much to increase people's access to jobs in the formal economy. As articulated in national urban policy in the late 1980s and early 1990s, the policy challenge was therefore to 're-connect', to open up routes for residents to the 'mainstream' economy. The metaphor of 'connecting to the mainstream'[2] tended to be constructed in terms of 're-admitting' the excluded and marginalised, through equipping them with new skills and capacities.

1 In this respect, there are parallels with the situation in the nineteenth century when the health hazards of inadequate water supply, sewerage and drainage threatened rich and poor alike.

2 See Davoudi and Healey (1995) for the way this metaphor was used in the City Challenge Initiative. US ideas were influential here (Bendick and Egan 1993).

Partnership in Cruddas Park

In the mid-1980s, a new idea emerged in the circles of Tory politicians, senior business people and some civil servants. This revived a pre-welfare state image of benevolent business, the city fathers and the Victorian ideal of patronage, and attached it to notions of the value of partnership between the public and private sectors in dealing with the challenge of urban regeneration. The *Business in the Community* initiative arose from this, as a national, business-led contribution to urban governance. Behind this idea lay a complex amalgam of political objectives – national government challenged local government for its 'failure' to address urban problems adequately; neo-liberal strategies sought to overcome the divide engendered by both the welfare state and the Unionist–Labour conception of the opposition between public and private sectors, community versus capital. Politicians and civil servants were also looking for ways of reducing costs and sought to 'lever-in' contributions from the private sector, and 'off-load' tasks to both business and the community. Business, and to an extent neighbourhood residents, were also concerned about the increasing fragmentation of governance agencies as a result of the vigorous organisational restructuring of government promoted by the Tory administration. 'Partnership' was in principle attractive as a way at least of getting close to government, for both business, often in search of good access to grants, and residents of areas dependent on government for housing and welfare (Bailey 1995).

It is in this context that the idea of a 'three-way partnership' between government, business and community emerged in English urban policy in the late 1980s.[3] The objective was to provide a time-limited programme of activities within a neighbourhood, designed to open up pathways for residents to the formal or 'mainstream' economy and to empower residents to take charge of setting their own governance agendas, by participating in defining and implementing the programmes. Implicitly, the role of civil servants and business in this empowerment was 'to lead not to learn' (Simey 1995, p.1).

The Cruddas Park Initiative was an experiment in three-way partnership which has a considerable influence on the government's City Challenge programme, started in 1991. The Initiative began as an idea in a business-led city promotion group in 1988, and was strongly supported by central government. The location selected for the Initiative was a neighbourhood of around 850 dwellings, nearly all in local authority ownership. The estate was located near the city centre, and adjacent to a new business park being developed by the local Urban Development Corporation. The dwellings were spacious and well-designed, mainly in tower blocks. But 40 per cent of economically active men were unemployed, as were 25 per cent of women; 70 per cent had no car and 22 per cent suffered from long term illness.[4] Crime rates were high.

Building Links for Community Governance: Multiple Objectives

> The idea was to get businessmen and civil servants on board … but have local people dictate the programme, a 'bottom up' approach. Professional people such as bankers, accountants and lawyers sat down with residents and helped them in the initiatives. They worked effectively together … it was heart-warming. People in the community who had seen themselves as wage slaves without a job (developed) a relationship of equals with (the business sector). (A civil servant[5])

The Initiative was to pioneer partnership between the business sector and the community in overcoming problems of social deprivation and exclusion. In parallel, civil servants would work to build links across departmentalised government programmes, to target resources and activities in ways which would be driven by what residents saw as their priorities. But this strategic objective came up against its own contradictions as it developed in the Cruddas Park context.

What, for example, could it mean for 'local people' to 'dictate' the programme, and why should they do this? Some people saw the problem as 'within the community'. The Initiative was therefore to help people

3 See Imrie and Thomas (1993), Davoudi and Healey (1995).

4 Census 1991.

5 Civil servants in this account work for central government.

help themselves, to encourage them to move away from a 'dependency' relation towards government. Others, as in the above quote, saw the issue in terms of exclusion. They used the word 'disenfranchised' and sought to 're-connect' local people to the mainstream of economic and governance activity. Another view was that the issue was empowerment, which meant the confidence to sort things out independently, and to challenge and capture power from the wider system which made life difficult for local people. Some also saw such initiatives as attempts either to off-load government functions onto local people, as a sort of 'privatisation' measure, to reduce government responsibilities, or as part of the continuing struggle for power between central and local government.

Such differences in interpretation of the purpose of the Initiative created a complex context in which to seek to build a 'partnership of equals'. In any case, people were not equals. They had different experiences, different priorities and wielded differential power resources in the wider world. In effect, they came from different cultural worlds. So building up partnership relations was shot through with tensions and inequalities in power relations. What to some looked like a partnership of equals because they sat round a table together, for others seems like patriarchal patronage or just plain terrifying.

The big tensions centred on the links which people hoped to develop through the Initiative. For many, the objective was to 'build routes' or pathways between local people and business in order to help people back into a job. Through such routes residents could be re-connected to the wider world. Several people in the government sector were also aware of the need to make links across the various sectors of government action, to make their delivery more relevant to local people as they set out on the path of 're-connecting to the mainstream economy'. Civil servants sought to make links across the increasingly fragmented landscape of government agencies, to put bits of policy action together in ways which made sense for their own tasks, and for local people, as the civil servants saw their needs.

Many local people, however, while concerned about these wider links, were more preoccupied with the conditions of daily life in the neighbourhood. For them, the key links were with each other, to deal with tensions within the neighbourhood, and to build up local social support systems. Conflict often surfaced about which were the priority links – those between local people and the outside world, those between sections of government, or those within the community. People were also aware that emphasis on one sort of link could undermine the others.

Whatever objectives people had, the Initiative implied organising governance in different ways to the traditional sectoralised and top-down approach characteristic of both British central and local government. The challenge of the Cruddas Park Initiative was therefore to pioneer new forms of governance in one small area where local people experienced, at the extreme end of difficulty, the contemporary economic, social and governance problems of our society.

Central Government Departments: Pro-active Mandarins

The pioneering and experimental nature of the Cruddas Park Initiative was well-understood by senior civil servants. Civil servants in Britain have traditionally had the status of a 'mandarin' class, clustering around government in London, serving the patriarchal relations of Britain's centralised state. In the mid-1980s in the North East, there came together a team of forward-thinking civil servants in the regional offices of several departments of government. These understood the positive side of the re-structuring of the state under the Thatcher administrations. British government was too functionally compartmentalised and too divorced from both the people and the business sector. What was required were initiatives in co-ordination and partnership.

This led to an enthusiastic commitment to developing the *Business in the Community* initiative. The result was the formation in Newcastle of *The Newcastle Initiative (TNI)*, a business-led grouping, which included civil servants, the leader of the City Council and some other local notables (Blackman 1995, Bailey 1995). It was in this arena that civil servants raised the possibility of an inner city neighbourhood initiative, influenced by estate regeneration projects in Scotland. This proposed:

- the focus on a particular inner city neighbourhood

- the co-ordination of government departments, services and local business to help local people back to work
- a division of responsibility whereby government would focus on training if local businesses would help by linking training to jobs
- people in the area, seen as 'tenants' of the local authority, would 'drive the process'.

The actual selection of Cruddas Park as the focus for the Initiative was made from a list supplied by the local authority, Newcastle City Council, which was represented on the TNI Board. It was an inner-city estate near a new business park on a former industrial site and also near the city centre, so city centre firms could reasonably be expected to be a source of potential jobs, thus demonstrating that the benefits of property-led initiatives did, with a bit of extra effort, 'trickle down' to unemployed people nearby.

The development of the initiative was then passed to the chair of TNI, at that time Chief Executive of a major bank in the North, and to civil servants, local authority officers and others. These set about working up the ideas and building links with local people. It soon became obvious that the project needed an anchor person, to hold it together operationally. An experienced community development worker was employed first as a consultant on an initial feasibility study and then took on a key leadership role in the development of the Initiative. Civil servants felt that, with these personnel, they had the 'right kind of people' for this difficult initiative.

A good deal of time and commitment was then spent in mobilising inter-organisational co-ordination and in finding ways of bringing in resources from the private sector. But those involved were also learning through the experience. Through this learning, some of the contradictions and complexities of such an Initiative became apparent to them. One lesson related to the division of responsibilities between people and government:

> Asking people to do the work that professional people with good education and years of training have been doing for them (is a big demand). It needs a lot of training and hand holding. The (Initiative) … has a great deal of responsibility and we expect people who have never done anything before to be able to run the show themselves. I've heard people say they were set up to fail. (Civil servant)

A second lesson pointed nearer to home, and the political environment of government programmes which developed in the 1980s. In the urban policy field in particular, there was a continual replacement of one initiative with another, with competition between and within each initiative for the funding available. Success in funding depended on being able to keep initiatives high profile with politicians and central government officials in London. The result was that Cruddas Park was catapulted into the political limelight, with visits from ministers and princes. In this climate, it was difficult for either the civil servants or local people to spend much time in the delicate institution-building work of 'getting to know each other' and to learn new skills in governance.

> When money is available for initiatives, you have to hit the ground running. … (something) has to happen quickly. There isn't time to do the type of research that it needs. Not enough was understood about the area, the people or what we were getting into, but if we had waited to think it through, we would have lost the opportunity. (Civil servant)

There were nevertheless a number of practical lessons which were transferred from the Initiative into the general practices of the government departments and agencies:

> If you get people at the right level in the organisation who are the 'doers', who are the ones who get round the table and say: how are we going to do this, and then through that they learn about each other's organisation – what was actually happening was that there was a greater understanding of things that would be preventing an organisation from doing things; also a way round how you could use the organisation in a different way. Because we were learning about the organisation. This helped Employment Services a great deal. We built up very strong links with other organisations, understanding what was going on. So from that aspect, Cruddas Park was very useful. (Civil servant)

The challenge was to break out of functionally compartmentalised departmental organisational cultures and to re-orient from producer-driven to consumer-sensitive conceptions of service delivery.

Through co-ordinated working on the Cruddas Park Initiative, practices built up among the various agencies which have since been adopted generally in the region. But transforming such practices required time, commitment and resources. It involved building up trust and understanding through which organisational learning could take place. Extending them more widely than a particular initiative is problematic with limited staff resources in the government agencies concerned.

The Business Community: The New City Fathers

The private sector in the North East, as anywhere, consists of a wide array of firms. As in most regional economies these days, they are not particularly strongly integrated with each other, either in a business sense or in a political or social sense. Nevertheless there are well-developed and long-established networks among the larger, older firms in the region.

> The North East Network works well – some people have called it a large village. There is a network of around 50–100 people who are fortunate to hold the power strings. They can pick up the phone and get help and advice. (Civil servant)

It was this network which was activated to persuade people to become involved in the Cruddas Park Initiative, just as it had been activated to get involved with TNI to begin with. But the network does not include many small firms or newer businesses in the region and is difficult to break into.

What was mobilised in the network were not merely financial resources. Business people provided their skills as well, in giving advice, particularly about the community businesses. They provided contacts, to sources of funds and further advice. Business people offered their time, their knowledge, and in particular, their commitment 'to care'. Several of the business people involved talked about their moral obligations. A senior manager in one firm felt that a large local employer should help neighbouring areas in need. Others talked about the duty to 'give something back' to the local community.

The Bank Chief Executive, as the prime mover from the business side, saw himself as the focal point for the private sector's involvement:

> Because of my contacts, I could be in a position to direct resources to Cruddas Park, mostly cash and people. Then I began to realise that there was a benefit in having the contacts that I had managed to get to know over the years, in trying to shortcut the problems that occasionally got bogged down in bureaucracy, and through layers and layers, because, just occasionally, – I never did this except sparingly – I could just pick up the phone and have a word with the leader of the City Council and say look: we are not getting anywhere with this. Would you mind just having a look at it, push things forward … That's where I saw myself. Increasingly I got to know the people … I do still care about the place. I don't go there that often, probably once a month, but I know a lot of people, certainly, those involved with the Trust. The relationship is I think fairly easy now.

The business sector provided considerable financial help to the Initiative, in cash and in kind, especially in the early stages. Small amounts of 'matching funds' were often provided for community projects. The advice of business people was also valued, although some commented that the directors of large firms did not always understand the problems of smaller enterprises and community businesses. In return, those business sector people who got involved learned a good deal about the situation of people in the area. Specifically, they came to realise the capacities of the people and the extremity of the difficulties they faced. They met real people, much like themselves, as people.

They also discovered their own limits. The specific case here was the 'jobs pledge'. This idea had been explored in senior business and government circles in the late 1980s. But it assumed that there were the jobs to pledge to local people and that legally they were allowed to do this. The directors of some firms committed their companies in principle but then found that there were no new jobs being created, and that their personal sections had concerns about such selectivity, partly because of EU legislation, and partly because of conflicting selection principles.

But there were limits to what business people could be expected to deliver. They preferred to perform specific tasks, rather than engage in 'talking shops'. They tended to have a distinctive perspective on what it takes to get things done. They emphasised the importance of having the 'right people' doing things, of having people who were 'project champions', the 'movers and shakers' who could get things going. They had a relatively low toleration for bureaucratic processes, and found themselves puzzled by, and critical, of the politics and bureaucracy of central and local government, even though they experienced bureaucratic processes in their own firms. There was a general unease about the distorting effect of the high political profile which surrounded the Cruddas Park Initiative in the early years.

There was also a low appreciation of the systematic nature of some of the issues addressed. Problems were more likely to be identified as due to specific people in specific places than to do with questions of institutional structure, or the 'way things go on'. It is one thing to help a particular business get started and to help particular people into jobs. It is another to think about the institutional capacity that needs to be in place to enable such things to happen more easily. For many in the business sector, the Cruddas Park Initiative was a one-off project, a form of 'business sponsorship', designed to add value. It was not about systematic institution building.

Those involved from the business field learned a good deal through the Initiative about the way others live. The Initiative also helped to build networks between key regional players in the public and private sectors, and to help governance think in business terms, which in turn could benefit business. More business people find themselves on the boards of governance agencies these days. But no sustainable long-term relationship could be said to have built up between the local 'community' and business as a result of the Initiative. Like people in Cruddas Park, the business players look to someone else to bother about building up the structures within which the 'right people' and the 'right networks' can work. There was thus a considerable implicit difference between the institutional vision of the civil servants and the perspectives of the business people involved.

The City Council: Defensive Pragmatists

Layered over this tension was a history of central–local government antagonism in Britain, particularly in the metropolitan areas. In Newcastle, this was compounded by an ideological divide between the Union–Labour culture of the city council at the time, and the neo-liberal strategies of the Thatcher administrations. While the city council had a history of acting pragmatically to capture whatever resources were going for the city, these tensions were never far from the surface. Further, city councillors and officers were in direct competition with civil servants and those from the business sector over governance roles. Each considered that they had a legitimate governance role with respect to the neighbourhood, as local elected representatives, national elected representatives and as business patrons. Further, central government and local government each blamed the other for the criticism of lack of co-ordination and excessive bureaucracy directed at them by business, residents and Tory politicians.

The City Council was already involved in Cruddas Park in many ways. It owned the shopping centre and was the landlord of most of the residents. It delivered key services, particularly education (a nursery and primary school were in the area), recreational amenities and community services. Cruddas Park was also covered by one of the City's *Priority Area Teams*.[6] This aimed to provide a channel between local people and council programmes. There were also two Tenants Associations, linking to the local authority's role as housing landlord. Cruddas Park was on the city council's list of estates to refurbish, and the council needed the co-operation of central government to provide the funds for this work.

Officially, then, the Council welcomed the Initiative. It was on the list of possible locations which they put forward to TNI. But councillors and officers saw its justification in economic terms. The business involvement was expected to deliver jobs. The housing and social issues were matters which were properly the responsibility of the City Council. This position was made very clear to the business people:

6 This had been set up under government urban policy of the late 1970s.

> We had a meeting with the leader of the Council and several senior local authority officers and politicians. He said, quite candidly, your success will be based solely on the basis of jobs … I said there is rather more to it than this, but I accept that people will judge us to a large extent on jobs. (Bank Chief Executive)

The dominant interpretation among councillors and most officers was that the problems of the area were economic, the lack of the jobs which had once been available. If these could be available again, then social conditions would improve. Empowering residents to 'drive' the initiative was a deflection from the real problem. In any case, the City Council had the machinery in place to deal with the social and housing issues.

> Politicians and Newcastle City Council were understandably resentful that the kudos of (the Initiative) would be taken by TNI and thus the private sector, when in reality the private sector in anywhere like Cruddas Park is minimal. (Council officer)

Thus the City Council had little interest in the institutional capacity-building and empowerment dimensions of the Initiative. Further, they tended at the time to underplay the complex interconnections between 'getting a job' and the social conditions which residents had to negotiate in getting a job. 'Jobs' were seen as material benefits which could be 'handed out' and 're-distributed'. Nor initially did councillors and officers respond pro-actively to the demands from articulate community representatives for more co-ordination of city council services and more respect and partnership in governance. The tradition of the City Council had been as a provider, a service deliverer. Yet they were under pressure from their communities, particularly their tenants, for a more interactive relationship with the Council.

> It used to be the community making representations about what they wanted, and hoping the council would respond. Now the community is making demands. Ward councillors these days are tested much more, with demands for reasons for why things have and haven't been done. (Local resident)

The functionally compartmentalised, service delivery approach militated against a 'holistic' focus to a person's problems or the problems of an area. It was difficult to interlink the various social dimensions of the Cruddas Park Initiative in an equivalent way to the links built up around employment issues. Central government agencies dominated the governance role in the economic arena, and co-ordination was vigorously supported by senior civil servants. There was no such impetus for departmental co-ordination in the City Council at the time. Residents and individual groups in Cruddas Park continued to make their own individual links with the City Council, rather than through any organisational channels representing the residents as a group. This served to fragment the relations between government and the people of Cruddas Park. Some people acknowledged that this was because there had been no previous concentrated effort in building up internal community relationships in the area.

There was a further dimension to the ambiguity felt by city councillors and officers to the Initiative. Many rejected the Tory attack on the welfare state and could see the damaging consequences this was having for the work they did and their own personal situations. Bringing business into government challenged people's political ideologies. Labour voters and public sector workers had been accustomed to view business in terms of an opposition between bosses and workers, capital and community. The private sector was thus understood as part of the problem, not the solution. Those who got involved soon dropped their reservations about the Initiative itself, as the emphasis shifted to the actual work going on. But such thoughts remained in the minds of many in the area and the Council.

The city council was on a slow trajectory of change of its own, moving towards greater co-ordination and community responsiveness. The Cruddas Park Initiative was tangential to this, and the high ideological profile surrounding it helped to generate hostility. This impeded co-ordination at the level of the 'doers', despite formal agreement to co-ordinate at the strategic level in the TNI arena. So, instead of a basis for building good relations between people and their city government, the Initiative tended to increase the distrustful and competitive relations between the city council and other governance activities.

The City Council also faced two other difficulties. Firstly, they had to recognise claims from all their neighbourhoods, many with as complex difficulties as Cruddas Park. It was thus politically hard for them to decide to focus their attention on a particular area. Secondly, throughout the time of the project, the Council was subject to severe budget restraints by central government. As a result, staff time was squeezed, and resources for investment and management within the neighbourhoods were cut. This was not a favourable climate for a positive partnership with central government and the business community.

The structural tension between central and local government in Britain over responsibilities and resources thus spilled over into the micropolitics of neighbourhood governance. Building partnership in such a context required major reserves of commitment and conviction.

Other Agencies: The Expanding 'Third Sector'

To add to the complexity of governance agencies surrounding the Initiative, several other organisations got involved. There exist an array of agencies nationally in the social welfare field which provide support through the charitable organisations, such as *Age Concern*, the *Save the Children Fund*, the *Prince of Wales' Volunteer Trust* and the *Council for Voluntary Service*. These have local offices in Newcastle. There are also local bodies co-ordinating support for the voluntary sector, such as *The Tyne and Wear Foundation* and *ENTRUST*. A number of key services are delivered by regional agencies, some accountable to the local authorities, such as *Northumbria Police* and some to national government, such as the hospital service. The main utility services are now privatised regional companies. In addition, there are special development and training agencies set up under the aegis of central government, such as the *Tyne and Wear Development Corporation* and the *Tyneside Training and Enterprise Council*.

This agency proliferation adds to the problems of delivering co-ordinated services to a neighbourhood such as Cruddas Park. Although there is considerable overlapping membership between the various agencies and organisations such as TNI, each agency has its own pre-occupations and criteria to deliver to. Some were drawn in by the Initiative, and some were already involved and saw in the Trust a vehicle through which to 'add value' to what they were doing.

For example, Northumbria Police were already involved. In 1987/88, they had participated in a national initiative studying community-based approaches to crime reduction. The idea was to replicate this model in a number of urban areas. Cruddas Park had already been chosen for such an initiative, the *Multi-Agency Crime Prevention Initiative* (MACPI). This too focused on co-ordinating 'sectoral' actions.

The Tyne and Wear Development Corporation was in a similar position. It had been created in 1986/87 with a great fanfare of publicity and had to get a good project underway early. The *Newcastle Business Park* fitted the bill. There was little time for building up a relation with the local community, although in all later projects, considerable attention has been given to doing just this (Robinson, Shaw and Lawrence 1994). Instead, the Corporation contributed resources, estimated at £80,000 in total, and encouraged companies on the park to consider employing local labour. There were no officer links with TWDC, but the Bank Chief Executive was on the TWDC Board.

The involvement of all these agencies added to the multiple lines of responsibility cutting across the project, and the diversity of organisational cultures and operational criteria with which participants assessed what was done and how to do it. It also meant that the management task for the community development facilitator was much more that of a co-ordinator and broker than a line manager. The problem for all these agencies was how to 'make links' with each other in ways which were both relevant to their own 'missions' and of value to their target 'clients', without getting in each other's way. Their problems in doing this reveal clearly the costs of the fragmented structure and dynamics of contemporary urban governance in Britain.

The Neighbourhood: Living with Powerlessness

For local residents, the Initiative came from the outside. Yet they had no cohesive organisation within the neighbourhood to represent their own views. Residents and outsiders described the neighbourhood as having many different groups. They spoke of tensions and conflicts between groups. There were tensions between families, with the potential for violent conflict over control of parts of the neighbourhood territory and social networks, and struggles for control of more formal organisations, such as the Tenants' Associations. These conflicts often cut through the work of the Initiative.

The one relation with government which most of them shared was their position as local authority tenants. It was to the Tenants' Associations that the promoters of the Initiative turned in the search for points of contact with residents. In retrospect, this was considered unfortunate:

> You can see why they did it that way ... (the Associations) were strong groups and they should have been consulted and included in the process. But they became the main channel of dialogue ... it alienated other groups and those who were not involved in groups or with the Tenants' Associations. Yes, they were strong groups but still, they didn't represent the majority of the people of the area ... there should have been the community work base there first and then that could have been the channel for dialogue and getting it going. (Local resident and community worker)

As a result, not only was the Initiative perceived among residents as coming 'from the outside', it was often seen as being 'captured' by particular interests and forceful activists. In the early stages, tensions and outright conflicts often broke out over priorities and over how to organise.

The residents were challenged by the Initiative with the offer of 'driving' the programme, and the possibilities of 'empowerment'. But their capacity to respond to this challenge, and to redefine it in their own terms, was limited. Partly, this was due to the lack of community cohesion and the multiplicity of groupings and networks in the area. This meant not only that information and opportunities did not easily 'filter through' to residents, it meant that some groups sought to capture them for their own group. The very creation of group cohesion could seem excluding to newcomers. Talking about a cookery group, one resident said:

> It is mainly women though we also encourage men and boys to come and two women brought their partners when these men were unemployed. It's hard to get new people to join because they walk into a room and it's full of people who know each other and while they're a great bunch of girls it can be intimidating for someone new.

In some cases, attempts to build social meeting points among residents found themselves caught up in a competition among families for the control of neighbourhood territory. This kind of situation exacerbated suspicion between neighbours, creating a climate of lack of trust in each other. But the Initiative itself did not initially generate confidence, as many local residents were also suspicious of the motives of the outsiders. They wanted to know 'what was in it for them'. Thus morally-motivated business people confronted residents who judged them in terms of individualistic material gains.

Because of the lack of co-ordination within and between the various central and local government departments, and the other agencies, local residents were accustomed to diverse 'routes' through to governance. In such a context, access to councillors, officers, civil servants, or other key players in formal systems gives a bit of power and the possibility of material benefit. Residents were thus implicitly competing among themselves for the small resources that were available, and increasingly aware that government agencies were competing for their attention. Thus the structure and practices of government itself contributed to the fragmentation of neighbourhood politics.

Many residents recognised that, unless they came together to form their own cohesive 'polity', any outside initiatives would be dominated by professionals rather than themselves. Some of those who got involved were strongly motivated to build up such community cohesion, to deal with things that government was neglecting or to make more articulate demands on government:

> Somebody has to do it. If we all sat back and didn't get involved we wouldn't have a life at all. You can't rely on the Council to do things, they don't care, not about the things that matter in life, not like communities and real people. (Local resident)

> The community needs to take control from the paid workers. Often the community's views seem to be overruled by the paid workers. (Local resident)

But this was not easy for residents to achieve. In their dealings with these outside worlds, they struggled with their own sense of inadequacy. This was a multidimensional burden, built up through layers of experience – as school failures, as workforce rejects, as struggling mothers, as unknowledgeable people. Many, and especially women, saw these experiences as personal failings. Slowly, through involvement in the Initiative, several came to re-interpret their situation, to find value in themselves and to build up confidence. This, though, made them look anew at the contextual situation which had produced the oppressing feelings of inadequacy and failure which they had experienced. This could lead to personal rejection of previous circumstances. One woman left her 'menfolk' (husband and sons) behind to move away from the area, after the personal development which occurred through the Initiative, because she had had enough of their anti-social behaviour. Others talked of the confidence-building that had come about through getting involved with the Initiative. But for many, the world of 'systems' remained daunting and terrifying, and they felt that it had neglected and rejected them.

There were also problems over ways of working. The outside world of business and government came to meet the residents but did not initially question their own ways of doing things. Local residents had a rich store of organising experience – organising gatherings of kin and friends, banding together to provide community support services, campaigning for particular causes, involvement in the Union of a Labour Party branch, involvement in Tenants' Associations.

These organising capabilities largely went unnoticed by the outsiders. Yet, residents noticed the ways the 'outsiders' organised themselves and often felt alienated by them. This noticing touched deep feelings about lack of power. For many women, their experience led them to see the world of business and of politics as a male-dominated environment, closed to their participation. It was a world of 'men in suits' and formalised organisational 'rituals' and ways of speaking which was far from their informal organising experience.

> There is something about men in suits which makes me fall silent. When I go there I feel that I can't speak because who will back me up, will they agree or just put me down. (Local resident)

Some were more hostile, seeing business people and the various external professionals as patronising professionals or patriarchs. Such terms expressed the feelings of domination and exclusion that residents experienced in their encounters with the outside world of governance. But they were knowledgeable about it too. They saw that this external world was full of people with particular perceptions and interests, and as liable to failures of co-ordination and understanding as their own efforts. At the same time, they appreciated the concern and commitment of the outsiders as individuals. Where the business people spoke of 'the right people', the residents saw 'good people'. The ability to know who to trust and how to judge people as individuals is of course a key capability needed for daily life survival in tension-ridden neighbourhoods such as Cruddas Park. That residents came to meet business as real people who could be judged and distinguished in the way we all do in our social relations was as important in building up trust and understanding for the residents involved in Cruddas Park as it was for the business people who got involved. Compared to the business people, however, residents had a deep consciousness of the structural nature of the forces which surrounded and limited their lives.

The Cruddas Park Initiative thus came to play a small role in building up community cohesion, though this progressed slowly, and was always undermined by other channels through which local residents sought to get access to wider governance influence and resources. People who got involved made new contacts which were useful to them and developed their own confidence and skills. They also learned new organising skills, for example, how to bid for funds in the 'multi-funding' climate of contemporary public funding. But there was no momentum which emerged from among the residents which led to a locally-based leadership which could be widely trusted.

Fragmented governance thus confronted a socially fragmented neighbourhood. In this context, the ambitions of the civil servants to let local people drive the project were naïve and unrealistic. The professional community leader sought to build up a community base to enable some 'empowerment' to take place, but most people acknowledged that this work should have preceded the high powered short-term initiative envisaged by both civil servants and the TNI. Local residents in any case were not certain how much governance responsibility they wanted. They often demanded a say in policy making, but taking over the running of programmes and the running of business was a different question:

> How can anyone begin to say there is ownership? People here can only just manage their own lives, they have enough to contend with. You can't expect them to take over this level of self-management … or government, I suppose it really is … they want to be consulted, they want a means of getting their views and needs heard, but they don't want to run the show … otherwise they'd all be in politics and working for the Council, wouldn't they? (Local resident and community worker)

Beyond these questions to do with roles and responsibilities, there were also tensions about the content of the Initiative and about the style of its processes. With respect to content, many people echoed the City Council's concern and thought the emphasis should have been on building up co-ordination and participation on the social side. They recognised more clearly than the councillors that the problems people faced were much more than about jobs, but about the social fabric of their lives, about children, partners, street life, welfare benefits, feelings of failure.

Through the Cruddas Park Initiative, therefore, some local residents became involved in building up internal cohesion within the neighbourhood, and in making links with the outside world. Others built up confidence. Many also learned a good deal about the nature of contemporary governance in British cities. But there were always going to be limits on the degree of citizen control and empowerment which an 'outsider'-led initiative could bring to an area with an array of problems of its own to deal with. Residents learned skills in communicating with the systems of business and government. But they already had capabilities in 'living with difference' which many of the outsiders lacked. They understood that social space had to be created within which the different groupings in the community could work together, if daily life was to be rid of conflicts and tensions.

Thus while the promoters of the Initiative thought to help local people learn about governance, residents already had much of the understanding and skills needed to work 'across divides'. Their problem was a practical understanding of formal organisational routines and a sense of inadequacy. It was the people from the formal systems who had most to learn about the practical reality of living and working with real differences, and about how their own practices could impede the development of 'a relationship of equals'.

Dilemmas of Collaborative Governance

This paper has been about the micropolitics of relation-building work. At one level, it is a story about grand strategic conceptions of new approaches to governance, within which business shares with the state the social responsibility for people and places experiencing economic difficulty. At another, it shows a struggle by good, capable and committed people trying to build relationships across the divisions and fragments of contemporary urban governance in Britain, motivated by the aim to make life better for people in disadvantaged areas and to make government more sensitive to people's life experience in such places. It is also a tale about individual learning, by residents, business people and officials, about their own capacities and limits, about the social worlds in which others live, about how to build understanding across these social worlds, about the differences in the ways people see things, and how to learn new conceptions, new roles, and new ways of performing old roles.

Permeating this relation-building work are transformative struggles; the transformation of personal identities; of the social relations of neighbourhood life; of organisational cultures and practices; and of the relations of governance. Different people interpreted the Initiative in terms of their particular perspectives and trajectories and sought either to integrate it into their relational worlds, or to limit its impact on them. Many individuals learned much through involvement with the Initiative and this learning spread out in small ways into wider relationships.

The 'initiating force' behind the Initiative came in this case from people 'inside the system', powerful players in the political and economic 'mainstream', as they conceptualised it. But the problem with the Initiative was not just that it was a 'top-down' project, made available through benevolent paternalism to the people of

Cruddas Park. The very conceptualisation of the social relations of both the political and economic system and the worlds of people in Cruddas Park was problematic. This emphasised a dominant mainstream economy and society from which some people were excluded, or marginalised, and a process of re-inclusion which emphasised providing individuals with better assets in the market place of jobs, and building routes along which they could progress to get them, rather than addressing the systematic re-building of the social relations which constructed for people in Cruddas Park the multiple oppressions and dominations which they faced.

The central government players saw the transformative task in terms of 'levering in' the private sector in a collaborative venture to haul people back on board who had fallen off the ship of the 'mainstream economy'. Their efforts were directed at improving the efficiency and effectiveness of the way the 'lifeline' was deployed. For some of those involved in the delivery of this task, it became much more an effort in the design of the lifeline, as they began to test its relevance and strength. Building pathways to jobs was initially seen in terms of providing enough training and making sure that there were job possibilities after training. It was about delivering 'things'. But those involved in delivering training and job access services began to see it in much more relational terms, as they came to recognise the importance of dealing with all the other problems people faced in order to make going for a job possible. For out there in the harsh seas of life 'on the edge', people had made little lifeboats of their own. It might not always make sense to them to make the effort to climb up into the 'mainstream' ship, merely to be thrown off again, when economic forces made the ship have to re-balance again, by which time the lifeboat they had made before, usually with a great deal of social and psychological effort, might have disappeared.

For other participants, it was the whole conception and design of the ship itself which was the problem. Instead of being a wide and stable flat-bottomed raft, with nooks and crannies for all ways of living, it was a high-sided imperial machine, if somewhat battered by global competition. The ways the world worked inside the ship were dominated by principles of hierarchy, class and rational order. Outside, people were surviving in different ways. The problem was thus one of cultural difference, as well as political and economic power.

The notion that the problem is the shape of the ship, that is, the systemic structure, is widely understood. Analysts differ, however, in conceptions of structuring dynamics. The approach used in this paper has been to emphasise the relational ways in which agency mobilises actions and meanings in the context of structuring constraints, and in particular the significance of ways of thinking and ways of organising in reinforcing and changing structural power.[7] If this is so, then the policy response needs to focus not on projects and schemes but on 'relation-building work', on how to build the trust and understanding between people which can allow 'mutual social learning' to take place.[8] Transformative work which changes a society, which changes systems, needs to engage in social learning at the level of individuals in social relations with others and to generate 'flows of learning' which permeate around into the 'nooks and crannies' of organisations and cultures.

The stories of learning in Cruddas Park show the potential for this, as people 'learned' new understandings and developed new capacities. But the expansion and flow of such learning was limited by the institutional context within which the Cruddas Park experiment existed. Its protagonists hoped that the 'project experiment' would act as a catalyst for building up new ways of governance. There are in Britain a myriad of such partnership initiatives. But such experiments do not change systems. They merely serve to highlight the institutional barriers to systemic 'learning' and 'transforming' (Duggan and Ronayne 1991, Taylor 1995, Thake 1995).

What is lacking in British urban policy is a coherent long-term project of institutional transformation, to demolish the ship and strengthen the raft. The raft would not be like the welfare state of old, which in Britain delivered services but without much sensitivity and respect for people's circumstances. It would build up new ways of recognising and giving support to diversity. It would aim to provide a stable, institutional glue, a set of key relational resources, for both business and citizens, directed by a moral requirement to have respectful regard to all those with a 'stake' in an issue[9] and to share resources to help people avoid the danger of

7 This draws on an institutionalist understanding of social relations, see Healey et al. (1995), Healey (1996a), and a communicative understanding of governance processes (see Healey 1992, 1996b).

8 See Friedmann (1987), for a discussion of the concept of social learning. It is well-developed within the context of management, but Friedmann has developed it in a transformative context.

9 For the concept of 'stakeholders' in a governance context, see Bryson and Crosby (1992).

drowning. Such a project of institutional transformation would emphasise the relational work of building links, rather than conducting experiments. It would focus on making connections rather than promoting individual success. It would seek to foster collaborative practices, rather than promoting competition between areas and among initiatives in capturing the 'scraps' of resources available for community development. It would emphasise processes as much as products, and focus on process outputs as much as project outputs. It would evaluate governance from the point of view of everyday life, and the world of conducting economic activity, rather than from the traditional functional organisation of governance. And it would operate in respectful ways towards all citizens, rather than expecting everyone to fit into the categories of official systems.

The story of Cruddas Park, and of many other such partnership experiments in Britain, is that, with sufficient effort, goodwill and commitment, people can build relations across wide cultural divides. But these efforts will continue to be fragmented and undermined so long as the institutional context emphasises the paradoxical mixture of division, fragmentation, competition and hierarchy which is evident in the story of the Cruddas Park Initiative. Setting up partnerships and experimental projects in community development is thus not the answer to social transformation, although through expanding these activities, the pressure for change may accelerate. What is needed is attention to the systemic *institutional capacity* of urban governance (Amin and Thrift 1995). For the next round of institutional change in urban policy in Britain, many are therefore now arguing for more attention to building the systemic institutional relations for sustainable community development and urban regeneration (Thake 1995, Taylor 1995, Healey 1995, Wood et al. 1995). This paper suggests that important dimensions of such relation-building work will be developing mutual appreciation of the experiential differences, life strategies, modes of thought and ways of organising, of those with 'stakes' in the conditions of everyday life in places such as Cruddas Park.

References

Amin, A. and Thrift, N. et al. (1995), 'Globalisation, "institutional thickness" and the local economy', in P. Healey (ed.), *Managing Cities*, London: John Wiley.

Atkinson, R. and Moon, G. (1994), *Urban Policy in Britain: The city, the state and the market*, London: Macmillan.

Bailey, N. (1995), *Partnerships in Practice*, London: UCL Press.

Bendinck, M. and Egan, M.L. (1993), 'Linking business development and community development in inner cities', *Journal of Planning Literature*, 8: 3–19.

Blackman, T. (1995), *Urban Policy in Practice*, London: Routledge.

Bryson, J. and Crosby, B. (1992), *Leadership in the Common Good: Tackling public problems in a shared power world*, Jossey Bass: San Francisco.

Campbell, B. (1993), *Goliath*, London: Methuen.

Davoudi, S. and Healey, P. (1995), 'City Challenge: sustainable process of temporary gesture', *Environment and Planning C: Government and Policy*, 13: 79–95.

Duggan, C. and Ronayne, T. (1991), *Working Partners? The state and the community sector*, Dublin: Work Research Centre.

Friedmann, J. (1987), *Planning in the Public Domain*, Princeton, NJ: Princeton University Press.

Gibson, M. and Langstaff, M. (1982), *An Introduction to Urban Renewal*, London: Hutchinson.

Haughton, G. (1990), 'Targetting jobs to local people: The British urban policy experience', *Urban Studies*, 27(2): 185–98.

Healey, P. (1992), 'Planning through debate: The communicative turn in planning theory', *Town Planning Review*, 63(2): 143–62.

Healey, P. (1995), 'The institutional challenge for sustainable urban regeneration', *Cities*, 12(4): 221–30.

Healey, P. (1996a) *Collaborative Planning: Shaping Places in Fragmented Societies*, London: Macmillan.

Healey, P. (1996b), 'The communicative turn in planning theory and its implication for spatial strategy formation', *Environment and Planning B: Planning and Design*, 23(2): 217–34.

Imrie, R. and Thomas, H. (1993) (eds), *Urban Policy and the Urban Development Corporations*, London: Paul Chapman Publishing.

Jessop, B. (1991), 'The welfare state in the transition from Fordism to post-Fordism', in B. Jessop and K. Nielse (eds), *The Politics of Flexibility*, Aldershot: Edward Elgar.
McArthur, A.A. (1993), 'Community business and urban regeneration', *Urban Studies*, 30(4/5): 849–73.
Robinson, F., Shaw, K. and Lawrence, M. (1994), 'It takes two to quango: UDCs and their local communities in the North East', *Northern Economic Review*, 21: 47–60.
Simey, M. (1995), 'The politics of social change', *Social Work*, 66(4): 1–5.
Taylor, M. (1995), *Unleashing the Potential: Bringing residents to the centre of regeneration*, York: Joseph Rowntree Foundation.
Thake, S. (1995), *Staying the Course: The role and structure of community regeneration organisations*, Joseph Rowntree Foundation, York: York Publishing Services.
Wheelock, J. (1990), *Husbands at Home: The domestic economy in a post-industrial society*, London: Routledge.
Wood, J., Gilroy, R., Healey, P. and Speak, S. (1995), *Changing the Way We Do Things Here: The Cruddas Park Initiative*, Department of Town and Country Planning, Newcastle University, Newcastle.

Chapter 16
The Treatment of Space and Place in the New Strategic Spatial Planning in Europe*

Patsy Healey

The Resurgence of Strategic Spatial Planning in a European Context

This article examines the way concepts of place and space are being used in the new wave of strategic spatial plans in Europe, in relation to the intellectual debates in the social sciences and humanities on these concepts. In the 1980s, the practice of spatial or territorial planning in many parts of Europe had deserted conceptions of the strategic development of cities and regions. Instead, the emphasis was on large projects of renewal and transformation of urban landscapes, justified through arguments about the need to break out of strategic spatial organising ideas locked into the urban plans of an earlier era (Healey et al., 1997; Salet and Faludi, 2000; Albrechts et al., 2001; Balducci, 2001). By the end of the millennium, however, strategic spatial plans, frameworks and perspectives were back in fashion among Europe's planning policy communities, and were actively being promoted by European Union initiatives (CSD, 1999; Salet and Faludi, 2000; Albrechts et al., 2001; Faludi, 2002; Faludi and Waterhout, 2002).

There are many reasons for the resurgence of interest in strategic spatial planning. These include: the persistent problem of coordinating public policy in particular localities; the search for ways of making urban regions more economically competitive by developing their collective 'asset base'; a parallel search for spatial forms and relationships with the potential to promote the (often diffuse) objectives of 'sustainable development'; and, most recently, a concern to redress the unequal distribution of access to opportunity across urban regions among the many groups now recognised as coexisting within localities.

Strategic spatial planning may also have a political role in strengthening the voice of municipal government or regional bodies within the 'multi-level' governance landscape widely recognised within Europe by the end of the twentieth century (Hooghe, 1996; Cooke et al., 2000). Articulating a strategic orientation with a spatial dimension may have direct material benefits in capturing resources from a higher government level. It may also help the formation of active coalitions among an array of small municipalities, or mobilise active stakeholder groups important to an area's development who can move perceptions (and hence actions) from just 'being in an area' to a recognition of an area as having an identity (a city or region 'in itself'), and beyond this, to having the capacity to act 'for itself' (Beauregard, 1995; Healey, 2002). Episodes in strategic spatial planning may thus be linked to the processes of institutional 're-scaling' identified in the European context (Brenner, 1999; Macleod, 1999), linked to attempts by urban and regional political and policy communities to reposition the relations of urban regions within interactions between global forces and local dynamics.

Strategic spatial planning efforts are demanding in terms of the institutional processes of their articulation and there has been much discussion of these processes (see, for example, Healey et al., 1997; Vigar et al., 2000; Albrechts et al., 2001; Salet et al., 2003). In this article my particular focus is on the way cities and regions – the spatial organisation of territory and the qualities of places within territories – are represented. These issues of representation are an important part of the persuasive capacity of strategic planning. They

* Previously published in the *International Journal of Urban and Regional Studies*, 28(1): 45–67 (2004). An earlier draft of this article was presented at the EURA Conference on Urban and Spatial European Policies in Turin, April 2002, and at seminars at Cardiff and Newcastle Universities. I am grateful for helpful discussions at these events and thoughtful subsequent comments from Ole Jensen, Enrico Gualini, Wil Zonneveld and Brendan Murtagh, as well as two anonymous referees.

have impacts in carrying framing concepts from the arenas of policy articulation to the arenas where decisions are made about specific investments and regulatory norms and permits. It is therefore important to examine both the concepts of space and place mobilised in strategic spatial planning episodes and the institutional work they perform (Fischler, 1995; Healey, 2002).

There has been much debate about the meaning of 'spatial planning', a term which does not easily translate between European languages (Williams, 1996; Faludi, 2002; Faludi and Waterhout, 2002). My understanding of 'strategic spatial planning' refers to self-conscious collective efforts to re-imagine a city, urban region or wider territory and to translate the result into priorities for area investment, conservation measures, strategic infrastructure investments and principles of land use regulation. The term 'spatial' brings into focus 'the where of things', whether static or in movement; the protection of special 'places' and sites; the interrelations between different activities and networks in an area; and significant intersections and nodes within an area which are physically co-located. *Strategic* is sometimes used to mean a higher level of administration, or a more general or abstract level of policy. But it is also used to mean an overview, or more specifically, a framework. It implies selectivity, a focus on that which really makes a difference to the fortunes of an area over time. *Planning* (or 'development') also highlights a developmental movement from past to future. It implies that it is possible to decide between appropriate actions now in terms of their potential impact in shaping future socio-spatial relations. This future imagination is not merely a matter of short-term political expediency, but is expected to be able to project a transgenerational temporal scale, especially in relation to infrastructure investment, environmental management and quality of life. The term 'planning' also implies a mode of governance (a form of politics) driven by the articulation of policies through some kind of deliberative process and the judgement of collective action in relation to these policies.

None of the above elements of the concept of 'strategic spatial planning' are easy to imagine, to address technically, to argue about deliberatively, and to translate into policies and programmes. It is for this reason that much of the discussion about strategic spatial planning has focused on *process*, on how significant stakeholders can be mobilised to develop strategic agendas in a 'diffused power' context, and become cohesive enough to develop 'collective actor' power (Healey et al., 1997; Salet and Faludi, 2000; Albrechts et al., 2001; Fuurst and Kneilung, 2002; Salet et al., 2003). There has been much less analysis of the nature of the concepts of place and space being deployed. Some analysts have examined the consciously-used 'images' deployed in planning episodes, but their emphasis has primarily been on their mobilising and coordinating power (Fischler, 1995; Faludi, 1996; Neuman, 1997). Others have explored the selectivity of spatial policy articulation, but from the point of view of competing economic, environmental and social agendas (Jensen and Richardson, 2000). Jensen and Richardson have analysed the tensions between a discourse of 'places' and of 'flows' in European-level policies for transnational transport networks and in the *European Spatial Development Perspective* (ESDP), a document prepared by the European Commission and Member States during the 1990s which has promoted strategic spatial planning in the context of European integration and cohesion (CSD, 1999; Jensen and Richardson, 2000; 2002; Faludi and Waterhout, 2002). Zonneveld has highlighted the contradictory discourses within the ESDP itself (Zonneveld, 2000; Bengs and Zonneveld, 2002). There has been very little analysis of the nature of the spatial vocabulary being used in these episodes.

In this article I examine the frames of reference, organising concepts and metaphors used in three recent episodes of strategic spatial planning, all of which are regarded as explicitly mobilising spatial concepts and all of which are to some degree contested. In each case I locate the spatial vocabulary in the institutional context which generates the episode and in which the vocabulary takes on a meaning in relation, potentially, to understanding place qualities and the spatiality of relations, to mobilising activity (in coordination, investment, new regulatory approaches, etc.), to legitimising action, and to dreaming about possible futures (a better quality environment and place to live and do business, a different kind of politics, etc.). In the next section, I set up an evaluative frame with which to analyse the concepts of space and place being used in the three examples which follow. I then consider the institutional work the concepts are being used to perform, before turning to the three cases. The article thus uses the tools of interpretive policy analysis to analyse the discourses and practices of episodes of strategic spatial planning.

Policy Discourses of Space and Place

In the geography and planning literature, there has been a longstanding critique of mid-twentieth century planning concepts of spatial organisation (Boyer, 1983; Dovey, 1999; Liggett and Perry, 1995; Graham and Healey, 1999; Graham and Marvin, 2001). Planners and plans have been criticised not merely for trying to 'order' the dynamic and inherently disorderly development of cities and regions. The concepts that have been used, from notions of central place hierarchies, to distance-decay models of urban regions and movement patterns within cities, and to ideas about reducing the need to travel by making cities physically 'compact', are seen to reflect a view of geography which assumes that objects and things exist objectively in contiguous space and that the dimensions of this space can be discovered by analysis, that physical proximity is a primary social ordering principle and that place qualities exist objectively, to be found by analysis and made by physical development and management projects (Gregory, 1994; Graham and Healey, 1999).

This so-called essentialist, 'Euclidean' geography is under heavy challenge from an alternative, relational conception which sees space as an inherent spatiality in all relations, whether social, ecological or biospherical, and which understands place as a social construct, generated as meanings are given in particular social contexts to particular sites, areas, nodes of intersection, etc. (Allen et al., 1998; Allen, 1999; Dematteis, 1994; 2001; Friedmann, 1993; Graham and Healey, 1999; Gregory, 1994; Thrift, 1996). In this non-essentialist, relational conception, 'places of the mind' are as significant as physical objects and flows, with a continual co-production of 'things' and meanings. This conception means that, rather than searching for some inherent 'natural' qualities of place to mobilise into spatial concepts for strategic purposes, meanings of place are likely to be diverse and contested. Articulating the spatial vocabulary for a spatial strategy is therefore a highly political process, involving struggle and selectivity, not just between different interests and power blocs but within the terrain of the mode of analysis and representation of the spatiality of phenomena.

The new relational geography also challenges notions of inherently coherent, integrated 'territory-based' systems of relations. Significant relations affecting the qualities of territories may stretch in many directions and link to many and different scales. Spatial effects cannot be analysed merely in terms of variations in physical proximities but may occur 'at a distance' as well as nearby. The social relations which transect a specific piece of territory may each have a different spatial reach, just as they may have different temporalities. They may or may not intersect as they pass 'over' or 'under' each other. They may coexist in a specific physical site or institutional arena without infusing the site with meaning as a 'place', still less mobilising that meaning to do political work. The qualities of places exist both as experienced materialities and as mental constructs related to the construction of individual and collective identities. By implication, 'development' follows not one trajectory through a common time dimension, but occurs in multiple timescales, follows many, often conflicting, pathways, which may be 'folded' and 'circular' as well as linear (Amin and Thrift, 2002). This new relational geography is proving particularly fruitful in developing understanding of the time/space implications of the information technology revolution (Mitchell, 1995; Graham and Marvin, 2001). It also aids in developing understanding of the multiple ways people and firms use and experience the 'city' in their diverse daily lives, and in grasping the multiple meanings and values with which 'places' become infused (Healey, 2002).

As the academic articulation of these ideas has gathered momentum, the criticism of spatial planning practices as being trapped in an out-worn essentialist geography has mounted.

> Within this contemporary urban world … the modern infrastructural ideal founders. Its essentialist notions of Euclidean space and Newtonian time, of functional planning towards unitary urban order, of single networks mediating some 'coherent' city, are paralysed. It is largely incapable of dealing with the decentred, fragmented and discontinuous worlds of multiple space-times, of multiple connections and disconnection, of super-imposed, cyborgian filaments, within the contemporary urban world. (Graham and Marvin, 2001: 215)

Episodes in strategic spatial planning in Europe in the 1990s have therefore faced a paradigmatic shift in geographical imagination. Are the new geographical conceptions having any influence on conceptions of spatiality and place qualities? Is an old geography being rolled forward into new contexts? Or are the two geographical discourses co-evolving as both seek to make sense of the reality emerging around them? And what difference does it make which geography underpins the spatial vocabulary deployed in a strategic spatial planning episode?

This raises the question of how to analyse the spatial vocabulary mobilised in a spatial planning framework and the discursive struggles which surround its articulation. Drawing on the geographical literature (especially Lefebvre, 1991; Dematteis, 1994; 2001; Gregory, 1994; Massey, 1994; Thrift, 1996; Brenner, 1999), I develop the criteria presented in Figure 16.1 and discussed below. In Figure 16.1, I contrast an essentialist and a relational approach against each criterion.

The first criterion relates to the treatment of *scale*. This highlights the critical distinction between scale conceived as a nested hierarchy from global to very local, and scale understood in terms of the 'reach' of a relationship in time and space, which may connect many discontiguous sites (such as those of a family with relatives in different countries), or may bring together intense interactions with a global range (as in the financial cores of London, New York and Tokyo). The new geography emphasises the potential multiple scales in play at any site of interaction (Thrift, 1996; Brenner, 1999).

Criterion	Essentialist conception	Relational conception
Treatment of scale	Nested hierarchy	Relational reach in different networks
Treatment of position	Hierarchy and borders	Different positions in different networks
Regionalisation	An integrated, differentiated physical fabric	Fragmented, folded conceptions of space; multiple networks coexist
Materiality and identity	A material physical future can be built, meshed with social relations in an integrated way	Materialities are co-existent with conceptions of identity and iconographies of space/place
Concept of development	An integrated linear trajectory	Multiple, non-linear, continually emergent trajectories
Representational form	Material metaphors of functional integration, expressed in maps	Metaphors of movement and ambience, expressed in multiple ways.

Figure 16.1 Criteria for evaluating concepts of space and place

The second is the treatment of the *position* of a relation or a site of intersection in a wider context. The essentialist approach emphasises both hierarchical organisation and the organisation of space into distinct areas, with clear boundaries and borders. It is primarily concerned with internal organisation. Connections to areas outside are governed by transport routes with simple distance-decay characteristics. Such a conception is reflected in attempts to identify the boundaries of distinct regions and settlements, or areas with distinct 'landscape characteristics'. A relational approach, in contrast, focuses on the way places and sites are positioned in particular relational networks, and how near and far they are, in relational terms, from nodal points in relevant networks. Position is thus not a geographical point, but an institutional site with an angle of vision. In a relational geography, a city has potentially multiple positions, depending on the site of observation and the relational webs within which an institutional site is situated (Amin and Thrift, 2002; Healey, 2002).

The third criterion relates to the 'internal differentiation' of a locale or place, that is, following Giddens (1984), its *regionalisation*. If the previous criterion of position refers to how those involved conceptualise 'where they are' and what they are related to, this criterion focuses on the internal spatial organisation of a locale. Making this step implicitly means that a locale is becoming recognised as a 'place', rather than merely a 'point in space'. The contrast here is between cities and regions understood in classical geography in terms of an 'integrated' physical fabric, from which the social organisation of a territory can be 'read-off'. In the classical model, geographical areas are divided into zones of activity and related property values, ordered so that more intense and higher value-generating activities are at the core and lower ones at the periphery. This structure is supported by integrated infrastructure systems. The relational geography talks, in contrast, about the fragmentation and splintering of social relations and of the physical fabric, of 'warps' and 'folds' (Graham and Marvin, 2001; Amin and Thrift, 2002), of 'bits' (Mitchell, 1995), and of the coexistence of multiple relational layers across a physical area (Massey, 1994). Nodes and borders, in this conception, are not derived from some clear model of socio-spatial organisation but are continually emergent, as nodes are actively constructed by mobilisation effort and boundaries established by mental maps of place qualities.

The fourth criterion focuses on the treatment of the *materiality* of spatial relations and place qualities, their imaginary content and the ontological role of images of space and place in constructions of *identity*. The relational approach adopts a social-constructivist perspective which recognises that, however real are material objects and needs, our recognition of them is always filtered by how we perceive them. In developing understandings and dreams about the future of places, this approach recognises that the imaginative content of strategic spatial planning episodes inevitably organises the way the materialities are thought about. 'Visions' and iconographic images are understood as significant in mobilising attention, an imaginative effort which builds from and contributes to shaping conceptions of identity. An essentialist geography, in contrast, focuses on the 'objective' material dimensions of cities and regions. Dreams about the future are anchored in the idea that the physical future can be built according to plan, and that social relations can be 'read off' from physical relations. The future can thus be concretely 'built'. A relational geography emphasises the dynamic complexity of the relations which shape the material flows through which physical objects and patterns are brought into being, and the significance of mental constructs in shaping actions which contribute to this complexity. The future is understood as continually emergent and unknowable, but yet shaped by the interaction between imaginative work and materialisation. Thus, the creation of material objects and the construction of conceptions of objects are co-generative processes. Hence the formation of the spatial patterning of the materialities of social relations and place qualities is co-emergent with the 'naming' of these spatialities and qualities.

The fifth criterion relates to the conception of '*development*'. Essentialist geography and its planning manifestation treated time as linear, and development as a linear trajectory from less-developed to more-developed states. Such a conception has by now received much criticism in the environmental and development literature for its privileging of certain place qualities (such as large western metropolises) and certain societies (especially affluent western countries) as the apex of development (Gregory, 1994). In contrast, a relational geography emphasises that what are recognised as place qualities are shaped by multiple forces, producing multiple development pathways, with different places having different options and potentials because of the specific interplay between local histories and wider relationships. Further, what becomes the trajectory of a place from past to future is continually contested and hence continually emergent, both through local struggles over meanings and values of place and place qualities and the changing positioning of 'places' in the materialities and mentalities of social relations whose locus of power lies far away from the site in question (Thrift, 1996).

The final criterion relates to the manner of *representation* of the ideas of spatiality and place qualities (Beauregard, 1995). This is partly a matter of the metaphoric content of the spatial vocabulary. The metaphor of 'flow' can be inserted into a traditional essentialist geography of gravity-based traffic models which assume that people's movement in space is governed predominantly by proximity principles. Or it can be part of a narrative of multiple social relations with multiple space-times, sometimes flowing co-terminously but not necessarily with any integration. Consequently, a slice of, say traffic flow, may be made up not merely of different types of journeys in terms of spatial and temporal reach, but used as a route in lives with quite different times and spatialities. However, the issue of representation is not only about the symbolic structure within a narrative. It also involves consideration of the chosen form with which to express a narrative. In the traditional planning approach, the privileged format was the two-dimensional map or three-dimensional perspective, within which each parcel or zone was allocated a function and even a physical shape within the integrated whole (Fischler, 1995). The relational approach opens up a wide possibility of expressive forms, from text, icons, pictures (still and moving) to musical expressions and fragrances. Thus the sights, sounds and smells of places recorded in artistic works become available as aids to understanding, mobilising and dreaming (Sandercock, 2003).

The Institutional Relations of Strategic Spatial Planning 'Episodes'

'Episodes' of explicit strategy formation around spatial issues arise in many different ways and in different institutional contexts. They involve complex multidimensional interactions between the institutional context of a planning episode and the creative force of agency in realising it. They result, if successful in arriving at some kind of conclusion, in new frames of reference embodied in governance practices. These frames may find expression in documents called plans, guidance documents or development strategies of some kind. Such frames are deeply shaped by the specific policy relations of their production and the purposes to which they are directed. This recognition directs attention to the institutional contexts in which concepts of place and space are mobilised and the institutional tasks these concepts are being called upon to perform. Does it matter what kind of geographical imagination is evoked in the spatial vocabulary used in a strategic spatial planning 'frame'?

It was traditionally assumed that the primary function of strategic spatial frames and plans was to direct the state's investment and regulatory power. This reflected an authoritarian conception of power, the ability of a government agency to command certain actions and control their implementation. A strategy sought to achieve specific material outputs, such as better living conditions, or property market stability, and served to legitimate the investment and regulatory actions of state bodies. But power may also be exercised in a generative way (Dyrberg, 1997; Giddens, 1984; Gualini, 2001), to release potentialities and to innovate. A strategic planning process and a strategic frame may generate new identities and new kinds of investment proposals from unexpected directions. The stabilising force of an enduring spatial frame may suppress some tendencies (for example, urban sprawl), but it may also nurture others (reduce uncertainties about land and property for small firms, for example). In some contexts, efforts at strategic spatial planning may help to generate a different kind of politics, focused around struggles over different kinds of issue. Most strategic spatial planning episodes which are able to accumulate sufficient power to have significant effects combine both authoritative and generative aspects of power, often in a complex tension between stabilising and restraining forces which re-mould the spatial relations of territories and releasing and innovating forces with the objective of transforming these relations. In the analysis which follows, I aim to draw out both the way the concepts of space and place used arise from the institutional context of their use and the mix of authoritative and generative force which the strategic spatial planning episode seeks to mobilise.

Developing New Vocabularies of Spatiality and Place in Strategic Spatial Planning

The emerging experience of strategic spatial planning in Europe since the mid-1980s provides an increasingly rich resource for the analysis of the spatial vocabulary of planning episodes. An early forerunner was the French urban region plans produced under the impetus of decentralisation in 1983. The plans for Lyons and Lille are the most well-known, but these were part of a general movement in French planning at this time (Motte, 1995). By the 1990s, city regions in Germany were also reworking spatial strategies in new ways (Salet et al., 2003).

These experiences fed into the preparation of the *European Spatial Development Perspective* (ESDP), finally produced in 1999 (CSD, 1999; Faludi and Waterhout, 2002). This influential advocacy document moves between the scale of the city region and the overall spatial organisation of the European territory. The concepts developed within it have interacted with the articulation of national and regional spatial strategies in several parts of Europe (Faludi, 2001), including two of the examples discussed below. The ESDP promotes a number of spatial concepts, notably 'polycentric development', 'a balanced spatial structure', 'dynamic zones of integration' and compact settlements and corridors. However, its final version avoids any diagrammatic expression of these ideas except in the form of iconic sketches, because of political disagreement over the content and manner of spatial representation (Faludi and Waterhout, 2002).[1] The application of the ESDP has been promoted by the EU's INTERREG programme, which has resulted in well-developed strategic 'visions' for several transnational

1 For a fine collection of spatial representations of the European territory in EU-level debates in the 1990s, see Faludi (2002: 21–36).

regions (see, for example, Doucet, 2002). At the sub-national level, in the UK the regional devolution impetus has been linked to the production of spatial strategies for Northern Ireland and Wales, and proposals for their production in all the English regions as well as in Scotland. At the city region level, by the late 1990s examples were appearing in many parts of Europe (Albrechts et al., 2001; Salet et al., 2003).

In this article I examine three experiences, moving from the level of a smallish country in the economic core of Western Europe to a region with a difficult geopolitical position, and to a large metropolis with a buoyant economy, the dynamism of which is slipping away from its urban core to the surrounding areas and municipalities. All three cases illustrate deliberate attempts to transform the spatial vocabulary used in planning practices and to mould a new kind of planning 'politics'. My sources for these cases are plan documents, interviews with some key actors, critical evaluations by local commentators and subsequent discussion with key actors and commentators.

- *Netherlands Fifth National Policy Document on Spatial Planning 2000–2020*, approved by government subject to parliamentary ratification in January 2001 (VROM, 2001) (referred to in what follows as the Fifth NPD);
- *Shaping our Future: The Regional Development Strategy for Northern Ireland 2025*, approved in September 2001 (DRDNI, 2002) (referred to in what follows as the Northern Ireland RDS);
- *Ricostruire la Grande Milano: Documento di Inquadrimento delle Politiche Urbanistiche Communal* (Reconstructing Greater Milan: Framework Document for Municipal Planning Policies), approved in June 2000 (Comune di Milano, 2000) (referred to in what follows as the Milan Framework Document).

The Netherlands Fifth National Policy Document on Spatial Planning 2000–2020[2]

The Dutch have produced national spatial planning policy documents since 1960. Spatial planning has a strong and well-developed tradition in the Netherlands, and spatial organising concepts have had an important leverage on national, provincial and local policy (Faludi and van der Valk, 1994; Hajer and Zonneveld, 2000; de Vries and Zonneveld, 2001). The Fourth National Policy Document on Spatial Planning was produced in 1990 (with a supplement in 1991, and widely referred to as VINEX). This focused on concentrating development in the Mainports (the port of Rotterdam and Schipol Airport, Amsterdam) and the Randstad ring of urban centres in the centre of the Netherlands. It contained firm policies to promote 'compact city' development across the country. In this way, the promotion of economic competitiveness at a European and global scale was combined with deeply-embedded planning concepts of urban form (de Vries and Zonneveld, 2001). By 1999 these policies were coming under sustained criticism (see, for example, WRR, 1999; Hajer and Zonneveld, 2000). The growth had been accommodated, but the growth locations identified lacked the supporting investments and quality of development hoped for them. In the meantime, 'the dynamics of today's society appear to be transcending the bounds of conurbations' (VROM, 2001: 10). Planners from the National Spatial Planning Agency (VROM) had become increasingly conscious of the importance of locating Dutch spatial policy in a broader European perspective. The *Fifth National Policy Document on Spatial Planning 2000–2020*, in the version approved by government in January 2001, was based, as is usual in the Netherlands (Woltjer, 2000; de Vries and Zonneveld, 2001), on intense discussion between levels and sectors of government. It sets out a new spatial development approach and strategy. This focuses on criteria for 'spatial quality', but centred in a strong concept of the country's spatial development.

The Fifth NPD, even in its English summary version, is extraordinarily rich in its spatial content, expressed verbally and visually.[3] The spatial ideas build on the Dutch spatial planning tradition, and the contribution which the Netherlands National Spatial Planning Agency was making to the ESDP (VROM, 2000). These ideas reflect a deliberate effort to link analyses of the changing dynamics of the economy and society understood in terms of the relational geography of a 'network society' to an appropriate spatial strategy. But the impact of the analysis in its translation into policy concepts has been limited by both the

2 The main source document for the Fifth NPD is the English version (VROM, 2001).

3 Its spatial concepts have developed and changed significantly from an earlier discussion paper produced in 1999 (VROM, 1999).

power of traditional Dutch planning concepts and by the weakening ability of spatial planning concepts to influence infrastructure development. I first discuss the concepts developed and then review the coexistence of a new geography discourse and an old planning one in the spatial vocabulary of the Fifth NPD text.

The search for a spatial planning approach is grounded analytically in an organising concept of spatial structure based on layers, or 'strata'. In one sense, this concept is used to break with notions of the integrated development of the national territory. The focus is shifted to the different space-time and transnational dimensions of the key relationships. But when the layer concept is translated into terms to describe the main 'spatial structure' of the country, it becomes more hierarchical:

> The morphology of the Netherlands has been divided into the primary stratum, the network stratum and the occupation stratum. The primary stratum pertains to the natural conditions, the most important of which are altitude and hydrology. The network stratum is formed by the dense and diverse network of infrastructure links. The occupation stratum depicts the physical pattern of land use, most importantly the nature and size of that land use. The picture these features form is completed by the most notable features of the North Sea: the busy shipping routes, the offshore platforms and outflow of fresh fluvial water into the sea (the coastal rivers). These elements construct the structurally formative elements of the Dutch landscape. (VROM, 2001: 15, Explanation to Map D)

The Fifth NPD then moves away from spatial expression to define criteria of spatial quality. These are: spatial diversity, economic and social functionalities, cultural diversity, social equality, sustainability, attractiveness and human scale. Such criteria are intended to frame the making of specific plans, investment programmes and regulatory principles by provinces and municipalities. But the Fifth NPD goes further and presents a clear and spatialised development framework across the country. This is elaborated through four themes. The first positions the Netherlands in a 'transnational policy perspective', in a Western European landscape of metropolitan relationships and land and sea connections. This is used to emphasise the importance of cross-border policy coordination. The following three themes are presented as development images. The first is the 'city and the country'. This focuses on the interaction between the 'red' and the 'green', the red representing six types of development: inner city, outer city, green urban areas (development on sites currently undeveloped), village centres and rural villages, and specific work environments (i.e. business parks). Rural areas are understood as 'green', intermediate or in need of special protection. The red-green distinctions become the basis for a 'contour policy'. The contours define the areas. The red cannot spill out of its contours unless 'demand' cannot otherwise be satisfied (VROM, 2001: 29). Here the 'compact city' idea of the *Fourth National Policy Document* is recast over a larger and more 'disjointed' terrain. The focus in the green areas is preservation and protection, but allowing for the dynamics of landscape and nature. Urban and rural areas remain as separate domains, reflecting traditional Dutch planning concepts.

The second development image presents 'urban networks', groups of cities to be promoted in an integrated way. The discussion of the urban network concept is located explicitly in the new relational geography. This emphasis seeks to overcome the criticism of the Fourth NPD that the focus on the compactness of individual towns and cities failed to appreciate the complex interactions between cities in a polycentric and highly urbanised landscape:

> The urban network concept is intended to promote an urban character in the network society and to render or keep cities suitable for the network economy. (VROM, 2001: 33)

But developed into policy concepts, the network idea is translated into a hierarchy of centres. This produces six major groups, an international network consisting of an enlarged Randstad, now called 'Delta Metropolis'; five other national urban networks, and some smaller regional ones. Within these 'urban networks' attention is to be paid to improving transport and providing a coherent 'green structure' within the urban realm. This approach has already generated a politics of classification among municipalities as they struggle to capture national funds for urbanisation projects (de Vries and Zonneveld, 2001).

The third development image, 'Going with the flow', refers to the country's main water systems. This section focuses on the need to 'give water room to find a new balance', especially with respect to the flow channels of rivers, with water quality management and rising sea levels in mind. This is complemented later by a special section on the North Sea, and the management of shipping lanes and the protection of dune systems. A concept of 'development corridors' envisaged in early ideas for the Fifth NPD (VROM, 1999) and promoted in the *European Spatial Development Perspective*, is not followed through into the January 2001 document. De Vries and Zonneveld (2001) suggest that this was because Dutch planning practitioners equated the corridor idea with urban sprawl, undermining long-standing concepts of contained urban development.

The Fifth NPD provides more detailed elaboration of its approach within four schematic regions (north, east, south and west) and contains an interesting discussion about how the strategic ideas should shape regulatory powers, investment and agreements, both trans-nationally and between levels of government within the country. The elements are drawn together into the key national spatial policy decision, which was approved by government subject to parliamentary ratification in early 2001 (see Figure 16.2).

As Zonneveld comments,[4] the Fifth NPD is a 'hybrid' document. Its analysis is clearly located in an understanding of the multi-scalar relations of the new relational geography. There is a sophisticated attempt to draw out the spatial implications of different kinds of relations, through the 'layer' approach. This could have been developed further in the concept of urban networks. However, there is a disjunction in the document between the analysis and the policy development. The ambition of the former has become interpreted into the more traditional spatial vocabulary of Dutch planning. De Vries and Zonneveld suggest that this has happened through the processes of intensive consultation. In effect, the Dutch spatial planning community reinterpreted the meanings of 'layer' and 'network' into their established spatial vocabulary and buttressed them with a reworking of clear divisions between 'urban' and 'rural' areas through the use of the red/green contour idea. Meanwhile, the power of the National Spatial Planning Agency (VROM), once very strong and able to impose its spatial frames on other government departments, is now much weakened (WRR, 1999; Hajer and Zonneveld, 2000; Wolsink, 2003). As a result, major investment departments, and particularly those concerned with major infrastructure projects, are less prepared to accept the new relational concepts and their transnational implications. De Vries and Zonneveld (2001) suggest that this has reinforced a retreat in policy content in the Fifth NPD back to a narrow spatial planning agenda of managing the location of urban development. Thus, the innovative spatial vocabulary of the analysis has been captured, reinterpreted and positioned back into the established spatial planning policy discourse.

The Fifth NPD in its January 2001 form has not, therefore, been able to shift this discourse significantly. However, critical debate about its content has continued, feeding into the major upheavals in Dutch politics in 2002. The National Spatial Planning Agency (VROM) has been divided up into a policy department and a more independent think tank. The Fifth NPD failed to achieve parliamentary ratification before the dramatic overturning of the Wim Kok Labour government in May 2002 and was abandoned by the new government. By late 2002 a new version was proposed, combining concepts from the January 2001 version with a more recent plan for green spaces/rural areas. Whether the spatial vocabulary in the policy concepts will retain their traditional geography in this new context remains to be seen.

Shaping Our Future: Regional Development Strategy for Northern Ireland 2025[5]

This strategy for the province of Northern Ireland is the first approved example in what is a new wave of strategic regional spatial planning in the UK (Marshall, 2002). However, it has grown out of a very distinctive local context in which a key priority is promoting social cohesion, in a situation riven in the past by violent sectarian struggles which played out across a well-understood sectarian geography. Institutionally, it is positioned in the transfer from direct government by the UK national government to devolved elected government resulting from the Peace Process which led to the 'Good Friday' agreement (April 1998). The development strategy was initiated by the Northern Ireland Executive in Northern Ireland in 1997 but approved by the Minister for Regional Development in Northern Ireland's own government in September 2001. It was produced through an intensive consultation process across the province (McEldowney and Sterrett, 2001),

4 Personal communication, email (3 December 2002).

5 The source material for the concepts is DRDNI (2002), supplemented by comments from key players in July and November 2002.

Figure 16.2 The Netherlands Fifth NPD: National Spatial Policy
Source: VROM, 2001: 48, Map L, with kind on-line permission of the Ministerie van VROM

oriented by the urgent political necessity of distributing development opportunity and public investment, while recognising the economic pressures to promote the 'competitiveness' of the regional economy at a European and global level. The first logic emphasised a spatially dispersed pattern of development, the second a concentration on the core metropolitan city of Belfast and its local region.

In developing this strategy, the planning team drew heavily on the spatial vocabulary of the *European Spatial Development Perspective*. In contrast to the predominant emphasis elsewhere in the UK on economic development and building regional competitiveness in the 1990s, the strategy adopts the ESDP emphasis on 'balanced development', that is, the 'balance' between focusing on 'globally competitive integration zones' and distributing development impetus and opportunity widely across the European territory (CSD, 1999). The ESDP also encompasses social cohesion among its concerns, in contrast to the preoccupation in the UK with combining economic imperatives with environmental considerations. However, the driving force behind the Northern Ireland RDS was grounded in the local political, social and economic context, which used the ESDP concepts to help deliver a distinctive local approach.[6]

The Northern Ireland RDS is refreshing in the clarity of its presentation and in the way it avoids much of the technical jargon in which spatial and land use planning issues became enmeshed in the UK in the 1990s. Instead, it tries to express organising concepts to help a wide array of stakeholders in Northern Ireland's future to think about the spatiality of the area. It is a strategy intended to be trans-sectoral and multi-level in its influence on the location of investment and the use of regulatory powers. Its language and argumentation speak directly to particular audiences across the province.

6 Personal communication, B. Murtagh (November 1992).

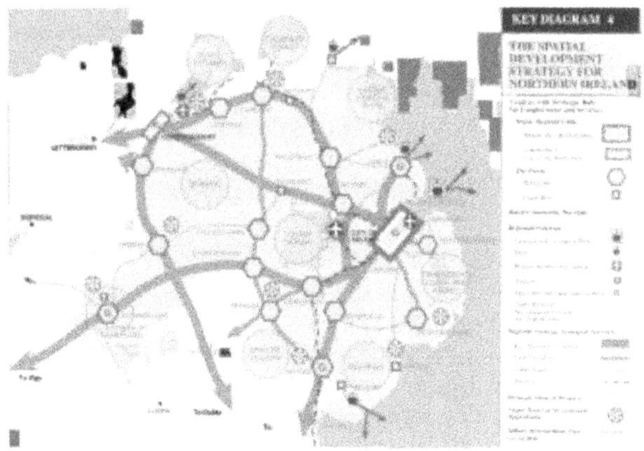

Figure 16.3 Northern Ireland Spatial Development Strategy
Source: DRDNI 2002: 45, Key Diagram 4, with kind permission of Department of Regional Development, Northern Ireland

It aims to provide 'an overarching strategic framework, to help achieve a strong, spatially balanced economy, a healthy environment and an inclusive society' (p. 1). It provides a thoughtful analysis of the main forces driving change in the region, organised into social, economic, transport and environmental forces. It develops a Vision, Guiding Principles and a Spatial Development Strategy (SDS), which wrap around the more traditional land use planning content. The Guiding Principles emphasise the importance of a people and community-focused approach, achieving a more cohesive society, achieving competitiveness and an integrated approach to future development. The SDS is introduced as a way of drawing these principles together to 'promote a balanced and equitable pattern of sustainable development across the region' (p. 41). The argumentation emphasises that all parts of the region have a contribution to make, helping to create a territory of diverse local identities around a sense of place (p. 22). Within this mosaic of diversity, all parts are to be encouraged to maximise their potential. This leads to a division of the area into the Belfast Metropolitan area, Derry/Londonderry[7] as a 'regional city' in the North West, and the small towns and rural areas in the rest of the region. These are interlinked, as in the Dutch National Policy Document, into a concept of urban networks. But rather than linking to the fashionable 'network society' concept, the Strategy develops its own concept, expressing 'living together'. The towns form:

> a 'family of settlements' which provide a fairly well-balanced hierarchy of employment and service centres across the Region. The extensive network of urban hubs is evenly spread and well placed to serve a strong rural community living either in villages, small settlements, or in distinctive patterns of dispersed dwellings in the open countryside. (DRDNI, 2002: 11)[8]

The intention is to strengthen intra-regional articulation while at the same time positioning the area in a 'web' of external linkages, including to South West Scotland and across the province's national land border with Ireland, where a National Spatial Strategy was also in preparation from 2000 (DoELG, 2002).

The Strategy articulates these spatialised conceptions into a strong vocabulary of elements:

> The SDS for Northern Ireland is based on a framework for the future physical development of the Region based on urban HUBS and CLUSTERS, key and link transport CORRIDORS and the main regional GATEWAYS of ports and airports. (DRDNI, 2002: 43) (see Figure 16.3)

7 The City is referred to as Derry by its city council, and Londonderry by the provincial government.

8 This 'distinctive pattern' recognises the old Celtic settlement pattern, in contrast to the traditional 'English' concept of a rural landscape of contained villages dating from the enclosures of the eighteenth century.

In contrast to the Dutch concept of layers and strata, the SDS uses the concept of transport corridors as a 'skeletal framework' for future physical development (p. 43). But it too seeks to encourage some degree of compact development, through the concepts of 'decentralised development' and 'a polycentric network of hubs/clusters' to act as 'growth poles' across the region. These hubs, which could be a town or a 'cluster of towns', are encouraged to develop as counter-magnets to the growth dynamic of the Belfast Metropolitan Area. The problems generated by the hyper-concentration of growth in Dublin in the Irish context are very much in mind here.

The intention in this strategy is not to provide too precise a specification of what a hub, corridor or gateway would look like, as this is envisaged as something that will evolve as key players (notably groups of local authorities) develop the ideas in a more specific way in different parts of the province. Instead, it aims to provide a conceptual frame and a mobilising vocabulary to focus development initiatives across the province and encourage some integration of the development effort across government sectors and among the various governance arenas, from rural community development projects to strategic planning efforts in Belfast. Despite criticisms of its processes and its potential real influence (Neill and Gordon, 2001), the Northern Ireland RDS reflects a good example in the UK of the 'endogenous development' of a spatial development strategy at the regional scale, well-informed by, but not too heavily borrowed from, European and UK exemplars. In particular, it uses spatiality as a key policy tool in a complex and very difficult political context. The ambition is to develop a new politics of spatial allocation to contribute to displacing the geography of the old sectarian politics. But despite the language of flows and webs used in the Strategy, the Northern Ireland RDS still uses traditional notions of the relation between physical structures and social and economic development, and uses physical metaphors. Transport is a 'skeletal framework', development spreads out primarily in a 'distance-decay' manner, smaller settlements have less important linkages to the national and international context than larger ones, settlements are somehow integrated among themselves.

By late 2002, and despite the political difficulties which resulted in the suspension of devolved government, there were many signs that the new policy discourse of regional spatial development was being used by other government departments and had the support of influential devolved government politicians. Local authorities and partnerships across the province were also increasingly interested in seeing the strategy translated into 'delivery mechanisms'.[9] Whatever the outcome, the *Northern Ireland RDS* is a brave effort, both in the specific governance context of Northern Ireland and as a contribution to the development of a stronger spatiality to planning thought in the UK context. It is also seen within the European Commission as an exemplar of the local development of the ESDP concepts.[10]

Ricostruire la Grande Milano: Documento di Inquadrimento delle Politiche Urbanistiche Communali[11]

At the heart of this planning episode is a struggle to develop a more strategic and metropolitan scale approach to major development investments and land use regulation, in the context of political upheavals and accelerating metropolitan decentralisation (Balducci, 2001; Gualini, 2003). The Milan Framework Document was approved by the Comune di Milano in June 2000, and published in January 2001. The objective was to initiate new procedures for land use regulation as well as a new way of thinking about planning strategy in the Comune di Milano. By the 1990s, Milan was facing pressures of congestion as well as decentralisation, with the wider region capturing much of the growth dynamic which used to locate within the city boundary. But the Comune had found it very difficult to realise proposals for major area redevelopment schemes which would have created more space within the city boundaries (Balducci, 2001). The Comune was also seeking ways of moving beyond its history of corrupt government, in which communal projects and planning regulations became opportunities for extracting the political bribes that became known as 'tangentopolis' practices.

In 1999 the *Assessore* for the Comune's planning function, advised by his officials, asked the well-known planning academic and consultant, Luigi Mazza, from the Politecnico di Milano, to assist in developing a new approach (Balducci, 2001). The product of this work was the Milan Framework Document. This

9 These comments are based on discussions with local actors in Belfast and Derry in November 2002.
10 Personal communication, R. Neissler (April 2002).
11 The text for the spatial concepts is Comune di Milano (2000).

introduces both a 'strategic frame of reference' and a new procedure for land use regulation. Rather than the traditional Italian 'urbanistic approach' to urban plan-making, which articulated a detailed specification for the future physical form and spatial organisation of a town, the Milan Framework Document offers a more flexible approach, based on strategies, policies and criteria.[12] The crucial innovation is the simplification of the city's zoning regime and the introduction of a unified project evaluation procedure. This allows the spatial organising concepts expressed in the Framework Document, which has advisory status only, to be brought to bear on regulatory judgements about individual projects. This procedure has now been operating for two years with significant results in terms of generating built space in locations long-recognised as critical development nodes for the city (Pomilio, 2003).[13]

In these comments on the Framework Document, I focus less on the new procedure and concentrate on the spatial concepts within the strategic framework. However, in evaluating these, it will be important to keep in mind the institutional work the framework is called upon to do with respect to the new procedure. The strategic concepts are left relatively general. The aim is to frame argumentation and encourage debate which can lead to elaboration among the different stakeholders interested in the development of the city and the wider Greater Milan region. The Framework Document is thus put forward as a necessary step towards the formation of a strategic plan for the Milan urban region.

The strategic concepts are presented in a rich and dense argumentation provided by Mazza.[14] At the heart of this argumentation is a plea for a strategic approach, in contrast to the past focus on projects, many of which did not get built. This is linked to an attempt to shift the spatial conception of the city from a traditional view of a hierarchically-organised city centred on a single historic core, to one based on the actually emerging new major axis of development interest. Infusing the Framework Document is a recognition that a major reason for the failure of the Comune of Milan to attract and retain people and firms is the lack of adequate development opportunities for commercial activities and the decreasing environmental quality of many residential areas across the city, including the city centre itself.

In articulating a new strategic orientation for the city's development, the Framework Document emphasises a 'strategy of relations':

> Recent transformations and future tendencies in the economic structure of Milan define the city as the major service centre in Italy; Milan lives through its relations, in particular its external relations. A general strategy for Milan is above all a strategy capable of reinforcing and developing the links between Milan and the world, both near and far. (Comune di Milano, 2000: 61, para. 91, author's translation)

In this context, the Framework Document argues that Milan has to continue to attract investment, and in particular commercial activity. This has been compromised in the past by the failure to provide suitable opportunities for property development and investment.[15] A major purpose of the new approach is to remedy this situation, while at the same time shaping the nature and location of market opportunities in relation to strategic objectives, through linking the development opportunities to nodal points in the physical infrastructure. A new model of spatial organisation for the city is proposed with the objective of creating new multi-nodal patterns of land and property value, shifting conceptions from the classical view of a land value pyramid focused on the city centre (p. 8) to create new values in more peripheral locations. The Milan Framework Document reviews development tendencies in the city in recent years and identifies that in practice, the land value pyramid has been altered by decentralisation dynamics. Any strategy for the future, it is argued, should build on these actual tendencies and not retain idealised models of city form. Yet a firm spatial 'model' is needed to create stability and generate visibility over time to the city's development trajectory. This is provided by

12 See Mazza (2001; 2002). This approach is, of course, already well-developed in the Netherlands and the UK.

13 I acknowledge the help given to me in understanding the context and effects of the Framework Document by Alessandro Balducci, Luigi Mazza, Gabriele Pasqui and Filomena Pomilia of the Politecnico di Milano, and Giovanni Oggioni, Paolo Simonetta and Paolo Riganti of the Comune di Milano.

14 This builds on his work in promoting an alternative conception of strategic planning (see Mazza, 1994; 1996; 1997).

15 Milan attracts considerable international interest in its urban property market. Key actors make comparisons between Milan and other European cities in terms of how the market works (see, for example, Magalhães, 2001).

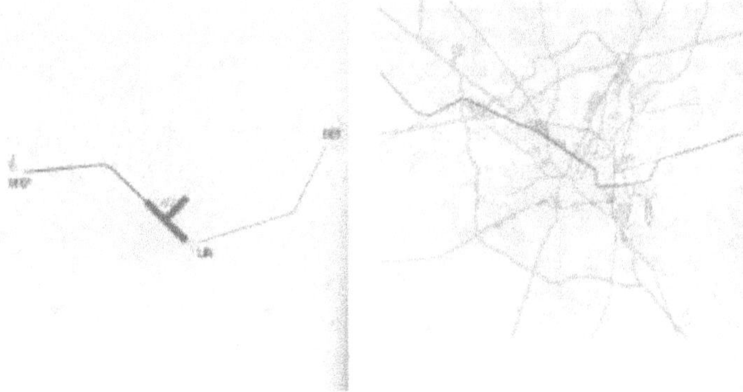

Figure 16.4 The Milan Framework Document: (a) the new urban 'dorsale' and (b) the urban spatial model

Source: Comune di Milano, 2001: 72, 76, with kind permission of the Progetto Pianificazione Strategica, Comune di Milano

the concept of an 'inverted T' ('*T rovesciata*') which consists of a principal urban 'backbone' ('*dorsale*'), with 'axes' ('*assi*') and a 'heart' ('*cuore*'), the heart being the city centre, with connections across the whole Grande Milano.[16] This 'dorsale' is intended to provide an efficient relationship between the city and the airport system and to provide new, large development areas with good accessibility and more competitive prices than are available in the central area. These spatial ordering concepts are not new and have been developed within the Comune since the 1980s, although new terms have been used for them (Balducci, 2001; Gualini, 2003). What is new for Italian planning practice is the highly selective and minimalist mode of expression of a spatial planning idea (see Figure 16.4a).

Within this structuring device, the Framework Document then emphasises the importance of improving urban and environmental quality. This concern centres on fostering a greater mix of activities in different locales, so that residential activity will return to the city centre, and peripheral areas will become more diverse. The emphasis on quality also focuses attention on areas where quality has been degraded and needs social and physical interventions to prevent further deterioration. It is supported by an emphasis on the importance of green spaces and a '*sistema a corona di parchi*' (ring of parks) (p. 10) in the periphery, with '*percorsi verde*' (green routes) connecting the '*corona*' to the centre. The Framework Document also expresses the hope that investments in education, health, sport and leisure facilities will be made in line with its criteria. Finally, it emphasises the importance of promoting accessibility generally across the city, while relieving congestion and reducing the damage which excess traffic produces to the residential environment. The Document concludes that the actual volume of traffic on the streets cannot be reduced, so what is needed is a better interlinked package of transport investments. These various considerations are pulled together in a 'new model of spatial organisation' (Figure 16.4b), which is little more than a sketch. This is in line with the conception of the development of a spatial strategy as a process, which should conclude with a spatial organising idea rather than start with one.

As with the two previous plans discussed, the Milan Framework Document is very specific to its context. Its key quality is the emphasis on argument, the argument for a strategic focus and for a change in conception of a city and its dynamics. Strategy formation is understood as a non-linear, emergent process. The Document not only recognises that Milan is a major commercial and financial city in the European context, but that, especially in such cities, the operation of the land and property market matters and needs to be viewed strategically as an asset, rather than as a way in which land and property owners capture land rents from the efforts of workers and the profits of firms. This recognition alone draws the spatial approach away from conceptions of the precise ordering of space. The emphasis instead is on providing opportunities, with criteria

16 By mid-2002, in ideas for developing a spatial strategy, these axes were being referred to as 'lines of force'.

to safeguard wider interests, and encouragement for new initiatives to emerge, even outside the broad spatial concepts. Just as the city is developing now 'outside' its formal planning framework, so the Framework Document takes an emergent view of urban development. It also emphasises the importance of linkages and relationships. But beyond noting their significance, there is little attention to the specific spatialities of different kinds of relationships and how these might develop and conflict. Exactly what it means to recognise a 'strategy of relations' is not well-developed.

The innovation of the Framework Document is in the approach to regulating development. Rather than lodging the regulatory power in the legal instrument of the plan, major developments in Milan now pass through an evaluation process, in which the ideas of the Framework Document are put to use, with recommendations made to the Comune's elected councillors. The approach has attracted much critical attention.[17] Some believe that the flexibility merely allows major property market actors to extract yet more benefit as well as creating opportunities for corrupt practice. Others are concerned about the lack of a broad participatory debate on the qualities of the city or any connection to discussions about urban qualities going on elsewhere in the Comune. There are also coordination problems within the Comune, especially as regards land use regulation and transport investment. Most seriously, there appears to be little political will to construct a strategic coalition within Milan and with actors in the wider region, to carry the strategic ideas forward. As regards the spatial concepts themselves, despite the emphasis on networks and relationships, the '*T-rovesciato*' image and its link to spatial development nodes conveys a strongly physical conception of relations. What is new in the Italian context, however, is the representation of these ideas in a very tentative, minimalist way, with the continual caveat that the representation is intended to initiate a thinking process which may ultimately result in a more developed spatial strategy. The new practices for land use regulation grounded in the Framework Document may also slowly lead key actors towards recognition of the value of a more strategic approach to their role in urban development.

Conclusions

Spatial strategies achieve their effects, if they get to have any leverage over future conceptions and actions, over the long term. They do this by influencing agendas of projects and schemes for physical development, and by shaping the values with which the qualities of places are promoted and managed. Shifting such agendas and sets of values is a complex institutional project which unfolds over time and unevenly, as these three examples illustrate. In each case, the struggles are not merely over the specific content or concepts of the strategies. They are part of wider political struggles over the approach to area and territory development and over modes and arenas of policy development and delivery. These struggles and their outcomes are important because, if the strategic spatial organising concepts influence government investment (especially in infrastructure and urban redevelopment) and help to shape land use regulation practices, then there will be significant material effects on emergent socio-spatial dynamics. In evaluating these three plans, it is only possible to consider their potential for exercising such power and, if they should realise this potential, the consequences of the effective mobilisation of the concepts of place and space contained within them. In these conclusions, I consider the kind of geography expressed in these plans, the institutional work being performed by the plans and the role of the spatial concepts within them. I conclude with a comment on the significance of the geographical imaginations developed in strategic spatial planning episodes.

Geographical Concepts

Figure 16.5 summarises the spatial concepts deployed within the three plans, using the evaluative criteria outlined in Figure 16.5. An obvious difference can be found in the strategy documents themselves. In terms of representational form, the Fifth NPD has a strong emphasis on managing the spatial evolution of territory, while the Milan Framework Document, which provides a single spatial concept, tentatively expressed visually,

17 See Balducci (2001), Salzano (2002), Mazza's reply (2002) and the special volume of *Urbanistica* (Volume 119, 2002).

recognises the emergent nature of spatial organisation. The Northern Ireland RDS lacks the sophistication of the spatial account in the Fifth NPD, and seems to lie midway between the other two plans, using a set of spatial metaphors as organising devices with which to grasp the territory and shape its future development.

Netherlands	**Fifth NPD 2001**	**Northern Ireland RDS 2001**	**Milan Framework Document 2000**
Treatment of scale	*Networks, not hierarchy* (but hierarchy in responsibility for detailing)	A hierarchy of hubs and gateways, *but interlinked through hub networks.*	*Very loose, the Framework concept to be inserted at the urban region scale, the rest to emerge.*
Treatment of position	*Emphasis on position in West* European flows *economic* and water movements.	*Internal differentiation and external positioning*, but the linkages are not strongly developed except through the hubs/gateways (dominant concern).	*External positioning critical, internal differentiation to evolve in relation to the Framework concept.*
Regionalisation	*Emphasis on cores, not borders*, but concept of strata and red/green holds the approach in a traditional vice.	Hubs, gateways and corridors 3 zones (BMA, Londonderry, rural areas).	A physical backbone, *structuring responses which will differentiate, shaped by policy Criteria.*
Materiality and identity	Strong emphasis on managing spatial futures.	*Family of settlements; multiple place identities in a cohesive society*; balanced development.	Focus on the material physical infrastructures and property market opportunities. *Re-imagining the City in a different way.*
Concept of development	Assumes continued affluence, but little articulated.	*Social and political dimensions dominant balance and cohesion as key concerns.*	*Positioning multiple relations in European context* (a strong economic emphasis). *An emergent concept of urban development.*
Representational form	Strong emphasis on mapped description as an anchor for mapped future strategy. Key concepts are about space. Organisation	Strong emphasis on spatial concepts, in text and diagrammatic maps. Key representations of physical flow and settlement organisation.	*Strong emphasis on argument in a textual statement.* A simple spacial concept.

Italics = more relational approaches; Normal = more traditional approaches

Figure 16.5 Analysis of spatial concepts

Figure 16.5 also illustrates the conflicting 'geographies' which coexist within the planning documents. All these examples combine relational and traditional geographies in some way. This reflects a tension between fixity and mobility, and between openness and closure. In this perspective, and despite its analytical emphasis on networks and flows, the Fifth NPD tends to emphasise the fixities, the land, the physical spaces, 'layered' over each other. While avoiding definitions of borders, even of the national territory, focusing instead on core locales (urban centres), the notion of red and green areas and contours pulls the conception back to physicalist notions of activity separation and environmental qualities. A relational geography developing in the analysis of urban and regional economic and social relations has become enmeshed in a traditional physical planning landscape.

The Northern Ireland RDS, like the ESDP on which it draws, reflects what Jensen and Richardson describe as a tension between a discourse of 'places' and a discourse of 'flows' (Jensen and Richardson, 2000). The RDS absorbs the notion of networks and understands the importance of place qualities and identities, but ties these down into rather traditional concepts of spatial organisation, with a hierarchy of settlements and corridors between them (the hubs and corridors), grounded in notions that what is physically 'near' represents a more significant linkage than what is 'far'. The gateway metaphor opens this up a little, suggesting openness and connectivity. Where the RDS innovates is in its concept of a 'family of settlements' with potentially multiple identities. The Milan Framework Document explicitly emphasises a relational approach, but this is not then developed. Instead, it is left implicit that a geography will emerge from many relationships and, apart from the 'dorsale' as the structuring device, its patterning cannot be defined in advance. Where the Fifth NPD and the Northern Ireland RDS confidently present their ideas in multi-colour maps, the Milan Framework Document is visually silent.

The Netherlands Fifth NPD and the Northern Ireland RDS in their specific development of a spatial vocabulary for policy development purposes are more traditional in approach than the Milan Framework Document, but *all* the plans are influenced by relational concepts. What is most striking, however, is that none of these strategic discourses appears to engage consciously with a new way to articulate its geographical understanding. Instead, they seem to exist in a discursive melting pot, full of metaphors from a traditional, essentialist past combined with images emanating from the new relational geography. These metaphors – like 'balanced development', hubs, networks, corridors – float around in an increasingly European planning policy discourse, to be pulled out and used to accrete meanings, older and newer, in specific planning contexts. In this usage, the metaphors may act transitively, to carry meanings from one institutional site to another. But they may also be co-opted to serve very specific local purposes. This highlights the importance of understanding the institutional context of the development and use of spatial concepts and vocabularies.

Institutional Work

Each of the three planning episodes discussed arises from a distinctive context and is driven by specific intentions. In the Netherlands context, the objective appears to be to insert a recast spatial conception into a traditional policy context where a high degree of spatial ordering has been a key government activity, underpinned by the negotiation of consensus among the main government actors. The critical question is whether such an 'authoritative' ordering approach can survive the more complex relations between state and society, between many stakeholders and between levels of government now emerging in the Netherlands, and whether conceptions of spatial dynamics more relevant to the Dutch society and economy can emerge from the grip of traditional Dutch planning 'doctrine' (Hajer and Zonneveld, 2000).

The Northern Ireland RDS and the Milan Framework Document reflect a quite different institutional reality. In both cases, the planning exercise has a mobilisation intention. This is much more clearly articulated in the Northern Ireland RDS, reflecting the delicate political context within which it was produced, given confidence by the extensive consultation process through which the RDS evolved. It is the outcome of a long and quite broad process of articulation. Its institutional ambition was primarily generative, to mobilise new ways of thinking about the territory, with which to focus and justify investment. It sought to stabilise and build more territorial coherence in a divided society, build democratic opportunity, capture investment opportunity and create place qualities for both citizens and investors.

The Milan Framework Document was intended as an input to an emergent process of strategy formation. It sought to create spaces for emergent property market opportunities, to mobilise and change a planning culture and its practices, and to improve place qualities for citizens. These ambitions too are focused by a generative conception of the power of a strategy. Whereas the Fifth NPD may be seen in terms of a way of maintaining the position of a well-defined set of actors, at a time when power has been slipping away, the other two strategies may be seen as contributing to an active process of developing a new constituency and politics around territorial development, rather than sectoral or fragmented policy agendas.

The spatial concepts used in the Fifth NPD may therefore be interpreted as searching for a different understanding of spatial order, to insert into a well-established (but now challenged) tradition of authoritative spatial ordering. The Northern Ireland RDS in contrast uses spatial concepts, albeit largely drawn from an old geography, to help mobilise policy attention across a governance landscape around a new way of thinking about place quality and identity and thereby contribute towards a hopeful dream of a less troubled future. The Milan Framework Document combines specificity about major infrastructural investment with a deliberate spatial vagueness, intended to open an imaginative space within which new ideas and dreams of the future can be articulated and mobilised. But as yet, there is no sign of that imaginative space being used.

Shifting Planning Discourses

Spatial concepts and vocabularies not only carry strategic ideas from the arenas of their articulation to these sites of material and imaginative use. They also affect the structuring of political debate and struggle over the impacts of projects, the distributive justice of investments and regulatory principles, and over imagined futures. Spatial vocabularies also provide a currency for local environmental politics as elected representatives seek to acquire or prevent particular types of development for their constituents. These effects in turn influence the behaviours of market actors as they acquire sites and develop project proposals. Strategic spatial planning episodes which aim to change established spatial concepts and vocabularies therefore face a challenging task. The established vocabularies represent significant political and intellectual capital embedded in ongoing political and market processes. The political and market implications of new concepts and vocabularies are often not easy to assess and/or may be interpreted in crude ways, as in the elision of 'development corridors' into 'urban sprawl' in the Dutch debates. This suggests that shifting a planning discourse will be hard without other supporting shifts in the institutional context which makes a new discourse more welcome. Such a moment of opportunity was clearly evident in the Northern Ireland case, but has so far been missing in the Milan case.

If it is so difficult to change the spatial content of a planning discourse, is there any merit in seeking to shift the geographical imagination from traditional, essentialist conceptions to the new relational geography? If the old concepts of urban form are easy to grasp, why not continue to use them? There are two main reasons why the struggle to shift the geographical imagination of strategic spatial planning discourse needs to be maintained, whatever the institutional resistances. The first is that the new dynamic relational conceptions capture the real material experience of many people and firms as they operate, intersect and interact in the space of urban regions. If planning concepts fail to absorb these material realities, the struggle between planning 'rules' and the demands and needs of people and firms will merely escalate, in the end undermining any remaining legitimacy of the planning concepts. The second is that a relational understanding of the spatiality of urban and regional dynamics in a world of multiple relations with diverse space-time dimensions and driving forces provides a more relevant way to understand the capacity locally to moderate and shape the power of external forces as they impact on territorial development, which in turn should lead to more effective ways to promote local capacities and values.

However, there is no easy answer to developing a locally-relevant relational spatial conception and vocabulary. There is always a danger that the vocabulary of a relational geography will be borrowed from some authoritative report or exemplar practice and imposed on a locality. Such borrowing is already occurring with the concepts in the ESDP, which has a very ambiguous geographical imagination (Zonneveld, 2000; Jensen and Richardson, 2000). However, concept 'transfer' is likely not only to weaken the power of a spatial vocabulary, making concepts easily open to capture by more traditional understandings. Such an imposition may also fail to recognise important local relations, networks, nodes and identities. The key to developing a locally-relevant and powerful relational spatial vocabulary is to encourage a relational geographical imagination with which to explore the many material relations and mental images of place and place quality which are locally-important, and the interaction between these and the wider relations of which they are a part. Such an imagination may also be easier to develop in discussion with the array of stakeholders concerned with place quality and territorial development, as it relates more to their daily life meanings and activities than the urban form concepts of traditional physical planning. It is the locally-specific realisation of a relational geographical imagination that could result in spatial strategies which have the potential to deliver real progressive effects for the quality of life in specific places.

References

Albrechts, L., J. Alden and A. da Rosa Pires (eds) (2001) *The Changing Institutional Landscape of Planning*, Aldershot: Ashgate.

Allen, J. (1999) Worlds within cities. In D. Massey, J. Allen and S. Pile (eds), *City Worlds*, London: Routledge/ Open University Press.

—, D. Massey and A. Cochrane (1998) *Rethinking the Region*, London: Routledge.

Amin, A. and N. Thrift (2002) *Cities: Reimaging the urban*, Cambridge: Polity Press.

Balducci, A. (2001) New tasks and new forms of comprehensive planning in Italy. In L. Albrechts, J. Alden and A. da Rosa Pires (eds), *The Changing Institutional Landscape of Planning*, Aldershot: Ashgate.

Beauregard, B. (1995) If only the city could speak: the politics of representation. In H. Liggett and D. Perry (eds), *Spatial Practices*, Thousand Oaks, CA: Sage.

Bengs, C. and W. Zonneveld (2002) The European discourse on urban-rural relationships: a new policy and research agenda, *Built Environment* 28(4): 278–89.

Boyer, C. (1983) *Dreaming the Rational City*, Cambridge, MA: MIT Press.

Brenner, N. (1999) Globalisation as reterritorialisation: The re-scaling of urban governance in the European Union, *Urban Studies* 36(3): 431–52.

Committee for Spatial Development (CSD) (1999) *The European Spatial Development Perspective*. European Commission, Luxembourg.

Comune di Milano: Assessore allo Sviluppo del Territorio (2001) *Ricostruire la Grande Milano: Documento di Inquadrimento delle Politiche Urbanistiche Communali*, Milan: Edizione Il Sole 24 Ore.

Cooke, P., P. Boekholt and F. Todtling (2000) *The Governance of Innovation in Europe*, London: Pinter.

Dematteis, G. (1994) Global and local geographies. In F. Farinelli, G. Olsson and D. Reichert (eds), *Limits of Representation*, Munich: Accedo.

— (2001) Reti globali, identita territoriali e ciberspazio. In P. Bonora (ed.), *Comcities: geografie delle comunicazione*, Bologna: Baskerville.

Department of the Environment and Local Government (DoELG) (2002) *The National Spatial Strategy 2002–2020*, Dublin: The Stationery Office.

Department for Regional Development (DRDNI) (2002) *Shaping Our Future: Regional Development Strategy for Northern Ireland 2025*. Department for Regional Development, Northern Ireland Government, Belfast.

de Vries, J. and W. Zonneveld (2001) Transnational planning and the ambivalence of Dutch spatial planning. Paper presented at the World Planning Congress, Shanghai, July.

Doucet, P. (2002) Transnational planning in the wake of the ESDP: the Northwest Europe experience. In A. Faludi (ed.), *European Spatial Planning*, Cambridge, MA: Lincoln Institute of Land Policy.

Dovey, K. (1999) *Framing Places: Mediating power in built form*, London: Routledge.

Dyrberg, T.B. (1997) *The Circular Structure of Power*, London: Verso.

Faludi, A. (1996) Framing with images. *Environment and Planning B: Planning and Design* 23: 93–108.

— (2001) The application of the European Spatial Development perspective: Evidence from the North-West Metropolitan Area. *European Planning Studies* 9(5): 663–75.

— (ed.) (2002) *European Spatial Planning*, Cambridge, MA: Lincoln Institute of Land Policy.

— and A. van der Valk (1994) Rule and order in Dutch planning doctrine in the twentieth century, Dordrecht: Kluwer Academic Publishers.

— and B. Waterhout (2002) *The Making of the European Spatial Development Perspective*, London: Routledge.

Fischler, R. (1995) Strategy and history in professional practice: planning as world-making. In H. Liggett and D. Perry (eds), *Spatial Practices*, Thousand Oaks, CA: Sage.

Friedmann, J. (1993) Towards a non-Euclidean mode of planning, *Journal of the American Planning Association* 59(3): 482–4.

Fuurst, D. and J. Kneilung (2002) *Regional Governance: New modes of self-government in the European Community*, Hannover: ARL, University of Hannover.

Giddens, A. (1984) *The Constitution of Society*, Cambridge: Polity Press.

Graham, S. and P. Healey (1999) Relational concepts in time and space: Issues for planning theory and practice, *European Planning Studies* 7(5): 623–46.
—— and S. Marvin (2001) *Splintering Urbanism*, London: Routledge.
Gregory, D. (1994) *Geographical Imaginations*, Oxford: Blackwell.
Gualini, E. (2001) *Planning and the Intelligence of Institutions*, Ashgate, Aldershot.
—— (2003) The region of Milan. In W. Salet, A. Thornley and A. Kreukels (eds), *Metropolitan Governance and Spatial Planning*, London: E&FN Spon.
Hajer, M. and W. Zonneveld (2000) Spatial planning in the network society – rethinking the principles of planning in the Netherlands, *European Planning Studies* 8(3): 337–55.
Healey, P. (2000) Planning in relational time and space: responding to new urban realities. In G. Bridge and S. Watson (eds.), *A companion to the city*, Blackwell, Oxford.
—— (2002) On creating the 'city' as a collective resource, *Urban Studies* 39.10, 1777–92.
—— , A. Khakee, A. Motte and B. Needham (eds.) (1997) *Making Strategic Spatial Plans: Innovation in Europe*, London: UCL Press.
Hooghe, L. (ed.) (1996) *Cohesion Policy and European Integration: Building multi-level governance*, Oxford: Oxford University Press.
Jensen, O. and T. Richardson (2000) Discourses of mobility and polycentric development: a contested view of European spatial planning, *European Planning Studies* 8(4): 503–20.
—— and T. Richardson (2002) Being on the map: the new iconographies of power over European space. Paper presented at the AESOP Congress, Volos, Greece.
Lefebvre, H. (1991) *The Production of Space*, Oxford: Blackwell.
Liggett, H. and D. Perry (eds.) (1995) *Spatial Practices*, Thousand Oaks, CA: Sage.
Macleod, G. (1999) Place, politics and 'scale dependence': exploring the structuration of Euro-regionalism, *European Urban and Regional Studies* 6(3): 231–54.
Magalhães, C. de (2001) International property consultants and the transformation of local markets, *Journal of Property Research* 18(2): 99–122.
Marshall, T. (2002) The re-timing of English regional planning, *Town Planning Review* 73(2): 171–95.
Massey, D. (1994) *Space, Place and Gender*, Polity Press, Cambridge.
Mazza, L. (1994) Piano, progetti, strategie, *Critica della Razionalita Urbanistica* (CRUU) 2: 50–55.
—— (1996) Difficolta della pianificazione strategica, *Territorio* 2: 176–82.
—— (1997) *Trasformazione del piano*, Milan: Franco Angeli.
—— (2001) Nuove procedure urbanistiche a Milano, *Territorio* 16: 53–60.
—— (2002) Flessibilita e rigidita delle argomentazioni urbanistica, *Urbanistica* 118: 149–59.
McEldowney, M. and K. Sterrett (2001) Shaping a regional vision: the case of Northern Ireland, *Local Economy* 16(1): 38–49.
Mitchell, W. (1995) *City of Bits: Space, place and the Infobahn*, Cambridge, MA: MIT Press.
Motte, A. (ed.) (1995) *Schema directeur et project d'aggtomeration: l'experimentation de nouvelles politiques urbaines spatialisees 1981–1993*, Paris: Les Editions Juris Service.
Neill, W.V.J. and M. Gordon (2001) Shaping our future? The Regional Strategic Framework for Northern Ireland, *Planning Theory and Practice* 2(1): 31–52.
Neuman, M. (1997) Images as institution builders: metropolitan planning in Madrid. In P. Healey, A. Khakee, A. Motte and B. Needham (eds), *Making Strategic Spatial Plans: Innovation in Europe*, London: UCL Press.
Pomilio, F. (2003) Il 'Documento di Inquadrimento delle politiche urbanistiche' di Milano: un caso anomalo di pianificazione strategica? In R. Pugliese and A. Spaziente (eds), *Strategie per la citta: piani, politiche, azioni: una rassegna do casi*, Milan: Franco Angeli.
Salet, W. and A. Faludi (eds) (2000) *The Revival of Strategic Spatial Planning*, Amsterdam: Koninklijke Nedelandse Akademie van Wetenschappen.
—— , A. Thornley and A. Kreukels (eds) (2003) *Metropolitan Governance and Spatial Planning*, London: E&FN Spon.
Salzano, E. (2002) Il modello flessibile a Milano, *Urbanistica* 118: 140–48.
Sandercock, L. (2003) *Cosmopolis II: mongrel cities in the 21st century*, London: Continuum.

Thrift, N. (1996) *Spatial Formations*, London: Sage.

Vigar, G., P. Healey, A. Hull and S. Davoudi (2000) *Planning, Governance and Spatial Strategy in Britain*, London: Macmillan.

VROM, Netherlands National Spatial Planning Agency (1999) *Planning the Netherlands: Strategic Principles for a New Spatial Planning Policy*, Ministry of Spatial Planning and the Environment, The Hague.

— (2000) *Spatial Perspectives in Europe*, Ministry of Housing, Spatial Planning and the Environment, The Hague.

— (2001) *Making Space, Sharing Space: Fifth National Policy Document on Spatial Planning 2000/2003.* (English Summary), Ministry of Housing, Spatial Planning and the Environment, The Hague.

Williams, R. (1996) *European Spatial Policy and Planning*, London: Paul Chapman Publishing.

Wolsink, M. (2003) Reshaping the Dutch planning system: a learning process? *Environment and Planning A* 35: 705–23.

Woltjer, J. (2000) *Consensus Planning: The relevance of communicative planning theory in Dutch infrastructure networks*, Aldershot: Ashgate.

WRR: Netherlands Scientific Council for Government Policy (1999) *Spatial Development Policy*, The Hague: SDU.

Zonneveld, W. (2000) Discursive aspects of strategic planning: a deconstruction of the 'balanced competitiveness' concept in European Spatial Planning. In W. Salet and A. Faludi (eds), *The Revival of Strategic Spatial Planning*, Amsterdam: Koninklijke Nederlandse Akademie van Wetenschappen.

Chapter 17
The Universal and Its Others: Reflections on Twenty-first-Century Planning

Ananya Roy

Global Mayoralties

A new urban optimism is in full swing. 'Mayors should rule the world', declared Benjamin Barber, a political scientist, in a recent TED talk. Indeed, from Enrique Peñalosa of Bogota to Rudolph Giuliani of New York, former mayors are today's planning gurus, disseminating their ideas and interventions to worldwide audiences. The most recent mayor to join this pantheon of planning innovation is Michael Bloomberg who is set to launch a 'global mayoralty', specifically a 'a high-powered consulting group to help him reshape cities around the world long after he leaves office' (Barbaro 2013). Funded by Bloomberg's personal fortune, and staffed by many of his former deputies who held key urban planning posts in New York, the organization is meant to showcase how urgent human problems, from public health to environmental sustainability, can be tackled in cities.

In keeping with Patsy Healey's (2013) analysis of the 'transnational flow of planning ideas and practices', these global mayoralties must be understood as instances of what she has termed 'circuits of knowledge and techniques'. At least four issues are at stake in these circuits. First, such transnational formations of expertise involve a distinctive methodology of urban problem-solving. At first glance, what seems to be at work is what Peck and Theodore (forthcoming), in their analysis of global regimes of poverty policy, have described as 'model power', rapidly ascendant and transnationally diffused policy models, 'orchestrated from global centres of persuasion, enacted through expert networks, sustained by narratives of best practice, and mediated by the rationalistic lore of experimental evaluation'. Peck and Theodore draw our attention to an 'emergent class of model peddlers' who wield such model power. And indeed, a plethora of hegemonic models of urban innovation now abound, from the Bogota model to the Singapore model. However, the recent launch of the Bloomberg global mayoralty suggests a methodology of urban problem-solving that departs from 'model power'. In his headline-making article for the *New York Times*, Barbaro (2013) describes the process thus: 'The organization, to be called Bloomberg Associates, will act as an urban SWAT team, deployed at the invitation of local governments to solve knotty, long-term challenges, like turning a blighted waterfront into a gleaming public space, or building subway-friendly residential neighbourhoods'. The metaphor of urban SWAT teams is a troubling one for Twenty-first-century planning and yet it may be quite accurate in depicting the global mayoralties through which planning ideas and practices are being produced and disseminated. It is not surprising that such regimes of planning involve what Patsy Healey (1997: 273) once characterized as the 'new city fathers', powerful business people who take on the task of reforming and improving cities.

Second, it is worth considering which 'knotty, long-term challenges' are addressed by such forms of planning. Amanda M. Burden, the director of city planning in the Bloomberg administration, and now a member of Bloomberg Associates, claims: 'We have heard this huge demand and need from other cities to learn from New York City … New York is the epitome that cities look to of how to get things done' (Barbaro 2013). But what precisely are the problems that the Bloomberg administration claims to have tackled (obesity, bike lanes) and those for which it is less stridently confident (income inequality, homelessness)? Ironically, the new mayoral administration in New York, led by Bill de Blasio, and elected on a wave of outrage about sharp socio-economic inequality in the city, is unlikely to draw on the expertise of Bloomberg Associates. In other words, global mayoralties must be understood as distinctive technologies of policy innovation and circulation, those that are focused on the problems that can be identified and solved through urban SWAT teams, often through familiar formats of environmental determinism. And once again, it is worth noting the rupture with 'model power'. My research on hegemonic models of microfinance (Roy 2010) and that of Peck and Theodore on conditional cash transfers demonstrate that global policy models have often been concerned with intractable social problems, notably that of poverty. Of course, these powerful models do not necessarily mitigate these persistent problems, and they often share the behavioural and technical interventions favored by global mayoralties, but the 'knotty, long-term challenges' they seek to tackle are of a different order than those that can be fixed through the ban of trans-fat foods or the painting of bike lanes.

Third, the global mayoralties represent a rescaling of problem-solving to cities. Such urban experiments are not novel. From Hellenic cities to medieval cities to Fordist cities to colonial cities, urbanism has been a key site and scale for the organization of the economy, government, and citizenship. What is distinctive about today's urban innovations is that they seek to address global problems in the context of cities. While globally interconnected through circuits of knowledge and techniques, such methodologies of problem-solving thus obscure what Popke (2007) describes as the 'politics of connectivity', the ways in which cities are 'locked into a multitude of relational networks of varying geographical reach' (Amin, 2004: 43). It is with this politics of connectivity in mind that Doreen Massey (2007) has called for an understanding of city life shaped by 'extended responsibility', one that foregrounds the global relationships of exploitation, dispossession, and dependency through which prosperous cities function and thrive. Her transnational ethics of place imagines a 'local internationalism' where world cities extend responsibility to the often invisible hinterlands on which they draw upon for labour and resources. Does the urban optimism of our times, the rehearsed faith that mayors can save the world, propagate a bounded rather than extended sense of responsibility?

In this chapter, I draw on Patsy Healey's analysis of transnational flows of planning to interpret such circuits of knowledge and techniques. In conversation with her essay, 'The Universal and the Contingent', I explore two key concepts presented in that work: the idea of planning as a 'contingent universal', and the idea of planning not as a technology but instead as produced by a 'community of inquirers' (Healey 2011). Healey argues that while the 'modernization myth' depended on the notion that planning is a 'universal good technology', this 'travelling capacity' is limited by the contingencies and complexities of local histories and geographies. In putting forward such an argument, Healey continues her long-standing interest in the relationship between planning and place (see for example, Graham and Healey 1999). But she also marries this interest with her articulation of a deliberative approach to planning. In the place of model power or urban SWAT teams, Healey (2011: 191) imagines the methodology of planning as a 'contingent universal' that is the 'product of the ongoing debates we have as a "planning community."' Building on Healey's important contributions, I take up these puzzle of the relationship between planning and place, as well as that of a 'community of inquirers' through which deliberation on planning problems can proceed. Such puzzles, I suggest, point to an analysis of, and critical intervention in, the political economy of global mayoralties and thus in the present history of planning practice. What is at stake in such an intervention is Healey's (2012: 19) ambitious call to 're-enchant democracy', the project of 'democracy-in-action', one inevitably shaped through 'contingent particularities and struggles'.

My interest in reactivating these key ideas drawn from Healey's work is two-fold. First, in various efforts to decentre the (EuroAmerican) canon of urban studies and to craft 'new geographies of theory' (Roy 2010), I have warned against the production of a universal theory of urban form and process. Most recently, I have noted that to study urbanism in its global reach and variegation does not, and should not, entail a universalizing concept of the urban (Roy forthcoming). But to reject a universal theory does not imply rejecting the study of universals. Thus, in her important ethnography of 'global connections', Anna Tsing (2005: 2) makes the case for the study of '*universal* dreams and schemes' (emphasis in the original). However, she conceptualizes the universal not as a 'self-fulfilling abstract truth' but rather as 'charged and enacted in the sticky materiality of practical encounters'. Her call to explore a 'practical, engaged universality as a guide to the yearnings and nightmares of our times' bears striking resemblance to Healey's concept of 'contingent universals'. In short, I am interested in how planning theory can study universals, 'not as truth or lies but as sticky engagements' (Tsing 2005: 6). What is at stake in such sticky engagements is the encounter with difference, the constitutive outside of the universal, the 'others' of universality. In this chapter, I argue that we need to carefully consider whether our theories, especially their geographical imaginations, are attentive to the stickiness of difference or whether they work with universalizing and smooth narratives of space and place.

Second, I expand Healey's notion of a 'community of inquirers' by placing it in relationship with the concept of 'communities of risk'. I borrow this latter concept from feminist philosopher, Nancy Fraser, and her meditations on justice. Such a relationship explicitly places Healey's project of 're-enchanting democracy' in an unequal world and suggests that the notion of the 'planning community' has to be expanded to take account of the urban poor as a community of inquiry. In other words, if we understand 'democracy' as a contingent universal, communities of risk constitute the grounds of the 'sticky engagements' of such universality. They embody difference. This type of analytical and political work requires attending to the grounds of contingency but more important to the grounds of universality. Such universality implies commonality, be it Healey's notion of a community of inquiry or Fraser's notion of global humanity as a community of risk. But it also requires a conceptualization of difference, specifically deep difference or radical alterity.

Through such work, I hope to demonstrate how Healey's scholarship not only shapes the conceptual architecture of planning theory but also enables useful lines of conversation between planning and other fields of inquiry, such as anthropology, history and philosophy. This is especially important because Healey seems unwilling to reify an instrumentalist or even institutionalist understanding of planning, instead expressing interest in conceptual frameworks that provide a 'broader meaning to the idea of planning' (Healey 2011: 198).

Planetarity

In an essay on 'relational concepts of space and place', Graham and Healey (1999: 623) argue that planning practice is 'grappling with new treatments of place, based on dynamic, relational constructs, rather than Euclidean, deterministic, and one-dimensional treatments inherited from the "scientific" approaches of the 1960s and early 1970s'. However, they lament that planning theory 'has so far failed to produce sufficiently robust and sophisticated conceptual treatments of place in today's globalizing world'. I interpret this lament as one concerned with the ontology, epistemology, and ethics of planning in a world constituted through global interconnections. Globalization is not new, but as Healey has often noted, what is perhaps distinctive about the present historical conjuncture is the crisis of the modernization myth, especially its tropes of universality and teleologies of progress. In this post-modern world, how should the politics of global connectivity be conceptualized and enacted in planning ideas and practices?

In previous work (Roy 2011), I have argued that it is necessary to forge counter-practices of critical transnationalism, those that can interrupt the global traffic of expertise. In doing so, I have relied on Healey's (2010: 8) argument that such critical work requires 'looking at what happens when planning concepts and instruments "travel" from place to place'. Healey argues that this 'helps not only to identify the implicit "pasts" that we still carry with us, feeding our dreaming in negative and positive ways; it also helps in understanding what could be the 'contingent universals' of our present global era'. While Healey argues that the idea of planning must be understood as a contingent universal, I am interested in how a more specific ethical concept, justice, often closely associated with planning, is also a contingent universal. Justice, in our present global era, not only requires an ethics attentive to the politics of connectivity but also new ontological and epistemological conceptualizations concerned with such global connectivity. In this manner, such conceptualizations of justice serve as useful templates for a (re)conceptualization of planning. An appropriate example is the case of climate justice.

We live in an era of super-storms. Each weather event seems more catastrophic than those that have preceded it. In November 2013, Typhoon Hainan hit the Philippines with devastating force, killing over 6,000 people and injuring and displacing thousands more. International agencies and national governments rushed to provide aid to the Philippines for disaster relief and reconstruction. But the super-storm also ignited an important debate about 'climate injustice' (Myers and Kulish 2013). At a United Nations conference on climate change held in Warsaw a month after the storm, representatives of global South countries held rich countries responsible for the climate crisis. Noting that the worst consequences of climate change are often borne by countries and peoples who have had the least role in producing the emissions that impact the world's climate, they demanded a new calculus of responsibility and compensation. But rich countries remained wary about the push for compensation. The US State Department's envoy on climate issues noted that not only was the US and other developed countries in the midst of a financial crisis but also that 'lectures about compensation, reparations and the like will produce nothing but antipathy among developed country policy makers and their publics'. The Warsaw conference thus concluded with what observers saw to be an 'uncomfortable shuffling of feet' (Readfern 2013). The conference recommended that a new branch of the United Nations Framework Convention on Climate Change be set up to deal with 'loss and damage' from climate change especially to help poorer countries deal with 'extreme events and slow onset events'. Such a recommendation acknowledged that extreme weather events can be linked to human-caused climate change but stopped short of placing liability on the rich nations – especially those in the North Atlantic and Arabian Gulf – with high levels of emissions per capita.

The debates about climate injustice highlight the significance of a relational concept of space and place, one that is dynamic and extended rather than static and bounded. It also suggests that while there are numerous interesting innovations to be launched at the urban scale, problems such as the climate crisis require a different scale of inquiry and action. No urban SWAT team can fix these problems. Perhaps most intriguing is the question this raises about what types of planning communities – communities of inquirers, to use Healey's specific phrase – are implicated in the deliberations about such problems.

In her work on 'scales of justice', philosopher Nancy Fraser (2009) poses the important question: Who counts as a subject of justice? She notes that while in the Westphalian world order, citizens of bounded nation-states counted as subjects of justice, in a global world such conceptions of justice are inadequate. Territoriality, Fraser emphasizes (2009: 5), cannot be 'the sole basis for assigning obligations of justice, given patently transterritorial problems, such as global warming or genetically modified agriculture'. Such transterritorial and transnational problems, she argues, require a reconsideration of 'communities of risk' (Fraser 2009: 5). If global humanity is considered a community of risk, then the very scale of justice, Fraser insists, must be reframed. With this in mind, Fraser (2009: 64) puts forward a new principle of justice, the all-subjected principle, whereby 'all those subject to a given governance structure have moral standing as subjects of justice'.

I cite the example of the all-subjected principle because it speaks directly to the two puzzles I derive from Healey's work on transnational planning: how to engage with contingent universals in a global era and how to ground the deliberation of such contingent universals in communities of inquiry. Fraser's conceptualization of scales of justice suggests a relationship between communities of inquiry and communities of risk, a point to which I will return in the next section of this chapter. Let me first turn to the matter of contingent universals in a global era. While Healey is concerned with the idea of planning as a contingent universal, and while I have interpreted Fraser's framework of justice as a contingent universal, I also want to pay attention to the relational geographies that are implicated in such concepts. Put another way, if planning theory is to 'produce sufficiently robust and sophisticated conceptual treatments of place in today's globalizing world' – Graham and Healey's (1999) mandate – then what are the geographic signifiers of such theory?

In an essay titled 'Rethinking Planetary Futures', Tariq Jazeel (2011) interrogates the geographically innocent signifiers through which our literatures are organized. In particular, Jazeel pays attention to a specific geographic signifier: planetary. Jazeel contrasts two frameworks of planetary futures. The first is Paul Gilroy's invocation of the planet as the basis of cosmopolitanism, a planetary consciousness which is meant to depart from the concern for the sovereign territory of national states. But the god-trick of Gilroy's vision is the view of Earth from outside its orbit by the Apollo spacecraft in 1972. As Jazeel (2011: 80) notes, the planet thus becomes the 'ground that unites humankind, a geocommon to all, but one that can only be glimpsed through moments of willing transcendence of that ground'. Needless to say, the view of Earth from the Apollo spacecraft was itself an effect of a very specific geopolitical struggle, the Cold War. Jazeel presents a second view of planetarity which draws upon the work of Gayatri Spivak and other postcolonial theorists. He presents these as 'alternative, less certain, and less avowedly 'cosmopolitan' imaginations that have the capacity to engage difference in non-assimilatory terms' (Jazeel 2011).

This second approach to planetarity has important implications for the relational understanding of space and place which Graham and Healey (1999) seek to encourage. It is also an approach that rejects the geographic signifier that is the 'global'. Spivak (2003: 72), for example, argues that the framework of globalization gives us 'an abstract ball covered in latitudes and longitudes, cut by virtual lines' and implying the same system of exchange everywhere. By contrast, the planet, she argues 'is in the species of alterity … we inhabit it, on loan'. This is not a view of the planet as an undivided 'natural' space – a geocommon – but rather as differentiated political space.

Spivak's disavowal of the global as a geographic signifier of critical theory is shared by many other postcolonial theorists. For example, Radhakrishnan (2005: 184) argues that worldliness must not be mistaken with globality. For him, 'globality is a condition effected by the travel of global capital', while worldliness is the state of 'being in the world'. Worlding is 'a perennial process of a lived and immanent contingency', while globality is a 'smooth and frictionless surface where oppositions, antagonisms and critique cannot take hold'. In my own work with Aihwa Ong, I have deployed the geographic signifier, worlding, to explain contemporary urbanism (Roy and Ong 2011). Our use of the term, worlding, is meant to break with the parochial but dominant framework of the global city, which as Jennifer Robinson (2002) has noted, increasingly functions as a 'regulating fiction' for hegemonic formats of urbanism.

The concept of planetarity, as put forward by postcolonial theorists, is also decidedly post-humanist. Unlike the transcendence of the planet imagined by Gilroy, these theories are interested in the articulation of human and non-human worlds. Thus, in his 'The Climate of History: Four Theses', Dipesh Chakrabarty (2009: 212) argues that to think about the planet means to think about anthropocene and that this in turn requires us to, in his words, 'put global histories of capital in conversation with the species history of humans'. What is at stake here in the shift from 'global' to 'planetary' is thus a renewed engagement with alterity. A relational understanding of space and place, such a geographic signifier requires a consideration of difference that extends beyond human ontologies. Thus, Spivak (2003: 72) argues: 'If we imagine ourselves as planetary subjects rather than global agents, planetary creatures rather than global entities, alterity remains underived from us'. Building on her earlier work, Spivak (2006: 107) clarifies this argument: 'By 'planet-thought' I meant a mind-set that thought that we lived on, specifically, a planet. I continue to think that to be human is to be intended toward exteriority. And, if we can get to planet-feeling, the outside or other is indefinite'.

In in its emphasis on alterity, specifically on difference that cannot be assimilated or even fully known, the philosophers of planetarity present an important provocation. When taken up in planning theory, there are many ways of understanding this provocation. For example, Vanessa Watson's (2006) conceptualization of 'deep difference' is in keeping with the postcolonial emphasis on alterity. But more broadly, Chakrabarty's statement that 'to be human is to be intended toward the other' is a reminder that all universals, including the universality of global humanity or of cosmopolitan responsibility, are predicated on visible or invisible others. If we are to consider the idea of planning as a contingent universal, what then does it to mean to take account of the forms of alterity that are constitutive of this universal? Planetarity, I am arguing, might be a more precise signifier of these relational geographies than the common invocation of globalization.

Indeed, in urban studies there is a new turn towards the concept of planetarity. Brenner and Schmid (2014: 743) have recently made a case for 'planetary urbanization' as a worldwide urban fabric, what they describe as 'the relentless dynamism of the urban phenomenon under modern capitalism: its endemic tendency to explode inherited morphologies of urbanism at all spatial scales; to create new, rescaled formations of urbanized territorial organization'. And they suggest that such processes entail a new ethics of habitation, that of 'a shared planetary existence'. Merrifield (2012) goes a step further, asking what sorts of politics is enabled or disabled by planetary urbanization. I am drawn to postcolonial theorists of planetarity because they insist that 'a shared planetary existence' is necessarily articulated through a politics of difference. Such frameworks bear striking resemblance to the work of planning theorists who draw on poststructuralist analysis to reconsider the epistemology, ontology, and ethics of planning. An important example is the work of Jean Hillier (2009: 640) whose Deleuzoguattarian analysis of 'assemblages of justice' emphasizes 'an oscillation between relations of presence and absence'.

I am arguing then that Graham and Healey's (1999) call for 'dynamic' and 'relational' 'treatments of place' serves as a mandate for planning thought and practice to consider the geographic signifiers through which the discipline and its literatures are organized. Inspired by the philosophical debates about justice in a global era, I suggest that contingent universals require careful attention not only to the grounds of contingency but equally to the grounds of universality. Such universality implies commonality, be it Healey's notion of a community of inquiry or Fraser's notion of global humanity as a community of risk. But it also requires a conceptualization of difference, specifically deep difference or radical alterity. As Jazeel (2011: 84) notes, the problem of universality is this: 'how can we conjure a notion of common humanity without normalizing the very idea of humanity itself?' I suggest that recent postcolonial articulations of 'planetarity' provide useful analyses of difference and thus stand in necessary contrast to the 'travelling capacity' of global mayoralties and other global regimes of power.

Learning

In their recent analysis of global policy regimes, Jamie Peck and Nik Theodore (2010) pinpoint the temporalities of expertise. Characterizing these transnational worlds as 'fast policy', they argue that the rapid transfer of ideas is a key characteristic of such circuits of knowledge. The global mayoralties and their technique of urban SWAT teams, which I described earlier in this chapter, can be understood as hyper-fast policy. In such a context, it is worth posing the second puzzle derived from Healey's work on transnational planning: who is the community of inquirers able to appropriately put ideas into motion and to translate and localize universals into contingent universals? Can such a community of inquiry interrupt and transform the hyper-temporality of fast policy?

In previous work (Roy 2011), I have argued that the transnational worlds of fast policy must be countered with a critical transnationalism that at least partially rests on what Colin McFarlane (2006) calls 'learning'. Interested in the 'creative possibility of learning between different contexts', McFarlane (2006: 1413) calls for forms of learning that are both ethical and indirect: 'ethical because it transcends a liberal integration of subaltern knowledge, and indirect because it transcends a rationalist tendency to limit learning to direct knowledge transfer between places perceived as "similar."' While McFarlane reads such forms of learning as an interrogation of the 'North-South divide', I also read it as a call for new temporalities of knowledge production, those shaped by slow learning rather than by fast policy. That slow register of time is perhaps most appropriately designated as 'ethnographic'. By the ethnographic, I mean deep immersion in the making of meaning. But following Tsing (2005: 4), I also specifically mean the capacity to be attentive to the 'friction' associated with the circulation of universal ideas: 'the awkward, unequal, unstable, and creative qualities of interconnection across difference'. Such friction is what I think Healey means by contingency and complexity as she seeks to question the travelling capacity of planning as a 'universal good technology'.

But in this final section of this chapter, I want to put forward the provocation that the types of ethical and indirect learning advanced by McFarlane are grounded in a distinctive type of community of inquiry: poor people's movements. McFarlane's (2011) own research underlying his important treatise, *Learning the City*, is concerned with slumdweller federations and the types of 'incremental urbanism' (32) that they are able to craft. For McFarlane (2011: 1) such forms of urban inhabitation are not only methodologies of survival but also what he terms a 'spatial grammar of urban learning'. I draw attention to two lessons from McFarlane's conceptualization of learning. First, poor people's movements must be understood as communities of inquiry. This is perhaps an obvious analytical point but I emphasize it because too often planning thought and practice ignore such communities of inquiry and their particular forms of expertise. Especially important about such communities of inquiry is that they coincide with communities of risk, a phrase I carry over from Nancy Fraser's deliberations of justice. Second, the spatial grammar of urban learning foregrounded by McFarlane involves a distinctive spatio-temporality, one that exists in relation to but also in contrast with the worlds of fast policy and urban SWAT teams. In his seminal article on the urban politics of the global poor, anthropologist Arjun Appadurai (2002: 30) provides a clue to such a temporality when he describes the everyday lives of slum dwellers as the 'tyranny of emergency'. In contrast, he designates the work of slum federations and more broadly poor people's movements as the 'politics of patience' (Appadurai 2002: 27). When the transnationality of planning is understood as the work of globally interconnected poor people's movements, when these transnational communities of inquiry coincide with transnational communities of risk, then the grounds of universality are quite different than those asserted by the travels of planning as a universal good technology.

Healey's insistent critique of planning as a universal good technology requires a reconsideration not only of the grounds of universality but also of the idea of the good city itself. In particular, her conceptualization of planning as a 'general concept contingently evolving through the experience and debates we engage in as a "community of inquirers"' (Healey 2011: 188) suggests a relationship between the epistemology of generalization, the ontology of a deliberative community, and the ethics of planning. This is perhaps most evident in her recent work on democracy where she has explored 'the potential for progressive governance practices which could help to sustain and promote more people-centred, open-minded, pluralist polities' (Healey 2012: 20). She acknowledges that such a participatory polity 'will always be "incomplete" and emergent', 'a direction to be struggled for' rather than a truth to be asserted (Healey 2012; 35). Once again, such conceptualizations of political community can be greatly enriched by close attention to the discourses and practices of poor people's movements, particularly their struggles to enact participation. This is evident in Appadurai's (2002: 22) concept of 'deep democracy' which foregrounds 'visions of emancipation and equity' put forward by 'grassroots political movements' and that are often 'at odds with the nationalist imagination' Most significant, according to Appadurai (2002: 24–5), are 'efforts to reconstitute citizenship in cities' that deploy 'new horizontal modes for articulating the deep democratic politics of the locality'. Appadurai (2007) reformulates such formations as the 'politics of hope'. In turn, he aligns the politics of hope with the 'politics of participation' (Appadurai 2007: 30). With a focus on mass politics and the empowerment of the poor, Appadurai (2007: 32) signals a 'great ethical shift' that is concerned with 'global poverty', i.e. 'human suffering and misery'. But it is 'participation' that remains central to Appadurai's politics of hope:

> Put in the sharpest terms, the new currency of democratic politics, stimulated by the emergence of poverty as a primary and measurable social ill and by the worldwide growth in mass agitational politics, makes participation the path to capacity rather than the reverse ... This reversal places a new value on the politics of hope, since it promises that mass participation in democratic politics can provide a more direct route to economic equality than the path of gradually improved qualifications for citizenship. (Appadurai 2007, 30–31)

Appadurai (2007:33) once again locates participation in the 'galactic explosion of civil society movements' that have 'brought into being a whole new range of practices that allow poor people to exercise their imaginations for participation'.

I am arguing then that the idea of the 'planning community' and the (incomplete and emergent) ideal of a 'participatory polity' have to be located in an unequal world. Such a location inevitably raises the question of what participation means in the context of extreme and persistent poverty. The question is even more vexing when we realize that the urban poor constitute the urban majority in the cities of the world where the global urban future is most rapidly unfolding. I am drawn to Appadurai's work on poor people's movements and their role in shaping what McFarlane has termed the 'spatial grammar of urban learning' because it appropriately broadens concepts of participation, planning, and political community.

It is important to note that I do not intend for the urban poor to be understood as the bearers of unique forms of solidarity and learning. Indeed, the practices of economic solidarity of the urban poor that Appadurai describes – slum-dwelling, microcredit – are saturated with the movement of capital. They are the emergent frontiers of 'poverty capitalism', where transactions are thoroughly commodified and where practices of exploitation and accumulation are prominent. In earlier work, I have argued that such economies are instances of 'neoliberal populism', a valorization of the people's economy that facilitates new circuits of market rule (Roy 2010). Similarly, formations of deep democracy can also be 'infrastructures of populist mediation' (Roy 2008). Here, the 'horizontal modes' of political organizing celebrated by Appadurai consolidate new forms of sovereign power and inaugurate new forms of governmentality that shape how political subjects imagine themselves.

It is perhaps more useful then to consider poor people's movements as producing templates and grammars of participation and learning. This is political community understood not as a stable and coherent ontology but rather as incomplete and emergent, congealing in solidarity only through struggle and mobilization. This is what Ashwin Desai (2002) depicts in his poetic and provocative narrative of how shackdwellers in Durban came to define themselves, across race lines, as 'poors'. As Desai (2002: 44) notes, identities were rethought in the context of struggle. It became possible, at those moments, to constitute a political community: 'We are the Poors'. I am arguing that such formations of political community must also be understood as planning communities or as communities of inquiry and that to do so makes possible an understanding of the sticky engagements of contingent universals.

These debates indicate that there is much more work to be done in planning thought and practice on the question of communities of inquiry and their relationship to planning. If, following Healey, planning is understood not as a universal good technology but instead as a general concept contingently evolving through the experiences and debates of such communities of inquiry, then there is even more urgency to such a task. Drawing on emerging strands of postcolonial theory, I have argued that a community of inquiry cannot be conceptualized as a 'geocommon'. Instead, as suggested by the theorists of planetarity, it is vital to consider the radical alterity through which global humanity is constituted. How then can a community of inquiry be asserted? Building on Fraser's philosophy of scales of justice, and inspired by poor people's movements, I have argued that communities of inquiry must be seen as coinciding with communities of risk. Instead of generic notions of participation, this type of social ontology bears the potential of what McFarlane has suggested are new spatial grammars of urban learning. In an era dominated by fast policy and SWAT-style global mayoralties, such forms of learning might just be the 'critical antennae and sensibilities' that Healey (2011: 188) insists upon to navigate and transform a world of 'travelling ideas'.

References

Amin, A. 2004. 'Regions Unbound: Towards a new politics of place' *Geografiska Annaler* 86B: 33–44.

Appadurai, A. 2002. 'Deep Democracy: Urban governmentality and the horizon of politics' *Public Culture* 14(1): 21–47.

Appadurai, A. 2007. 'Hope and Democracy' *Public Culture* 19(1): 29–34.

Barbaro, M. 2013. 'Bloomberg Focuses on Rest (as in Rest of the World' *New York Times,* December 14 http://www.nytimes.com/2013/12/15/nyregion/bloomberg-focuses-on-rest-as-in-rest-of-world.html?_r=0 (accessed December 14, 2013).

Brenner, N. and C. Schmid. 2014. 'The "Urban Age" in Question' *International Journal of Urban and Regional Research* 38(3): 731–55.

Chakrabarty, D. 2009. 'The Climate of History: Four theses' *Critical Inquiry* 35, 197–222.

Desai, A. 2002. *We are the Poors: Community struggles in post-apartheid South Africa.* New York: Monthly Review Press.

Fraser, N. 2009. *Scales of Justice: Reimagining political space in a globalizing World.* New York: Columbia University Press.

Graham, S. and P. Healey. 1999. 'Relational concepts of space and place: Issues for planning theory and practice' *European Planning Studies*, 7(5): 623–46.

Healey, P. 1997. 'City Fathers, Mandarins, and neighbours: Crossing old divides in new partnerships' in Kaltorp et al. (eds) *Cities in Transformation: Transformation in cities*. Aldershot: Avebury, pp. 266–88.

Healey, P. 2010. 'The universal and the contingent: Some reflections on the transnational flow of planning ideas and practices' Paper prepared for the 24th AESOP conference, Finland, July.

Healey, P. 2011. 'The universal and the contingent: Some reflections on the transnational flow of planning ideas and practices' *Planning Theory* 11: 188–207.

Healey, P. (2012) 'Re-enchanting democracy as a mode of governance' *Critical Policy Studies*. 6(1): 19–39.

Healey, P. 2013. 'Circuits of knowledge and techniques: The transnational flow of planning ideas and practices' *International Journal of Urban and Regional Research* 37(5): 1510–26.

Hillier, J. 2009. 'Assemblages of Justice: The "Ghost Ships" of Graythorp' *International Journal of Urban and Regional Research* 33(3): 640–61.

Jazeel, T. 2011. 'Spatializing difference beyond cosmopolitanism: Rethinking planetary futures' *Theory Culture Society* 28: 75–97.

Massey, D. 2007. *World City*. Cambridge: Polity Press.

McFarlane, C. 2006. 'Crossing borders: Development, learning and the North–South divide' *Third World Quarterly* 27(8): 1413–37.

McFarlane, C. 2011. *Learning the City: Knowledge and translocal assemblage*. Oxford: Wiley-Blackwell.

Merrifield, A. 2012. 'The urban question under planetary urbanization' *International Journal of Urban and Regional Research* 37(3): 909–22.

Myers, S.L. and N. Kulish. 2013. 'Growing clamor about inequalities of climate crisis' *New York Times*, November 17 http://www.nytimes.com/2013/11/17/world/growing-clamor-about-inequities-of-climate-crisis.html?_r=0 (accessed on November 17, 2013).

Peck, J. and N. Theodore, forthcoming. 'Paying for good behavior: Cash transfer policies in the wild' in A. Roy and E. Shaw Crane (eds) *Territories of Poverty*. Athens: University of Georgia Press.

Peck, J. and N. Theodore. 2010. 'Recombinant workfare, across the Americas: Transnationalizing "Fast" social policy' *Geoforum* 41: 195–208.

Popke, J. 2007. 'Geography and ethics: Spaces of cosmopolitan responsibility' *Progress in Human Geography* 31(4): 509–18.

Radhakrishnan, R. 2005. 'Globality is not worldliness' *Gamma: Journal of Theory and Criticism* 13: 183–98.

Readfern, G. 2013. 'How rich countries dodged the climate blame game in Warsaw' *The Guardian*, November 25 http://www.theguardian.com/environment/planet-oz/2013/nov/25/climate-change-warsaw-rich-countries-blame-paris-deal (accessed November 27 2013).

Robinson, J. 2002. 'Global and world cities: A view from off the map' *International Journal of Urban and Regional Research* 26(3): 531–54.

Roy, A. 2008. 'Civic governmentality: The politics of inclusion in Mumbai and Beirut' *Antipode* 41(1): 159–79.

Roy, A. 2010. *Poverty Capital: Microfinance and the making of millennial development*. New York: Routledge.

Roy, A. 2011. 'Placing planning in the World: Transnationalism as practice and critique' *Journal of Planning Education and Research* XX(X): 1–10.

Roy, A. and A. Ong. 2011. *Worlding Cities: Asian experiments and the art of being global*. Chichester: Wiley-Blackwell.

Simone, A. 2004. 'People as infrastructure: Intersecting fragments in Johannesburg' *Public Culture* 16(3): 407–29.

Spivak, G. 2003. *Death of a Discipline*. New York: Columbia University Press.

Spivak, G.C. 2006. 'World systems and the Creole' *Narrative* 14(1): 102–12.

Tsing, A. 2004. *Friction: An ethnography of global connection*, Princeton, NJ: Princeton University Press.

Watson, V. 2006. 'Deep difference: Diversity, planning, and ethics' *Planning Theory* 5(1): 31–50.

Chapter 18

Making Places Better: Institutional Design for Local Plan-Making

Geoff Vigar

Introduction

This chapter focuses on the significance of institutions in making places better: how do different institutional configurations affect what is possible or what is achieved in places, and how do we analyse them? The significance of institutions as a focus of study and at times an explicit sociological institutional approach to such questions run through much of Patsy Healey's work. I then review some of this work and look at a specific example of institutional design for very local plan-making, neighbourhood planning, in the English context. The chapter concludes by noting the value of an institutionalist approach to designing plan-making processes.

Institutions and Institutionalism

Institutionalist approaches to studying politics and public policy remerged in the 1980s. This 'second wave' of institutionalist accounts was posited as having emerged with three strands – historical, rational choice and sociological (Hall and Taylor 1996). Patsy Healey's work around this time was informed by and was also part of an emerging sociological institutionalism.

Institutionalism appealed as an analytical tool as it had close links to Giddens' structuration theory, frequently cited as a direct inspiration by Healey. Giddens saw institutions as systems of meaning which play a vital role in structuring the agency of actors in specific contexts (Giddens, 1984). Institutions are not organisations, although they can be, but rather are forms of social organisation (Lowndes and Pratchett 2013), typically associated with established norms and rules of conduct. There is much debate on exactly what forms of social organisation should be defined as significant but for planning research institutional analysis provides most critically a way of highlighting how norms and routines, in organisations and through professional acculturation for instance, are important in understanding how planning is enacted: outcomes cannot be explained by sole reference to the laws, power of interests, individuals and organisations for example. There is a tendency among institutionalists to interchange, and fail to define, terms such as habits, rules, conventions, norms, customs, practices, routines, procedures etc., which frustrates some critics (Fleetwood 2008). Indeed to borrow Wildavsky's (1973) phrase if institutions are potentially everything then maybe institutionalism is nothing? For these reasons perhaps Healey has tended to describe later work as relational in its focus and to talk more about interpretive approaches (e.g. Healey 2006, 2012). Nevertheless the institutionalist contribution is arguably epistemological, providing a way of conceptualising what is important, of what to look for, when engaging with research on planning practice (Vigar et al. 2000).

We can see how this focus on the significance of forms of social organisation in shaping outcomes would resonate with Healey's early work as the 1979 paper reproduced in Part III.2 highlights. Here Healey critiques influential work of the time, especially that derived from operational research (Friend et al. 1976). The nature of the critique is fascinating in that such research is taken to task for being 'power blind': the irony being supplied by later critiques of her own work, to the workings of capital in particular, but also the motivations of planners (e.g. Tewdwr Jones and Allmendinger 1998). Healey's critique of Friend et al. rested more on the failure in their research to account for the structuring power of institutions. Power is thus reinserted through recognition of the constraints on actors afforded by such institutions. The value of such a focus is powerfully present in the 1997 paper where a principal contribution is to show the feebleness of well-intended initiatives that fail to recognise and build wider structural foundations to change the context for action. In failing to attend to institutional design, which in this case might address wider structural conditions, leads the initiative studied to have limited impact. We see this expressed as a need for wider network and relation-building here too, a focus that can be found in other papers reproduced in this volume.

An institutionalist approach to planning research thus enabled both formal and informal rules and practices to be considered. It also allowed for a consideration of how such rules and practices embody the values of society more widely, the significance of individual property rights in the UK context for example. The sociological institutionalist variant developed by Healey and others cohered into an empirical focus on: identifying the stakeholders relating to a particular issue, who is present and who is not and what are the networks, both formal and informal that bind them together; what are the arenas and fora where policy is discussed and how does their construction affect who deliberates and how; and, what storylines and discourses are mobilised to do policy work in such places. There were elements adopted here from work in other fields. The work of Maarten Hajer (1989, 1995) proved significant as did a general focus in the 1990s on ideas of policy networks (e.g. Marsh and Rhodes 1992), and learning through such networks (e.g. Sabatier 1993).

Taking each of the three elements of a sociological institutionalist account in turn, we see in many of the chapters in this section how such an approach emerged. First, there is a strong interest present throughout on identifying the range of actors in play, and particularly those with a stake who are not present or have very limited presence, and the significance of the communities and networks that might bind them together. In the 1979 paper we see the emergence of ideas of stakeholder mapping, a prominent feature of later work, alongside the significance of planners acting as 'nodes' between networks of such stakeholders. Emphasis is also present here of the role of planners as co-ordinators, and of planning work as being in part about co-ordinating between policy fields, as well as contributing specific expertise. Here policy communities were expressed as 'interagency networks' (p.58), reflecting a greater dominance at the time of the formal machinery of government and bureaucracies, although Healey notes the 'increasing interpenetration' of public and private sectors. An enduring feature of Healey's work is also present here as place is identified as a key way to achieve coordination, in part due to its significance as a site of meaning to citizens.

Second, we can see the emerging focus on the institutional sites where planning work is conducted and debated, and the distinction between such sites centred on the formal rules deployed within them, leading to the idea of arenas, fora and courts (Bryson and Crosby 1992). The empirical work underpinning the 1997 paper shows the value of such a focus. In particular it demonstrates how the construction of deliberative fora can influence the work done in them and who garners authority and power. In describing the presence of disadvantaged residents from a deprived neighbourhood being thrown into partnership arrangements with members of the business community and policy-makers Healey is able to call upon an empathic, ethnographic sensibility, '*what to some looked like a partnership of equals because they sat round a table together, for others seems like patriarchal patronage or just plain terrifying*' (Healey 1997:270). In addition, the quote highlights the attention to the micropolitics of such encounters which remains a central feature of her work.

Thirdly, analysing policy discourses is the central thrust of the 2004 paper where three examples show the path dependent nature of planning and how the presence of long-standing institutions with histories of innovation shape how space is conceived and how place might be shaped. But Healey observes that this too can be a trap and sometimes a shift in institutional context is needed to shift the discourse. What it takes to do so, and thus achieve transformative change, has become a central feature of the so-called third generation of institutionalist analysis (Lowndes and Roberts 2013) and remains a key challenge for those seeking to shift planning practices. Indeed following the limited success in implementing changes to the hard institutional infrastructure of planning law and policy in the UK in the 2000s, the need for attention to the soft infrastructure of practices, norms and routines was alighted upon by government. This became known as the need for 'culture change' (Shaw and Lord 2007).

Discourse was also strongly associated, both in Healey's work and elsewhere across the policy sciences, with ideas of policy and social learning in the 1990s. Her 1992 paper explicitly raises the question of how far the configuration of institutions and deliberate attempts at institutional design might facilitate or hinder learning. Many of the emerging contextual issues that framed such questioning, such as the marketisation of planning through trends such as outsourcing, and the widespread assault on the idea of public sector-led regulation as being capable of making places better, have accelerated since these writings. There is a great deal of research to be done on how the changes in such contexts continue to affect the ability of planning to make places better when the space for learning, among public sector employed actors in particular in the UK and elsewhere, are dispersed to the private sector or lost in a focus on managerialism (Clifford and Tewdwr Jones 2013; Gunn and Vigar 2012). I now turn to speculate on these possibilities before looking at how the framework of a sociological institutionalist approach might be used to analyse a plan-making initiative, very much framed by these new conditions, and very different in its institutional design as a result of them.

Institutional Design Work in the Context of Marketisation and Austerity

Much of Healey's work, including the papers in this section, is highly prescient in identifying emerging trends in society, planning practice and theory. Healey notes in a number of papers in the 1970s and 1980s of the increasing amount of planning work done in the private sector. By the 1997 article we also see how the idea of business being at the heart of shaping planning interventions was well-established. Ideas of collaborative planning embodied in the 1992 article are not specific about who might be in control of a communicative planning exercise, however how collaborative planning exercises are mobilised and by who remains a critical question. I examine two issues below that have particular traction in the UK in the early 2010s: the outsourcing of planning activity; and localism and the role of civil society in plan-making.

Planning in an Age of Marketisation and Austerity

The 1979 paper shows the significance of actors in formal government, local and central. The populating of the governance landscape with quasi non-governmental organisations and the opening up of governance processes to put business in particular into a steering role are some way in the distance here. The increasing role of private sector actors doing planning work is noted, but it remains at the margins of UK planning practice. This situation still holds in many nations where a strong local state steers and enacts planning. Here institutional design will have a different focus than in places where local government is relatively emasculated, one actor among many.

The 1992 paper is written after an assault on planning and the public sector under successive UK governments led by Margaret Thatcher. Given the growing marketisation of planning activity, communicative rationality is proposed as a way forward for planning given the narrow emphasis on economic evaluation in this emergently dominant tradition, and conceptual problems in four others identified in the Healey paper. Such economic pressure gathered momentum in the UK and elsewhere, often bracketed as a roll out of neo-liberalism. Within such an overarching discourse, the UK version privileged business voice as before, but also sought to put private sector practices at the heart of public sector work in two ways. First, managerialism, influenced by ideas of New Public Management. While slow to penetrate planning, by the 2000s targets and steering mechanisms imposed by central government were strongly influencing what was done and how it was done in local planning departments in the UK (Gunn and Vigar 2012). The second process was one of outsourcing planning work, often whole local authority functions and departments, to the private sector. How collaborative planning might be exercised, where open-ended dialogue is encouraged, yet where the institutional context is shaped by discourses of neo-liberalism and practices which focus on getting things done, typically through contracts of one kind or another, isn't clear from reading literature. So while in theoretical terms a critique was readily mobilised, the power of neo-liberal discourse and its associated discourse coalition affected the ability of 'planners' to mobilise processes embedded in ideas of communicative rationality. Indeed arguably the scientific rationalism that provides the basis for the design of the bureaucracies that linger on today is better suited to deliver the pared-down practices envisaged by the neo-liberals.[1] So while the 1992 paper focuses on the need for planners to mobilise alternative argumentation [p.159],

1 Bevir and Rhodes (2010) suggest that the state as currently composed is a nineteenth-century bureaucracy with a series of governance mechanisms oriented around knowledge creation, market regulation and so on, grafted into an increasingly incoherent and unsuccessful muddle.

little appears to have changed in the past 20 years to suggest this is happening, with a modus operandi of 'decide-announce-defend' very common. Indeed Healey's 2004 paper shows how little these ideas have changed the practices and institutions of contemporary planning in many instances with plans in particular hostages to short term political expediency.

Into this domain in the UK, and mostly in England, came in 2010 a Coalition government using the window of opportunity associated with the Recession of 2008 onward to implement wide ranging cuts in local government. While this was spatially patchy with areas controlled by the Labour Party unfairly hit, in such areas particularly the search was on to do things very differently.

The context for such change was prefaced in the decade previously. As the 2000s progressed, all UK political parties were looking to civil society to take up an enhanced role in delivering public services. For the Left this meant turning toward its cooperative roots focused on ideas of mutuality, rather than the state as solution. For the Right, mobilising civil society to bypass local government and its armies of 'untrusted welfare-bureaucrats' provided its possible solution. The result was that many policy areas such as education became a contract between an individual service delivery point i.e. a school, and central government, with little role for the local state. In planning, central government used the Localism Act 2011 to mobilise new tools, notably Neighbourhood Plans, for civil society to use which largely bypass the local state. The problem in such places was in dealing with the lack of capacities available often to do such work or contribute to debates. In this sense Healey's 1997 paper should be required reading for any politician thinking that this is an easy solution. Here Healey demonstrates how hard it is for civil society to take on roles formerly done by a professionalised, experienced public sector; and how people living difficult lives in hard circumstances will be systematically disadvantaged if there isn't support to build institutions in such places. An interesting point to make here is that Healey has been spending part of her retirement mobilising to do some of the bridging and knowledge-building work associated with such initiatives [http://www.theguardian.com/local-government-network/2013/jan/08/working-lives-local-professor]. Latterly indeed Healey has been working on a Neighbourhood Plan for the place where she lives. What then can thinking institutionally offer to such an exercise?

'Local plans for local people': Institutional Design for very Local Planning

Neighbourhood planning has a long and rich history. However, historically planning at this scale, with the exception of some (local authority-led) Action Area Plans and (community-led) Village/Town Design Statements has had little statutory force. Non-statutory plans led by citizens, such as parish plans, have however achieved a great deal in community development terms and also occasionally on planning issues. The new Neighbourhood Plans in England are different in that statutory authority is granted to a community-led plan. What is different in process terms is that it is indeed intended to be community-led. This is not about 'participation' in an initiative led by the state (although it is of course *central* state-enabled); rather the local state plays a supporting role, advising where called upon dependent somewhat on its interpretation of a 'duty to support', and it is up to citizens to construct the processes that may be assessed as being more or less democratic.

The demotion of local government's role in shaping local planning agendas and processes also renders some of the planning literature on collaboration and participation – with its assumption of a 'planner' at the heart of the process – only tangentially useful (Watson 2009). While the aim of much participation literature is for some form of co-production, with collaborators including the public, the public sector employed planner typically sits at the centre in a coordinating role. Neighbourhood plan-making offers a chance to test Healey's ideas of the significance of place as a mobilising tool for the community and to see what 'nodes' might develop new networks focused on neighbourhood plan-making.

Neighbourhood Development Plans (NDPs), as they are formally known, can be seen as part of two of the UK Coalition Government's central discourses, 'Localism 'and the 'Big Society'. They also represent a wider concern to devolve power to communities, expressed rhetorically at least in UK planning since the inception of the 1947 system. There is broad political agreement about the desirability of communities having a greater influence over what happens to the places where they live, work and play, mirroring the emphases in Healey's work. Indeed Healey's work was cited in the 2009 open source white paper on planning with a specific section on 'collaborative planning', that paved the way for the Plans (Conservative Party 2009). There are concerns, however, in relation to the specific forms that processes of NDP preparation might take – the institutional design. How can the sociological institutionalist tools of stakeholders, networks, arenas and discourses shed light on this new mechanism?

In relation to stakeholders the main concern is over who comes forward to make such plans and relatedly the potential dominance of various forms of self-interest. With regard to self-interest, two forms are often talked of: first, the potential for corporate takeover. Here NDPs can provide apertures for big capital to exploit local communities' powers to achieve corporate ends, through potentially funding and contributing to delivery of a neighbourhood plan. This appears to be largely unrealised to date but the potential for it to occur should be noted. More positively this fits with greater direct engagement between developers and civil society that seems to be an emergent feature in the UK, typically at pre-application stage. This feature results from the support given in national guidance through the National Planning Policy Framework but also as many private sector planners come to see the benefits of working in this way (LGA 2014).

The second area of self-interest lies in potentially narrow sectional interests in communities' taking control. Looking at who is participating in neighbourhood plan-making reveals a mixed picture. Wealthier neighbourhoods were more likely to initiate a neighbourhood plan (Vigar et al. 2011). Looking in more detail the activists in such places also tended to be the 'usual suspects' although a wider mix of people was engaged in the broader networks contributing to plan-making (Vigar 2013).

But a further question arises as to who has the chance to participate. Looking at the arenas for plan-making provide a window on to this. Many neighbourhood plan-making exercises have proved to be very open, with extensive use of online and offline mechanisms to inform wider and engage wider communities. That said the composition of the arenas – pin up boards in draughty village halls for example – has done little to widen participation in some instances (Vigar 2013). But more broadly, the mechanism of neighbourhood plans provide the arenas for debate and decision that Healey is so keen to emphasise, where issues of place futures can be discussed and agreement reached as to ways forward. The collaborative process can lead to creative solutions and a commitment to implementation shared among diverse stakeholders, although there is also evidence that community members avoid more agonistic practices that may lead to more inclusive and/or creative solutions (Vigar et al. 2011).

And what of the storylines and discourses that emerge from such exercises? The fear with neighbourhood plans was that giving communities more power would result in them using such powers to resist development, both of contentious Locally Unwanted Land Uses (LULUs) such as travellers' sites, homeless hostels, waste recycling plants etc. and of development such as housing in general. For this reason the 2011 Localism Act was clear that NDPs should be permissive – it would be difficult to use them to resist LULUs which should be considered at local plan level. And NDPs can only allocate the same amount or more housing than is contained in the wider local plan. This may be seen as 'necessary' framing to get houses built, but it runs counter to the idea of NDPs having a genuine devolved authority which many suggest is needed, or should be expected, in order to encourage citizens to participate (e.g. Mouffe 2000). That said, the framing is based on a likelihood that communities may indeed opt for less not more development, running counter to a widespread consensus (at a regional rather than very local level) of the need for more housing, at least in the Greater South East of England. So far there are examples of neighbourhood plans proposing the same or more housing than allocated in local plans. Typically the distribution of the housing varies from the local plan determined by local government. Neighbourhoods appear to want to disperse it more widely, sharing the 'burden' of housing across a neighbourhood. This redistribution does make it harder for developers and regulators to provide and pay for infrastructure and accommodate the increased transaction costs associated with many rather than few sites. Resistance from planners in public and private roles may arise for these reasons alone.

At the time of writing over 1000 NDPs had been launched and over 30 adopted. The adopted ones cover very different population sizes from around 1000 to over 10,000. Even with a significant increase in the rate of adoption only a small proportion of the population of England will be covered. So, are they irrelevant? Many professional planners are sceptical but where NDPs have been implemented the effects on daily life, on community capital, on relation-building, as well as some 'better' planning outcomes – where local knowledge has shaped policies and actions with greater community buy-in and arguably with a richer, citizen-informed evidence base – can be identified (Vigar 2013). The preparation process might divide some communities but it has developed capacity and goodwill in others. NDPs also look rather different from the perspective of the citizens that develop them than from the professional planning community i.e. the process for citizens might solve a small, but to them significant problem, or they may get to know their neighbours better with possible later consequences. Looking at the storylines of some of the handful of plans adopted so far suggests they are broadly progressive, for example recognising and addressing problems such as the need for more affordable housing. They can thus fit with a desire for institutions that mediate between the harsh winds of globalisation and their impacts locally on issues such as affordable housing, better local high streets and more local jobs. They could be havens in a marketised planning world where austerity prevents the public sector from conducting such fine-grained community level work.

Working at this smaller scale means they may have a greater chance of creating institutions that make people feel valued, an important point made in Healey's 1997 paper and a subsequent focus on 'respect' in governance work (e.g. Healey 2012). Resurrecting the idea of institutions, particularly civic ones, being there to make people feel cared for seems a long way from the objectives of the new public management theorists and many of the other public policy initiatives of recent times (see Hoggett et al. 2008). In this way localism and austerity may, for some at least, lead to something better and rather unexpected. For communities such as those in Healey's 1997 paper the outcomes are likely to be much less positive, and this is highly significant.

Capturing these benefits more widely requires resources, both financial and human, going into NDP, or more likely other, 'localist' processes. It is clear that poorer neighbourhoods are much less likely to engage in NDP than wealthier ones (Vigar 2013). Whether this is due to the lack of capacities noted in Healey's 1997 paper or the irrelevance of a land use plan in many such areas is unclear. But greater justice in NDP could be achieved by a more proactive local state and/or through skilled intermediaries. In Bristol for example poorer neighbourhoods get more resources, such as the time of local authority planners, for community-led initiatives including local planning (Cowie et al. 2014).

A critical institutional design issue here in generating, rather than undermining, community capital is the presence of intermediaries with good facilitation skills. As Healey's 1992 paper notes, changing the systems is typically much easier than changing the historically-embedded practices of the system operators i.e. the planners (p.158). Awareness of the potential of community engagement is central to this need. Planners on the ground have been rather hostile to NDPs. Such hostility may be a professional defensiveness reflecting the loss of power that NDPs imply. Conversations with planners reveals more pragmatic reasons, with the value of such plans questioned given the high costs of developing them due to their statutory nature compared to the small scale of the issues existing in some smaller places and the process costs associated with turning 'citizens into planners'.[2] But some planners can be ignorant of the potential of deeper community engagement (Ploger 2004). Citizens are seen as annoying 'consultees' who create problems in the midst of rambling processes of plan production. The attitudes and directions of local politicians may also make engagement exercises futile. Some planners are also bruised through encounters within poorly designed community engagement processes, as well as clinging on to a perceived superiority of technical-rational knowledge (Campbell and Marshall 2002). Planners' tend to prioritise their own values and expertise over 'lay' knowledge. The idea of shifting from a techno-bureaucratic mind-set and disposition towards a more techno-democratic one (de Suza Santos 1998 cited in Krantz 2003) centred on collaborative rationality, from a democratic elitist model to a participatory democratic one, is thus far from prominent in English planning practice. One hope is that NDPs accelerate it a more participatory democratic ethos into being and indeed some planners have enjoyed engaging in such very local plan-making activity, but this seems far from a mainstream view (Vigar et al. 2011).

So the intermediaries may come from beyond the typical local authority and they may not be planners. While the focus on local authority support for communities is centred on technical planning knowledge, a good facilitator may be more crucial. Planners often do not have the requisite skills or mind-sets (of planning *with* communities, not *for* them) to do this work well. Flair and creativity in engagement is essential, as are mediation skills. The package of such skills and the drive to deploy them is rarely found in local planning departments and may not be well taught in planning education. This requires bureaucrats (and planners) to change, both their outlook and culture, but also their *institutions* from within.

To sum up, NDPs are popular with a Government keen on reducing resources granted to the local state because it gets citizens to do the work. But the virtual withdrawal of the state is problematic for two reasons:

2 I am grateful to Jules Brown for this idea.

first, capacity-rich neighbourhoods are much more likely to come forward (Vigar 2013); and second, as the discussion above suggests, the state, in the form of the expert resources that lie in its workers, is useful as is its 'distance' from the highly personal planning issues attendant at this micro-scale. Such collaborative and technical resources are likely to be more needed in areas less likely to come forward for NP thus creating a double-bind in relation to matters of social justice. The dependence of English local government on central government for resources, and the general climate of public spending cuts explain this to a large degree.

Conclusions

To conclude, Healey's papers in this Part are a fascinating and diverse collection that serves to remind us of a number of important things in conducting planning research. They highlight the importance of a rich, qualitative, often ethnographic examination of planning work. They provide a historical account and trajectory of place governance work and provide many lessons with contemporary relevance. They show that some issues continue to recur; but also that the way governance is enacted has changed considerably. The language of planning as a public sector activity in the 1979 paper mutates to the more diffuse power world of the 1990s in the two papers from the 1990s. By the 2004 paper we see a public sector trying to recapture planning, through enabling and coordinating roles, but with planners struggling to both conceptualise their contemporary world, and develop the power to make things happen, to make places better.

A focus on institutions, particularly on the social relations, the micropolitics, of what is going on in planning work is vital to the understanding of these broad shifts. Attention in practice and research, to the institutional contexts planners and others find themselves can lead reflexive practitioners to make a real difference. But Patsy's work also shows us that we need to know the broad context for action, the systems in play, and the historically-situated institutions that shape the context for actors, interests and organisations to do their work. While the focus of empirical work on planning practice might change, Patsy has given us a great platform on which to build theoretically and empirically informed research that can make places better.

References

Bevir, M. and Rhodes, R.A.W. 2010, *The State as Cultural Practice*, Oxford: Oxford University Press, 2010.
Bryson, B.C. and Crosby, J.M. 1992, *Leadership for the Common Good*, San Francisco: Jossey Bass.
Campbell, H. and Marshall, R. (2002), Values and professional identities in planning practice. In P. Allmendinger and M. Tewdwr-Jones (eds) *Planning Futures. New directions for planning theory*, London: Routledge, 65–92.
Clifford, B., Tewdwr-Jones, M. 2013, *The Collaborating Planner*, Bristol: Policy Press.
Conservative Party, 2009, *Open Source Planning*, London: Conservative Party.
Cowie, P., Healey, P. and Vigar, G., 2014, *Great Planning Successes*, London: Royal Town Planning Institute.
Fischer, F., 2009, *Reframing Public Policy: Discursive politics and deliberative practices*, Oxford: OUP.
Fleetwood, S. (2008) Institutions and social structures, *Journal for the Theory of Social Behaviour*, 38(3): 241–65.
Friend, J.K., Power, J.M. and Yewlett, C.J.L. 1974, *Public Planning: The intercorporate dimension*, London: Tavistock.
Giddens, A., 1984, *The Constitution of Society. Outline of the theory of structuration*, Cambridge: Polity Press.
Gunn, S. and Vigar, G. 2012, Reform processes and discretionary acting space in English planning practice, 1997–2010. *Town Planning Review*, 83(5): 533–52.
Hajer, M. 1995, *The politics of environmental discourse*, Oxford, OUP.
Hajer, M. 1989, *City Politics: Hegemonic projects and discourse*, Aldershot: Avebury.
Hall, P., and Taylor, R. 1996, Political science and three new institutionalisms, *Political studies*, 44: 936–57.
Healey, P. 2012, Re-enchanting democracy as a mode of governance. *Critical Policy Studies*, 6(1): 19–39.

Healey, P., 2006, *Urban Complexity and Spatial Strategies: Towards a relational planning for our times*, Abingdon: Routledge.
Hoggett, P., Mayo, M. and Miller, C. (2008) *The Dilemmas of Development Work: Ethical challenges in regeneration*, Bristol: Policy Press.
Krantz, R.S. 2003, Cycles of reform in Porto Alegre and Madison. In A. Fung and E. Olin Wright (eds) *Deepening Democracy*, London: Verso.
Local Government Association, LGA, 2014, 10 commitments for effective pre-application engagement, London: LGA.
Lowndes, V. and Roberts, M. 2013, *Why Institutions Matter*, Basingstoke: Palgrave.
March, J.G. and Olsen, J.P. 1989, *Rediscovering Institutions*, New York: Free Press.
Marsh, D. and Rhodes, R.A.W. 1992, *Policy Networks in British Government*, Oxford: Oxford University Press.
Mouffe, C. 2000, *The Democratic Paradox*, London: Verso.
Petts, J. and Brooks, C. 2006, Expert conceptualisations of the role of lay knowledge in environmental decision-making: Challenges for deliberative democracy, *Environment and planning A*, 38(6): 1045.
Ploger, J. 2004, Strife: Urban planning and agonism, *Planning Theory*, 3(1): 71–92.
Sabatier, P.A. and Jenkins Smith, H. 1993, *Policy Change and Learning: An advocacy coalition approach*, Boulder, CO: Westview Press.
Shaw, D. and Lord, A. 2007, The cultural turn? Culture change and what it means for spatial planning in England, *Planning Practice and Research*, 22(1): 63–78.
Tewdwr-Jones, M. and Allmendinger, P. 1998, Deconstructing communicative rationality: A critique of Habermasian collaborative planning, *Environment and Planning A*, 30(11): 1975–89.
Vigar, G. 2013, Promoting fairer neighbourhood planning. In K. Shaw and J. Blackie (eds) *New Directions in Planning: Beyond localism*, Newcastle: Northumbria University Press.
Vigar, G., Brooks, E. and Gunn, S. 2011, The innovative potential of neighbourhood plan-making, *Town and Country Planning*, July/August.
Vigar, G., Healey, P., Hull, A.D. and Davoudi, S. 2000, *Planning, Governance and Spatial Strategy-Making in Britain: An institutionalist analysis*, Basingstoke: Macmillan.
Watson V. 2009, Seeing from the South: Refocusing urban planning on the globe's central urban issues, *Urban Studies*, 46(11): 2259–75.
Wildavsky, A. 1973, If planning is everything then maybe it is nothing, *Policy Sciences* 4: 127–53.

Chapter 19

From British City Centre to British Columbia's Central Coast: The Transferability of the Institutional Capacity Development Framework

Janice Barry

Introduction

Once referred to as a 'bi-continental Euro-American enterprise' (Friedmann 1998: 245), planning theory is now very transnational in scope, with planning researchers and practitioners actively seeking to import and adapt new ideas and new ways of approaching complex problems (Healey 2012, reproduced in Part IV). As Healey argues (2013; 2012), this exploratory and translatory nature of planning is not new. Planning history is littered with examples of attempts to recreate innovative policy approaches and modes of urban and regional development, from the second century attempts to export Roman ideas about urban form to more contemporary attempts to apply Brazil's participatory budgeting approach – to name just two examples cited in Healey's articles. Yet, specific models of urban and policy development are not the only ideas that 'travel' within the field of planning. Planning scholars are engaged in their own 'mining expeditions into the universe of knowledge' (Friedmann 2008: 255), on the lookout for concepts that will increase understanding of both the normative basis and 'realpolitik' (Flyvberg 2002) of contemporary planning.

These conceptual explorers' 'specific contribution to theory is to return from these expeditions to home base and translate their discoveries into the language of planning where they will either take root or be unceremoniously forgotten' (Friedmann 2008: 255). Although Friedmann was thinking about intellectual expeditions into the theoretical canons and emergent bodies of literature in other allied disciplines, his 'expedition' analogy works equally well within the planning field. It also applies to instances where planning scholars are drawing on conceptual models and analytical frameworks developed in response to cases and contexts elsewhere. What Friedmann neglected to mention is that the danger of such theoretical expeditions is not simply that some ideas will fail to 'take root', but that some ideas may reveal themselves to be entirely inappropriate tools for understanding the particularities of the new context. Pushing this analogy of transplanting a conceptual 'sapling' one step further – is there not also a risk that some ideas may become invasive species: captivating and imported for a specific desire or purpose, but also contain the potential to obscure if not completely choke out the unique characteristics of the new context?

This chapter uses my own research on the evolution and subsequent institutionalization of a new planning arrangement between Indigenous peoples and the Government of British Columbia (Canada) to reflect on travelling ideas in planning theory. More specifically, it looks at the transferability of Healey and her colleagues' Institutional Capacity Development (ICD) Framework (de Magalhães et al. 2002; Healey 2007; Healey, 2006; Healey 1998; Healey et al. 2003; Healey et al. 1999): a set of ideas about how collaborative planning can contribute to broader processes of institutional change. Although the framework grew out of the study of city centre regeneration in one of Britain's larger and rapidly changing post-industrial cities (Newcastle-upon-Tyne), I used it to interrogate Indigenous-state relations during the development of the Central Coast Land and Resource Management Plan (CCLRMP). On one level, this case bears little similarity to Newcastle in that there was absolutely no overlap in the substantive planning issues. The British Columbia (BC) case was not about city centre regeneration, but rather the ecological sustainability of one of the province's most isolated and least populated regions. Furthermore, the BC planning context is profoundly shaped by the presence of distinct Indigenous rights: a governance characteristic for which there is no equivalent in the British system. Yet, like Newcastle, it was shaped by a wide range of governance actors exerting different forms of power and influence over a planning process. In both cases, lasting and potentially transformative change was the result of these actors' ability to mobilize different kinds of institutional resources. Given these broad similarities, the ICD Frameworks' way of conceiving and analytically interrogating institutional change was a foundational concept in my research.

Now several years on from the completion of this work, I wish to critically reflect on the appropriateness and potential limitations of transferring the ICD Framework to such a wildly different context. My reflections are developed with reference to Healey's own work on the ways in which ideas travel amongst planning scholars and practitioners (2013; 2012). In her 2012 paper, Healey calls attention to how the rich narratives of situated learning and pragmatic responses to context-specific factors tend to get left behind when ideas travel from one locale to another. This omission is seen to open the door for distorted and potentially damaging applications of travelling ideas and practices: a danger that equally applies to planning theorists and practitioners. In this chapter, I will use my own attempts to apply the ICD Framework to draw attention to the converse problem of what factors and experiences are not adequately picked up when a travelling idea lands in a new locale. For as it will be shown, the ICD Framework was a useful tool for understanding the ways in Indigenous and provincial officials were able to generate new relational networks, new forms of knowledge and new organizational structures to solidify their emergent planning relationship. However, it was ill equipped to handle some of the unique characteristics of planning with Indigenous peoples. This limitation forces critical reflection on the possibility of stretching a conceptual framework beyond its natural limits: an issue that will be returned to at the end of the chapter.

The chapter begins by outlining the core components of the ICD Framework and some of the key features of the empirical context to which it was developed in response. It then looks at how the framework has travelled to new locales, including BC's Central Coast. The presentation of my own work on Indigenous peoples' involvement in the CCLRMP begins with a discussion of the evolution of this planning process and concludes with some reflections on how it can be read as an example of institutional capacity development. I then turn to aspects of the case that do not sit easily within the ICD framework and that prompted one rather inelegant and potentially inappropriate alteration of the original framework. My challenge to apply the ICD framework to such a radically different planning context is then used as fodder for a more general discussion of how planning theory travels from one context to another.

The Institutional Capacity Development Framework

The ICD Framework was developed to enhance understanding of the potentially transformative power of a more deliberative approach to spatial planning that promotes greater coordination amongst different governance actors. As such, the framework emerges out of Healey's longstanding interests in the communicative nature of planning (1992, 1997b), the coordinating role of social and policy networks, (1979), and the crossing of organizational and cultural divides through alliances and partnerships (1997a) – key examples of which are reproduced in this Part. The ICD Framework also speaks to Healey's interests in the development of a relational conception of strategic spatial planning (2004b, reproduced in this Part): one that calls attention to the intersection and possible co-existence of multiple conceptions of space and place and that explores how divergent perspectives can be brought together within a specific planning 'episode'. At its core, the framework (Figure 19.1) is a 'conceptual device' intended to help tease apart the strategic importance of 'three elements of social interaction' (de Magalhães et al. 2002: 54). These elements are: knowledge resources (K), relational resources (R), and mobilization capacity (M), which are conceptualized as follows:

> The first [element] is the flow of knowledge of various kinds between stakeholders in a locality, and the learning processes that take place as knowledge is exchanged. These are the *knowledge resources*, the frames of reference, creativity and knowledgeability, the concepts of place and identity relevant to governance. The second is the nature, reach and quality of relational networks brought into the governance process by the stakeholders. These constitute the *relational resources*, the resources of trust and co-operation contained in such networks, the nature of the bonding elements in them and their reach. The third element is the ability of stakeholders and their networks to draw resources, rules and ideas into the effort of collective action. This is the *mobilization capacity*,

> the capacity of stakeholders to mobilize knowledge and relational resources to act collectively at the level of the city/region/neighbourhood for some common goal (de Magalhães et al. 2002: 54).

The framework posits that collaborative processes have the potential to alter and enhance the capacities that exist within the wider governance system. It is through the reconfiguration of existing resources, along with the development of new forms of knowledge, new relationships, and new ways of mobilizing those resources that the governance system gains the institutional capacity (IC) needed to 'nurture and sustain new forms of collective action' (Gualini 2002: 35).

Yet, as Healey (2007) also notes, collaborative processes need to be understood as but one level in a nested hierarchy of governance interactions. These levels range from the episodic convening of diverse governance actors in defined decision-making arena, to the formation of policy agendas and rules, to the articulation of broader cultural norms and values. Meaningful and sustainable change is said to occur when all three of these levels shift in a similar direction (Healey 2007) and when there is a 'coming together of various dynamics to create a moment of opportunity larger than the sum of the parts' (Healey 2004a: 99). These moments of opportunity are at least partially the result of external forces, which shape a specific collaborative 'episode' (Healey 2007) and provide the necessary 'shock' for the development of new institutional capacities. Institutional capacity development is, therefore, a product convergence of external forces and internal evolutions. 'In the language of sociological institutionalism, [institutional capacity development] expresses the complex interactions between 'structuring' driving forces and the active work of agency in inventing ways of going on' (Healey et al. 1999: 124).

In this way the framework also extends Healey's earlier work on the 'hard and soft infrastructures' of collaborative planning (1997b) by exploring not only the structural and interactional conditions that make a more collaborative approach possible, but also the ways collaborative processes can contribute to broader processes of institutional change. Given its interests in structure/agency dynamics, the ICD Framework also sits within and shares many commonalities with other new institutional perspectives in planning theory (see, for example: Verma 2007, as well as the analytical framework developed by Van den Broeck 2011; Van den Broeck et al. 2013).

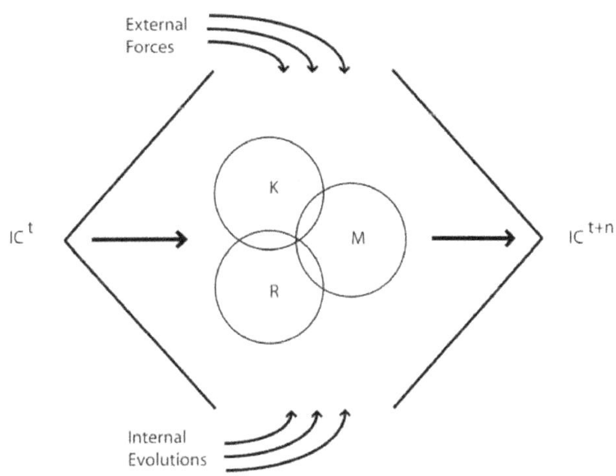

Figure 19.1 Institutional capacity development and the three elements of social interaction – knowledge resources (K), relational resources (R), and mobilization capacity (M) (Re-drawn from: Healey et al. 2003, p.64)

As a Conceptual Response to the Governance of City Centre Regeneration in Newcastle, UK

Returning to the question of the transferability of ideas and the importance of attending to the unique contextual factors that led to their development, what is the Institutional Capacity Development Framework's 'origin narrative' (Healey 2012: 195)? What is the story of how it 'came to life' (Healey 2012: 195)? Admittedly, Newcastle-upon-Tyne is not a city I am familiar with – or, indeed, one that I have ever visited. But Healey and her colleagues' work on the ICD Framework (de Magalhães et al. 2002; Healey 2007; Healey 1998; Healey et al. 2003; Healey et al. 1999) does contain numerous nuggets of contextual information that allow for the reconstruction of the origins of this way of conceptualizing the processes of institutional change.

First and foremost, urban regeneration in Newcastle was seen an example of multi-level governance in that the development of the entire metropolitan area is informed by the actions of the constituent metropolitan districts (of which the City of Newcastle-upon-Tyne is one) and various regional agencies. Various forms of local government were involved in and/or exercised responsibilities over a variety of functions, including education, social service delivery and planning. Central government also played a significant role by not only creating the over-arching policy and planning framework, but also by providing funds to the metropolitan districts. Yet, as Healey and her colleagues also noted when first developing the framework, regional bodies were playing an increasing role, particularly in the area of economic development, and were being used to channel funding (Healey 2006, reproduced in Part IV): just one example of a 'complex and cluttered' formal governance landscape with 'many areas of overlapping responsibility' (Healey 2006: 308). According to Healey's account, the informal governance landscape was no less complex. Within the City of Newcastle itself, there were ongoing efforts to widen stakeholder involvement, though often with a focus on major business groupings (Healey 2006). Yet, this landscape was also described as changing, with new networks and arenas arising both within and beyond the official regeneration process allowing different stakeholders to build new linkages to one another and to the formal centres of government power (Healey et al. 2003). The ICD Framework was, therefore, an attempt to make sense of this changing governance landscape. It was a way to begin to explore how a collaborative urban regeneration initiative might trigger shits in alliances, agendas and political influence, potentially opening up new spaces for a more deliberative form of planning and urban policy development.

Alternate Uses: A Brief Overview of the Existing Literature

Despite the frequent citation (and seemingly wide appeal) of Healey's work on collaborative planning (1997a), the Institutional Capacity Development Framework does not have the same prominence in the literature. The articles that first present the framework are frequently cited and can be connected to other socio-institutional approaches to the study of spatial planning (see, for example: Van den Broeck 2011; Van den Broeck et al. 2013). However, there are still only a handful of authors that have used the ICD Framework to structure empirical analyses. That said, the various elements and sub-components of the ICD Framework have been found to be a useful analytical tool in a wide range of planning contexts, from British regional waste planning (Davoudi and Evans 2005), Swedish sustainability planning (Khakee 2002), Portuguese spatial planning (Breda-Vazquez and Oliveria 2008), Finnish forestry planning (Saarikoski et al. 2012a) and Dutch energy planning (Breukers and Wolsink 2007). The ICD Framework's treatment of mobilization capacity has also informed an investigation of Belgian agricultural planning (Van Herzele et al. 2011). Despite the geographic and substantive differences that exist between these various planning contexts, all seem to be united by a much more universal phenomenon. To one degree or another, all the cases speak to the need to find new capacities for responding to complex governance situations in which knowledge, trust and influence are dispersed amongst a wide range of actors operating in multiple arenas and across multiple scales of policy development.

As will be explored, my own research on Indigenous peoples' experience of natural resource planning in BC's Central Coast is situated in an equally complex governance landscape. Different layers and forms of government (including Indigenous governments) exert different kinds of influence over a planning process, acting amongst an increasingly wide array of stakeholders. Non-governmental and corporate actors were also amassing and making strategic use of a wide range of existing political, social and intellectual resources to make their voices heard and to advance the development of new modes of forestry management. The development of new ways of addressing Indigenous peoples' distinct rights and title became of particular importance to many of these actors, not the least of which were the Indigenous governments themselves. The next section elaborates on the development of the Central Coast Land and Resource Management Plan and its role in the formation of a new 'government-to-government' approach to working with Indigenous peoples.

Government-To-Government Planning for BC's Central Coast

Over the past 25 years, the Province of British Columbia has built up a considerable amount of experience with the development of more collaborative forms of spatial governance. These developments have predominantly occurred within the context of large-scale natural resource planning. Constitutionally, the planning and management of public lands and resources is an area of provincial jurisdiction. But in a province like BC, which has a long history of intense conflicts between different user groups, these decision-making processes have been opened up to a wide variety of actors. Since the early 1990s, BC has pursued a 'shared decision-making approach' (Owen 1998) to regional and watershed-level planning. The provincial government creates the overarching policies and principles, as well as specific planning scenarios used to generate discussion within multi-stakeholder forums. These forums would develop consensus-based recommendations on appropriate land use zones and resource management strategies (BC IRPC 1993; O'Riordan 2005). Provincial planners would then transform those recommendations into a comprehensive planning document, which would be brought to the provincial cabinet for approval.

Despite its initial framing as an exercise in multi-stakeholder decision-making, BC strategic planning has been confronted with an additional set of issues related to increased recognition of Indigenous rights and title. Unlike other parts of Canada, treaties were generally not signed in BC and it was settled on the assumption that the introduction of European law and systems of property ownership simply extinguished Indigenous rights and title (Harris 2002; Tennant 1990). This assumption has begun to unravel through a series of court cases, including the highly influential 1997 *Delgamuukw* decision. The cases not only asserted that Indigenous rights and title had not been extinguished, but that both the federal and provincial governments have an ongoing duty to engage Indigenous peoples in land use decision-making. BC natural resource planning has been particularly impacted by these rulings, which triggered the introduction of a 'government-to-government' (G2G) approach to working with Indigenous peoples (Barry 2011). G2G planning is based on the recognition of Indigenous peoples as unique orders of government. In BC, it is now seen to include joint process delivery and funding provisions to support Indigenous research, travel and skill development (BC ILMB 2009). It also tends to include bilateral negotiations on the draft land use plan (just prior to cabinet submission), with the results of these negotiations either becoming or providing a strong basis for the final land use plan. Therefore, although the Province formally maintains final decision-making authority, G2G planning could accurately be described as culminating in a de facto bilateral approval process.

Although the approach is of relevance and has been subsequently applied to other natural resource planning processes, this chapter reflects on a single case study: the Central Coast Land and Resource Management Plan (CCLRMP). Located approximately halfway up Canada's western coastline, the 4.6 million hectare forest management region became the subject of one of BC's longest running and most complex natural resource planning processes. It was through this process that the idea of G2G planning evolved from a vague policy commitment to a concrete governance framework (Barry 2012; 2011). As will be discussed, the ICD Framework was used to identify the external forces and internal evolutions that contributed to the development and mobilization of the knowledge and relational resources needed to sustain this newfound collaboration between Indigenous and provincial officials.

Institutional Capacity Deficits and Initiating the CCLRMP

'Known for a large concentration of unlogged watersheds and the presence of the genetically distinct white spirit bear' (Barry 2012: 215), the Central Coast had been a long-standing area of interest for wilderness campaigners. The area also had a long history of forest harvesting, though the industry was under a great deal of stress due to declining stocks, changing government policy and market volatility (Holman and Terry

2001). From a government perspective, this potential for conflict underscored the importance and potential benefits of a more collaborative approach. Yet, key stakeholders were reluctant to participate. In fact, BC's major environmental groups disengaged from the process entirely, unable to accept the lack of a moratorium on logging activities while land use zones and management prescriptions were being developed (Saarikoski et al. 2012b). Approximately one year into the CCLRMP process, they took their discontent one step further by calling for an international boycott of coastal forest products. Many of the affected Indigenous groups also opted out of the process. Those that did participate described their involvement as being 'under duress' (CCLRMP Completion Table 2001), since past experiences had taught them that major decisions regarding their traditional territories were going to be made, with or without them. Yet, unlike some of the other actors, Indigenous peoples' reluctance to engage was not based on substantive concerns alone; they were equally, if not more, concerned with the governance arrangement and seeming lack of recognition of Indigenous rights and title. This lack of recognition was not unique to the CCLRMP process, but rather a systemic failure of early attempts at strategic natural resource planning.

By the time the CCLRMP was initiated in 1996, the BC government had approximately five years of experience with large-scale collaborative processes but had yet to secure any meaningful Indigenous involvement (Barry 2011). Many of the policy documents and discussion papers that guided these processes made reference to the need for a government-to-government relationship, but there was a great deal of uncertainty as to what that relationship would entail and how much of it fell within the remit of natural resource planning. The BC and federal governments had already launched an extensive process of modern treaty-making and there was real hesitancy to confuse these negotiations by attempting to address issues of Indigenous governance within a more narrowly defined planning process (Barry 2011). Indigenous groups did not generally support this separation (Commission on Resources and Environment 1993) and, as a result, their participation in these early planning processes was either completely non-existent or extremely intermittent (Barry 2011).

Indigenous groups that elected to engage in the CCLRMP arguably fared better, in that were invited to participate in the multi-stakeholder forum, as well as the province's own inter-agency caucuses and planning teams. The initial terms of reference also contained clear recognition of Indigenous groups' 'distinct standing as a separate government' (BC 1997a) and a commitment to ensure that their land use interests were taken in account through additional opportunities to review the final plan (BC 1997b). At the same time Indigenous representatives report racist comments from some of the stakeholders and, as a result, felt a constant need to defend their rights and title during the first 18 months of the CCLRMP process. Though nowhere near as blatant, provincial planners also expressed a certain amount of resentment towards the Indigenous representatives involved in the process. For example, one internal document describes the first few years of the process as having been 'regularly bogged down with rights and title issues, government policy issues, trust issues and rhetoric' (BC LUCO n.d.: 2). As I have argued elsewhere (Barry 2012; 2011), these tensions can be read as an institutional capacity problem. The Indigenous and provincial governments not built sufficient relational resources and lacked a shared frame of reference. As suggested by the framework, overcoming these deficiencies required sufficient external pressure, but it was also predicated on the strategic actions of the provincial and Indigenous representatives most involved in the day-to-day work of the CCLRMP process.

External Changes and the Development of New Institutional Capacities

New legal decisions on the nature and scope of provincial responsibilities towards Indigenous rights and title provided the first and arguably most obvious external force. However, the shift away from the defensive strategy initially adopted by participating Indigenous groups was not so much a product of the formal legal requirements arising out of these decisions, but rather how the stakeholders perceived and responded to them. At the request of Indigenous representatives, provincial officials organized an extensive training session on the *Delgamuukw* decision, which reaffirmed the Indigenous participants' role as designated representatives of distinct orders of government that possess distinct political and territorial rights. These sessions allowed the stakeholders to begin to see Indigenous rights as less of a threat and more as another set of 'variables in the area that have to be managed' within the planning process (personal interview, 25 March 2009). In the language of institutional capacity development, one of the key frames of reference used in planning process (a component of knowledge resources) had undergone a tremendous shift, opening up space for additional processes of institutional capacity development.

The shift in the actors' perception of the planning objectives, as well as each other's overall roles and political identities helped open the door for new networks and relationships. No longer feeling the need to constantly defend their rights and title, the Indigenous representatives were able to start 'learning the business' of the different stakeholders (personal interview, 25 March 2009). More importantly, they were able to make strategic use of this new information when engaged in bilateral negotiations with the Province. They were beginning to formulate solutions that not only protected their rights and title, but also spoke other actors' interests. For example, it allowed them to re-define the idea of 'certainty', which was a key concern of many of the stakeholders and the forest industry in particular. As they successfully argued, a comprehensive land use plan with rigid zones and management directions was not the only mechanism for ensuring the economic and ecological certainty of the region; certainly could also be achieved through the creation of a clear and comprehensive governance structure. In sum, and as argued elsewhere (Barry 2012), legal changes provided a catalyst for altering the discourses and frames of reference used in the planning process, which led, in turn, to the development of new relational resources. These relationships not only facilitated the transfer of strategic information, but the intensity of the bilateral negotiations helped further solidify Indigenous groups' political identity as a collaborating government.

The actions and political fallout from the boycott of coastal timber products provided the second major 'shock' to the CCLRMP process. Unable to ignore numerous cancelled contracts and a tarnished public image, six major timber companies took the unprecedented step of entering into private discussions with four large environmental groups. Working under the banner of the 'Joint Solutions Project', these organizations sough to negotiate an end to the boycott by formulating possible forest management options (Clapp 2004; Smith and Sterritt 2007) – all completely outside of the official CCLRMP process. Not wanting to have decisions being made outside of the main planning forum, the Province began to use its legal responsibilities towards Indigenous people as a way to re-orient attention back to official process. Given that only the Province could exercise the duty to consult Indigenous peoples, all proposals would need to be filtered through its multi-stakeholder forum.

Indigenous groups that were willing to continue working with the Province in a close collaborative relationship were rewarded with additional funding to conduct their own research and to augment their skills and training. The development of new knowledge resources amongst the participating Indigenous groups had knock-on effects in terms of their relationships with provincial planners, who began to see them more as colleagues who understood their field of practice and who they could speak to on a professional level. The participating Indigenous groups were able to use their deepening relationship with the provincial planners to access higher orders of government and to build strong relationships with elected officials as well. Again, these changes can be interpreted as examples of institutional capacity development (Barry 2012), in that the financial resources that precipitated out the perceived threat of the Joint Solutions Project enabled the development of new knowledge resources amongst the Indigenous participants as well as more complex and multi-layered relations with the provincial government. With these relationships in place, Indigenous and provincial government officials were able to more towards the development of an entirely new government-to-government arrangement for plan implementation, in which both parties would share the responsibility for overseeing the long-term development of the region.

Reflections on the Transferability of the ICD Framework

Overall, there are several aspects of the CCLRMP process and its role in the development of a G2G approach that sit easily within Healey and her colleagues' view of institutional capacity development. Traditional power structures were challenged, resulting in the creation of new relational resources that could be mobilized to support the creative alteration of existing knowledge resources surrounding both the processes and outcomes of planning. More specifically, key legal decisions helped facilitate the shift towards a more proactive style of engagement, which ultimately led to a re-framing of the over-arching goal of the CCLRMP and the emergence of a shared policy agenda. Over time, the desired outcome shifted from the creation of a clear and certain land use plan to a clear and certain governance arrangement. The idea that external shifts provide the catalyst for a whole series of internal evolutions is further supported by the analysis of the second major shift in the internal dynamics of the CCLRMP process. The perceived threat of the ENGO-industry coalition triggered an identifiable shift in the depth and types of relationships between the Indigenous and provincial actors, which allowed the parties to build rapport and to transfer substantive and strategic information. Perhaps most importantly, it got the parties to the point where they were willing to commit to a long-term, more open-ended planning and policy-development framework.

Yet, as I have argued elsewhere (Barry 2012; 2011), Healey and her colleagues' original formulation of the ICD Framework does not adequately address the political resources that are evoked and play a powerful role in shaping the governance dynamics of a complex collaborative process such as the CCLRMP. Following Sorensen and Torfing, these political resources were seen to include an actor's 'level of access … to decision-making processes (endowment); their capacity to make a difference in these processes (empowerment); and their perception of themselves as political actors (political identity)' (2003: 610) – all of which had a major impact on the development of new planning relations between the provincial and Indigenous governments. For example, legal changes such as the precedent-setting *Delgamuukw* decision formally endowed the Indigenous governments with additional levels of access to the decision-making process, while the provincial desire for a closer relationship with Indigenous governments precipitated further empowerment through increased funding opportunities and an increased role in information and process management. Both these factors helped solidify Indigenous governments' changing political identity from a particular kind of 'stakeholder' (a label Indigenous groups have long rejected) to a 'long-term governance partner' (Barry 2012: 227). The ICD framework was subsequently altered to account for these more overtly political factors. 'Political resources' were added as a third type of institutional resource (along with knowledge and relational resources) that can be strategically mobilized by the different actors involved in a collaborative process (see: Barry 2012; 2011).

Other aspects of the CCLRMP process proved less easy to accommodate through alterations to the original ICD framework. Unlike the Newcastle case, where the multiple levels of government have arisen out of a broadly similar history and cultural worldview, the provincial and Indigenous governments involved in the CCLRMP cannot be assumed to possess a broadly similar understanding of spatial planning. The small but growing body of literature on Indigenous planning has already made note of the Eurocentric underpinnings of contemporary planning and has drawn attention to the fact that Indigenous peoples often plan from an entirely different epistemological and ontological foundation (e.g. Jojola 2008; Matunga 2012; Porter 2010). Working from a broadly socio-institutional perspective (see: Healey 2007; 1997b), I saw these ways of knowing and being as a part of the informal 'institutions' that shape Indigenous action and decision-making in their traditional territories. I went on to suggest that collaborative planning between Indigenous peoples and the state therefore needed to be understood as a potential example of 'inter-institutional capacity development', given that there were multiple formal and informal institutional structures at play (Barry 2012; 2011). This admittedly cumbersome phrase provides some indication that the processes of institutional capacity development that I had identified during the CCLRMP were qualitatively and politically different than those observed in Newcastle-upon-Tyne. But, at the same time, my attempts to make only minor adjustments to the language and conceptual syntax of an existing way of understanding how institutions evolve and change also potentially has the unintended impact of obscuring one of the core and defining features of the reconciliation of Indigenous-state relations: the recognition of Indigenous sovereignty.

Although sovereignty is not often discussed in planning theory, it is absolutely central to the development of a more just relationship between Indigenous peoples and state-based planning authorities (Dorries 2012; Hibbard et al. 2008; Lane and Hibbard 2005; Porter 2004). Lane and Hibbard's 2005 work on Indigenous peoples' changing relationship to planning draws on political theory to offer the following definition of sovereignty. 'Sovereignty is fundamental to any people's independent existence. The concept of sovereignty has been given many meanings in political philosophy but at its core is the idea of autonomy, a people's capacity to guide and control their own fate' (2005: 173). Drawing on the work of American sociologist Stephen Cornell, Lane and Hibbard go on to connect this broad definition to the specific struggles of Indigenous peoples. Indigenous sovereignty is seen to include 'three interlocking matters: how to maintain or regain control over resources, especially land; maintain particular sets of social relations and more or less distinct cultural orders; and have some measure of political autonomy' (2005: 173).

Dorries (2012) develops the argument about the importance of attending to issues of sovereignty even further by calling attention to Indigenous peoples' political, cultural and historical understanding of the term. Indigenous sovereignty is seen as a call for the restoration of Indigenous independence, but also to explore their interdependence with settler governments. In Canada, the two-row wampum provides an often cited image of Indigenous sovereignty. The two-row wampum is a beaded belt with two rows of purple separated by rows of white that has long been used as a symbol for the respect and recognition of Indigenous sovereignty. The conventional interpretation of the two-row wampum is that the purple rows represent the vessels of the settler and Indigenous governments as they travel side-by-side down the river of existence (the white beads). The fact that the purple rows never meet is often seen to suggest that the two cultures should not interfere with each other's journey, while the placement of the two vessels on the same river is seen to underscore the importance of peace, trust and friendship (Borrows 2000; McGregor 2004; Peters 2003). While not refuting the conventional reading of the two-row wampum, Borrows' interpretation (2000) places greater emphasis on the shared waters of the vessels, noting that Indigenous sovereignty also needs to include the development of mechanisms for coexistence, conflict resolution and joint strategy-making in the river of shared existence.

This discussion of importance of attending to deep and fundamental questions about Indigenous sovereignty raises serious questions about the appropriateness of using a British theory to study the particular challenges and opportunities that may arise when Indigenous peoples and a state-based planning agency work towards the development of a long-term governance arrangement. Putting aside the very important issue of how British planning approaches and methods have been used to dispossess Indigenous peoples of their land and livelihood (see: Porter 2010), contemporary British planning practice has simply not had to confront the co-existence of sovereign nations, each exerting distinct forms of political authority over the same territory. Pushing the argument one step further, there are those that would assert that collaborative planning – and, by extension the ICD Framework – is an insufficient, if not politically dangerous, tool for the study of Indigenous-state planning relations (Dorries 2012; Porter 2010; Porter 2004). For as Porter (2004) forcefully asserts, Indigenous-state planning relations are not about 'inclusion' but about the co-existence of sovereign nations and, therefore, cannot be interpreted accorded to the normative standards inherent in theories of collaborative planning.

At the same time, Indigenous sovereignty is not well recognized in the Canadian context and many Indigenous peoples are forced to work within legal and political conventions established by the nation-state (Barry and Porter 2012). Perhaps then, my application of the ICD Framework might be best understood as an analytical tool, as opposed to a normative one. It offers a way of unpacking how Indigenous governments negotiate their way around and strategically mobilize institutional resources to pursue their own political goals and aspirations in an environment that continually attempts to limit the expression of their political sovereignty. For as the analysis of the CCLRMP suggests, the processes of institutional capacity development were messy, conflict-ridden and often bore little resemblance to the collaborative ideals of public deliberation, mutual learning and collective action (Innes and Booher 2010, Forester 1999, Healey 1997b). In fact, it often bore a much closer resemblance to the agonistic view of planning processes advocated by Pløger (2004) and Hillier (2003). Yet, the ICD framework was still helpful in terms of unpacking the external forces and internal evolutions that led to a different kind of relationship between Indigenous and provincial officials – perhaps not a sovereign relationship, but still a significant step forward.

Conclusions

So what, then, can be learned from this analytically useful yet somewhat normatively fraught application of the Institutional Capacity Development Framework? In many respects, the lessons that might be drawn from this case further underscore Healey's own arguments about the potentially damaging consequences of transplanting planning ideas from one location to another. Healey's article stresses the importance of ensuring that a travelling idea does not leave home without its baggage (so to speak): the 'origin narrative' that 'identifies the situation in which it came to life, and describes its particular history' (Healey 2012: 195), as well as some of the transformations it has undergone in its various journeys. I, on the other hand, would stress the importance of a different kind of narrative. Planning theorists have an equal (if not greater) responsibility to call attention to the elements that could not be neatly squeezed into the suitcase. For me, Indigenous sovereignty was one of those elements: a historical, political and cultural concept that has very particular implications in the Canadian context and that may demand a complete re-thinking of the appropriateness of applying a conceptual framework developed elsewhere.

Following my own advice, what kind of story would I tell about transferring the ICD framework to the study of Indigenous-state relations? How would I describe my intentions, challenges and omissions? My intention was to find an effective way of explaining the rather dramatic shift from an almost complete lack of recognition of Indigenous rights and title to the formation of a government-to-government approach to regional land use planning. As an analytical tool, the ICD Framework served my purposes well, drawing attention to the catalytic influence of external factors and critical importance of the highly strategic actions of individual actors and groups of actors working within the CCLRMP process. Yet, the collaborative norms that underlay and inform the ICD Framework also presented the unintended risk of obscuring, underplaying or misrepresenting Indigenous peoples' unique and sovereign authority. Although I would not go so far as to say that this 'conceptual sapling' (imported from Britain) became invasive by choking out key elements of the Canadian planning context, it did require some careful tending. While far from complete, my own process of making room for the concept of Indigenous sovereignty required letting go of the ICD Framework's normative baggage. I also needed to create space for a much more agonistic view of the processes of institutional capacity development that were occurring as Indigenous groups struggled to assert their sovereignty. In terms of the underlying moral of this story, these reflections point to possibility of disentangling the analytical power of a conceptual tool (such of as the ICD Framework) from the normative principles that surrounded its creation. But, they also speak to the importance of being explicitly clear about what aspects of a conceptual framework you are packing up for travel and what you are intentionally leaving behind to make room for the unique characteristics you anticipate encountering along the way.

Acknowledgements

I would like to thank Patsy Healey, who unwittingly provided the impetus for this chapter by challenging me in my PhD thesis defence to reflect on the appropriateness and potential limitations of transferring conceptual ideas developed in a radically different planning context.

References

Barry, J. (2011) Building Collaborative Institutions for Government-to-Government Planning: The Nanwakolas Council's Involvement in Central Coast Land and Resource Management Planning. PhD Dissertation. University of British Columbia.

Barry, J. (2012) Indigenous-state planning as inter-institutional capacity development: The evolution of 'government-to-government' relations in coastal British Columbia, Canada. *Planning Theory and Practice* 13(2): 213–31.

Barry, J. and Porter, L. (2012) Indigenous recognition in state-based planning systems: Understanding textual mediation in the contact zone. *Planning Theory* 11: 170–87.

Borrows, J. (2000) 'Landed' Citizenship: Narratives of Aboriginal Political Participation. In: W. Kymlicka and W. Norman (eds) *Citizenship in Diverse Societies*. New York: Oxford University Press, pp. 326–44.

Breda-Vázquez, I. and Oliveira, C. (2008) Coalition-building in Portuguese spatial planning: Is there a Southern European context? *European Planning Studies* 16(6): 761–84.

Breukers, S. and Wolsink, M. (2007). Wind energy policies in the Netherlands: Institutional capacity-building for ecological modernisation. *Environmental Politics*, 16(1): 92–112.

British Columbia [BC] (1997a) *Central Coast Land and Coastal Resource Management Plan: Process Structure and Membership – October 30, 1997*. Province of British Columbia.

—— (1997b) *Central Coast Land and Coastal Resource Management Plan: Workplan – October 1, 1997*. Province of British Columbia.

BC Integrated Land Management Bureau [ILMB] (2009) *First Nations Engagement throughout the SLRP Planning Phases*. http://www.ilmb.gov.bc.ca/slrpfnengagement/fnandslrp.html [accessed 5 June 2009].

BC Integrated Resource Planning Committee [IRPC] (1993) *Land and Resource Management Planning: A Statement of Principles and Process*. http://ilmbwww.gov.bc.ca/lup/policies_guides/lrmp_policy/ stmt. htm [accessed 11 April 2007].

BC Land Use Coordination Office [LUCO] (n.d) *Central Coast LCRMP Chronology*. Internal government document provided by a research participant.

Central Coast Land and Resource Management Plan [CCLRMP] Completion Table (2001) *Minutes of the Meeting, Airport Inn, Port Hardy, December 6, 2001*. http://archive.ilmb.gov.bc.ca/slrp/lrmp/nanaimo/ cencoast/plan/reports/ccpasrpt/CompletionTableAdoptedMinutes_Dec_6_2001.pdf [accessed 5 August 2010].

Clapp, R. (2004) Wilderness ethics and political ecology: remapping the Great Bear Rainforest. *Political Geography* 23: 839–62.

Commission on Resources and Environment (1993) *Discussion Paper: Aboriginal Participation in Regional Planning Processes*. Victoria, BC: Commission on Resources and Environment.

Davoudi, S. and Evans, N. (2005) The challenge of governance in regional waste planning, *Environment and Planning C: Government and Policy* 23(4): 493–517.

Delgamuukw v. British Columbia, [1997] 3 SCR 1010. http://canlii.ca/t/1fqz8 [accessed 7 July 2010].

de Magalhães, C., Healey, P. and Madanipour, A. (2002) assessing institutional capacity for city centre regeneration: Newcastle's Grainger Town. In: G. Cars, P. Healey, A. Madanipour and C. de Magalhães (eds) *Urban Governance, Institutional Capacity and Social Milieux*. Aldershot: Ashgate, pp. 45–62.

Dorries, H. (2012) Rejecting the 'False Choice': Foregrounding indigenous sovereignty in planning theory and practice. PhD Dissertation. University of Toronto.

Flyvbjerg, B. (2002) Bringing power to planning research: One researcher's praxis story. *Journal of Planning Education and Research* 21: 353–66.

Forester, J. (1999) *The Deliberative Practitioner.* Cambridge, MA: MIT Press.

Friedmann, J. (1998) Planning theory revisited. *European Planning Studies* 6(3): 245–53.

Friedmann, J. (2008) The uses of planning theory: A bibliographic essay. *Journal of Planning Education and Research* 28: 247–58.

Gualini, E. (2002) Institutional capacity building as an issue of collective action and institutionalisation: Some theoretical remarks. In: G. Cars, P. Healey, A. Madanipour and C. de Magalhães (eds) *Urban Governance, Institutional Capacity and Social Milieux*. Aldershot: Ashgate, pp. 29–44.

Haida Nation v. British Columbia (Minister of Forests), [2004] 3 S.C.R. 511. http://www.canlii.org/ en/ca/scc/ doc/2004/2004scc73/ 2004scc73.html [accessed 24 November 2013].

Harris, C. (2002) *Making Native Space: Colonialism, resistance, and reserves in British Columbia*. Vancouver: UBC Press.

Healey, P. (1979) Networking as a normative principle with particular reference to local government and land use planning. *Local Government Studies* 5(1): 55–68.

Healey, P. (1992) Planning through debate: The communicative turn in planning theory. *Town Planning Review* 63(2): 143–59.

Healey, P. (1997a). City fathers, mandarins and neighbours: Crossing old divides in new partnerships. In: O. Kalltorp and I. Elander (eds) *Cities in Transformation: Transformation in cities*. Aldershot: Avebury.

Healey, P. (1997b) *Collaborative Planning: Shaping places in fragmented societies*. Basingstoke: MacMillan Press.

Healey, P. (1998) Building institutional capacity through collaborative approaches to urban planning, *Environment and Planning A* 30: 1531–1546.

Healey, P. (2004a) Creativity and urban governance. *Policy Studies* 25(2): 87–102.

Healey, P. (2004b) The treatment of space and planning in the new strategic planning in Europe. *International Journal of Urban and Regional Research* 28(1): 45–67.

Healey, P. (2006) Transforming governance: Challenges of institutional adaptation and a new politics of space. *European Planning Studies*, 14(3): 299–320.

Healey, P. (2007) The new institutionalism and the transformative goals of planning. In: N. Verma (ed.) *Institutions and Planning*. Rutgers, NJ: Center for Urban Policy Research, pp. 61–87.

Healey, P. (2012) The universal and the contingent: Some reflections on the transnational flow of planning ideas and practices. *Planning Theory* 11(2): 188–207.

Healey, P. (2013) Circuits of knowledge and techniques: The transnational flow of planning ideas and practices. *International Journal of Urban and Regional Research* 37(5): 1510–26.

Healey, P., de Magalhães, C. and Madanipour, A. (1999) Institutional capacity-building, urban planning and urban regeneration projects, *Futura* 18: 117–37.

Healey, P., de Magalhães, C., Madanipour, A. and Pendlebury, J. (2003) Place, identity and local politics: analysing initiatives in local governance. In: M.A. Hajer and H. Wagenaar (eds) *Deliberative Policy Analysis: Understanding governance in the network society.* Cambridge: Cambridge University Press, pp. 60–87

Hibbard, M., Lane, M. and Rasmussen, K. (2008) The split personality of planning: Indigenous peoples and planning for land and resource management. *Journal of Planning Literature* 23(2): 136–51.

Hillier, J. (2003) Agon'izing over consensus: Why Habermasian ideals cannot be 'real', *Planning Theory* 2: 37–59.

Holman, G. and Terry, E. (2001) *Central Coast Land and Coastal Resource Management Plan (LCRMP) Phase 1 'Framework Agreement' – Socio-Economic and Environmental Assessment: Final Report.* http://archive.ilmb.gov.bc.ca/slrp/lrmp/nanaimo/cencoast/docs/central_assessment_SEfinal.pdf [accessed 7 July 2010].

Innes, J. and Booher, D. (2010) *Planning with Complexity: An introduction to collaborative rationality for public policy*. New York: Routledge.

Jojola T. (2008) Indigenous planning: An emerging context, *Canadian Journal of Urban Research* 17 (Supplement): 37–47.

Khakee, A. (2002) Assessing institutional capacity building in a local agenda 21 process in Göteborg, *Planning Theory and Practice* 3(1): 53–68.

Lane, M. and Hibbard, M. (2005) Doing it for themselves: Transformative planning by indigenous peoples. *Journal of Planning Education & Research* 25(2): 172–84.

Matunga, H. (2013) Theorizing indigenous planning. In: R. Walker, T. Jojola and D. Natcher (eds) *Reclaiming Indigenous Planning.* Montreal: McGill-Queen's Press-MQUP, pp. 3–32.

McGregor, D. (2004) Traditional ecological knowledge and sustainable development: Towards coexistence. In: M. Blaser, H.A. Feit and G. McRae (eds) *In the Way of Development: Indigenous peoples life projects and globalization.* London: Zed Books, pp. 72–91.

O'Riordan, J. (2005) Protecting natural capital in British Columbia: The role of land and resource management planning. *Dialogues*. http://www.cwf.ca/V2/files/DialoguesWinter%202005.pdf [accessed 11 April 2007].

Owen, S. (1998). Shared decision-making: A case study. In: B. Wharf and B. MacKenzie (eds) *Connecting Policy to Practice in the Human Services.* Don Mills, ON: Oxford University Press, pp. 81–95.

Peters, J. (2003) Aboriginal perspectives on planning in Canada – Decolonizing the process: A discussion with four Aboriginal practitioners. *Plan Canada* 43: 39–41.

Pløger, J. (2004) Strife: Urban planning and agonism, *Planning Theory* 3: 71–92.

Porter, L. (2004) Unlearning one's privilege: Reflections of cross-cultural planning research with indigenous peoples in South-east Australia. *Planning Theory and Practice* 5: 104–8.

Porter, L. (2010) *Unlearning the Colonial Cultures of Planning*. Farnham: Ashgate.

Saarikoski, H., Åkerman, M. and Primmer, E. (2012a) The challenge of governance in regional forest planning: An analysis of participatory forest program processes in Finland. *Society & Natural Resources*, 25(7): 667–82.

Saarikoski, H., Raitio, K. and Barry, J. (2012b) Understanding 'successful' conflict resolution: Policy regime changes and new interactive arenas in the Great Bear Rainforest. *Land Use Policy* 32: 271–80.

Smith, M. and Sterritt, A. (2007) *From Conflict to Collaboration: The story of the Great Bear rainforest.* www.forestethics.org/article.php?id=1809 [accessed 4 August 2008].

Sorensen, E. and Torfing, J. (2003) Network politics, political capital, and democracy, *International Journal of Public Administration* 26: 609–34.

Tennant, P. (1990) *Aboriginal Peoples and Politics: The Indian Land Question in British Columbia, 1849–1989*. Vancouver: UBC Press.

Van den Broeck, P. (2011) Analysing social innovation through planning instruments: A strategic-relational approach. In: S. Oosterlynck, J. Van den Broeck, L. Albrechts, F. Moulaert and A. Verhetsel (eds) *Strategic Spatial Projects. Catalysts for change*. London and New York: Routledge, pp. 52–78.

Van den Broeck, P., Abdelwahab, M., Miciukiewicz, K., and Hillier, J. (2013) On analysing space from a strategic-relational institutionalist perspective: The cultural park for children in Cairo. *International Planning Studies* 18(3–4): 321–41.

Van Herzele, A., Dendoncker, N. and Acosta-Michlik, L. (2011) Mobilisation capacity for agri-environmental management, *Journal of Environmental Management* 92: 1023–32.

PART IV
Transformation Processes

Introduction to Part IV
Transformation Processes

Jean Hillier

'The association of planning with change is frequently stated as an article of faith by planners' (Healey, 1974: 176), often linked to the concern of planning practitioners and planning systems with future states: with 'shaping futures and promoting hope' (Hillier and Healey, 2008: xxiii). Patsy Healey (2010a: 45) distinguishes between processes and events of change and transformation. She appreciates that change is continuous in a place, organisation or institution. What may appear outwardly unchanging may actually be changing subtly (Chia and Tsoukas, 1999), such as tenants change in a rented house, planning staff leave a department or consultancy, others arrive, and so on. Places, organisations and institutions are thus constantly-evolving complexities with perpetually changing sets of relations (Hillier, 2007). Acts of organisation, such as planning practice, are acts of 'stabilising and simplifying what would otherwise be the irreducibly dynamic and complex character of lived experience' (Chia, 1988: 362), or, as Patsy Healey writes (2010a: 45), of 'shaping the routine flow of on-going changes to the built environment' (I add natural environment too). In her 1974 article, 'Planning and Change', Healey refers to the negentropic, structural stability-building, tendencies of planning frames (planes) of reference and practices. This reference reflects the systems theoretic concerns of the early 1970s planning theorists and prefigures today's concerns with complexity theories (see, for example, Read, 2012).

Healey explains that such shaping of change tends to be the fine-grained work of development assessment planning. In contrast, she regards transformation in the planning sphere as a deliberate activity, 'either through creating new or redeveloped parts of cities or landscapes, or by promoting a new mixture of economic, social, cultural and environmental dynamics in an over-arching strategic effort' (2010a: 45). Whilst the issue of transformational intent may be debatable, especially with regard to what Patsy Healey terms 'big events' (2015: 5), such as the 2008 global financial crisis, or Hurricane Katrina in the Gulf of Mexico in 2005, no one would doubt the role of spatial planning practitioners in preparing for and/or managing transformation (Sundermann et al., 2013).

Healey (1997: 4) writes that 'the understanding and practice of planning is thus at the interlocking of the study of the dynamics of urban and regional change and the study and normative practice of governance'; succinctly expressed in the title of the editorial introduction to the volume *Managing Cities* (Healey et al., 1995), 'The City: Crisis, Change and Invention'. As Leonie Sandercock points out in Chapter 23, Healey believes that the planning project of managing our co-existence in shared spaces inherently involves doing so for transformative purposes; reflecting her interest in how planning can initiate and proactively facilitate change. Transformation incorporates shifts in cultural norms and expectations, as Sandercock illustrates, including those of planning and governance cultures and their institutional capacities.

This Part, Transformation Processes, includes articles by Patsy Healey and chapters from invited authors which discuss issues of change and transformation. Healey's work has always been concerned with issues of how planning can contribute to wider societal change, how planning practice changes and also how planning ideas and concepts change/are changed by their environment. These three strands of concern resurface at various times throughout her body of work, from her 1974 article, 'Planning and Change' onwards. From her earliest work, as André Sorensen states (Chapter 25), Patsy Healey has framed her research in terms of questions about the possible roles of planning in shaping urban change, the roles of structures and agencies in shaping planning action and the possibility for governance processes to contribute to 'transformative change', as well as questions of how and why planning and governance systems and processes are themselves changed and transformed. With regard to 'transformative change', Sorensen points out that Healey emphasises two modes of change in particular: the role of collaborative governance activity in working out policies and practices and, through such activities, the development of new institutional capacity, new forms of social capital, new ways of thinking (cognitive templates) and new cultures of place governance. At the same time, Healey always remains conscious of factors slowing down or opposing such change, including vested interests ('resources'), entrenched ideas, 'powerful exogenous forces' and so on: the situated practices of Anthony Giddens (1984).

Reflecting both Patsy Healey's international research and the international audience for her work, chapters in this Part include examination of, and reflection on, the ways in which ideas may travel between places and institutions. Authors recognise the three themes of what constitutes the planning project, theory-practice relations and spatial dynamics and place qualities, addressing them through case studies of Burns Lake in British Columbia, Canada (Leonie Sandercock) and Cairo, Egypt (Zeinab Tag-Eldeen). These case narratives complement those from Patsy Healey's reproduced articles, of Valencia, Venezuela and Cali, Colombia (Healey, 1974), Newcastle-upon-Tyne, England (Healey, 2006) and the vignette of Porto Alegre, Brazil (Healey, 2012).

In the remainder of this contextual Introduction, I briefly introduce some of the main elements which recur in Patsy Healey's work on transformative processes and which the authors develop in their contributions below. Firstly, the issue of whether there is, or are, a planning culture or cultures. Culture is inevitably linked to institutionalisation and to governance and institutional capacity. Transformations in economic, social, political and environmental conditions generate pressures for the transformation of formalised ways of 'doing government' (Coaffee and Healey, 2003: 1979–1980), which includes 'doing planning'.

Whether the changing politics of an integrated Europe (Healey, 2006), shifts to public-private partnerships through urban entrepreneurialism and infrastructure provision (see, for example, Stoker, 1998; Skelcher, 2005; Hodge and Greve, 2007; Peck, Theodore and Brenner, 2009), or the evolution of strategic planning into a more project-driven practice (Albrechts, 2006; Oosterlynck et al., 2011), empirical case studies are extremely useful in helping us understand what is taking place, how and why. Many of the shifts mentioned above include ideas which spread, or travel, from one jurisdiction to another across the globe (Healey, 2012, 2013; Healey and Upton, 2010). This aspect of travelling provides the final key element introduced below.

Planning Culture

How people in an organisation think, feel, value and act are guided by ideas, meanings and beliefs of a socially shared nature; i.e. a culture (Alvesson, 2002). Coaffee and Healey (2003: 1982) suggest that a 'culture' comprises 'taken-for-granted assumptions, habits and routines' embodied in accepted modes of governance, their embedded values and formal and informal structures for policing discourses and practices (Coaffee and Healey, 2003; Healey, 2007a). Schein (1992) identifies three distinct characteristics of an organisational culture: artefacts (such as language, style of clothing, products), espoused values (a sense of what ought to be) and basic underlying assumptions (unconscious taken-for-granted beliefs, perceptions and so on), all of which may be found in spatial planning organisations. Institutionalised cultures, therefore, structure the way that people think, feel and act (du Gay, 1996: 151).

Is there, then, a planning culture? Following Bish Sanyal's (2005: xxi) understanding of planning culture as 'the collective ethos and dominant attitudes of planners regarding the appropriate role of the State, market forces, and civil society in influencing social outcomes', we can see that planning cultures will differ significantly around the world. Cultures may also refer to processes or substantive issues of planning practice. In some countries, there is a process culture of instrumentalist expertise, in which planners 'know best'; in others there is a more 'people-centred' planning culture; in yet others, there is a 'development' culture of planning equated with facilitation or encouragement of economic growth and progress (Hillier, 2010). With regard to substantive issues, early emphases on cultures of urban design in Western countries were challenged in the late 1960s and after by conceptualisations of social justice, sustainability and so on. This was observed by Healey and Underwood (1978) in their study of planning practice in London Boroughs, manifest as a partial shift from planning as urban design towards 'planning as policy'.

Back in the mid-1970s, Cleland and King argued that there was a lack of planning culture in the USA which was deleterious to the development of the profession: 'to 'go through the motions' of planning without inducing a planning culture' is to make the error of 'making oneself the victim of the planning of others'

(1974: 71). By 1988, Healey et al. suggested that there existed a planning culture in Britain of the management of change. Whilst this broad cultural idea has not changed much over the years, its constituent ethos and attitudes have been challenged on several occasions. For instance, by Prime Minister, Tony Blair's, (in)famous 'practical pragmatism' of 'what counts is what works' (1997). In the early twenty-first century, the English government Department responsible for spatial planning called for 'a culture change' in planning practice away from perceived 'failings' of sluggishness, rigidity and opacity, towards a flexible, proactive, transparent culture organised round achievement of key performance indicators and targets (DTLR, 2001; ODPM, 2002). But, as authors such as Healey (2004a, 2004b, 2013), Shaw and Lord (2007) and Inch (2010, 2012) illustrate, planning and governance processes and cultures cannot be changed by formulae.

As Patsy Healey observed in 1974, we can identify several conceptions of planning at any one time, in any one place, as different actors and their values encounter each other. As her research in the early 1970s in Venezuela and Colombia discovered, planning concepts may represent an alternative set of values to those prevailing in wider society. Developing an operational culture of planning in situations of very rapid change, where the notion of planning is not well understood, or accepted is, therefore, extremely difficult. Healey's work with Jacky Underwood (1978, reproduced in Part II) reported not dissimilar conclusions from England following a major restructure of the planning system in 1968. Traditional ways of thinking and acting may be difficult to shake, leading to path dependence (see Sorensen, Chapter 25). According to Roland (2004, cited in Sorensen, Chapter 25), culture is a prime example of a slow-moving institution. It is to governance and institutional capacity that I now turn.

Governance and Institutional Capacity

In Chapter 25, André Sorensen, outlines what has been called 'the institutional turn' in theory and also the trajectory of Patsy Healey's work on institutional capacities. Healey regards institutions as social structures: 'a set of repeated events of interactive behaviour, the forces leading to this repetition defining or giving cohesion to the structure' (1974: 148). The family is thus an institution, development assessment sections are institutions, and so on. There are clearly strong links between 'culture' and institutionalisation of frameworks of formal and informal norms, rules and practices which structure action (Healey, 2007a). As Sorensen explains, Healey's sociological institutional approach differs from what he terms an historical institutionalist framework, which he then explores further in relation to questions of land development planning.

'Institutionalists focus on interactions, not decisions' (Coaffee and Healey, 2003: 1982). A key difference, for Patsy Healey, is that sociological institutionalists emphasise the interactions between structuring dynamics and micro-politics. Concern is not simply with path dependence or 'institutional inheritance' and how this might shape future possibilities, but, in particular, on how new institutional capacities get built; the power to learn new practices and create new capacities (Healey, 2007).

Healey (Healey et al., 1997) identifies seven key elements of institutionalist thinking which are useful in analysing spatial planning practices. Recognition of interrelationships between structure and agency (after Giddens, 1984) is crucial to an understanding of knowledges and values as social constructs articulated through social interaction. As such, actors are not autonomous, but rather part of various, often interlinked and overlapping, assemblages in which new discourses and ideas are actively constructed, generating what Judith Innes (1994) refers to as institutional capital, comprising intellectual capital (knowledge resources), social capital (social network or relational resources) and political capital (mobilisation or power resources). Together these resources and capitals generate institutional capacity, the social infrastructure of relationships between governance, citizens and companies (Healey, 1998: 1541). Healey explains that the concept of institutional capacity refers to 'the overall quality of the collection of relational networks in a place' (1997: 61) which can build on new systems of meaning, new organising routines and styles and so on (1997: 285).

Rich institutional capacity facilitates flows of knowledge/intellectual capital and enhances the abilities of stakeholders to 'improve their power to "make a difference" to the qualities of their place' (Healey, 1998: 1541). Institutional capacity is intrinsically linked to governance processes and 'cultures'. As Healey (2006: 302) explains, in sociological institutionalism, the term 'governance' has a double connotation. It can describe all the processes for the regulation and mobilisation of social action and it can also refer to the particular context in which it performs institutional work, such as the shift in implication from govern*ment* to govern*ance* in Europe in the late twentieth century (see Pierre and Peters, 2000; Bevir, 2013). Healey is especially interested in both the potential regressive and potential progressive implications of particular institutionalisations of governance, their territorial imaginations and actor capacities (2006).

The nature of the institutional context in which practices of 'making a difference' might flourish has long been of importance to Patsy Healey. In her work in Venezuela in the early 1970s, for example, she illustrates how municipal planning efforts were unable to generate sufficient institutional capacity 'to direct, or influence the direction of, any emerging new structures' (1974: 222). Planning reference frameworks were unable to accommodate to the rapidly changing and unpredictable forces in wider Venezuelan society at the time, and, in fact, were generally incompatible with them (1974: 223–4). Healey augments her discussion of the situations in Venezuela and Colombia with a brief look towards how planning concepts had achieved acceptance/institutionalisation in the UK. She concludes with a warning, however, that 'institutionalisation, by introducing bureaucratic values, may inhibit rather than promote the development of imaginative thinking' (1974: 230), an issue which she takes up on several occasions in her later work on professionalisation (see Part II).

With regard to development of imaginative thinking, Healey is concerned with what kinds of mobilising efforts and institutional contexts might help to build capacity for releasing 'creative energies' (2004a). She again refers to the blockages of 'following rule-bound bureaucratic procedures which impede(d) innovation' (2004a: 88–9) and suggests that a key task for planning is to serve as a 'project for sustaining the capacity for innovation' (2004a: 89). This would require an institutional capacity which could understand and accept that innovations are inherently experimental, and that experiments can – and often do – fail as well as succeed. Unfortunately, a capacity for taking 'risky bets' (Healey, 2004a: 91) such as these, is not something which is easy to create in an organisational world of performance indicators and targets, where many planning practitioners would rather play safe than risk 'failure' (Gunn and Hillier, 2013). This is far from the qualities of governance culture which Healey (2004a, Box 2) advocates in order to promote creativity: an emphasis on performance not conformance, encouragement of open-minded tolerance and sensitivity, and so on.

In Chapter 23 Leonie Sandercock offers an inspirational example of these qualities. Whilst not working at the broader scale of governance culture per se, Sandercock's action research with the Ts'il Kaz Koh First Nation in British Columbia, Canada, needed to find resonance with cultural assumptions. She and Giovanni Attili designed a planning process which created a safe and therapeutic space for conflicting parties to meet and speak without fear of dismissal, attack or humiliation: 'a new space of recognition in which historic injustices are acknowledged, as a necessary prelude to addressing contemporary conflicts'. Through the episode of the film, *Finding Our Way* (2010), Sandercock and Attili created an effective catalyst for productive conversation between Native and non-Native residents of Burns Lake. This triggered new ways of seeing and new ideas about real governance partnerships for managing their co-existence across the different groups. Sandercock and Attili's film generated what Patsy Healey (2007b) terms 'transformative force', destabilising established ways of thinking and acting and creating the momentum and potential for thinking and acting otherwise. The film reconfigured the institutional terrain of Burns Lake, through its achievement of generalised support for the idea that 'things need to change' (Healey, 2004c).

Zeinab Tag-Eldeen (Chapter 24) also deals with the notion that 'things needed to change' in Egypt in order to transform planning processes from a one-way transfer of knowledge to the joint creation of innovative ideas. Her example of the Zifta Demonstration project illustrates how institutional capacity was created between formal and informal sectors which, in turn, mobilised local resources and assisted program delivery of local refurbishment and other initiatives in the *Zifta Vision 2022*. As with Sandercock and Attili's film, it was the innovations in process design which generated capacities for change in Zifta. In both cases, the processes empowered people and motivated them to collaboratively think about new and different ways of achieving change.

Empirical Studies

Leonie Sandercock and Zeinab Tag-Eldeen offer us strong empirical cases, as Patsy Healey has done throughout her career. The benefits of in-depth, empirical research through case studies in the field of spatial planning are well known (see, for example, Flyvbjerg, 2006; Watson, 2009; Birch, 2012; Silva et al., 2015). Universals cannot be found in the study of planning practices which evolve in deeply situated and contingent manners (Healey, 2007b: 291). Rather than 'best practice' summaries or systematised typologies (Healey, 2012: 196), the power of context-dependent knowledge is extremely valuable to researchers and practitioners who can learn from thick case narratives in many different ways, including that of comparative reflection (Robinson, 2011; Healey, 2012).

Patsy Healey's early work in Venezuela and Colombia (reproduced in this Part) challenged the generalised applicability of institutionalised planning activity (1974: 229). Context-appropriate practice is critical, as Leonie Sandercock's and Zeinab Tag-Eldeen's cases illustrate. All the authors in this Part would agree with the importance of 'innovative planning' practice (Friedmann, 1967) which has so inspired Patsy Healey, especially where planning initiatives have been introduced into climates of values very different from those in which they originally evolved (Healey, 1974: 177); an issue which prefigures Healey's later work on 'travelling'.

Empirical case studies are of value in a world of 'fluidity and experimentation' (Healey, 2004c) when generalisation is of little utility compared with rich experiences of the situated diversity of ideas and practices. Healey (2004c: 7) writes that 'it is important to look at specific examples, in order to generate and test hypotheses about the contribution of experiences of strategic spatial planning to transformation dynamics and trajectories'. Her own rich case studies (Healey, 2004a, 2004c, 2005, 2007b) illustrate change and transformation processes in the cities of Amsterdam – where planning practitioners attempted to transform themselves from 'introverted' dominant players to becoming intellectually and politically 'nimble' 'multi-faceted' facilitative players in strategic planning; Cambridge – where practitioners struggled with development of broad collaborative processes in a hierarchical government structure with an oppositional culture of policy debate; and Milan – where planners were eager to transform practice from technical conformance-checking to collaborative approaches to strategic spatial planning.[1] Healey concludes that in each of her three cases, 'little transformation potentials' were discovered, though they were different in each case and their outcomes could not be predicted.

Probing ideas about present 'problems' and future potentials must be context-dependent, as Healey has demonstrated through her empirical work in South America and in several European countries, including in the English London Boroughs and in Newcastle-upon-Tyne. As she (2007b: 294) perceptively states, cases should not become 'exemplars of "good practice", or archetypal cautionary tales', to be copied or avoided elsewhere. Rather, they are 'stories of situated experiences' which might offer 'probes to imaginative endeavour' for practitioners to consider reflexively. It is in this spirit that Leonie Sandercock's and Zeinab Tag-Eldeen's stories are offered below.

Travelling

Theory and practice ideas have long 'travelled' within individual countries and around the world. It is known, for example, that William James, Albert Einstein, Sigmund Freud, Pablo Picasso, Henri Poincaré, Émile Durkheim, Marcel Mauss and Henri Bergson, among other prominent philosophers, scientists, sociologists and psychiatrists, attended talks at the Société française de philosophie (Chimisso, 2008; Hillier, 2012) and Bergson's lectures at the Collège de France in Paris.

Such 'circuits of knowledge' (Healey, 2013) were, and are, both national and international in nature, spread by travelling scholars and practitioners and the globalisation of electronic rapid information transfer. Patsy Healey has investigated how various innovations may 'travel' from particular episodes and institutional sites to become translated (or not) into 'mainstream' practices (2004a, 2006). In these analyses, Healey was influenced

1 See Balducci et al. (2011) for updated information on Milan.

by Lukes' (1974) concept of dimensions or levels of power, as reworked by Giddens (1984) and Dyrberg (1997). Her hypothesis is that, to have transformative effects, governance institutions must 'shift level' from explicit formation episodes to arenas of investment and regulatory practices. In order to endure, they need to become institutionalised in the routines of governance practices and of mainstream governance culture (Healey, 2006, this Part). New ideas and practices, therefore, have to 'challenge and shift an array of already routinized governance processes, with their complex mixture of conscious and taken-for-granted modes of practice' (2006: 305). This means that new ideas must 'jump' boundaries and 'break through' resistances.

As Healey states, for 'a real quantum transformation' to be identified, all three levels – specific episode, governance practices, governance culture – need to change significantly (2004a: 94). Her work on partnership processes in Newcastle-upon-Tyne (e.g. Healey, 1998, 2006; Healey et al., 2003; de Magalhães et al., 2002) for instance, demonstrated that several innovations at the episodic level of interaction between actors were not taken up into wider governance processes because of the 'deliberate strategies and embedded discourses and practices of ... key organisations' (Coaffee and Healey, 2003: 1983).

Healey's analytical frame has subsequently been referenced extensively in fields as diverse as tourism studies, climate adaptation and cleaner production, as well as in urban governance, by international authors tracing the mobility, or otherwise, of initiatives (e.g. Minnery, 2007; Beaumont and Dredge, 2010; Carmin et al., 2012; Nevens et al., 2013). Initiatives have been tracked, not only from specific episodes into accepted processes/practices and governance cultures, but also from attempted government-influenced macro-scale culture-change to processes and practices and the micro-scale of specific episodes (e.g. Gunn and Hillier, 2012).

Policy ideas and practices have 'travelled' around the globe with significant consequences for spatial planning, from the grid plan of Hippodamus onwards (Peck and Theodore, 2010; McCann, 2011; McCann and Ward, 2011; Peck, 2011). As Patsy Healey's work in Venezuela and Colombia exemplifies, often, for countries in the Global South, 'the earliest type of planning to arrive was urban planning derived from European urban design traditions' (1974: 177). As she explains, 'great faith was put in planning' (1974: 178), which was attributed 'almost magical powers' (Friedmann, 1965, cited in Healey, 1974: 178) by those who implemented the imported ideas, sometimes rather blindly.

Healey (1974) explains that Le Corbusier visited Bogotá, Colombia, in 1947 when he was asked by the municipality to undertake the re-planning of the city. Le Corbusier produced a 'Pilot Plan' which would be expanded by CIAM members Sert and Wiener (from the consultancy, Town Planning Associates, in New York) into a major town plan: the 'Regulating Plan'. The specially created 'Regulating Plan Office' in Bogotá prepared all the necessary information for the Pilot Plan for a meeting with Le Corbusier, Sert and Wiener in Le Corbusier's holiday home in the Côte d'Azur, where the Regulating Plan was born (Fondation Le Corbusier, nd). Written in French, not only was the Regulating Plan for Bogotá, therefore, produced by internationally-based architects and urban designers not overly familiar with the city, but it was largely written overseas in Paris, sight (and site) pretty-much unseen. It contained designs for blocks of housing in typical Corbusier style. Not surprisingly, Healey's research (1974: 226) found that the planning reference framework was 'incompatible' with the environment in which it was used. The American consultants also produced physical plans for Medellin and Cali in Colombia, Cidade dos Motores in Brazil, Maracaibo in Venezuela, Lima and Chimbote in Peru and Havana in Cuba.

The invitation of named, international architects and consultants to provide planning and design expertise to a 'host' city has been a feature of urban development for several centuries, from the time of European colonisation in Africa, South America, Australasia and Asia onwards, to the recent and current work of 'overseas' consultants in cities such as Dubai, Hong Kong and Beijing. It was in the mid-twentieth tentury, however, that international transportation of policy ideas appeared to peak, justified, as Healey (2012: 191) explains, by 'the belief in a single, "universally valid" pathway for human social development'. Ideas, from the USA, France, Germany and the UK in particular, were transported around what was regarded as the 'developing world' as marks of 'progress' and 'modernisation', often with little relevance to most of the local inhabitants (Nasr and Volait, 2003; Healey, 2010b, 2012).

With regard to importation of A-list architects and also of policy innovations, 'the principal appeal ... remains the expectation of comparable results [to those in the Global North], despite obvious differences in institutional arrangements and political alignments between places of policy innovation and the zones of policy emulation' (Peck and Theodore, 2010: 170). International transfer of cargo cult-like, so-called 'best practice' initiatives, such as New Urbanism, imported from the USA (McCann and Ward, 2010), or the 'Barcelona model' of culture-led regeneration (Gonzales, 2011) may, however, promote a new form of 'colonisation', crowding out the potential for local invention (Healey, 1974, 2011). The spread of such models owes much to globe-trotting 'gurus' such as Le Corbusier, Richard Florida and Andreas Duany, to international consultants and to students, academics (Healey, 2008, 2010b) and practitioners who transcend borders both physically (attending conferences) and in cyber-space. As the collection by Healey and Upton (2010) illustrates, policy transfers often legitimise Western expertise and ideas, abstracted from the socio-economic-political contexts which made them 'practically useful' in the first place (Healey, 2013: 1511), dropped into 'technical systems', at the expense of local knowledges (Hillier, 2014), where they are in danger of causing real harm (Healey, 2012), blind as they are to the dynamics of very different cultures (Healey, 2013).

Not all ideas are copy/pasted exactly as they left their place of origin, however. Many 'mutate' (Peck and Theodore, 2010), are translated and/or adapted by prescient practitioners for their new contexts (Tait and Jensen, 2007). As demonstrated in Healey and Upton (2010), ideas and their journeys are complex, ranging from 'crude impositions and emulations' to 'sophisticated and selective' processes of 'learning, challenging, adapting and inventing' linked to locally specific, social, political and economic dynamics (Healey, 2010b: 6).

It is these issues which the stories by Zeinab Tag-Eldeen and Leonie Sandercock illustrate so well. Tag-Eldeen's case study illustrates the complex process of transferring Patsy Healey's collaborative planning 'model' to the Egyptian context of Zifta City. A cultural milieu of mutual roles and responsibilities of individual, familial and societal relationships (what Tag-Eldeen calls a 'we' consciousness) served to facilitate patterns of communication and consensus-building, such that the author is able to conclude that the collaboration processes contributed to the transformation of power relations between the different actors and the restructure of institutional settings. Crucial to its success, as Tag-Eldeen claims, was a new interpretation of collaborative planning appropriate to the cultural values rooted in, and inherited from, the local society.

In her case narrative, Leonie Sandercock reflects on how collaborative planning and transformative governance were useful in the very different cultural context of Burns Lake, where Indigenous First Nation Ts'il Kaz Koh residents encounter those of 'pioneer stock' heritage. The five-year collaborative project of making and discussing the film, *Finding Our Way* (Attili and Sandercock, 2010), so engaged the communities who eventually came to recognise each other 'in their full humanity rather than as "other"'. Was this an act of transformation, however? As Sandercock explains, there was transformation as evaluated 'through the lens of Western planning theory'. But, as she then points out, this is a conceptualisation which continues to frame Indigenous people as 'stakeholders' seeking inclusion in Western political processes. It, therefore, fails to appreciate the unique status of Indigenous peoples as original owners of the country. It fails to move forward the important struggles of Indigenous Nations to achieve recognition and rights, as Sandercock tellingly writes, 'involving the return of lands that were never ceded'.

Copy/pasting strategies and ideas unthinkingly (or 'stupidly', Hillier, Part II, Chapter 5), from dominant Western cultures into the living and waking spaces of Indigenous Nations would be to court disaster. As Healey (2012: 200) suggests, when concerned with 'the challenge of co-existence in the shared spaces of complex urbanised societies', there can be no 'absolute universal good technology' of planning and its practices. There is much work still to be done on what 'co-existence' might mean in the context of Indigenous recognition and rights in colonised societies around the world (Sandercock, Chapter 23; Barry, Chapter 19; Barry and Porter, 2012; Howitt and Lunkapis, 2010; Porter, 2013).

Conclusions

As André Sorensen writes, one of Patsy Healey's particular contributions to planning theory and practice 'is to show convincingly that governance processes around shared spaces are a particularly important venue for … transformative democratic practices'. Yet, as she has always taken care to emphasise, it is impossible to define 'planning' simply as a universal, holistic conceptualisation when thick case narratives point to the contingency of ideas and practices.

From her early work in South America, right through her trajectory of deep empirical research, Patsy Healey has demonstrated that the manner in which ideas and practices 'travel' is linked to the institutional capacity of the 'host' area. Her conceptualisation of travelling or mobility (Healey, 2004a, 2006; Coaffee and Healey, 2003) offers an important investigatory frame useful at national and international scales. 'Mapping the institutional terrain' (Healey, 2007b: 196) is crucial if new ideas are to be context-relevant and to gain momentum and acceptance: 'doing place governance work in a planning way involves capabilities in situated, pragmatic learning' (Healey, 2012: 201). In addition, as Leonie Sandercock's and Zeinab Tag-Eldeen's case narratives illustrate, trajectories of urban governance transformation are not contingent only on the local institutional landscape, but 'are shaped by the imaginations which frame the mobilisations and struggles of transformation processes' (Healey, 2004a: 95).

References

Albrechts, L. (2006) 'Bridge the gap: From spatial planning to strategic projects' *European Planning Studies*. 14(10): 1487–500.

Alvesson, M. (2002) *Understanding Organizational Culture*. London: Sage.

Attili, G. and Sandercock, L. (2010) *Finding Our Way*. 90 min documentary. Vancouver: Moving Images.

Balducci, A., Fedeli, V. and Pasqui, G. (2011) *Strategic Planning for Contemporary Urban Regions – City of Cities: a project for Milan*. Farnham: Ashgate.

Barry, J. and Porter, L. (2012) 'Indigenous recognition in state-based planning systems: Understanding textual mediation in the contact zone' *Planning Theory*. 11(2): 170–87.

Beaumont, N. and Dredge, D. (2010) 'Local tourism governance: A comparison of three network approaches' *Journal of Sustainable Tourism*. 18(1): 7–28.

Bevir, M. (2013) *Governance: A very short introduction*. Oxford: Oxford University Press.

Birch, E. (2012) 'Cities, people and processes as planning case studies'. In: R. Crane and R. Weber (eds) *The Oxford Handbook of Urban Planning*. Oxford: Oxford University Press, pp. 259–84.

Blair, T. (1997) 'Labour's New Economic Approach for Britain', Speech at the Corn Exchange, City of London. 7 April 1997. www.prnewswire.co.uk/news-releases/labours-new-economic-approach-for-britain-by-tony-blair-156238675.html [accessed 26 March 2014].

Carmin, J., Anguelovski, A. and Roberts, D. (2012) 'Urban climate adaptation in the Global South: Planning in an emerging policy domain' *Journal of Planning Education and Research*. 32(1): 18–32.

Chia, R. (1988) 'From complexity science to complex thinking: Organisation as simple location' *Organization*. 5(3): 341–69.

Chia, R. and Tsoukas, H. (1999) *On Organisational Becoming: Rethinking organisational change*. WP99/12. Colchester: Department of Accounting, Finance and Management, University of Essex.

Chimisso, C. (2008) *Writing the History of the Mind: Philosophy and science in France, 1900 to 1960s*. Aldershot: Ashgate.

Cleland, D. and King, W. (1974) 'Developing a planning culture for more effective strategic planning' *Long Range Planning*. 7(3): 70–74.

Coaffee, J. and Healey, P. (2003) '"My voice: my place": Tracking transformations in urban governance' *Urban Studies*. 40(10): 1979–99.

de Magalhães, C., Healey, P. and Madanipour, A. (2002) 'Assessing institutional capacity of centre regeneration: Newcastle's Grainger Town'. In: G. Cars, P. Healey, A. Madanipour and C. de Magalhães (eds) *Urban Governance, Institutional Capacity and Social Milieux*. Aldershot: Ashgate, pp. 45–62.

Department for Transport, Local Government and the Regions (DTLR) (2001) *Planning: Delivering a fundamental change*. London: DTLR.

du Gay, P. (1996) 'Organizing identity'. In: S. Hall and P. du Gay (eds) *Questions of Cultural Identity*. London: Sage, pp. 151–69.

Dyrberg, T. (1997) *The Circular Structure of Power*. London: Verso.

Flyvbjerg, B. (2006) 'Five misunderstandings about case study research' *Qualitative Enquiry*. 12(2) : 219–45.

Fondation Le Corbusier (nd) *Urbanisme, Bogota, Colombia, 1950*. www.fondationlecorbusier.fr/corbuweb/morpheus.aspx? [accessed 31 March 2014].

Friedmann, J. (1965) *Venezuela: From doctrine to dialogue*. Syracuse: Syracuse University Press.

Friedmann, J. (1967) 'A conceptual model for the analysis of planning behaviour' *Administrative Science Quarterly*. 12: 225–52.
Giddens, A. (1984) *The Constitution of Society*. Cambridge: Polity Press.
Gonzales, S. (2011) 'Bilbao and Barcelona "in motion": how urban regeneration "models" travel and mutate in the global flows of policy tourism' *Urban Studies*. 48: 1397–418.
Gunn, S. and Hillier, J. (2012) 'Processes of innovation: Reformation of the English strategic spatial planning system' *Planning Theory & Practice*. 13(3): 359–81.
Gunn, S. and Hillier, J. (2013) 'When uncertainty is interpreted as risk: An analysis of tensions relating to spatial planning reform in England' *Planning Practice and Research*. 29(1): 56–74.
Healey, P. (1974) Planning and Change. *Progress in Planning*. 2(3): 143–237.
Healey, P. (1997) *Collaborative Planning: Shaping places in fragmented societies*. Basingstoke: Macmillan.
Healey, P. (1998) 'Building institutional capacity through collaborative approaches to urban planning' *Environment & Planning A*. 30(9): 1531–46.
Healey, P. (2004a) 'Creativity and urban governance' *Policy Studies*. 25(2): 87–102.
Healey, P. (2004b) 'Towards a 'social democratic' policy agenda for cities'. In: C. Johnstone and M. Whitehead (eds) *New Horizons in British Urban Policy*. Aldershot: Ashgate, pp. 159–72.
Healey, P. (2004c) 'From collaborative episodes of institutional transformations: Experiences in city region spatial planning in Europe'. Paper presented to ACSP conference, 21–24 October. Portland, Oregon.
Healey, P. (2005) 'Urban region strategy-making in a relational world'. Paper presented to AESOP conference. July. Vienna.
Healey, P. (2006) 'Transforming governance: challenges of institutional adaptation and a new politics of space' *European Planning Studies*. 14(3): 299–319.
Healey, P. (2007a) 'The new institutionalism and the transformative goals of planning'. In: N. Verma, (ed.) *Institutions and Planning*. Oxford: Elsevier, pp. 61–90.
Healey, P. (2007b) *Urban Complexity and Spatial Strategies: Towards a relational planning for our times*. London: Routledge.
Healey, P. (2010a) *Making Better Places: The planning project in the Twenty-First Century*. Basingstoke: Palgrave Macmillan.
Healey, P. (2010b) 'Introduction: the transnational flow of knowledge and expertise in the planning field'. In: P. Healey and R. Upton (eds) *Crossing Borders*. Abingdon: Routledge, pp. 1–25.
Healey, P. (2012) 'The universal and the contingent: Some reflections on the transnational flow of planning ideas and practices' *Planning Theory*. 11(2): 188–207.
Healey, P. (2013) 'Circuits of knowledge and techniques: the transnational flow of planning ideas and practices' *International Journal of Urban and Regional Research*. 37(5): 1510–26.
Healey, P. (2015) 'Planning theory: The good city and its governance'. In: J. Wright (ed.) *International Encyclopedia of the Social and Behavioral Sciences* (2nd edition). http://dx.doi.org/10.1016/B978-0-08-097086-8.74027-X
Healey, P. and Hillier, J. (2008) 'Introduction'. In: J. Hillier and P. Healey (eds) *Foundations of the Planning Enterprise: Critical Essays in Planning Theory: Volume 1*. Aldershot: Ashgate, pp. ix–xxvii.
Healey, P. and Upton, R. (eds) (2010) *Crossing Borders: International exchange and planning practices*. Abingdon: Routledge.
Healey, P., Cameron, S., Davoudi, S., Graham, S. and Madanipour, A. (eds) (1995) *Managing Cities: The new urban context*. Chichester: Wiley.
Healey, P., de Magalhães, C., Madanipour, A. and Pendlebury, J. (2003) 'Place, identity and local politics: analysing partnership initiatives'. In: M. Hajer and H. Wagenaar (eds) *Deliberative Policy Analysis: Understanding governance in the network society*. Cambridge: Cambridge University Press, pp. 60–87.
Healey, P., Khakee, A., Motte, A. and Needham, B. (eds) (1997) *Making Strategic Spatial Plans*. London: UCL Press.
Healey, P., McNamara, P., Elson, M. and Doak, J. (1988) *Land Use Planning and the Mediation of Urban Change*. Cambridge: Cambridge University Press.
Hillier, J. (2007) *Stretching Beyond the Horizon: A multiplanar theory of spatial planning and governance*. Aldershot: Ashgate.

Hillier, J. (2010) 'Introduction'. In: J. Hillier and P. Healey, P. (eds) *The Ashgate Research Companion to Planning Theory: Conceptual challenges for spatial planning*. Farnham: Ashgate, pp. 1–34.

Hillier, J. (2012) 'Baroque complexity: "If things were simple, word would have gotten round"'. In: G. De Roo, J. Hillier and J. Van Wezemael (eds) *Complexity in Planning*. Farnham: Ashgate, pp. 37–73.

Hillier, J. (2014) 'Troubling the place of the border: On territory, community, space and place'. In: W. Steele, T. Alizadeh, L. Eslami-Andargoli and S. Serrao-Neumann (eds) *Planning Across Borders in a Climate of Change*. New York: Earthscan-Routledge, pp. 31–42.

Hodge, G. and Greve, C. (2007) 'Public-private partnerships: An international performance review' *Public Administration Review*. 67(3): 545–58.

Howitt, R. and Lunkapis, G. (2010) 'Co-existence: Planning and the challenge of Indigenous rights'. In: J. Hillier and P. Healey (eds) *The Ashgate Research Companion to Planning Theory: conceptual challenges for spatial planning*. Farnham: Ashgate, pp. 109–33.

Inch, A. (2010) 'Culture change as identity regulation: The micro-politics of producing spatial planners in England' *Planning Theory and Practice*. 11(3): 359–74.

Inch, A. (2012) '"Cultural work", spatial planning and the politics of renewing public sector planning professionalism in England' *Town Planning Review*. 83(5): 513–532.

Innes, J., Gruber, J. Thompson, R. and Neuman, M. (1994) 'Coordinating growth and environmental management through consensus-building'. Report to the California Policy Seminar. Berkeley, CA: University of California. www.escholarship.org/uc/item/6tgls896 [accessed 28 March 2014].

Lukes, S. (1974) *Power: A radical view*. London: Macmillan.

McCann, E. (2011) 'Urban policy mobilities and global circuits knowledge: Toward a research agenda' *Annals of the Association of American Geographers*. 101: 107–30.

McCann, E. and Ward, K. (2010) 'Relationality/territoriality: Toward a conceptualization of cities in the world' *Geoforum*. 4: 175–84.

Minnery, J. (2007) 'Stars and their supporting cast: state, market and community as actors in urban governance' *Urban Policy & Research*. 25(3): 325–45.

Nasr, J. and Volait, M. (eds.) (2003) *Urbanism: Imported or exported? Native aspirations and foreign plans*. Chichester: Wiley-Academy.

Nevens, F., Frantzeskaki, N. Gorissen, L. and Loorbach, D. (2013) 'Urban transition labs: Co-creating transformative action for sustainable cities' *Journal of Cleaner Production*. 50: 111–22.

Office of the Deputy Prime Minister (ODPM) (2002) *Sustainable Communities: Delivering through planning*. London: ODPM.

Oosterlynck, S., Albrechts, L. and Van Den Broeck, J. (2011) 'Strategic spatial planning through strategic projects'. In: S. Oosterlynck, J. Van Den Broeck, L. Albrechts, F. Moulaert and A. Verhetsel (eds) *Strategic Spatial Projects: Catalysts for change*. Abingdon: Routledge, pp. 1–14.

Peck, J. (2011) 'Geographies of policy: From transfer-diffusion to mobility-mutation' *Progress in Human Geography*. 35: 773–97.

Peck, J. and Theodore, N. (2010) 'Mobilizing policy: Models, methods and mutations' *Geoforum*. 41: 169–74.

Peck, J., Theodore, N. and Brenner, N. (2009) 'Neoliberal urbanism: Models, moments, mutations' *SAIS Review*. 29(1): 49–66.

Pierre, J. and Peters, G. (2000) *Governance, Politics and the State*. New York: Palgrave MacMillan.

Porter, L. (2013) 'Co-existence in cities: The challenge of Indigenous urban planning in the twenty-first century'. In: R. Walker, T. Jojola and D. Natcher (eds) *Reclaiming Indigenous Planning*. Montreal: McGill Queens University Press.

Read, S. (2012) 'Meaning and material: Phenomenology, complexity, science and "adjacent possible" cities'. In: J. Portugali, H. Meyer, E. Stolk and E.Tan (eds) *Complexity Theories of Cities Have Come of Age*. Dordrecht: Springer, pp. 105–27.

Robinson, J. (2011) 'Cities in a world of cities: The comparative gesture'. *International Journal of Urban and Regional Research*. 35(1): 1–23.

Shaw, D. and Lord, A. (2007) 'The cultural turn? Culture change and what it means for spatial planning in England' *Planning Practice and Research*. 22(1): 63–78.

Silva, E., Healey, P., Harris, N. and Van Den Broeck, P. (eds) (2015) *The Routledge Handbook of Planning Research Methods*. New York: Routledge.
Skelcher, C. (2005) 'Public-private partnerships and hybridity'. In: E. Ferlie, L. Lynn and C. Politt (eds) *The Oxford Handbook of Public Management*. Oxford: Oxford University Press, pp. 347–70.
Stoker, G. (1998) 'Public-private partnerships and urban governance'. In: J. Pierre (ed.) *Partnerships in Urban Governance: European and American experience*. London: Macmillan, pp. 34–51.
Sundermann, L., Schelske, O. and Hausmann, P. (2013) Mind the Risk: A global ranking of cities under threat from natural disasters. Zurich: Swiss Re. www.media.swissre.com/documents/Swiss_re_Mind_the_risk.pdf [accessed 26 March 2014].
Tait, M. and Jensen, O. (2007) 'Travelling ideas, power and place: The cases of urban villages and business investment districts' *International Planning Studies*. 12: 107–28.
Watson, V. (2009) 'Seeing from the South: Refocusing urban planning on the Globe's central urban issues' *Urban Studies*. 46(11): 2259–75.

Chapter 20
Planning and Change*

Patsy Healey

[Chapter 1] Introduction

The central concern of this study is with the nature and potentials of planning as a basis for action, examined in the particular context of situations experiencing the kind of changes associated with economic development. In other words, this study is about planning and change.

The emphasis of the study is not on developing a particular prescriptive solution to the urban problems of Latin America, but on examining the activity of planners and planning agencies in relation to these problems and to the wider society of which they are a part.

More generally, this study sets planning activity within the context of the social and political structure of society. This leads to a consideration of the terms and purposes of the evaluation of planning.

Planning

But an initial problem is to find an acceptable definition which will allow the exploration of such a broad approach to planning. Looking at the recent history of planning in this country alone, a variety of conceptions of planning can be identified, and challenged on the grounds of inadequacy as bases for action or as concepts (see Hebbert, 1973, Hart, 1973). For example, the rational decision-making model of planning, with its emphasis on process, is currently emphasised in this country [the UK] over earlier conceptions of planning as urban design or land use control. Yet this model, imported from the United States, can be criticised on many grounds,[1] and others are sought, especially in the States (e.g. Friedmann, 1973; Grabow and Heskin, 1973).

One might describe the history of planning concepts as an oscillation between models, modification or replacement of a particular model occurring as a result of challenges on the grounds of internal inconsistency or conflict with other conceptual, often ideological, principles. Accepting this variety in conception and consequent definition of planning, then planning and planning activity must be viewed more widely than in the simple functional sense of a tool which may be used in specified circumstances. Equally, it must be recognised that, despite the lack of clarity and diversity of definition surrounding the concept of planning, and the various evaluations of ineffectiveness and undesirability, planning activity has been introduced, encouraged and has persisted in many situations, i.e. its use and persistence may not be related to any functional ideal such as that of the rational planning process, or if such an ideal is used, it may serve other purposes than as a basis for the consistent evaluation of policy decisions.

Change

Rather than resolve the problem of definition at this stage, this study examines urban planning as defined by those doing it, i.e. self-acknowledged urban planners and planning agencies, and it will be found that several conceptions of planning can be identified within the group of planning agencies studied. But the particular

---

* This chapter contains selected extracts from 'Planning and Change', *Progress in Planning* 2(3): 145–237 (1974). The original text has been substantially reduced. Editorial inserts for the purposes of this partial reproduction are inserted in square brackets.

1 See Banfield (1959), Braybrooke and Lindblom (1963), Altshuler (1965), Rabinovitz (1969) and Stuart (1969).

characteristic of the urban planning investigated in this study is that as a concept and an activity it has been a recent innovation, usually associated with 'modernisation' and 'development', this innovation occurring in societies undergoing the kind and scale of changes implied by the concept of development.

This approach, of examining planning, a term considered to be difficult to define, in a context of change, may be justified for two reasons. Firstly, planning itself is often considered particularly appropriate for either inducing change or guiding it (see on, Chapter 5). This justification leads one to ask how far has 'planning' been an appropriate transfer or innovation into such situations. Secondly, to examine a phenomenon under extreme conditions, assuming that the nature of the changes occurring in developing countries represents extreme conditions for planning, may reveal that phenomenon in greater clarity. Using an ecological analogy, the nature of an organism's adaptive qualities are more obvious at the boundaries than at the core of its distribution range.

To talk about change and development, however, raises further definitional problems.

Since the concern with these concepts here is contextual to the central concern with planning, a more specific meaning has been sought at this stage. Some definition was needed which would distinguish changes in kind or state from changes in quantity, and for this purpose the concept of *structural change* has been adopted.

Structural Change

However used, 'change' is a relative term. Structural change can therefore be considered as such that a particular state is contrasted with a previous one in terms of the underlying structure of the two states. A situation *undergoing* structural change may similarly be contrasted with one characterised by structural stability or persistence; more specifically in the sociological field, with stability in institutional structures (Eisenstadt, 1968). Social structure, or social institutions, according to one writer, may be defined as the persistent form of the ongoing process of interactive behaviour (Geertz, 1957). There are a variety of ways in which such behaviour may be described, some emphasising the process, some the persistent form (Eisenstadt, 1968; Barth, 1967). But the general point is made that social structures – and here institutions, systems, forms, patterns may often be used synonymously with structure – consist of a set of repeated events of interactive behaviour, the forces leading to this repetition defining or giving cohesion to the structure. To change such a structure requires a breakdown in these cohesive forces. Similarly, spatial structures can be defined as a set of material objects linked by some cohesive force such as, for example, a transport system.

It is obvious that the identification of particular events as causing structural change will depend on what is identified as the particular structure or institution in question. The death of a father may cause structural change within the institution that was his family, but not necessarily in the institutions of the town where he lived. Sociologists have been concerned with the institutions which affect an individual's social behaviour; economists with the institutions which allocate material goods between competing demands; geographers with spatial structures. Thus the identification of structures may emphasise different dimensions by which phenomena may be described (social, political, economic, spatial). It may also stress different levels, local institutions rather than national institutions, urban spatial patterns rather than national spatial patterns, for example. It is important to raise this variety of definition, since when observing the relation of planning activity to change, it is necessary to consider not only what planning activity sought to change and did change, but what was accepted by planners as constant or continuous within the situation in which they were operating.

It must also be noted that, although structural change can be generated by pressures external to the structure, such change can also be generated from within. Eisenstadt (1968) maintains that 'the propensity to change is necessarily built into any institutional system', since any institutional definition conflicts with the interests of certain members of that system. This point is taken further by Geertz, who suggests that conflict may occur between the different structures or systems within which individuals operate. In particular, he contrasts the 'casual-functional integration' which he suggests is the basis of social systems and 'logico-meaningful integration' which is the basis of cultural systems. He argues that the tensions, dysfunctions, or inconsistencies, between these two organising principles by which such structures may be defined act as important generators of change, elaborating from examples to show that it is the perception of such an inconsistency that leads to changes in the structures based on behaviour with which he is concerned (Geertz, 1957).

Development

Using the concept of change in the way discussed above, it is now possible to distinguish between the nature of change, whether it is structural or not, and the rate of change which may or may not be associated with structural change. It has also been stressed that the concept of structure refers simply to phenomena held in persistent inter-relationship by some cohesive force or organising principle and that structures can be identified through a variety of such principles and at a variety of levels. Finally, any structure, it has been suggested, is inherently unstable, its own organising principles being inconsistent with certain of its elements.

Relating the discussion to the context of development, it is clear that the pursuit of development to which the governments of most Latin American countries are committed can be interpreted as the pursuit of structural change, in the national economy at the least. In the two countries on which most of the ensuing discussion is based, Venezuela and Colombia, wider social changes were also sought explicitly. Such development has been conceived in a variety of ways, an emphasis on linear progression from pre-industrial socio-economic societies to industrial ones giving way more recently to models in which one set of institutions seeks to expand and take over another pre-existing or peripheral set (Nettl, 1969; Ponsioen, 1968). Like change, 'development' is a relative term, locating a particular society or economy in relation to others, usually developed or more developed. This means that, apart from its descriptive use for academic purposes, the motivating force behind the pursuit of development by 'underdeveloped' or 'less developing countries' is the perception of an unfavourable relative condition, that is perception of the kind which Geertz suggests is necessary to provide structural change. The conscious attempt to produce such change can of course be equated with national planning defined in efficiency or effectiveness terms (Spicer, 1968). Hence in the examples presented in this study, some association of planning with attempts to induce structural change may be expected.

Change and Urban Areas

To provide some focus for what could otherwise become a very general discussion, this study concentrates on urban areas and urban planning: i.e. a level or aspect of a situation likely to be experiencing structural change has been emphasised. Since the innovation with which I am concerned in this study is urban planning, it is necessary to examine the urban areas to which this planning has been applied or thought applicable. In particular, some attempt has been made to identify the extent to which structural change can be said to be occurring in the different dimensions of what may for convenience be called 'the urban system'.

The consideration of urban areas, and the discussion of urban planning, has been centred around actual examples taken from Venezuela and Colombia, deriving from the author's research work in these two countries. This study is therefore presented in the following way. After some background notes on Latin America, and in particular Venezuela and Colombia, the urban phenomenon is considered by the presentation of two examples of large cities, one from each country, which relate to the examples of planning agencies discussed later. In these examples, spatial development, or the mechanisms for allocating land, have been especially emphasised since this has been a central concern of much of the urban planning activity associated with these areas. Some attempt is then made at interpreting what is described in the examples, in the terms of structural change as presented above.

In the second stage of the study, five examples of urban planning organisations are examined, with particular emphasis being put on their origins and development, their activities and effectiveness. These are introduced by a general discussion of the introduction and evolution of planning in the two countries, Venezuela and Colombia, from which the examples have been taken.

The third stage is a theoretical interpretation and discussion of some of the problems presented when attempting to evaluate the planning observed in the examples. This considers first planning as an institutional innovation in the situations described. This leads to a suggested definition of planning allowing for the problems which impeded definition initially. From this position it is then possible to return to a consideration of the general issue of planning and change, the juxtaposition which this study uses in an attempt to explore, and if possible elucidate, some aspects of the conceptual and organisational phenomenon: planning.

[The next three chapters are omitted from this extract. Chapter 2 provides a brief discussion about the Latin American development context, and about the rapid urban growth being experienced in Venezuela and Colombia. The two urban areas studied were Valencia in oil-rich Venezuela (popn 1971: 367,000) and Cali in Colombia (popn 1970: c.800,000). The chapter sketches the socio-political and spatial context of each country.

Chapter 3 provides a detailed account of the urban trajectory of each urban area from the 1940s to the 1970s. In Valencia, the main groups responsible for spatial growth were private urbanisers, public authorities and squatters. The account focuses on the increasing sophistication of each group's interest and method of operation as a land urbaniser. Each group came to produce different segments of the city. Land markets operated in the areas developed by both private urbanisers and squatters (who take land by occupation rather than by legal means), and both attempted to manipulate public agencies. But the two markets did not interact, and there was little direct conflict between the different land development interests, though there was a clear differentiation between high and low income areas. A variety of developer groups emerged, 'each with their own system for exploiting the new opportunities presented by rapid urban growth and increased prosperity' (p.173 of the original).

In Cali there had been significant industrialisation in the 1950s, followed by substantial immigration from rural areas due to prolonged national civil unrest. But by the 1960s, hopes of increasing prosperity were fading and many lived in extreme poverty. As in Valencia, there was a clear differentiation between high and low income areas. In contrast, however, the developers providing for the two groups often competed for the same land. Squatters had a much more limited role as land urbanisers, being pushed to marginal areas. Land urbanisation was dominated by traditional elite landowning families, who resisted squatting incursions and at times refused to allocate land for development. This led to acute housing shortage and political protest. In the late 1950s, some progressives in the landowning elite aligned themselves with the poor, sponsoring land invasions. In response, the national housing agency instituted an emergency housing programme, encouraging landowners to release land. In Cali then, urban development in the 1960s was dominated by private urbanisers and public agencies. As a result in Cali, 'major changes in spatial extensions and distributions, and rapid population growth, have been accompanied by strong efforts at maintenance of the status quo in the political context' (p.173 of the original).

Chapter 4 summarises key points from the comparison. In both, the urban spatial structure had changed dramatically, but only in Valencia was this accompanied by significant economic and political changes, while in Cali the competition for urban space was much more acute between groups. The analogy of a game is used to describe the development process in both cities, but with a focus on games in which the rules themselves are in evolution. Rather than overseeing the game, public authorities, including those charged with 'planning' urban development, were only one group among many involved in urban development processes.]

[Chapter 5] The Introduction of Urban Planning in Two Countries in Latin America: A Review

The association of planning with change is frequently stated as an article of faith by planners, sometimes simply because of planning's alleged concern with future states (e.g. Chadwick, 1966), sometimes because of a concern with societal transformation (e.g. Grabow and Heskin, 1973). Justifications for the need for 'planning' in situations such as those described in Chapters 3 and 4 fall into three broad groups: those emphasising system transformation, those concerned with partial reform, and those where the imposition of order in crisis is stressed. Friedmann is perhaps the most prolific exponent in English of the need for planned system transformation in Latin America. Defining planning as 'the guidance of change in a social system', he contrasts allocative and innovative planning. This latter he describes as 'a form of social action intended to produce major changes in an existing social system'. 'Innovative planning is especially relevant in rapidly changing social systems. It is in fact a method for coping with problems that arise under conditions of rapid change and will tend to disrupt existing balances' (Friedmann, 1967). Those justifying planning in terms of partial reform have been particularly concerned with management and public administration, with planning as a mechanism for improving the allocation of resources (e.g. the underlying assumptions in Boyce, 1969).

Many Latin American planners, especially in the early sixties, saw planning as a mechanism for establishing order, counteracting anarchy […].

Whatever the justification for planning, however, to provide the basis for action, or policy-making leading to action, all such concepts have had to be institutionalised in some form or other. It is the purpose of Chapter 6 to explore how such institutionalisation has occurred, and what opportunity or area of influence planning, or rather those promoting and carrying out 'planning', have been able to evolve, in five examples of urban planning offices, three in Venezuela and two in Colombia. To provide a context to these examples, it is helpful to look at the background to planning in both countries.

In Venezuela and Colombia such planning organisations and planners can be found at all levels in the public sector -national government or autonomous agencies, regional agencies, state and municipal governments – and they can be found pursuing several models of planning. And in both countries the introduction of planning as a government activity has been associated with three main objects – national economic development, the control of urban development and administrative reform – and the various organisations encountered have emphasised one, or a combination, of these objects. It should also be stressed that planning has been introduced as an already recognised professional activity evolved in countries considered 'developed', rather than evolving in response to particular local problems. It must therefore be considered as a package innovation, and one with close associations with 'modernisation'. This is not to say that planning has no antecedents. In the field of urban control, for example, the mechanisms for developing and expanding towns, with a grid and central nucleus of fixed proportions, derive from strict rules laid down in the sixteenth century, and apparently based on the example of Vitruvius and the Bastide towns (Martinez, 1967). But the complex background of building and health regulations against which urban planning evolved in the United Kingdom is lacking, nor is there, in most cases, an acceptance of the right of government to intervene in private affairs for the public good, i.e. planning has been introduced into a climate of values very different from that in which it evolved in the United Kingdom.

In both Venezuela and Colombia the earliest type of planning to arrive was urban planning derived from European urban design traditions. In Venezuela a strong local school of architects and urban designers, initially under French influence, had managed to establish an Urban Design Division in the administration of Caracas' Federal District by 1938, and this same group were also working with Banco Obrero, the National Housing Agency, at this time on the remodelling of the centre of Caracas. Urban design (Urbanismo) was seen as a way of improving the health and well-being of citizens, providing a form of city management and leading towards greater economic efficiency and a reduction of social ills, claims familiar for this planning tradition from both Europe and the United States (Lander and Rangel, 1970).

A plan for Caracas was developed under the guidance of Rotival, the French architect, although, apart from the work of Banco Obrero, it had little effect on development since there were no legal mechanisms through which land could be controlled. In any case, the plan greatly underestimated the city's future growth, the target population for the year 2000 being in fact reached well before 1960. Both this type of planning and the two others referred to were supported by the 1945–8 democratic government. Several ministries produced plans for developing the sectors of the economy for which they were responsible, both as a method of promoting development and for making their own activities more efficient. Urban design was furthered by the creation of the Comision Nacional de Urbanismo in 1946, whose main aim was to draw up urban plans, to be known as Pianos Reguladores or Pianos de Zonificacion, for all Venezuela's towns. The members of the Commission were drawn from the original group of urban designers, still under the guidance of Rotival and Villanueva, Venezuela's most famous architect to date. The Commission survived the fall of the first AD government [the Accion Democratica was followed by a dictatorship until 1958] and several urban plans were produced between 1950 and 1955. These plans had little more success in influencing the course of development than the first plan for Caracas. They were resisted by local municipal councils who were supposed to implement them, and frequently ignored by the Ministry of Public Works (MOP) and the autonomous agencies, whose activities the plans were supposed to coordinate. After making unwelcome protests to the dictator, Jimenez, whose grandiose building projects had reinforced the difficulties of plan implementation, the Commission was abolished in 1957, its functions and powers being transferred to MOP, where its technical office had always been based.

Since 1958 the main development has been the successful (from an organisational point of view) introduction of national economic planning. The leaders of the AD party, who came to power after the fall of the dictator, had, during their exile, been strongly influenced by the style of planning developed by the UN Economic Commission for Latin America and practised in Puerto Rico. This approach focused on the need for national development to break with international dominance, but within each country, planning was conceived by ECLA as a neutral tool for guiding development in a mixed economy towards goals on which everyone agreed. Its introduction into Venezuela, however, had strong political intentions, aimed as much at creating a consensus as to implement a consensus view (Lander and Rangel, 1970; Friedmann, 1965). Strong emphasis was put on coordination and upon rational government decision-making. The main organisation set up to implement this planning philosophy was CORDIPLAN (Oficina Central de Coordinacion y Planificacion), but planning offices were encouraged in all public organisations to promote administrative reform.

Great faith was put in planning at this time, and Friedmann has remarked: ‘it might be argued that unreasonable hopes were placed on the efficacy of planning. Its use frequently suggested almost magical powers that would instantly resolve the most intractable of problems. Such was the implicit faith in planning, whose internal structure was correctly understood by only a small number of insiders’ (Friedmann, 1965).

In Colombia no type of planning has received the amount of support which national economic planning did in Venezuela after 1958. As in Venezuela, physical planning appeared first. Bogota had a road plan by 1936 and a physical layout plan by 1945. In 1951, le Corbusier visited Bogota, leaving behind an outline which was expanded by Wiener and Siert of New York into a major road plan.[2] Wiener and Siert went on to produce a physical plan for Medellin and much later for Cali. During the 1940s legislation was passed in the national congress enabling the control of development and municipal planning. In 1947 all municipalities with a population of over 10,000 were required to have an office to approve construction in their areas, and all those with a budget of 200,000 pesos or more were to have a comprehensive development plan to control future growth. A more realistic modification to this law was approved in 1951, restricting the towns which had to draw up plans to the larger and richer ones, and requiring that Bogota, Medellin and Cali had plans approved by the local municipal council with a local Junta de Planeacion to deal with modifications (Kammerer, 1966 […]. During the sixties there has been an increased interest in urban planning, with several towns now having urban plans, usually drawn up by consultants. One of the main factors behind this activity has been the requirements of loan agencies that projects for which loans are requested should be part of a general development plan (S82).

National economic planning appeared first in the 1940s, Congress passing enabling legislation in 1945 […]. The first national plan was developed by the IBRD Mission under Lauchlin Currie, reporting in 1950 (Currie, 1950), and during the fifties a Direccion Nacional de Planeacion, with its associated Council, did various planning studies for particular states. A second plan was produced in 1962 (S44) under the new national front government, following, as in Venezuela, the UN-ECLA model. However, unlike Venezuela, the emphasis was more on developing the plan as a technical document, in which effort UN consultants worked closely with local professionals, rather than on establishing the planning offices and ideology required for the successful operation of ECLA-style planning. Consequently, although hailed at the time on technical grounds, this plan lapsed into government archives and has been criticised by Bird (1970) as unrealistic in the ‘difficult’ economic and political circumstances prevailing at the time. The pattern of erratic interest in national planning continues.

Planning as administrative reform has mainly taken the form of budget planning offices, which are found in many organisations, often associated with other functions, and usually related to the rational government planning model derived from the United States.

The five urban planning offices to be discussed in Chapter 6 are: in Venezuela, the Direccion de Planeamiento, descendant of the Comision Nacional de Urbanismo, FUNDACOMUN, a national agency concerned with municipal development, and the municipal planning office in Valencia; in Colombia, two large municipal planning offices, that of Cali and of Medellin, will be examined. One contrast between the two countries is worth re-emphasising at this stage. Although in Venezuela the public sector as a whole plays a much more important role in national investment, and in urban infrastructure investment, than in Colombia,

2 Le Corbusier reached Colombia over ten years after his original expedition to Latin America in the thirties.

the importance of the municipal level of government as an investor is much less than in Colombia. Municipal income in Venezuela represented only 8% of national government income in 1967, compared with 39% in Colombia.

[Chapter 6 presents case studies of five urban planning organisations, three in Venezuela and two in Colombia. In Venezuela, national level organisations provided a significant role in drawing up development plans for urban areas. The *Direccion de Planeamiento* was a department of the national Ministry of Public Works, within which responsibility for major public works and social housing provision was located. The *Direccion* had the power to influence urban spatial development through providing zoning ordinances to be approved by local municipal councils, through approving development projects referred to it by local municipal engineers, and by influencing the way national government agencies invested in physical infrastructure. The *Direccion* produced a plan for Valencia in 1965. (The chapter also comments on a more recent plan for Barquisimeto to illustrate a change of approach). However, its impact on emerging urban environments was very limited. It had little influence on physical investment projects and budgets, in part because there was little inter or intra ministerial co-ordination. Partly also, investment projects were strongly influenced by political considerations. There was also a criticism of the strict zoning approach. In competition for local influence was *FUNDACOMUN*, with a focus on community development and municipal growth, funded by national government and international aid, especially US-AID. *FUNDACOMUN* had a planning unit, which helped muncipalities set up planning offices, worked with officials and politicians to explain the value of urban planning, and developed models and techniques for plan-making and for assessing the work of planning offices. (The chapter discusses *FUNDACOMUN's* work with the municipality of Barquisimeto). In developing their approach, *FUNDACOMUN* used the experience of the Valencia Municipal Planning Office. This had been established in 1965, following some longstanding conflicts between national and local government over the location of industrial land. The municipality decided to seek its own advice from a French company. The result was a study which suggested some development locations, but also argued for administrative and legal reforms. One function of the new planning office was to introduce development guidelines, and a group of Venezuelan architects drew up a development plan, to challenge that produced by the *Direccion.* By 1968, drawing on both plans, a municipal zoning plan was produced. The office also promoted projects, such as a recreational park, and undertook studies of the squatter areas, providing some funds for community projects in these areas. Yet by 1971, it was clear that this planning office had little capacity to influence spatial development in Valencia. It was unable to influence national level investment in key highway projects. Its control over the private urbanisers was limited to discussions about layout and density, but not location. Meanwhile, the squatters were always ahead of any planning office idea to contain and regulate their activities.

In Colombia, municipalities played a more important role in urban development as they had significant infrastructure responsibilities. In theory, this could have allowed greater co-ordination between development projects, infrastructure investment and development regulation. Municipal planning offices also had more staff than those studied in Valencia. In Cali in the late 1960s, there was an energetic effort to produce a municipal development plan, but interest lapsed by the time it was completed in 1970. Politicians were involved in disputes among factions of the land-owning elite. Co-ordination with departments with investment budgets was limited. Staff in the planning office were not always interested in doing professional work. The Cali plan itself was very ambitious in its investment and regulatory ambitions. It was suggested in retrospect that the main reason for producing the plan was to capture loan finance potentially available with the Panamerican Games in mind. The chapter contrasts this experience with Medellin, which was, at the time, the centre of Colombia's earliest industrial region. The local elite, dominated by business rather than land-owning interests, was more focused on the efficient operation of public services and with the quality of the city. A planning office was created in the 1950s. By 1970, the office was expected to produce an 'integral development plan' with an emphasis on promoting equality of opportunity and an atmosphere of solidarity among citizens. A spatial plan had already been introduced in the 1950s, and did shape the pattern of subsequent physical development. Politicians generally supported the work of the planning office, which could operate effectively in a professional way.]

[Chapter 7] Planning Institutions, Planning Concepts and Planning Action

This study has been concerned with the introduction and evolution of an innovation, urban planning, which has the characteristic that it is both a concept and a specialist activity for which specialist institutions have been created. In the preceding chapter, five examples of this institutionalisation have been briefly summarised. Some attempt at evaluating this experience, in the light of local considerations and in view of the central concern of this study with planning and change, will now be made. The main question to be investigated in the first part of this chapter is the extent to which planning concepts and planning activity have been appropriate introductions into the local situation.

The Origins of Planning Offices

In the examples presented here, and in Latin America generally, planning offices as established were derived from prior models or precedents for planning activity, these models deriving either directly from international, especially developed countries, examples or from earlier introductions elsewhere in Venezuela and Colombia. In no case was an office set up out of a local inspiration to 'invent' planning as the solution to a particular need or problem.

Four kinds of reason seem to underlie the establishment of planning offices at the urban level in situations such as these:

a. Issue-oriented: i.e. where an office was created out of some direct concern to accelerate development, coordinate the development effort, modernise administration, or to regulate building and land use.
b. Promotional: i.e. where particular groups have promoted the idea of planning as an appropriate government activity. Thus in Venezuela architects and engineers actively promoted physical planning in the forties; FUNDACOMUN is an institutionalised promotion agency, advocating planning as one of a package of municipal activities.
c. Symbolic: i.e. where a planning office was set up to express meaning rather than carry out a specific technical function. Thus planning offices were set up to express a modern or progressive outlook (OPM Cali, 1969–70) or local independence (OTPM Valencia).[3]
d. Imitative: in one sense, all the planning offices encountered were imitative in origin, since all derived from the prior existence of planning offices. What is referred to here are the cases where offices were created because other agencies were doing so, as with several of the state planning offices in Venezuela and the departmental offices in Colombia, and OPM Cali in its early period.[4]

All four reasons can be found behind the setting up of planning offices in both Venezuela and Colombia, several offices originating from a combination of such influences. In Venezuela, however, the generally more positive attitude to development and to new ideas (see Chapter 2) is reflected in a more positive approach by politicians, local business interests and administrators to the possible usefulness of planning. Only in the first case can it be assumed that the organisational complex into which planning was introduced would have a direct interest in the technical potential of planning activity. On the other hand, the existence of a planning office provided a platform from which the members of the office could promote planning concepts if they chose. It also acted as a resource to be manipulated if desired for bureaucratic and political purposes by other interests in the organisation.

3 CORDIPLAN, the Venezuelan national planning office, is a major example of this.

4 Neither Venezuela nor Colombia had planning offices created through primarily legislative requirement as in the United Kingdom. Where such legislation was passed, in Colombia, while enabling planning offices to be set up, it could not compel their foundation since, in view of the resources available to municipal governments, this was an impossible task. The only comparable blanket legislation to that in the United Kingdom is the Venezuelan Regionalisation Decree of 1969.

The Subsequent Use of Planning Offices

Turning now to how planners and planning were used, the following ways may be distinguished:

1. A particular output required of a planning office, or the planning office itself, was used as a symbol; for example, representing 'modernisation', or 'progressive government', or as representing local capacity versus national government, as in the case of Valencia.
2. The planning office could be used as a receptacle for issues where the appearance of action was required; as a form of delaying tactics. Friedmann ascribes this function to CORDIPLAN (Friedmann, 1965). But the 'reference for study' procedure used in CORDIPLAN- and other examples can be found (OTPM Valencia, DAP Medellin) – can also be seen in the political terms used by Easton as a mechanism for issue reduction, with the planning office acting as a kind of filter to deal with items that the political system cannot immediately process (Easton, 1965).
3. More specifically, planning could be used both as a symbol of progressive development based on political consensus and a mechanism for achieving it. Planning was thus used politically, especially in Venezuela, for its symbolic expression of neutrality in political terms. In Colombia, technical approaches were more closely associated with particular political groups and the concept of technical neutrality was less tenable.
4. Finally, the planning office might find itself required to perform a specific service such as controlling physical development, especially squatter invasions (Cali or Medellin), or providing the necessary documentation for a loan application. Thus urban planning offices could find themselves subjected to a variety of demands. Political demands were the most obvious – demands for outputs to increase the status of a particular party or ruling group, or to promote the interests, in status or financial terms, of individuals in positions of power. Bureaucratic interests were most evident in resistance to the innovation that the planning office represented, for example in the resistance of Venezuela's Ministry of Public Works to attempts at the coordination of its activities by CORDIPLAN or the Direccion de Planeamiento, and the various municipal agencies and departments to coordination attempts by the Cali and Medellin planning offices.

But a fundamental problem facing the acceptance of planning as a concept-innovation was a difference in underlying values. Payne (1968) has said that Colombian politicians are more concerned with the acquisition of status than with the effective implementation of programmes. Parsons (1951), elaborating his distinction between universalistic and particularistic values, suggested that Latin American societies may be dominated by the latter. In contrast, planning conceptions lay a much greater emphasis on programme effectiveness, to use Payne's terms, and, using Parsons' distinction, derive from an acceptance of universalistic-achievement values.

In the light of this conclusion, it is possible to explain what is often referred to as widespread 'corruption' in Venezuelan and Colombian public organisations as a consequence of the juxtaposition of two sets of values about the function of these organisations. A planning programme allocating resources in one way is corrupted, i.e. its values and consistency are disturbed, by the simultaneous operation of a quite different system for allocating resources. Similarly, the symbolic use of planning concepts and planning offices can be interpreted as allocating means to planning not in the terms of planning concepts but in terms of other ideological positions, in particular, local political ones.

It must be concluded, therefore, that planning concepts have been transferred into environments dominated by alien values, and that their introduction has not been generated by any specific idea of its character in technical terms, or its possible uses, technical or otherwise. Rather than an immediately acceptable innovation, a change in bureaucratic and political values would be required before the activity of the organisation of which the planning office is a part could be carried out consistently with proposals for such activity derived from planning concepts.

Concepts and Reference Frameworks for Action

To justify this conclusion it is necessary to look more closely at the character of concepts as a basis for action and planning concepts in particular. Is this problem of incompatible values particular to planning concepts? and is the nature of planning concepts likely to affect their relevance as bases for action in conditions of structural change?

The word concept has been taken to mean more than a 'general notion' (OED). It is understood throughout this discussion to imply a *conceptual framework*, defining the way some phenomena are related and valued. Since the concern is with the use of concepts as bases for action, the term reference frameworks has been substituted for concepts, meaning the conceptual framework to which actors may refer to direct, justify or evaluate their activity. The assumption behind this connection between conceptual frameworks and activity is that actors use ordering concepts implicitly or explicitly to locate themselves in the environment in which they act. Obviously there are many possible such reference frameworks and there seems no reason in theory why there should be a limit to them. To simplify for the purposes of this analysis, it is helpful to extract three kinds of framework which have affected the activity of the planners and planning offices described in Chapter 6: technical, bureaucratic and political.[5]

By technical is meant reference to a set of tenets and procedures associated with a recognised body of knowledge or enshrined as a technical/professional activity by the society being studied. A central theme in technical reference frameworks is the concept that the function of an object or institution is directly related to its explicit or overt capacities.

By bureaucratic is meant reference to the survival requirements of organisations. By this it is not implied that the only function of organisations is to establish themselves, expand and maintain their position, but that such needs frequently motivate organisational activity.[6] Bureaucratic reference frameworks may be identified not only in the wider organisation to which a planning office is attached, but within a planning office itself.

By political is meant reference to the sources of power within the society in which the planning office is located. By power in this context is meant the ability to influence and control the distribution of valued resources, such as, in these examples, capital, land, labour, support, status.[7] Once again, organisations and planning offices may also use political reference frameworks when concerned with the distribution of power within the organisation or office.

In terms of a technical reference framework, activity might be directed at achieving the efficient allocation of resources; in terms of a bureaucratic framework, towards organisational stability; or politically, at acquiring control over the allocation of resources. One or more such frameworks could be in use by planners in determining appropriate activity and outputs, and such activity could be evaluated according to all three frameworks.

But unless a single reference framework is used, the resultant activity is a composite of several frameworks. Consequently, when evaluated, failure to satisfy technical criteria may represent success in satisfying bureaucratic criteria. What is being described is not a set of groups each with their own reference framework, but multiple reference frameworks in use by each group. Thus there can be no technical process, political process or bureaucratic process as such, distinguished by activity specific to it since any activity may be part of several 'processes', depending on the set of terms or reference frameworks used in description.

So far the term 'reference framework' has been considered as providing an actor with some means of interpreting the environment within which he acts. The basis for each framework is an initial structuring of a situation, some representational model. It is this underlying structure, or the organising principles of such a model, which provides the basis for selecting those aspects of the 'reality' of which the model is a representation to evaluate and the criteria for evaluating them. The concept 'reference frameworks' therefore both defines an actor's position and provides a basis for determining future actions. It can be considered both

5 These are similar to Levin's categories, used in discussing increasing commitment and specificity in decision-making (Levin, 1972). My definition of these groups differs from Levin's in some respects (e.g. political), and where he focuses on processes I am focusing, for definition, on reference frameworks.

6 See Meyerson and Banfield (1955) and Banfield (1957) for this view of bureaucracies. Much of the literature on bureaucracies assumes a technical viewpoint according to these definitions e.g. Simon (1961).

7 This definition derives from Easton's definition of the political aspects of a society as those concerned with the authoritative allocation of values (Easton, 1965).

analytically and normatively; to provide a *model* of the situation in which an actor finds himself and a *model* for his actions, a distinction which Geertz (1966) has made when discussing cultural models.

Considering 'reality' from the point of view of any actor as consisting of himself, an actor within an environment of other phenomena, including other actors, the first model defines where an actor fits into his environment and the second the bases for his intervention in this environment.

To take an example, the rational planning approach promoted by FUNDACOMUN in Venezuela both provided a model of where the planner was located in relation to the problems of urban areas, and a model for what actions he could propose for others and take himself. Disorganised urban development was interpreted as caused by irrational and uncoordinated community decision-making, the role of the planner as key agent in promoting greater coordination and efficiency/effectiveness.

Whether developed for analytical or normative purposes, such models, the basis for the reference frameworks, are simplifications or reductions of 'reality' and therefore are a distortion in one form or another of the situation which an actor in practice encounters. Thus the environmental totality in which an actor operates can be described in a variety of ways depending on the organising emphasis taken (e.g. technical, bureaucratic, political; or within the technical framework – economic or physical), and the position the actor accords to himself. Given that any actor has some implicit or explicit model of his environment and his position in it to which he relates his actions, and that such a model is a simplification or distortion of the totality of the environment, it is possible to explain the use of a variety of models by a single actor, and hence of a variety of reference frameworks, as an adaption to the inconsistencies produced by single model distortion. As a further elaboration, such models may be expected to vary over time and between individuals and groups.

In other words, planning activity will be evaluated differently depending on the evaluator's position and terms, or reference frameworks. Not only are there a variety of such frameworks, but there are many positions from which they may be constructed. Finally, those evaluating the activity of planners will not necessarily be concerned with planning concepts.

Planning Reference Frameworks

Given these general points about the conceptual reference frameworks to which an actor may refer, it is now necessary to define more precisely the particular qualities of planning reference frameworks, i.e. planning concepts. At the most general level, I suggest that planning as a technical reference framework is based on a belief in intervention, since all actors could be said to 'believe' in intervening or interacting with their environment. The emphasis is on the prior definition of that intervention, according to some plan, policy or programme. Associated with this belief is an acceptance that the prior definition of its actions is or should be a distinct function of a group, organisation or society, and consequently specialist practitioners are needed to perform this function.

Stated another way, planning is about making explicit models for intervention, which implies further that planning is a particular sort of technical reference framework which requires that the framework for any action or set of actions be established explicitly, in advance, and, by implication, consistently, with the general organising principles of the framework chosen. From this it can be derived that planning reference frameworks are based on a belief that action can and should be organised according to a single framework. It represents the imposition of organising principles not only on a perceived environment in order to describe it (model of), but through intervention, in order to bring the environment into line with the organising principles (model for). Obviously this definition could be limited further by taking particular sets of principles ('rational', 'balanced community') or particular environments (organisations, urban areas, regions, etc.).

However, all frameworks, or models, it has been suggested, represent a distortion of the totality 'reality'. Consequently, any planning reference framework will be in some way or other incompatible with the totality. This conclusion elucidates the oscillation among various models of planning referred to in Chapter 1, new models or types being put forward when earlier ones revealed their inevitable inadequacy in absolute terms. Of greater importance for the present study is the implications of this definition of planning – as a reference framework attempting to direct action according to a single set of terms – when applied in the context of structural change and rule evolution described in Chapter 4. To conclude this chapter before embarking on a consideration of these implications, an attempt has been made to summarise the types of planning reference frameworks encountered in this study.

Types of Urban Planning

It is only possible to identify the various types of planning reference framework in use from the various statements about what planning was for and could do, and the character of the outputs of planning offices, including any supporting statements about their particular purpose and content. None of these directly define a framework per se, and many documents and statements were produced with bureaucratic or political rather than technical considerations in mind.

Given these provisos, the summary in Table 6 [20.1; table modified for readability] has been attempted, limited to planning offices with directly urban concerns. Evidence was used about the emphasis; organising principles; measures of achievement or evaluation criteria expressed explicitly or implicitly; about the limitations to the area of planning concern; about the type of intervention envisaged; and the role or position of the planner in relation to planning as a function generally, or to the implementation of a particular proposal.

The fact that all the types of planning derive from models developed elsewhere suggests that when introduced into Venezuela and Colombia, these types of planning were either considered to have general application despite differences in the contexts to which they were applied, or that the contexts themselves, rapidly growing Latin American cities, were considered generally similar to cities, for example, in the United Kingdom or United States. Those types of planning concerned with development paths or progressions, such as rational decision-making or community development, appear to have taken the former position, with the additional value being attributed to this type of planning that it was of even greater usefulness in a developing situation than in a developed one. Physical/land use planning, on the other hand, which emphasised the provision of containers for expected development, was justified on the grounds that urban development had reached such a stage of complexity that a similar tool to that in use in developed countries was now appropriate. Thus all five types of planning have been considered appropriate transfers into a local situation.

A common theme to all statements about planning was its ability to impose integration and balance in a confused situation, to integrate and coordinate the at present diffuse development effort. In terms of the definition of planning reference frameworks considered above, the general attribute of planning is seen as its ability to impose, through intervention, a set of coherent organising principles on an environment which, because of the rate of change at all levels of society, is perceived as generally disorganised.

However, as we have seen above, although planning may have been perceived as an appropriate transfer, in practice it often proved an alien graft in the institutional environment in which it was implanted. To use Adams' terminology, the innovation had not reached the stage of secondary development (Adams, 1967) (see Chapter 2). In the next chapter I will discuss how far the planning reference frameworks used were appropriate to situations described as undergoing structural change characterised by incompatibilities between both continuities and changes, and between different aspects of change.

[Chapter 8] Planning in a Context of Structural Change

[This chapter discusses the role of planning concepts and activity in relation to the evolving rules for urban land allocation and then considers how these attempts to evolve rules played out in a context of disjointed, inconsistent structural change. Only the conclusion to this chapter is included here].

Conclusion

The conclusion from this [analysis] is that there is little area of opportunity for planning action consistent with or in terms of planning reference frameworks in the examples studied, or in any situation with similar characteristics. But is this how planning should be regarded? Should its purpose be considered as

Table 20.1 The planning reference frameworks encountered in the planning offices

Type	Emphasis	Evaluation Criteria	Organising Principles	Limitations	Intervention	Position of the Planner
Urban design	Spatial arrangement of urban areas	Functional relations between objects and spaces	Direct relation between functional and physical structures	Focus on an area or project	To impose both new arrangements and re-arrangements	Designer of a part of society
Physical land use planning	Framework for allocating land uses	Adequate provision to meet 'needs'	Land as the container for activities of various developer groups	Mixed economy development path assumed	Direct provision to meet needs; guidance of all physical development	Co-ordinator of administration related to urban physical development. Controller of how land used and developed
Community development	Provision for social equipment for 'modern' society	Acquisition of values, skills and means to satisfy immediate needs	Incorporation of marginalised groups into mainstream system	Structure of a 'modern' system assumed	To develop skills/ values via co-operation between intervener and 'community'	Social worker, linking the community with government resources
General development	Integrated development framework; multi-dimensional	Administrative efficiency/ effectiveness. General welfare	Economic, social and administrative processes integrated in the development effort	Mixed economy development path assumed	To change administrative behaviour	Co-ordinator of the overall development effort; especially within an administration
Rational public decisions	Behaviour of the administration	Administrative efficiency/ effectiveness. General welfare	Economic, social and administrative processes integrated in the development effort	Existing structure of political, social and economic relations assumed	Mixed economy development path assumed	Co-ordinator of administration

operationalising a reference framework as a basis for action, or can action directed by planning reference frameworks be seen to have any other purpose?

Planning has been recently introduced in the examples, and was an innovation into an environment different in many ways from that in which the models from which the innovation was taken had evolved. It was introduced for a variety of purposes, several of which were associated with a desire not simply to direct change but to induce it. In this context, the fact that planning was incompatible in concept in the above four senses need not be considered negatively.

As has been argued in many other situations, the awareness of conflicts and inconsistencies leads to the perception of the need for adjustment and from this perception to adaptive activity, to changes. If the introduction of planning increases the conflicts, increases the awareness of inconsistencies, then it can operate as a contributory generator of adaptive changes.

This is essentially Hirschman's thesis, that economic development can be stimulated by inputs whose consequences cannot be anticipated but the demands of which will generate new adjustments and the unpredicted mobilisation of new resources (Hirschman, 1967, 1971). Where, in these examples, planning has had any influence on behaviour at all, such adjustments can be identified. The clearest case of this is in administrative behaviour in Venezuela, where the demands and conflicts revealed in the administration when viewed in planning terms have led to changes, such as improved accounting, budgeting and at least some coordination. It t must be emphasised, though, that the introduction of planning itself was associated with the will to change, in contrast, for example, with the Cali situation.

It is thus possible to argue that the introduction and operationalisation of planning concepts into inappropriate situations can serve a constructive function as a change agent. The problem, as far as planning as a basis for action is concerned, is that while adaptive changes may be generated by the introduction of planning, such changes cannot be assumed to occur in 'foreseen' or 'anticipated' directions. Thus it is difficult to define planning as, for example, 'the guidance of change in a social system' (Friedmann, 1967), and apply this definition to situations such as those described here. Such definitions are based on holistic conceptions which, as a basis for action, require:

1. A society which can be described in holistic terms; a society which can be conceived as an integrated and defined whole, which it has been suggested is not the case in these examples.
2. The possession of comprehensive powers to implement such planning, which was not available to any bureaucrat or politician with whom planning was associated in these examples.

Nor is it possible to support the argument developed by Friedmann that planning may be particularly appropriate to societies in transition between one level of economic development and another, i.e. undergoing structural change, if it is concluded that the phenomenon of structural change inhibits coordinated action (Friedmann, 1966).[8]

Elsewhere, Friedmann contrasts developmental planning with adaptive planning (Friedmann, 1967). According to the argument developed in this study, developmental planning would not be possible until sufficient structural cohesion had been achieved to permit the comprehensive powers necessary to implement such planning. Once such structural cohesion exists, planning could only operate to maintain that structure, that is, modify or adapt it. Hence planning can only be adaptive in the functionalist sense of structural maintenance. Criticising Friedmann's 'sanguine expectations' for planning in Latin America, Morse (1971) notes that in any case there is a contradiction between the evolutionary view of development which Friedmann holds and the 'revolutionary strategy of social and institutional overhaul' which may be the 'prerequisite' for the innovations which Friedmann identifies with development.

The conclusion must be that while in situations such as these, planning activity has led to some changes – in spatial structure, in administrative behaviour – it is not in a position to induce directed changes. Planning as a concept cannot be operationalised as such under the conditions found in Venezuela, Colombia and elsewhere in Latin America.

8 Friedmann develops this argument in relation to regional planning, but the criticism applies to any form of planning.

To extend the case, in situations of rapid change affecting a whole society, where such changes are occurring incompatibly with each other, planning cannot be put into practice; planning concepts cannot be translated into or provide a coherent basis for action. This confirms the point made in socialist philosophy that planning is only appropriate after the revolution, once a set of rules and powers have been established. The converse of this is that it is only possible to operationalise planning under conditions of structural stability, i.e. where the forces producing the structure into which intervention is proposed can be assumed to be constant in character. In other words, planning can only operate when it is used functionally, to maintain an existing order, and one to which planning values are appropriate. Returning to Mannheim's distinction between ideologies and utopias, planning can only operate where it is used ideologically.

Nevertheless, in these examples planning institutions have been established under apparently inappropriate conditions. They have survived and expanded. As an organisational innovation, they have to some extent adjusted and diffused into a wider environment. Yet planning as a reference framework, it has been concluded, cannot operate within this environment. Each organisation, or centre of planning activity, therefore, had to evolve some accommodation to the contradictory situation in which it found itself.

One solution to the dilemma was to abandon the attempt at maintaining a planning reference framework. This seems to have happened for long periods in the [Cali municipal planning office]. Another alternative was to adopt a utopian position in some form, and accept the inconsistent character of planning. Two strategies seemed possible in this case: either to subvert or undermine an existing structure, or those parts of it which appeared an impediment, or to convert by demonstrating the viability of an alternative, i.e. to play an evangelistic role. Among the examples presented here, [Valencia municipal planning office] was possibly the most obvious example of this last approach.

What is of interest is that in none of the cases encountered was the objective of planning – a particular goal or problem – a dominant organisational concern, except possibly in Medellin. The core of technically oriented activity was establishing planning as a recognised and supported activity. Evaluated on the basis of success in achieving what they set out to do, the only successful office was Medellin's municipal planning office, and this success was due to a consciously limited conception of planning. On the basis of effecting changes, the Venezuelan offices which adopted evangelistic styles and which either concentrated their effort on administrative reform […] or on the promotion of projects […] may be considered the most successful.

It seems likely that planning offices that operated an evangelical strategy although often operating with a very uncertain and confused knowledge of planning concepts, were nevertheless likely to have a more fundamental effect on rule evolution in the long run than for example, the municipal planners in Medellin, concentrating on effective implementation and in Cali, emphasising a particular planning type. Derived from this, it may be suggested that the type of planning reference frameworks, as identified in Chapter 7, may be relatively unimportant where all planning reference frameworks are incompatible with the environment in which they are used and the main planning-directed effort of planners is towards effecting changes and propagating planning as an accepted activity.

[The final chapter considers the relevance of the findings of this study to other countries. It suggests that, to be effective, institutionalised planning activity of the kind promoted through the planning ideas used in these situations demands some degree of cohesion between the government context in which the ideas are applied and the planning concepts themselves. But planning ideas are not just about being effective. They may also represent an alternative set of values to those prevailing. The chapter concludes as follows:]

It is possibly this particular quality of planning, that it is associated with a desire for hopeful visions of the future, that makes it on the one hand so complex a phenomenon in the intellectual and practical sense and consequently, such an absorbing object of study. The study of planning is intimately linked with the history of ideas and the history of practical endeavour. What is called planning today – both the ideas and the practice – has appeared at other periods in history and may recur in different forms again if it does not persist in present ones. As a final question one may ask what drives a society to attempt the explicit prior definition of its actions and when does it look to hopeful visions that this may be possible?

References

[*Only those texts cited in this edited chapter are included.*]

Adams, R.N. (1967) *The Second Sowing: Power and secondary development.* San Francisco: Chandler.

Altshuler, A.A. (1965) *The City Planning Process: A political analysis*. Ithaca, NY: Cornell University Press.

Angell, A. (1966) Cooperation and conflict in Colombian party politics. *Political Studies*, 14(1): 53–71.

Banfield, E.C. (1957) The decision-making schema. *Public Administration Review*, 17: 278–85.

Banfield, E.C. (1959) Ends and means in planning. *International Social Science Journal* 11(3): 361–8.

Barth, F. (1967) On the study of social change. *American Anthropologist*, 69: 661–9.

Bird, R.M. (1970) *Taxation and Development: The lesson from Colombia.* Cambridge, MA: Harvard University Press.

Boyce, C. (1969) *Elementos de Planijicacion Urbana*. Caracas: Fondo Editorial Comun.

Chadwick, G.F. (1966) A systems view of planning. *Journal of the Town Planning Institute*, 52: 185–6.

Currie, L. (1950) *The Basis of a Development Programme for Colombia.* I.B.R.D.

Easton, D (1965) *A Framework for Political Analysis*. Englewood Cliffs, NJ: Prentice Hall.

Eisenstadt, S.N. (1968) Introduction. In *Comparative Perspectives on Social Change*, ed. S.N. Eisenstadt. Boston, MA: Little, Brown & Co.

Friedmann, J. (1965) *Venezuela: From doctrine to dialogue.* Syracuse: Syracuse University Press.

Friedmann, J. (1966) *Regional Development Policy. A case study of Venezuela.* Cambridge, MA: MIT.

Friedmann, J. (1967) A conceptual model for the analysis of planning behaviour. *Administrative Science Quarterly*, 12(3): 225–52.

Friedmann, J. (1973) The public interest and community participation: Towards a reconstruction of public philosophy. *Journal of the American Institute of Planners*, 39(1): 2–7.

Geertz, C. (1957) Ritual and social change. *American Anthropologist*, 59: 32–54.

Geertz, C. (1966) Religion as a cultural system. In M. Banton (ed.) *Anthropological Approaches to Religion.* ASA Monograph No. 3. London: Tavistock.

Grabow, S. and Heskin, A. (1973) Foundations for a radical concept of planning. *Journal of the American Institute of Planners*, 39: 106–14.

Hart, D.A. (1973) Ordering change and changing orders; A study of urban policy development. *Policy and Politics*, 2(1): 27–41.

Hebbert, M. (1973) Scale models, statistical models, models of man. In *Ideologies in Planning.* Papers of a conference held at Kingston Polytechnic.

Hirschman, A.0. (1967) *Development Projects Observed.* Washington: The Brookings Institution.

Hirschman, A.0. (1971) *A Bias for Hope.* New Haven, CT: Yale University Press.

Hobsbawn, E.J. (1963) The revolutionary situation in Colombia. *World Today*, 19: 248–58.

Kammerer, K.C. (1966) *Law and the Urbanisation Process in Colombia.* Bogota: CINVA.

Lander, L. and Rangel, M.J. de (1970) *La Planificacion en Venezuela.* Caracas: Soc. Venezolana de Planificacion.

Levin, P.H. (1972) Specificity and commitment in urban planning. *Town Planning Review*, 43: 93–115.

Martinez, C. (1967) *Apuntes sabre el urbanismo del Neuvo Reina de Granada.* Bogota.

Meyerson, M. and Banfield, E.C. (1955) *Politics, Planning and the Public Interest.* New York: The Free Press.

Morse, R.M. (1971) Trends and issues in Latin American research, 1965–1970. *Latin American Research Review*, 6(1): 3–52, No.2, 19–75.

Nettl, J.P. (1969) Strategies in the study of political development. In C. Leys (ed.) *Politics and Change in Developing Countries*. Cambridge: Cambridge University Press, pp. 13–34.

Parsons, T. (1951) *The Social System.* London: Routledge and Kegan Paul.

Payne, J.L. (1968) *Patterns of Conflict in Colombia.* New Haven, CT: Yale University Press.

Ponsioen, J.A. (1968) *National Development: A sociological contribution.* The Hague: Mouton.

Rabinovitz, F.F. (1969) *City Politics and Planning.* New York: Atherton.

Simon, H.A. (1961) *Administrative Behaviour.* 2nd edition. New York: Macmillan.

Spicer, E.H. (1968) Developmental change and cultural integration. In A. Gallagher (ed.) *Perspectives in Developmental Change*. Kentucky: Lexington, pp. 172–200.

Stuart, D.G. (1969) Rational urban planning: Problems and prospects *Urban Affairs Quarterly*, 5: 151–82.

Chapter 21
Transforming Governance: Challenges of Institutional Adaptation and a New Politics of Space*[1]

Patsy Healey

Urban Governance Transformation Processes

For the past 20 years, social science and public policy debate in Europe have focused attention on the changing governance of cities and urban regions. The reasons for this are broadly shared – the new politics of European 'integration', the impact of globalization on the relative importance of the nation state vis-à-vis sub-national governance arenas, the shifting dynamics of local economies in the face of globalizing tendencies, the changes in the forms of social welfare delivery and transfer payments, the changes in social expectations about policy agendas and modes of politics (Amin and Hausner, 1997; Cars et al., 2002; John, 2001; Le Gales, 2002; Pierre, 1998). However, there is much controversy about the nature and significance of the emergent governance forms which are arising. Some emphasize the 're-scaling' of governance arenas and networks (Brenner, 1999; MacLeod, 1999; MacLeod and Goodwin, 1999; Swyngedouw, 2005). Others are exploring the nature of the territorial focus emerging at new scales (Albrechts et al., 2001; Albrechts et al., 2003; Fürst and Kneilung, 2002; Healey, 2004, 2006; Novarina, 2003; Salet and Faludi, 2000; Salet et al., 2003). There are also debates about the extent to which new forms of politics are developing (Healey, 1997a; Hajer and Wagenaar, 2003) and whether these are progressive or regressive in their direction (Harvey, 1989; Imrie and Raco, 1999; Swyngedouw, 2005).

Earlier structural analysts presented cities as sites for the organization of 'collective consumption' (Castells, 1977; Saunders, 1979), the organization of welfare delivery. This could take many forms, from bureaucratic procedures, to rational policy analysis, to interest-based politics and clientelism. Shifts in urban governance were then seen as structured by contestation over the manner and distribution of welfare activity, over the balance between social and economic objectives and over the tendency for some governance insiders to pursue private interests under the cover of public authority. But by the 1980s, it became obvious that urban elites were deeply concerned about their local economies, especially in areas affected by the decline of traditional industries. David Harvey (1989) described the change as one from 'managerialism' to 'entrepreneurialism', as urban governing elites shift attention to the promotion of projects and strategies to re-configure and re-position their local economies. By the 1990s, regulation theorists had become interested in the extent to which changes in urban governance processes reflected a shift in 'mode of regulation', to parallel the shift in 'modes of accumulation' from Fordism to post-Fordism (Jessop, 2000, 2002; Painter, 1997). It is in this context that the 're-scaling' debate surfaced (Brenner, 1999; MacLeod, 1999). Others began to explore the extent to which urban processes were structured by markets (the logic of economic activity), hierarchies (the logic of the established twentieth-century welfare state) or networks (the logic of webs of social relations stretching in multiple space/times, rather than just hierarchically (Amin and Hausner, 1997; Castells, 1996). These and other contributions argued that a new form of politics was emerging for a new globalizing or 'network' society (see also Hajer, 2003).

But such analyses assumed a certain coherence in urban governance processes and their manner of transformation. In contrast, other analysts whose focus of attention was on the working out of particular policies and mechanisms in urban governance contexts highlighted the experience of 'fragmentation' in the

* Previously published in *European Planning Studies*, 14(3): 299–319.

1 This paper is a revised version of one given at the 54th Meeting of German Geographers in Bern, 28 September–4 October, 2003. Workshop FS26: New Urban Governance – Institutional Change and Consequences for Urban Development.

1980s (Cars et al., 2002; Le Gales, 2002; Pierre, 1998). Established procedures were perceived as breaking down, in different ways in different areas of governance activity. Systematic 'comprehensive' town planning was giving way to a project-driven practice which broke through the constraints of bureaucratically-delivered norms and standards. Poverty policies moved from a focus on the distribution of access to dwellings and welfare benefits to citizen empowerment and area regeneration projects. All kinds of experiments appeared – new kinds of project and new kinds of 'partnership' between actors within formal government and between formal government, the economy and civil society. Instead of nesting neatly in a hierarchical model of levels of government responsibility, new urban governance arenas and practices were introduced which drew in actors from a variety of different levels of government, from the level of the European Union (EU) to local neighbourhood organizations.

Whatever the assumptions about the driving forces of change, these empirical experiences raise questions about how to understand the dynamics of contemporary struggles to introduce new urban and regional governance agencies and new modes of governance. Do the experiences represent the dynamic evolution of new scales of governance articulation? Or the emergence of 'territory' as a new focus in public policy, to challenge deeply-embedded traditions of sectoral government policy communities? Or are they a sign of a new collaborative mode of policy formation, to challenge the traditional separation of politics and policy-making? Is the experience of 'fragmentation' to be understood as new social reality in itself, a more market-like, competitive mode of governance, in which authority and responsibility, actions and consequences, become hard to identify and therefore difficult to call to account? Or is it to be understood as a process of 'dis-integration', reflecting the weakness of a territorial focus of governance in the face of global networks and technologies, a sign of a disappearing urban collective actor? Or is such disintegration quite the opposite, a necessary destabilization which generates the 'moments of opportunity' for new governance modes, coalitions and agendas focused around a strong territorial sensibility able to generate new relations of 'integration' so that a city or urban region can once again become a strong collective actor locally and in the European arena? Is the process one of re-scaling arenas from city and 'nation state' to 'urban region' and 'neighbourhood', with the consequent challenge to create a territorial collective actor around a new arena and scale? Or is the 'fragmentation' there all the time, for any analyst to see who looks carefully enough at the fine-grain of urban governance processes? Should urban governance be understood as in 'continual transformation'?

To address these questions, some 'intermediate' analytical tools between the language of structural transformation and that of the experience of particular actors in particular places are needed.[2] Information is drawn in this paper from a particular strand of 'institutionalist analysis', particularly as developed in policy analysis and planning, to develop an approach to the analysis of urban governance transformation processes. In this paper, the aim is to summarize this approach, and then illustrate its use in relation to examples from different arenas of governance within an urban region, all of which in some way challenge established governance processes. From the approach and experience, comments are made on the themes raised at the start of this paper and the issue of how to understand contemporary experiences of fragmented urban governance.

An Institutionalist Approach to Governance Transformation

There are many academic frames of analysis which currently cluster under the 'institutional' and 'institutionalist' label. Broadly, they all share a recognition of the importance of 'institutions' in shaping social action. Any analyst of governance processes is inherently concerned with the 'institutions' through which social action is mobilized and regulated and hence is to some extent an institutionalist, as political scientists recognize (Peters, 1999). 'Institutionalists' differ, however, in the meaning given to institutions and in their ontologies and epistemologies. Within this array, the author's own work has developed in a social constructivist and non-positivist frame, related to a dynamic, relational view of social action (Healey, 1999, 2004a).

From this viewpoint, institutions are understood as the ensemble of norms, rules and practices which structure action in social contexts (Giddens, 1984; Powell and Dimaggio, 1991). Such ensembles are brought

2 The author and colleagues have been developing such intermediate tools through empirical research on planning and urban regeneration experiences (Healey and Barrett, 1990; Healey, 1992, 1998, 2004a; Healey et al., 2003).

into existence and given meaning through continual active effort, which reinforces and changes both meanings and materialities. Social action is constituted by people acting in relation to others, not just through the pursuit of an individualized 'rational calculus'. In such interaction, people draw on a (shifting) store of cultural resources through which meanings, values and 'knowledge' are shaped (Fischer, 2003; Hall and Taylor, 1996).

In this 'sociological' view of 'institutionalism', the term 'governance' has a double connotation. Analytically, it can be used to describe all the processes for the regulation and mobilization of social action. In this meaning, there are always governance processes (Cars et al., 2002; Le Gales, 2002). The research question becomes an empirical inquiry into modes of governance manifest in a particular time and place. The political question centres on which modes to pursue in a particular instance in relation to specific concerns. But as a sociological phenomenon, it is also necessary to look at the way the term 'governance' is being used in a particular context and the 'institutional work' its use performs. In Europe at present, the term is often deployed by analysts and political actors to describe a particular transformation, from modes of government associated with a bureaucratized welfare state focused on universal service delivery to the more 'entrepreneurial' modes of governance recognized by commentators such as David Harvey (1989) and Bob Jessop (1998, 2002).

In the author's own work, the general meaning of 'governance' is used to encompass all forms of collective action focused on the public realm (sphere) in one way or another, from those orchestrated by formal government agencies, to lobby groups, self-regulating groups and social campaigns and movements targeted at resistance or challenge to dominant governance relations. Analytically, there is no necessary reason why these diverse governance processes should be articulated around an 'urban' or any other territorial focus, although in Europe, as Le Gales (2002) convincingly argues, there are strong traditions of 'city government' and territorial identity which provide cultural and organizational resources for such a focus. In examining current transformations in urban governance, the author's specific current interest is in the extent to which the diverse actors and networks which coexist within a territory, urban region, city or neighbourhood cluster around a territorial focus, and create the capacity to act (collective actor capacity) for a territory. And there is interest in the potential regressive and progressive consequences of particular institutionalizations of such a territorial imagination and collective actor capacity. This interest connects to the wider debates in urban governance analysis about the 're-scaling' and 're-territorialization' of contemporary sub-national governance.

How then may these institution-building processes be examined? Traditional institutionalists in political science focused on the creation of formal competences, procedures and laws. Or they examined the 'overt' politics of battles for formal political control of key arenas and the resources attached to them. These days, analysts struggle with the recognition of the huge multiplicity and complexity of governance processes manifest within an urban area. There are the various and usually sectoralized agencies of higher tiers of EU, national and regional government; the different segments of a municipal administration and the professional and policy communities to which they are typically connected; the organizations of representative democracy, with the structures of municipal councils and neighbourhood districts, linked to the organization of political parties and electoral processes. There are a whole array of individual actors and social networks and lobby groups linking businesses and citizens to the formal arenas of government, intertwining in complex ways the worlds of the state, the economy and civil society. These linkages looking 'from outside in' to formal government are complemented by the linkages 'from inside out', ranging from formal initiatives in building partnerships or 'empowering' particular groups to participate in 'their' governance, to the informal networks which connect politicians and officials with firms and lobby groups, and in which they participate as citizens themselves. Thus the landscape of the 'everyday life' practices of governance in an urban region is not a clearly demarcated territory of levels of government and public versus private action, but a complexly intertwined social reality in which integrations and boundaries, cohesions and exclusions cannot be read off from simple 'maps' of organizational structures.

In the author's own work in exploring territorially-focused governance transformation processes, four strands of conceptualization are combined together: Giddensian concepts of the binding flows which link agency to structuring processes (Giddens, 1984; Bryson and Crosby, 1992); Hajer's notions of discourse structuration and institutionalization (Hajer, 1995); concepts of levels of power developed by Lukes and re-worked by Dyrberg (Dyrberg, 1997), and ideas about institutional capacity-building developed by Innes and Booher (2003) and Healey and colleagues (Healey, 1998; Healey et al., 2003).

Giddens' structuration theory addresses the interaction between external 'structuring' pressures and how actors-in-social-relations respond to these through an emphasis on the formative interaction between 'structures' and 'agency' (Giddens, 1984).[3] Giddens identifies key linkages through which this interaction flows, and which in turn shape the materialities and identities of actors and create the structural forces which they experience. The first is the flow of material resources (goods and assets, technologies, finance, labour power). The second is the flow of authoritative resources or regulatory power, the power to regulate the behaviour of others through formal and informal norms, codes, laws, etc. The third is the flow of ideas and frames of reference, the power to generate new imaginations and shape identities and values. These flows interweave with each other in different spaces and through different temporalities, producing the material distributions and mental expectations which generate the diversities and inequalities in life experiences. To the extent that they follow stable patterns, they generate the 'structural forces' that exert such a powerful influence on opportunities for particular people in particular places. But their patterning is also in continual formation, through shifts in their interactions and through the continual creative work of agency in both realizing and resisting the traditions and forces which bear down upon them. Transformations in urban governance driven by structuring forces would, following this approach, be identified through the power mobilized as resources are circulated, regulatory norms activated and concepts brought to life in arguments and justifications. Transformations driven by agency power in specific situations would be identified by their capacity to produce different patterning tendencies through the impact of the active work of, for example, a territorial collective actor arena.

But such an analysis would say little about how such dynamics were brought about and in effect could become little more than the traditional input–output analysis of Eastonian policy analysis which treated an institutional site of policy-making as a 'black box' (Easton, 1965). For many years now, it has been recognized that it is necessary to enter this 'black box' to examine the complex dynamics of actors and networks as they struggle over resource allocation, over regulatory practices and over framing ideas. Over the past 20 years, there has been an explosion of in-depth case studies of governance processes examining policy formation, policy change and policy 'implementation'.[4]

Initially, the focus of the stories of these cases was on themes such as the inadequacy of the formalized rational planning model to account for the processes described, or the power of particular groups to define what had been assumed to be the 'public interest' which government agencies should pursue. More recently, drawing on policy discourse analysis, there has been a stronger emphasis on how policy change is brought about. A primary inspiration here has been the work of Maarten Hajer (1995) who described the processes of 'discourse structuration' and its subsequent 'institutionalization'. In his work on the acid rain controversy, he looked at a period of policy contestation in a particular department of British government during which the structure of framing ideas was finally shifted by the efforts of activists and scientists who eventually won the argument over the causes and effects of British power station emissions on the acidity of lakes and rivers in north-west Europe. Once this argument was won at the centre of a centralized government policy system dealing with energy and pollution effects, it spread through an array of sub-national government arenas through the central power to define regulations and monitor regulatory practices. The new discourse thus became institutionalized, embedded in norms, practices and discourses, as a new normality. This raises the issue of how discourses and practices come to 'travel' from one arena to another and what happens to them as they do so.[5] Following these ways of analysing governance processes, urban governance transformation could be identified where a new discursive frame appears and diffuses to a range of arenas with sufficient effect to shift significantly the way resources are allocated and regulatory tools formulated and used.

3 There are many critiques of Giddens' structuration theory [see, for example Seidman (1998) and Thrift (1996)], but the basic conception of the mutual constitution of 'structure' and 'agency' remains a very helpful approach.

4 See, for example, Flyvberg (1998), Harding (2000), Healey et al. (2003), Hillier (2000), Murdoch and Abram (2002), Raco (2003).

5 See Innes and Booher (2003), Healey (2004a), Hajer (2003). In developing their concepts of 'institutionalization', Gromart and Hajer (2003) have recently been drawing on some of the concepts in the sociotechnical systems literature with its particular conceptions of actor-networks involved in the formation of knowledge and technology, and with the capacity of this knowledge to 'travel' from one institutional arena to another when 'black-boxed' into a discursive metaphor, a technique or a procedure (see also Murdoch and Abram, 2002).

Level	Dimension
Specific episodes	*Actors* – key players – positions, roles, strategies and interests *Arenas* – institutional sites *Ambiances and interactive practices* – communicative repertoires
Governance processes through which bias is mobilized	*Networks and coalitions* *Stakeholder selection processes* *Discourses* – framing issues, problems, solutions, interests, etc. *Practices* – routines and repertoires for acting *Specification of laws, formal competences and resource flow principles*

Figure 21.1 Levels and dimensions of governance: an analytical schema
Note: This figure is an adaptation of equivalents in Healey (2006) and Coaffee and Healey (2003).

In the author's own work on the power of ideas and practices generated in a partnership to diffuse more widely into 'mainstream' urban governance practices, significant barriers to such diffusion were found (Healey et al., 2003). This confirmed not only the experience of other similar episodes in innovation in area partnerships across the city that were examined, but echoed much of the British experience of urban regeneration partnerships (Carley, 2000; Hastings, 1996; Healey, 1997b; Healey et al., 2003; Lowndes and Skelcher, 1998; Peck and Tickell, 1994). In trying to grasp the nature of these barriers, there was a move away from a conception of the landscape of governance as merely an array of different institutional sites or arenas, articulated with each other and with the worlds of economic activity and civil society by dynamic webs of relations, all operating as sites of everyday practices. It was felt necessary to explore more explicitly the deeper structuring that was present in these practices but not always recognized by actors in the episodes in which they were engaged. The idea here is that transformative initiatives which succeed in 'institutionalization' need to have the capacity to 'travel' not just from one arena to another, but from one level of consciousness to another. By this, a translation is meant from the level of conscious actor invention and mobilization to that of routinization as accepted practices, and beyond that to broadly accepted cultural norms and values. To analyse this, a concept of 'levels of power', first articulated by Lukes in 1974, and reworked later by Giddens (1984) and Dyrberg (1997) is studied. Figure 21.1 summarizes this conception.[6]

Figure 21.1 expresses the hypothesis that, to have transformative effects, governance innovations (such as new discourses, new allocatory or regulatory practices, the formation of new arenas or networks) must move from explicit formation episodes to arenas of investment and regulatory practice. To endure, they have to become institutionalized in the routines of governance practices. At the level of specific episodes of experience, those actively involved may develop what they believe are new and relevant conceptions of issues, objectives and appropriate processes which guide their own practices. But it is hard to get these new conceptions and their implications to 'travel' to other arenas just through processes of 'snowballing' individual learning experiences. This is not just because there are other actors in other arenas seeking to resist such initiatives or to promote their own learning experiences. New concepts have to challenge and shift an array of already routinized governance processes, with their complex mixture of conscious and taken-for-granted modes of practice. New concepts have to 'jump' boundaries and 'break-through' resistances, involving implicit and explicit struggles. For initiatives seeking to create a new concept and arena for territorially-focused collective action, this may involve a complexity of relations between different departments of local government, between administrators/officials and politicians, between politicians/officials and citizens, between the state and all kinds of power elites and lobby groups, each with their own relation to allocative power, regulatory power and their own discursive frames. Sustaining and legitimizing both governance processes and specific episodes of governance are cultural assumptions about appropriate agendas and practices of governance held by different

6 Figure 21.1 in effect provides a way to express the structure/agency relation, often expressed by others in terms of linking macro-, meso- and micro-levels of power. See also Fischer (2003) for a similar conception of levels.

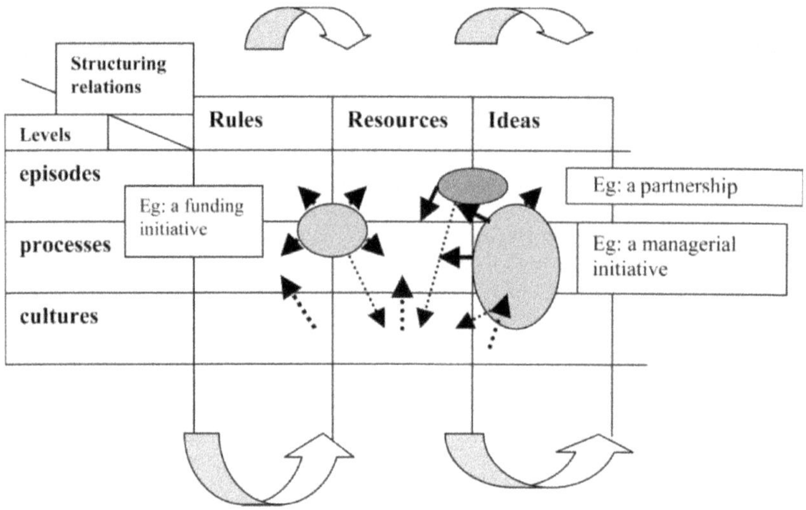

Figure 21.2 Transformation initiatives in governance dynamics
Note: Although laid out as separate levels and structuring dimensions, in any actual instance these all interact.

social groups in society generally. It is through these assumptions, as recognized by actors themselves and the 'media chorus' of critical commentary on their performance, and as evaluated in the formal procedures for challenging governance actions, that those involved in governance are held to account and their legitimacy judged. Shifts at this cultural level, promoted by long-term shifts in economic, sociocultural and political relations, influence both those involved in particular episodes of governance and those involved in routine governance practices. These 'prepare the ground' for new ideas and discursive vocabularies.

These levels are not separate realities but mutually constitute moments of governance activity and the everyday life experience of 'doing government work'. The reasons for an analytical separation is that the levels move according to different temporalities and respond to different driving forces. The pressure for transformation may come from any of the levels. Episodes of innovation may create pressures to change governance processes more generally, but there may also be mobilization efforts to initiate such changes elsewhere in governance systems. Shifts in cultural assumptions may put pressure for change on governance processes but provide resources for episodes of innovation. This conception of different experiential levels of governance firstly emphasizes the complexity, the multiplicity of interacting and often counteracting movements promoting and resisting change, the multiple timescales and the likely instability of urban governance transformation processes. Secondly, it stresses that significant transformation would have to affect the level of governance process at the least. To achieve this, initiatives would have to move from the level of an episode to the level of processes, and in some way find resonances with cultural assumptions to have any capacity to be seen as legitimate and to endure, that is, to 'institutionalize'.

Figure 21.2 links this conception of levels of governance with the earlier discussion of the Giddensian relation between structure and agency. Episodes of governance are shaped by structuring power embodied in the mobilization of rules and norms, material resources and framing ideas, but they may also be creative of them, implicitly by interpretive work or explicitly by direct challenge. Rules, resources and framing ideas may be vigorously sustained by mobilizing work at the level of governance processes. But these processes may also be open to pressures from broader cultural shifts and from the learning and mobilizing taking place in specific episodes. At the level of cultural assumptions, the authority of rules, the justice of resource allocations and the validity of framing discourses is under continual challenge and re-formulation, re-moulding the principles of legitimacy and accountability through which governance processes and episodes are judged.

In this approach, it is clear that what we take to be 'politics' and 'policy' are not articulated merely in either the formal arenas of council chambers or the informal processes in which politicians are involved. As Hajer (2003) argues, politics and policy-formation emerge from many institutional sites and processes. In this complex institutional terrain, building a new territorial arena of governance or a new mode of politics is likely to involve very challenging efforts in mobilization, in discourse formation and diffusion, and in aligning old practices with new discourses. This is widely acknowledged in the literature on the difficulties of establishing urban region or metropolitan governance where none existed before (Albrechts et al., 2001; Healey, 2004; Herrschel and Newman, 2002; Salet et al., 2003). Yet sometimes new territorial arenas have emerged as powerful collective actors. The structural arguments for 're-scaling' urban governance to the urban region scale suggest there are contextual pressures acting on all the levels and structuring dimensions of governance identified in Figure 21.2 which create a 'moment of opportunity' for such institution-building work to succeed.

This leads to the fourth strand of concepts, which focus on the building of 'institutional capacity' (Amin and Thrift, 1995; Innes and Booher, 2003; Healey, 1998; Healey et al., 2003). Some analysts focus on the qualities of 'leaders', building on notions of individual capacities ('human capacity'). This kind of approach leads to an emphasis on the qualities of actors such as City Mayors and key professionals who have had a role in transforming some city government agendas and processes. But enduring governance transformation rarely results from the actions of single individuals. Most transformatory episodes mobilize networks of political supporters and expertise, often linking across traditional boundaries and divisions, and with the knowledge to challenge concepts and assumptions embedded in established practices. This suggests that capacity building is a social process, generating, to use Innes and Booher's concept, intellectual, social and political capital (Booher and Innes, 2002). In the author's own work, the notion of 'resources', rather than 'capital' has been preferred.[7] This examined how and why the resources deployed and developed in an episode of governance largely failed to diffuse into the wider governance context (see the examples which follow). More generally, this work on institutional capacity-building emphasizes that urban governance transformation which leads to the institutionalization of a new territorial collective actor with significant authoritative and generative power needs to mobilize and build knowledge resources and relational resources (social networks) which not only help to consolidate power and legitimacy around the new arena but have the capacity to carry the new ideas, understandings and recognitions of opportunity and struggle through to a wide range of other arenas in the urban governance landscape where practices shape how resources flow and regulatory rules are exercised.

In the next section, the approach outlined earlier is illustrated through three brief examples. They have been selected from the author's own work in Newcastle as representative of attempted innovations, initiated from different positions in an urban governance landscape. The first is a city centre regeneration partnership.[8] The second is an attempt at a city-wide development strategy.[9] The third is an initiative arising from outside the range of formal government which has slowly been drawn into a more formalized relationship.[10] In each case, the initiative as an episode of governance activity in terms of its territorial focus of attention and its position in the governance landscape is outlined. Then, the potential of each initiative to institutionalize, using the analytical lens of Figure 21.2, is assessed. Finally, the implications of these experiences for the potential for governance transformation in the Newcastle governance context are commentated on.[11]

7 'Capital' suggests that different resources can be agglomerated into a unitary currency, which is not the author's view.

8 See Healey et al. (2002, 2003). The author would also like to thank collaborators, Claudio de Magalhães, Ali Madanipour and John Pendlebury, and to the ESRC, who financed project R000222616.

9 See Coaffee and Healey (2003). The author would also like to thank collaborators Jon Coaffee and ESRC Ph.D. student Lorraine Johnstone.

10 The author would like to thank colleagues Sara Gonzalez and Geoff Vigar, and EU F5 SINGOCOM project leader Frank Moulaert, for permission to use this material.

11 The Newcastle area was selected not because it has a striking history of governance transformation but because, through the various research projects conducted there by the author and colleagues in SAPL/ GURU, the author has a rich understanding of its governance dynamics.

The Lurching Pace of Innovative Momentum

The Newcastle Governance Landscape

As an urban region, the area is one of Britain's industrial conurbations, with a population of around 1.5 million in its travel-to-work area, which also includes rural areas, coasts and hills, and attractive small towns. Although considerable shifts in social composition are taking place, with an expanding middle class and an ageing and largely static overall demographic profile, it is an area still characterized by high levels of unemployment and social deprivation relative to national levels. There is currently no overall political-administrative government for the most urbanized part of the conurbation, which is divided into five large metropolitan districts. Politically, Newcastle is just one of those districts, although, with some political reluctance from neighbouring districts, Newcastle city centre is recognized as the 'regional centre' of the wider urban region. Metropolitan districts are responsible for education, social service delivery, planning and some aspects of transport provision, as well as environmental health, housing management (of what was, in Newcastle's case, once a large council-owned housing stock) and promotion of leisure, sport and cultural provision. Councils also engage vigorously in economic development initiatives, though their work in this field is on a much smaller scale than that of regional agencies. There are also important agencies operating at the regional scale dealing with economic development, health, tourism promotion, and arts and culture promotion, and some others operating at the scale of the conurbation, notably the Passenger Transport Authority which oversees public transport provision. The formal government landscape is complex and cluttered, with many areas of overlapping responsibility.

Local authority funds have come predominantly from central government and have been significantly constrained in recent years. Extra central government funds have been channelled through an array of special initiatives – for housing estate regeneration, area regeneration, overcoming social exclusion, education, healthcare, transport and safety and security. These have brought into conjoined actions in various arenas at different times agencies organized at metropolitan and regional scales. Central government activities are in part coordinated through a Government Office for the Region (GO.NE). With the possibility of regional devolution and an elected regional assembly in prospect, the region also has a Regional Assembly of elected representatives from the different local authorities.[12] Outside the organization of politics and administration, business groupings have consolidated what were dispersed chambers of commerce and lobby groups into a regional Chamber of Trade and Industry. Throughout all these recent changes, and despite some personnel changes, some kind of 'governing elite' can be identified, still largely associated with the old industrial politics, dominated by trades unions, a paternalist mode of governance, and focused primarily on poverty reduction and economic development.[13]

In terms of governance, metropolitan city councils in England remain major political actors within this complex governance landscape in which sectorally-organized policy communities and their networks encounter each other but remain driven by their own logics. City councils are themselves multifunctional organizations, with several departments linked to functional service delivery tasks, and with tensions in inter-departmental linkages. Within the Newcastle area, the councils vary in the extent to which it has been possible to develop and maintain a strategic and coordinated approach to city council organization. There have also been initiatives in all councils for devolution to sub-areas, a movement encouraged currently by central government's proposals for 'modernizing' local government, which advocates the creation of 'area committees' (DETR, 1998; Coaffee and Healey, 2003). The creation of a council-wide system of area committees, however, challenges traditions of ward politics and a whole array of special area development initiatives pursued in different neighbourhoods over the past 30 years. Politically, Newcastle City Council seeks to operate as a strong collective actor, with the territorial scale expressed primarily in terms of political and administrative legitimacy rather than any sense of the specific relations and qualities to be found in the city.

---

12 Under recent legislation, regions can set up political assemblies with powers 'in between' those available to Wales and Scotland, if a majority of local voters support this in a referendum, which will also convert all local authorities into a single administrative layer (i.e. abolishing either counties or districts where a two-tier system still exists). The first referendum was held in the North East region in November 2004 and was heavily defeated.

13 The opportunity of the recent and ultimately unsuccessful 'Cultural Capital of Europe' bid helped to enrich this traditional pre-occupation, but was continually justified in the language of job creation and social inclusion.

The three examples selected are all within the area of Newcastle City Council itself. In terms of economic and social life, the boundaries of the city council area have little meaning. But in terms of the connections between social and economic networks and political ones, the impact of a territorial boundary can be considerable. Access to many material resources allocated by the public sector and to negotiation over regulatory approvals requires some relationship with one or more parts of the city council. This is very obvious in the case of the Grainger Town Partnership.

Revitalizing the Historic City Centre: The Grainger Town Partnership

This Partnership was created to promote the physical, economic and social regeneration of the nineteenth century core of Newcastle city centre, which had been progressively undermined by the development of new shopping complexes to the north in the 1960s and 1970s and the Metrocentre in an 'edge city' location in neighbouring Gateshead, as well as riverside redevelopment for offices and leisure activities to the south. Initiated by conservation and urban design campaigners within and around the city council in the 1980s, the project grew into a regeneration project in order to attract substantial national government funding. In doing so, it had to widen the stakeholders involved to include business interests from the retailing, commercial and property sectors. This was done through the regional Chamber of Commerce and a special business promotion agency, The Newcastle Initiative. The funding streams to which it had access were in part channelled through regional bodies. But when the Partnership was formally created, it was in the form of an 'arms-length' agency of the city council. This meant that the projects and practices of the Partnership were under continual challenge from city councillors and officers who felt their own autonomy threatened by the Partnership mechanism and the introduction of strong business partners. For both councillors and officers, the city centre as an area focus had traditionally been of less interest than the poor neighbourhoods to the east and west, which were the heartlands of traditional working class support for councillors and the centre of the council's role as a housing provider.

The Partnership in its formal lifetime (1997–2003) achieved substantial physical improvements in Newcastle city centre, helped by the uplift in property markets regionally and the trend in city centre living among young professionals in the UK. Those involved, from different segments of the business community, those concerned with conservation and design as well as the increasing numbers of residents, not only came to appreciate the area focus of the Partnership but valued the consultative style pursued by the Partnership, which developed an array of mechanisms for consultation with different groups. The Partnership drew on and drew in a number of different networks into its governance arena. It also helped to create social-political networks. These could have some influence on the fine detail of programme development and delivery. The author's research also shows that the interchanges between actors located in different positions led to a considerably enhanced understanding of the dilemmas which had to be faced in managing the complex socio-physical interactions of the city centre public realm. But the knowledge resources and networks generated through the Partnership were not so easily diffused from the Partnership arena to the wider relations of local area management in Newcastle. Nor was it clear how much of the Partnership style and focus would survive as it was 'folded' back into the city council in 2003, as the main extra funding came to an end, a typical experience of the preceding urban regeneration partnerships in the city.

In terms of power dynamics, this episode depended on the nature of the funding available to the Partnership. Without the national funding, it is unlikely that a project of this scale would have taken place and an area-focused Partnership form for the city centre created. The project challenged the geographical imagination which councillors had of the city, which encompassed electorally-significant neighbourhoods in poverty which needed re-distributive attention, while the city centre was the territory of commercial interests which, it was assumed, could look after themselves and even generate funds for the city council. Significant regulatory responsibilities (planning and environmental health) remained in the hands of the city council, and these arenas were used to challenge Partnership proposals. In this case, the authority derived from governance culture and formal government process was used to maintain clear boundaries between the Partnership and the Council, generating all kinds of resistances to the flow of ideas and practices into the Council (see Figure 21.3). Other pressures were acting on the Council to introduce a new mode of politics, to encourage a much more open and consultative relationship with citizens generally and all kinds of stakeholders, generating a momentum which could not be avoided. But these pressures bore into the Council from different directions

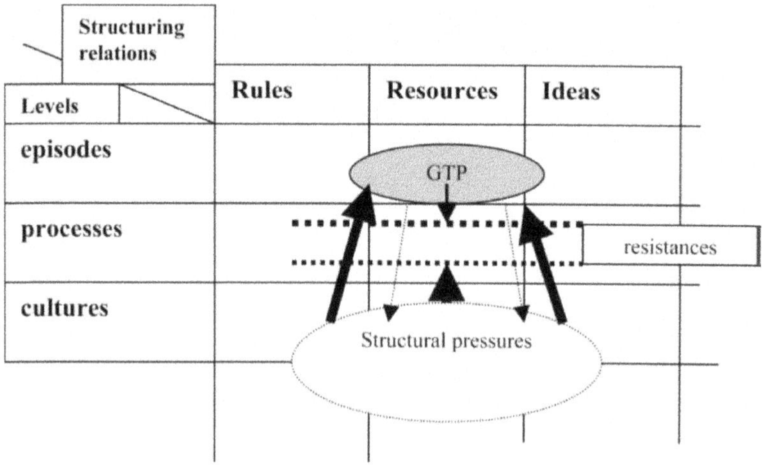

Figure 21.3 Transformation initiatives in governance dynamics: Grainger Town Partnership (GTP)

and cannot be attributed to the impact of the Grainger Town Partnership (see the next case). Thus the Grainger Town Partnership created an effective but contained and temporary arena in the Newcastle governance landscape. It innovated a strong focus on the area of the city centre and attempted to introduce a new kind of consultative politics, but in the end its power base was too weak to expand and endure. However, it has left behind tensions as many stakeholders valued its presence and argue for some continuation.

A Strategic Initiative: 'Going for Growth'

For many years, the city council (in its political and administrative activities) was held together by a strong pragmatic leader,[14] backed by a group of senior councillors and heads of departments. Council practices were structured by service delivery functions, in which councillors played an important role as 'patrons' to the residents in their wards, particularly helping to solve their housing problems. In the mid-1990s, the leader was forced out. In the past 10 years, the Council has struggled to develop a new direction and mode of operation. A new Chief Executive, with a background in management strategy, backed by some councillors, recognized the value of having a clear over-arching strategy for the city. This could provide a focus for political decisions, coordinate the activities of the various council departments and position the Council as in the forefront of the post-1997 national Labour government's agenda for 'modernizing' local government. Such a strategic initiative represented a break with the pragmatic past, when a clear strategy was considered a hostage to fortune in the shifting sands of national and local politics and funding opportunities.

The result was the production of a strategy known as 'Going for Growth' (NCC, 2000). This was articulated by a core group within the Council. It was based on the assumption that the problems the Council faced were well-understood, after years of small-scale regeneration attempts in different parts of the city. The city's population was falling. More affluent families were leaving for the surrounding areas, leading to a concentration of an already large number of poorer families, badly affected by a long history of temporary and then permanent unemployment. Council income was falling, while the problems it had to address were increasing. In addition, its large stock of 'council housing', along with other parts of the low cost housing stock, were increasingly affected by disrepair and abandonment. There was plenty of space, on 'brownfield land', within the city for housing to accommodate more people. But the social character of many parts of the city had undermined many

14 In the UK, the Leader of the Council has traditionally been the important political figure, the mayor being a ceremonial position. As part of the 'modernization' of local government, there is an encouragement for new, more powerful elected mayors, as in the London case.

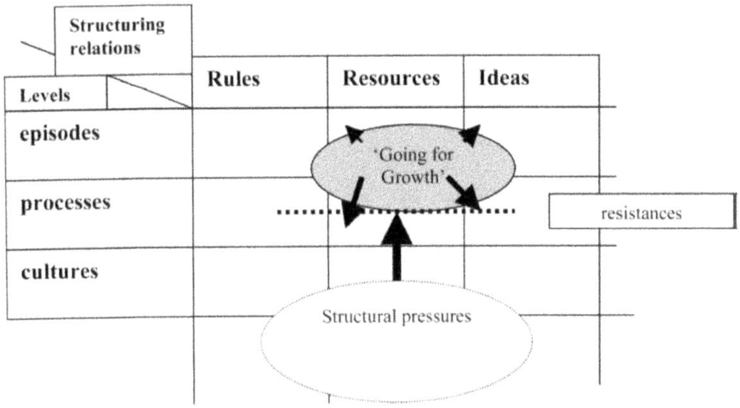

Figure 21.4 Transformation initiatives in governance dynamics: 'Going for Growth'

past attempts at improving and diversifying both the housing stock and the social life of neighbourhoods. The solution, in the words of the Strategy, was to turn around population decline and promote large scale redevelopment within the poorer neighbourhoods. This in effect meant demolition of some streets and the creation of new neighbourhood environments with a different (and more middle class) ambience.

The Strategy was discussed with some elite stakeholders[15] in the city, who were pleased to see a more strategic focus and interested in the potential for new development areas, especially if backed by national government funds. After a brief consultation exercise on the Strategy in 1999, the Council then commissioned 'master plan' studies for the two main areas for comprehensive redevelopment. These were produced in mid-2000. A more intensive consultation exercise followed. At this point, residents in these neighbourhoods, many of whom had been involved in council regeneration initiatives before and who expected a reasonably close relation with their ward councillors, exploded in protest (Byrne, 2000; Healey, 2002; Cameron, 2003). In part, this was due to the content of the master-plans. Residents were worried about road proposals which seemed to turn some streets into major highways. They were anxious about their houses, fearing demolition, especially if they had spent a good deal on interior decoration, despite the abandonment going on around them. In any case, the neighbourhoods were already very diverse with different qualities in different streets. But above all, they felt disconnected from the processes which had produced the master-plans. They too understood the problems. They were living with them on a day-to-day basis. What they demanded was to be 'part of the solution'. Especially, they resented the lack of consultation with them about their futures. They were challenging the Council's mode of governance as much as the content of the strategies.

Well-orchestrated citizen protest led to a fundamental re-think within the Council. The overall Strategy was retained, but the Strategy Team shifted attention to engage in very detailed consultations with residents through which new plans were produced and new ways of understanding the plans developed. Thus a new mode of politics began to emerge, with the Strategy used as the basis for obtaining substantial government funding for urban regeneration. But by 2003, many of the senior councillors and officers actors involved in the 'Going for Growth' initiative had left the Council. Some of the elements of the Strategy are rolling forward, backed by national government money. Some of the consultative practices have spread into other areas of the Council's work. But at this moment, the Council is once again in re-organization, the outcome of which is not yet clear. The struggle here lies within the Council, over modes of politics and, implicitly, over concepts of the city and over the priorities which should underpin Council strategy and programmes. Collective actor capacity in terms of relations with other stakeholders and with other government levels is continually undermined by this uncertainty (see Figure 21.4).

15 Especially from other public agencies and commercial interests.

The 'Going for Growth' initiative can be interpreted as an attempt to 'upscale' the horizons of councillors and officers of the city council, to focus attention on the city as a whole, and the wider picture affecting options for action and the coordination of effort across departments. It was very much encouraged by the wider climate of local government 'modernization' as well as the building up of an emphasis on the wider 'region'. Key actors in the Council sought to position themselves in this wider discourse and arena. But to do this, they had to change the mode of operation of local politics, particularly the relations between citizen and councillor, and the internal organization of the council, which had operated in the past very much through individual service fiefdoms. But in attempting to make this kind of shift, they initially neglected to consider the relation between citizen and state, particularly among the councillors' heartland voters in the poor neighbourhoods. Since 2000, the Council has set up Local Strategic Partnerships and Area Committees, following national guidance, and has sought a much more consultative relationship with residents, in effect the kind of mode of politics promoted in the Grainger Town Partnership and in earlier partnerships in the area. But how far this will protect the politicians from electoral challenge from Liberal Democrats,[16] and how far either party will continue to sustain an 'up-scaled' strategic approach and a deeply consultative relationship with residents, remains to be seen. Transformation in governance discourses and practices is a long process with many 'folds' and 'bends'.

Mobilization Outside the State: The Ouseburn Trust

Within the UK, the state in the form of national and local government bodies, is a very strong presence. This has meant that governance initiatives outside the state have had difficulties growing and surviving without finding a way to link to formal government in some way. This issue is currently being examined in relation to 'social economy' and related initiatives in an EU F5 project led by Frank Moulaert (SINGOCOM). The final example is drawn from this work (Gonzalez and Healey, 2005). The Ouseburn Valley is an area of the city between the city centre, poor neighbourhoods to the east, and an area of intensive riverside redevelopment initiated in the late 1980s. It is a typical 'in-between' part of the city, with old industrial buildings being used for various workshops, some ecological projects, and pubs which attracted a distinctive 'alternative' clientele, all clustered down the sides of a steep valley, making it a curious and attractive physical locale.

The initiative which turned this 'in-between' place into an active 'place-for-itself' was sparked off by mobilization 'against the state'. A group of local people, including someone from the local church, organized first to demand a voice in the major redevelopment of the riverside area, so that the Ouseburn Valley area would not be cut-off from the main riverside. Success in this mobilization gained them the status to become active participants in the ongoing consultation process which developed around the riverside regeneration projects. This focus soon widened out to include people concerned with industrial heritage and environmental issues, neither of which were well-articulated policy issues in the city council, where the focus had long been on 'housing' and 'jobs'. By 1996, these various networks around the Ouseburn area had become consolidated into a formalized Trust. They then widened out further to involve other community, business and local groups and the city council, forming a Partnership which obtained modest funding from central government for development activities. This funding finished in 2002, and the Trust now survives with a more permanent link to the city council, employing four officers seconded from the city council in the Ouseburn Resource Centre in the area, dealing with several development projects.

The trajectory of this initiative is not unlike the Grainger Town Partnership (see Figure 21.5). It started off through an activist campaign around neglected issues, developed a wider agenda in order to participate in government opportunities for consultation and funding, became more formalized, and then was absorbed more closely into the city council organization, where its future has to jostle with many other commitments and organizational arenas, and will depend on both the commitment of a few activists and how well the initiative 'fits' the prevailing interests of local and national politicians and officials. It has created a 'place identity' which has filtered into the consciousness of both government actors and local people. To safeguard as much autonomy as possible, the Trust has set out to acquire some land and buildings. It has also built links to a range of national agencies. It has been a local pioneer for a fine-grained, ecologically-sensitive and

16 It did not. The Liberal Democrats gained control of the city council in May 2004.

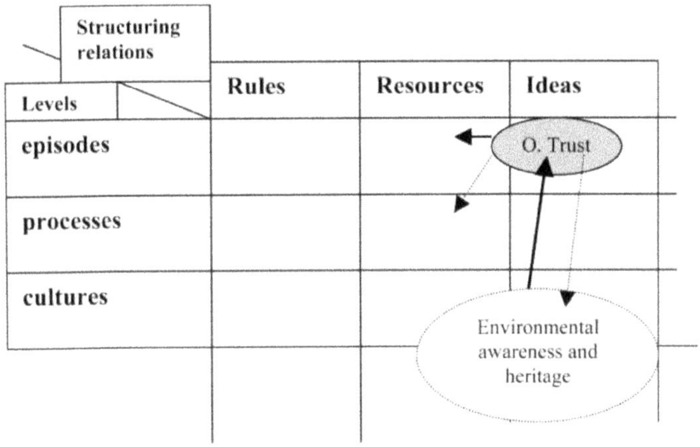

Figure 21.5 Transformation initiatives in governance dynamics: the Ouseburn Trust

consultative approach to project development, which is in line with much national thinking but challenges Council traditions. It thus provides an exemplar of a new way of doing things, but, as with the Grainger Town Project, has had little wider impact on Council discourses and practices, except perhaps as a demonstration project of 'downscaling' governance activity and what a new mode of local, socially-driven entrepreneurial development might look like. Nor has it made links to a wider 'movement politics' which could have enlarged its scope of influence.

Struggles for Change

These 'cameos' of governance activity in and around Newcastle City Council all seem to suggest the continuing capacity of a local council to resist change and to undermine the innovative potential of experiments in new government forms. Yet the content of these pressures and the initiatives, conflicts and struggles they give rise to are not the same as they were in the 1970s or even the neo-liberal 1980s. The context of power dynamics, funding, demands and expectations is different. The terrain of invention and struggle is different. The interpretation is that, drawing on Figure 21.1, the level of governance processes in Newcastle is 'in movement', particularly since the change of leadership in the mid1990s (Coaffee and Healey, 2003). What is unclear is the trajectory of this movement, and whether it will ever come to 'rest' in any kind of stable set of networks and coalitions, discourses and practices. Within this 'movement', there are some signs of a re-scaling of focus, from political wards and social housing neighbourhoods, and the promotion of the city's direct interests in land, property and its income base, both upwards to its wider role as a city in a region and downwards to neighbourhoods and locales within the city. The 'Going for Growth' experience shows that this double re-scaling cannot be achieved without equivalent attention to residents' concerns, particularly in the political heartland poor neighbourhoods. There are initiatives underway to develop a more systematic approach to absorbing residents' views into the shaping and delivery of Council policies (the Area Committees etc.). But this brings different arenas in different locales within the city into clear conflict with each other, whereas in the old days of councillor patronage, it was the councillors who struggled in the City Council Chamber on behalf of their ward constituents. The acute tension for city council politicians and managers is how to manage the shift going on in its broader socio-economic context, which is bringing many more middle class people to work and live in the city and thereby over time potentially changing both policy priorities and modes of politics. Can a city council transform itself without struggle and crisis? Or are these essential ingredients for 'unlocking' the inertia of 'old ways'? Certainly, Newcastle City Council has been experimenting in all kinds of ways, which at least provides an experience on which to draw if and when the various strands of initiative come together into a 'new way' of doing governance in the city. But inevitably, such a transformation will and should involve struggle, as it would re-distribute access to resources flows, regulatory practices and the formulation of policy ideas.

Conclusions

The rhetoric of the well-publicized initiatives in urban and regional governance 'modernization' in Europe provides many indications of an emphasis on new scales for the articulation of policy and politics, for the emergence of a territorial focus and for the development of new modes of more collaborative governance. But a fine-grained and institutionally-situated examination of such initiatives, such as those described in the previous section, suggests that whether and how far these practices develop is highly variable and contingent on local particularities (see also Le Gales, 2002; Swyngedouw, 2005).

In relation to re-scaling, the Newcastle governance case suggests that the old power of the city council is under serious challenge, with a strengthening possible at both the regional and the sub-municipal level. But the arena of the metropolitan district remains important, both in terms of the articulation of formal politics, and the organization of policy and administration. That this power is constrained by national government regulatory and resource allocation power has been a long-standing feature of central-local government relations in Britain, and shows few signs of abating as yet in England. It remains to be seen whether new regional levels emerge with real power to act 'for themselves' rather than operate as creatures of central government, or local government, or both (Jeffrey et al., 2002; Murdoch and Norton, 2001). The continuing strength of city councils in England lies in part in the range of functions they still perform and in part in their well-established legitimacy and authority. But it also lies in the expectations that others have of 'the Council'. Residents and business interests may be very critical, and may complain of the distance between local government and citizen, but yet they expect councils to 'do things' for them and often wrap together a whole array of government bodies as an imagined collective 'council'. Structural imperatives for both more efficiency and more open and participative democratic processes challenge traditional practices, but do not necessarily encourage the re-formulation of scalar relations.

This raises the question of territorial focus. Curiously, and despite the 'Going for Growth' initiative, the institutional arena of Newcastle City Council has a very uncertain conception of Newcastle as a place. This vagueness about territorial qualities is shared with many other organizations in the city (Vigar et al., 2005). The city council is clearly a powerful collective actor, but the city as a territory is only implicitly present. For politicians, councillors, citizens and other stakeholders, the Council is a collection of service departments, not an aggregation of distinctive locales or a locale in and for itself. This is probably true of many other city councils in the UK. In such a context, the rhetoric of 'joined-up government' and of 'integrated governance' promoted by the UK national government in the late 1990s and early 2000s finds difficulty in transferring to an actual practice. Where it does so, the focus tends to be on the needs of individuals and firms in locales, rather than on the qualities of places, that is, developing a voice for people-in-places. The current initiatives in integrated council strategy and in area/neighbourhood government may have the effect of creating new policy spaces in which a new politics with a strong 'territorial focus' will emerge, but there is no clear evidence for such a development at present in Newcastle.

The pressures for a more consultative and collaborative style of policy-making from residents is clear in the Newcastle case. In fact, the experience of area regeneration partnerships in the past had encouraged the demand for 'voice' in the poorer neighbourhoods (Coaffee and Healey, 2003). This is challenging the old paternalist and quasi-clientelist practices of councillors and the delivery modes of officials. But councillors and officers have also to adjust to the demands of central government to perform according to a 'best value' target-driven performance regime, which gives little space for transforming citizen-state relations in significant ways (Stewart, 2002). Caught in this structural tension, initiatives in more intensive resident consultation are driven by the search to underpin the authority of council attempts at strategy with the legitimacy of 'deep consultation', but residents are often well-aware that the council needs the consultation for its own purposes rather than as part of a shift in power relations between council and citizen.

In the Newcastle case, therefore, my analysis suggests that governance processes are 'on the move'. This conclusion is made not because of the array of new initiatives of all kinds, but because, at the level of government process (see Figure 21.1), the mixture of attempts to manage and resist the changes set in motion by the new initiatives and the general sense of uncertainty (reflected in regular re-organization attempts) suggests a diffused sense of the need to change. But what kind of trajectory will emerge remains very unclear, and is still strongly dependent on a central–local government power dynamic. In this dynamic, some business interests and residents in some neighbourhoods may gain, but a new inclusive 'politics of space' which could provide a voice for people across the city in relation to the different needs, lifestyles and their different locales of living seems a long way off. This case thus confirms Le Gales' general conclusion for urban governance in Europe that some kind of transformation is underway (2002), but that as analysts we should be careful of deterministic over-generalization.

A possible 'over-generalization' of the Newcastle case could be that the city council level is slowly 'dis-integrating' its established departmentalized organization and paternalist mode of politics. It may be losing power as a new regional scale begins to institutionalize. In which case, this could be an example of de-stabilization prior to a new and re-scaled territorial integration. Such an integration, in the particular context of the north-east of England, where there is a long-standing awareness of economic vulnerability, is likely to focus on territory as a 'container' of economic assets, favouring business interests directly in order to win 'jobs' to favour the traditional north-east worker. However, how far a new arena will be able to accumulate sufficient force to challenge the embedded practices of central–local relations and whether a concept of territory will emerge which is more than a 'container' remain very uncertain potentials at the present time. In 10 years' time, analysts may conclude that the Newcastle governance story is one of continual transformation, of an unending and 'restless search' (Offe, 1977) for a stable resolution to contradictory structural imperatives and historically-embedded discourses and practices. The analytical tools summarized in this paper are put forward to help both in assessing these evolutions of 'governance on the move' and as aids to those actively involved in struggles for transformation.

References

Albrechts, L., Alden, J. and da Rosa Pires, A. (eds) (2001) *The Changing Institutional Landscape of Planning* (Aldershot: Ashgate).

Albrechts, L., Healey, P. and Kunzmann, K. (2003) Strategic spatial planning and regional governance in Europe. *Journal of the American Planning Association*, 69(2): 113–29.

Amin, A. and Hausner, J. (eds) (1997) *Beyond Market and Hierarchy: Inter-active governance and social complexity* (Cheltenham: Edward Elgar).

Amin, A. and Thrift, N. (1995) Globalisation, institutional thickness and the local economy, in: P. Healey, S. Cameron, S. Davoudi, S. Graham and A. Madanipour (eds) *Managing Cities*, pp. 91–108 (Chichester: John Wiley).

Booher, D. and Innes, J. (2002) Network power for collaborative planning, *Journal of Planning Education and Research*, 21: 221–36.

Brenner, N. (1999) Globalisation as reterritorialisation: The re-scaling of urban governance in the European Union, *Urban Studies*, 36(3): 431–52.

Bryson, J. and Crosby, B. (1992) *Leadership in the Common Good: Tackling public problems in a shared power world* (San Francisco, CA: Jossey Bass).

Byrne, D. (2000) Newcastle's going for growth: Governance and planning in a post-industrial metropolis, *Northern Economic Review*, Spring/Summer, 30: 3–16.

Cameron, S. (2003) Gentrification, housing redifferentiation and urban regeneration: Going for growth in Newcastle upon Tyne, *Urban Studies*, 40(12): 2367–82.

Carley, M. (2000) Urban partnerships, governance and the regeneration of Britains cities, *International Planning Studies*, 5(3): 273–98.

Cars, G., Healey, P., Madanipour A., et al. (eds) (2002) *Urban Governance, Institutional Capacity and Social Milieux* (Aldershot: Ashgate).

Castells, M. (1977) *The Urban Question* (London: Edward Arnold).

Castells, M. (1996) *The Rise of the Network Society* (Oxford: Blackwell).

Coaffee, J. and Healey, P. (2003) My voice: My place: Tracking transformations in urban governance, *Urban Studies*, 40(10): 1979–99.

Department of the Environment, Transport and the Regions (DETR) (1998) *Modern Local Government: In Touch with People* (London: DETR).

Dyrberg, T. B. (1997) *The Circular Structure of Power* (London: Verso).

Easton, D. (1965) *A Systems Analysis of Political Life* (New York: John Wiley).

Fischer, F. (2003) *Reframing Public Policy: Discursive Politics and Deliberative Practices* (Oxford: Oxford University Press).

Flyvberg, B. (1998) *Rationality and Power* (Chicago, IL: University of Chicago Press).

Fürst, D. and Kneilung, J. (2002) *Regional Governance: New Modes of Self-government in the European Community* (Hannover: ARL, University of Hannover).

Giddens, A. (1984) *The Constitution of Society* (Cambridge: Polity Press).

Gonzalez, S. and Healey, P. (2005) A sociological institutionalist approach to the study of innovation in governance capacity, *Urban Studies*, 42(11): 2055–70.

Gromart, E. and Hajer, M. (2003) Is That Politics? For an inquiry into forms of contemporary politics, in: B. Joerges and H. Nowotny (eds) *Looking Back, Ahead – The (2002) Yearbook of the Sociology of Sciences* (Dordrecht: Kluwer).

Hajer, M. (1995) *The Politics of Environmental Discourse* (Oxford: Oxford University Press).

Hajer, M. (2003) A frame in the fields: Policymaking and the reinvention of politics, in: M. Hajer and H. Wagenaar (eds) *Deliberative Policy Analysis: Understanding Governance in the Network Society*, pp. 88–110 (Cambridge: Cambridge University Press).

Hajer, M. and Wagenaar, H. (eds) (2003) *Deliberative Policy Analysis: Understanding Governance in the Network Society* (Cambridge: Cambridge University Press).

Hall, P. and Taylor, R. (1996) Political science and the three institutionalisms, *Political Studies*, 44(5): 936–57.

Harding, A. (2000) Regime formation in Manchester and Edinburgh, in: G. Stoker (ed.) *The New Politics of British Local Governance*, pp. 54–71 (Houndmills: Macmillan).

Harvey, D. (1989) From managerialism to entrepreneurialism: The formation of urban governance in late capitalism, *Geografisker Annaler*, 71B: 3–17.

Hastings, A. (1996) Unravelling the process of partnership in urban regeneration policy, *Urban Studies*, 33: 253–68.

Healey, P. (1992) An institutional model of the development process. *Journal of Property Research*, 9(1): 33–44.

Healey, P. (1997a) *Collaborative Planning: Shaping places in fragmented societies* (London: Macmillan).

Healey, P. (1997b) Mandarins, city fathers and neighbours: Crossing old divides in new partnerships, in: O. Kalltorp, I. Elander, O. Ericsson and M. Franzen (eds) *Cities in Transformation:Transformation in cities*, pp. 266–88 (Aldershot: Avebury).

Healey, P. (1998) Building institutional capacity through collaborative approaches to urban planning, *Environment and Planning A*, 30(9): 1531–46.

Healey, P. (1999) Institutionalist analysis, communicative planning and shaping places, *Journal of Planning Education and Research*, 19(2): 111–22.

Healey, P. (2002) Place, identity and governance: Transforming discourses and practices, in: J. Hillier and E. Rooksby (eds) *Habitus: A sense of place*, pp. 173–202 (Aldershot: Avebury).

Healey, P. (2004) The treatment of space and place in the new strategic planning in Europe, *International Journal of Urban and Regional Research*, 28(1). 45–67.

Healey, P. (2006) The new institutionalism and the transformative goals of planning, in: N. Verma (ed.) *Planning and Institutions*, pp 61–87 (New Brunswick, NJ: Center for Urban Policy Research).

Healey, P. and Barrett, S.M. (1990) Structure and agency in land and property development processes, *Urban Studies*, 27(1): 89–104.

Healey, P., de Magalhães, C., Madanipour, A. and Pendlebury, J. (2002) *Shaping City Centre Futures: Conservation, regeneration and institutional capacity* (Newcastle upon Tyne: Centre for Research in European Urban Environments, University of Newcastle).

Healey, P., de Magalhães, C., Madanipour, A. and Pendlebury, J. (2003) Place, identity and local politics: Analysing partnership initiatives, in: M. Hajer and H. Wagenaar (eds) *Deliberative Policy Analysis: Understanding governance in the network society*, pp. 60–87 (Cambridge: Cambridge University Press).
Herrschel, T. and Newman, P. (2002) *Governance of Europes City Regions: Planning, policy and politics* (London: Routledge).
Hillier, J. (2000) Going round the back: Complex networks and informal action in local planning processes, *Environment and Planning A*, 32(1): 33–54.
Innes, J. and Booher, D. (2003) Collaborrative policy-making: Knowledge through dialogue, in: M. Hajer and H. Wagenaar (eds) *Deliberative Policy Analysis: Understanding governance in the network society*, pp. 33–59 (Cambridge: Cambridge University Press).
Imrie, R. and Raco, M. (1999) How new is the new local governance?, *Transactions of the Institute of British Geographers*, 24(1): 45–63.
Jeffrey, C., Mawson, J., et al. (2002) Special Issue: Devolution and the English Regions, *Regional Studies*, 36(7): 15–810.
Jessop, B. (1998) The narrative of enterprise and the enterprise of narrative: Place marketing and the entrepreneurial city, in: T. Hall and P. Hubbard (eds) *The Entrepreneurial City: Geographies of politics, regime and representation*, pp. 77–99 (London: John Wiley).
Jessop, B. (2000) Governance failure, in: G. Stoker (eds) *The New Politics of Local Governance*, pp. 11–32 (London: Macmillan).
Jessop, B. (2002) Institutional re(turns) and the strategic-relational approach, *Environment and Planning A*, 33(7): 1213–35.
John, P. (2001) *Local Governance in Western Europe* (London: Sage).
Le Gales, P. (2002) *European Cities: Social Conflicts and Governance* (Oxford: Oxford University Press).
Lowndes, V. and Skelcher, C. (1998) The dynamics of multi-organizational partnerships: An analysis of change in modes of governance, *Public Administration*, 76(2): 313–33.
MacLeod, G. (1999) Place, politics and scale dependence: Exploring the structuration of euro-regionalism, *European Urban and Regional Studies*, 6(3): 231–54.
MacLeod, G. and Goodwin, M. (1999) Space, scale and state strategy: Towards a reinterpretation of contemporary urban and regional governance, *Progress in Human Geography*, 23(4): 503–29.
Murdoch, J. and Abram, S. (2002) *Rationalities of Planning: Development versus environment in planning for housing* (Aldershot: Ashgate).
Murdoch, J. and Norton, A. (2001) Regionalisation and planning: Creating institutions and stakeholders in English regions, in: L. Albrechts, J. Alden and A. da Rosa Pires (eds) *The Changing Institutional Landscape of Planning*, pp. 109–32 (Aldershot: Ashgate).
Newcastle City Council (NCC) (2000) *Going for Growth: A citywide vision for Newcastle 2020* (Newcastle upon Tyne: Newcastle City Council).
Novarina, G. (Ed.) (2003) *Planet Projét: L'urbanisme en France et en Italie* (Paris: Anthropos).
Offe, C. (1977) The theory of the capitalist state and the problem of policy formation, in: L.N. Lindberg and A. Alford (eds) *Stress and Contradiction in Modern Capitalism*, pp. 125–44 (Lexington, MA: D.C.Heath).
Painter, J. (1997) Regulation theory, regime and practice in urban politics, in: M. Lauria (ed.) *Reconstructing Urban Regime Theory: regulating politics in a global economy*, pp. 122–43 (London: Sage).
Peck, J. and Tickell, A. (1994) Too many partnerships, *Local Economy*, 9(8): 251–65.
Peters, G. (1999) *Institutional Theory in Political Science: The new institutionalism* (London: Continuum).
Pierre, J. (Ed.) (1998) *Partnerships in Urban Governance: European and American Experience* (London: Macmillan).
Powell, W.W. and Dimaggio, P.J. (eds) (1991) *The New Institutionalism in Organizational Analysis* (Chicago, IL: University of Chicago Press).
Raco, M. (2003) Assessing the discourses and practices of urban regeneration in a growing region, *Geoforum*, 34(1): 37–55.
Salet, W. and Faludi, A. (eds) (2000) *The Revival of Strategic Spatial Planning* (Amsterdam: Koninklijke Nederlandse Akademie van Wetenschappen [Royal Netherlands Academy of Arts and Sciences])

Salet, W., Thornley, A. and Kreukels, A. (eds) (2003) *Metropolitan Governance and Spatial Planning: Comparative studies of European city-regions* (London: E&FN Spon).

Saunders, P. (1979) *Urban Politics: A sociological interpretation* (London: Hutchinson).

Seidman, S. (1998) *Contested Knowledge: Social theory in the post-modern era* (Oxford: Blackwell).

Stewart, M. (2002) Compliance and collaboration in urban governance, in G. Cars, P. Healey, A. Madanipour and C. de Magalhães (eds) *Urban Governance, Institutional Capacity and Social Milieux*, pp. 149–67 (Aldershot: Ashgate).

Swyngedouw, E. (2005) Governance innovation and the citizen: The Janus-face of governance-beyond-the-state, *Urban Studies*, 42(11): 1991–2006.

Thrift, N. (1996) *Spatial Formations* (London: Sage).

Vigar, G., Graham, S. and Healey, P. (2005) In search of the city in spatial strategies: Past legacies and future imagining, *Urban Studies*, 42(8): 1391–410.

Chapter 22

The Universal and the Contingent: Some Reflections on the Transnational Flow of Planning Ideas and Practices*[1]

Patsy Healey

Fate is the wind that blows the seed
Choice is the ground
Where it lands and blooms[2]

Travelling Ideas

This essay is addressed to the 'community of inquirers'[3] who make up the field of scholarship and practical engagement associated with planning as a project for shaping urban and regional/territorial futures. In this field, we are continually producing, critiquing, applying and circulating concepts, techniques and practice experiences from one place and time to another. What do such concepts carry with them, as they flow around? How do they fare and what are their consequences in a world of multiple contingencies? What kind of critical antennae and sensibilities should we cultivate as we encounter 'travelling ideas' and engage in this circulation? I start the paper with an example, a story of a contemporary urban governance experience, invented in very specific circumstances, which by now is vigorously 'travelling' internationally.

> Brazil since the 1980s has been a fertile ground for innovations in governance processes. The fall of a dictatorship and the struggles by workers and residents of urban peripheries provided a strong momentum for devolving power to municipalities led by city mayors, and for governance ideas which emphasize empowerment of all citizens, especially those marginalized by established formal government procedures (and practices). In one city, Porto Alegre, where the *Partidos dos Trabalhadores* (Workers' Party) came to power in 1989, the new administration had strong support from the neighbourhoods, mostly developed 'informally'. A key objective was to reverse the budget spend on projects which benefitted the small elite and development industry nexus, and distribute resources more fairly among the neighbourhoods. The challenge was to identify and prioritize a new project agenda in ways which were grounded in what residents actually felt were important. The new political leaders were keen to promote practical ways through which citizens could see that their voice had a real impact on what happened. They therefore focused on the processes for determining the city's investment budgets, although initially there was little such investment to redistribute. A process was invented through which people across the city got involved in identifying projects that were needed and in debating how to set priorities between the demands of different neighbourhoods. Citizens were encouraged to form neighbourhood groups, in loose, informal ways, to identify and prioritize needs. They had then to elect representatives, to carry their view to forums composed of groups of neighbourhoods. These then elected representatives to what became a municipal council

* Previously published in *Planning Theory*, 11(2): 188–207 (2012).

1 This paper was originally presented at the AESOP Congress, Helsinki 2010. My thanks to those who commented at the time, and to John Friedmann, Jean Hillier and Jonathan Metzger who commented on a draft version, and to Andrew Donaldson for his helpful editorial comments.

2 From a Folken Word presentation, the Sage Gateshead 15 May 2010, excerpt from a song 'When the orchard withers it will be spring' by Peter Tickell (music) and Carolyn Jess-Cooke (song).

3 This concept refers to those who engage in ongoing exploration and deliberation about a set of puzzles in relation to some phenomena. It is a central concept within pragmatic philosophy, derived from the work of the nineteenth-century philosopher Charles Peirce (see Bernstein, 2010).

> for setting the budget. At each level, projects were discussed and prioritized. After a while, groups involved in city-wide services, such as transport, education, health and welfare, also participated. Neighbourhood groups were asked to help in monitoring the progress of the agreed projects. These various groups and representatives did not become entrenched, as each year anyone could participate and each year new representatives were elected. As Brazil decentralized more functions and finance to municipalities, Porto Alegre municipality had greater resources to spend, and the process could be seen to have real effects in terms of providing improved conditions in the poor urban neighbourhoods of the city's expanding periphery that had hardly been considered part of the city until then. One commentator has calculated that, after seven years of operation, 14,000 citizens had an active involvement in the process, and significant budget redistribution occurred. The process was also not opposed by the middle classes and businesses. Instead, these groups appreciated the introduction of 'clean, efficient government' (Abers, 1998: 53). This experience is very interesting in itself. By the mid-1990s, clothed in the label of 'participatory budgeting', it was attracting a great deal of attention – among other Brazilian cities, elsewhere in Latin America, and across the world. By the late 2000s, it was circulating in Europe, and many projects were being promoted in the UK.[4] Yet careful observers and analysts of the Porto Alegre experience itself, and of attempts to replicate it elsewhere, underline the very particular circumstances in which the practice was invented, and the need to consider carefully how far such circumstances were available in other situations where experiments were being made in similar forms of 'participatory budgeting'.[5]

The attraction of the Porto Alegre story is that it creates hope that there are possibilities for shifting core governance activities, such as capital investment budgets, from one pattern to another, in ways which draw on broadly based discussion and participation in agenda-shaping in urban complexes with histories of massive inequality and injustice in resource distribution. But the lesson from research and experience is that the technique could not just be extracted from its context of invention, uprooted and 'planted' somewhere else. It arose from a particular ground and context, and might well not transplant easily somewhere else. This suggests that it is helpful to attach some kind of 'origin narrative' to planning ideas as they travel from place to place, to help others work out what could be learned from it of relevance to other situations and other contexts (Healey, 2013).

Today it sometimes seems, to those of us in planning academia and practice, as if we are in a whirlpool of new policy ideas, sometimes swirling about within national political and policy discourse, but often circulating vigorously transnationally and in global networks. There are concepts about spatial form (such as the 'compact city', 'gated communities', 'eco-towns') and about instruments (such as 'micro-credit', 'developer contributions', 'walking school buses', 'equity-share housing'). There are ideas about governance processes (such as 'participatory budgeting', 'community engagement', 'joined-up governance', 'multi-level' governance), or about analytical techniques (such as evaluation methods, local housing market analysis, sustainability indicators, impact assessments). Our theories also travel, shaping what people think the focus of attention in a situation should be (see for example, the influence of 'communicative planning theory', or 'new public management' concepts, or 'neo-liberal hegemony').[6] Yet our recent theorizing, as well as our increasing awareness of experiences of planning activity as practised in different parts of the world, has made us doubtful about the role of such 'travelling planning ideas' when they 'land' in particular places. Do they just promote a new form of 'colonization' or hegemonic domination, under the broader umbrella of globalization processes (Roy, 2010a; Watson, 2008, 2009)? Are they repressive technologies, crowding out the potential for 'local invention'? Are they little more than empty rhetorics, waiting to be filled with localized content and relevance? How could we tell?

In this essay, I reflect on these questions, drawing on work I have been doing recently in a book on the transnational flow of planning ideas and practices (*Crossing Borders*, Healey and Upton, 2010) and in work for

4 The Church Action on Poverty has been very active in this promotion (see www.participatorybudgeting.org.uk; see also work at the School of Public Policy in the University of Birmingham).

5 See Holston (2008) for the general Brazilian context. Abers (1998, 2003) and Baiocchi (2003) are my main sources for the Porto Alegre case, which is modified from Healey (2010: Chapter 3, Box 3.1). See also Souza (2001).

6 See Sager (2012).

the UN-Habitat Global Report on Urban Planning (UN Habitat, 2009).[7] In both projects, there is an underlying concern to promote better futures through some kind of development initiative – from the promotion of specific initiatives, to particular ideas about urban form, access to housing or finance, or transport management and general conceptions of territorial development. However, there has been a major shift in the past century in the planning academy and in the social sciences generally from a conception of a universally valid, linear pathway to economic and social development, linked to a set of technologies, to a recognition of the diversity of ways in which 'development' happens in particular places. In a linear development conception, the idea of planning was part of the package of forms of societal guidance which sought to keep development 'on track'. Planning concepts and practices could be and were borrowed from one place to another with few qualms. Once we recognize the complexity and contingency of development pathways, then the 'universal validity' of both the idea of planning and particular planning experiences are called into question.[8] In the following section, I look more closely at what is involved when a concept, technique or instrument is extracted from its place of 'origin', packaged up with a surrounding narrative, circulated around and translated into the particular circumstances of a different locale. More broadly, however, a concern with travelling concepts in the planning field raises questions about the idea of 'planning' itself. In the subsequent section, I therefore examine the 'travelling capacity' of the idea of planning itself. I conclude by suggesting that what we may claim as universal in the planning idea is the product of the ongoing debates we have as a 'planning community'. Rather than a technology, its existence is grounded in the tradition of deliberation of a 'community of inquirers'. Through such deliberations, we articulate what the 'planning idea' gets to mean and the work such an idea and the practices promoted by and through it in particular circumstances get to perform. In other words, I claim that the planning idea has the properties of a continually evolving 'contingent universal'.

Where We Were: Universal Pathways to Modernization

Back in the mid-twentieth century, a simple uprooting and transplanting transfer process could be justified by the belief in a single, 'universally valid' pathway for human social development.[9] Societies or countries were conceived as at different stages of development. Through the technology of planning, their development could be speeded up, to help them catch up with the 'most developed' societies. We know this now as the 'modernization' project. There has been some very interesting historical work in the planning field analysing what happened when such ideas – from the US, from France and Germany, and from the UK – were promoted in countries such as the Lebanon, India, Sri Lanka and China.[10] In many parts of the world, the imported practices of land-use zoning and forms of 'masterplanning' are now considered as alien implants by imperial colonizers which have become used and abused by elites to bolster their own positions, but have very little relevance to the very many living in informal settlements in rapidly expanding cities (Holston, 2008; Miraftab, 2009; Roy, 2009a; UN-Habitat, 2009; Watson, 2009). Such studies not only highlight the channels through which ideas were circulated, especially the linkage to the colonial networks of the early twentieth-century empires. They also show how local elites actively sought out such ideas as part of programmes for modernization.[11] Vidyarthi (2010a) shows very clearly the way ideas about urban form and the 'neighbourhood unit' were promoted in India by the first post-independence government, as if American ideals of urban living

7 UN-Habitat mobilized a considerable number of planning academics in the production of this report. See also Watson (2009), for reflections making use of this involvement.

8 See Friedmann (1973, 2011) and Friedmann and Weaver (1979) for accounts of this shift.

9 The debates were of course about the economic and political form best able to pursue that future. See Friedmann (1973, 1987) for planning thought in this period.

10 See Nasr and Volait (2003), the work of the International Planning History Society, and its Journal, *Planning Perspectives*, and several chapters in Healey and Upton (2010).

11 See several of the cases in Nasr and Volait (2003). Note that in the mid-twentieth century, the corpus of people trained in the planning field was very small, so there was a heavy reliance on consultancies from countries with such expertise. Slowly, communities of planners, mostly trained overseas, built up in other parts of the world. As an example of this process, see Vidyarthi (2010b), who shows how the neighbourhood concept was 'inappropriately appropriated' as it became institutionalized in Indian planning practice.

were somehow to be transposed into the very different socio-economic and political landscapes of a huge and very poor sub-continent.[12]

The modernization urge has not evaporated as social scientists in the West have become more cautious and context-sensitive in our development theories. Planners from Europe and North America have been called upon to advise governments energetically involved in promoting their countries or regions as key players in the current global political economy.[13] In China, a modernization dynamic is vigorously underway.[14] Along with specific ideas, the general idea that a 'planning approach' had value in shaping a development trajectory was also strongly evident in this modernization philosophy. Such an approach could be applied to overall societal development, or to territorial development on a sub-national scale – to regions and cities. It carried with it a variable mix of values, from accommodating the 'needs' of expanding masses with rising expectations about living standards to obtaining investment efficiencies by effective coordination. Within the 'modernization' umbrella, planning could be promoted as a universally beneficial policy approach.[15]

Yet, in the mid-twentieth century, planning practitioners and scholars lived primarily in 'national' policy cultures. Spatial planning systems in Western countries evolved with only limited exchange with other countries. Those working in the planning field knew little about the way the planning ideas and instruments they developed and used were shaped by the specific legal and political constitutions in which they were situated. The character of such contexts was often assumed as the 'natural' way of things. Thus our predecessors purveyed – through colonial and aid networks, and through other forms of international consultancy – locally developed planning instruments and processes as if they were 'universals'. While promoting greenbelts around cities to protect a 'green and pleasant land',[16] British planners not only imagined a cool and rainy country with a love of rurality, but tacitly assumed a system of individualized property rights, commercial investment and retail markets in land and property. US planners promoted a notion of 'individuals with interests', moving from one place to another to capture the best benefits and seeking a single family dwelling unit on a large land plot, accessed by wide roads for a car-based lifestyle.

There is still much work to be done to unpack in detail the way implicit assumptions and frames of reference such as these have got locked into all kinds of standards and instruments, which have then been circulated to other parts of the world. Although since Donald Schon's *The Reflective Practitioner* (1983), we have been encouraged to interrogate our own frames of reference, these 'built-in assumptions' are very difficult to bring to the surface.[17] Part of the value of working transnationally is that, as researchers and practitioners, our assumptions about 'how things go on' are continually challenged and unsettled by unexpected ways of thinking and acting (Watson, 2009).[18] Debates around theory have also helped us to surface the implicit in our thinking and acting. We have learned to use more probing, archaeological excavation in Foucauldian mode, and to undertake deconstructive analysis drawing on ideas developed in work in the humanities to enable us to 'see' these concepts and frames of reference buried deep inside our socially disciplined consciousness.

It is these recognitions and inquiries, pursued both conceptually and in empirical studies, which have undermined the idea of a linear development pathway to better futures. This in turn has made us very cautious in our claims about general principles which could provide a basis for acting in and evaluating specific instances of planning activity outside our own 'home ground'. We have become nervous of 'universal' claims and generalizations (see Campbell, 2006; Fainstein, 2010; Watson, 2008). Academics in particular these days emphasize the situated contingencies of particular instances of planning activity. We struggle to find the

12 Anthony King's pioneering study (King, 1984) of the circulation of the idea of the 'Bungalow', from India to many parts of the Anglophile Western world, is a fascinating study of the trajectories of a design idea.

13 Strategic planners Larry Beasley from Vancouver and academic Alessandro Balducci in Milan have both been asked to advise the Dubai authorities. See also Hardy (2008).

14 Abramson (2010) discusses his personal experience of this dynamic.

15 See Mannheim (1940), and for the debates at that time see Hillier and Healey (2008). By the 1970s, Castells (1977) was arguing that urban planning was the managerial arm of a capitalist socio-political order.

16 This phrase is from William Blake's famous poem against the industrialisation process in Britain (*Jerusalem*).

17 Work in planning history has been very valuable in this area, see for example Huxley (2006) and Vidyarthi (2010b).

18 Friedmann's landmark book, *Transactive Planning* (1973), develops the author's own experience of 'critical learning from transnational practice'.

appropriate concepts and relations upon which to focus our analytical work.[19] Our analyses of comparative experiences of instruments and approaches promoted by national governments and the EU have highlighted both the diversity of politico-institutional structures and processes and the significance this dimension makes to the fate of policy ideas.[20] We stress that every situation is different and that 'no one size fits all'. We argue for more attention to the contextualized political economy of development dynamics, or to the way planning ideas get caught up in struggles for hegemonic domination in the context of global flows (Roy, 2009a, 2010b). We emphasize the dynamics of complex overlapping systems, producing unpredictable conjunctions through which futures emerge (Hillier 2007, 2011a).

Meanwhile, however, the political and policy-making 'mood' by the turn of the century in the West came to celebrate 'innovation' – as much in policy ideas as in economic activity. In this climate of rejecting past experience, old ways of doing things have to be reformed and reconfigured, marked by the adoption of new concepts and vocabularies, and what seem to be new instruments. In this process, localized experiments and innovations are 'captured' and cut out of their local applicability, to act as symbols of innovative and reforming energy. They become part of a sea, or ether, of circulating ideas among policy and practice communities. 'Participatory budgeting' has experienced this fate. Taking another example, Roy (2010a) discusses how the practice of 'micro-credit' institutions has similarly been extracted from its Bangladesh origins, and then adopted and evangelized by the World Bank. Many officials and consultants these days seem to be living in bubbles of deracinated policy ideas and practices, which make it hard for them to see the local specificities of the situations they apply their ideas to. They are somehow experientially innocent, without a collective memory of previous endeavours. Such innocence is valuable at times when used to fuel a critical energy in unsettling and disturbing the taken-for-granted, and mobilizing attention to possible new directions. But it may also lead to the scattering of seeds on stony or ill-prepared ground, where they fail to flourish.

Moving Beyond the Contingent

Such 'restless' policy cultures carry many dangers, not just of ineffectiveness and inefficiency, but of causing real harm – failure to notice significant problems, voices and conflicts; unwitting capture by the vocal powerful; reinforcing the distrust of citizens and businesses in governance processes and expert advice. Echoing a question that reverberates through the *Crossing Borders* book, I think it does matter if planning ideas circulate in a deracinated way. But where do we 'stand' in making such criticisms? Does the discovery of the significance of agency and contingency just lead us down the path of extreme localism, where all we can say is that it is up to each site of collective action to go its own way and invent its own 'wheel', drawing on its own history and culture to create a future? Does this imply that the planning idea is itself contingent upon a specific conjunction of circumstances, and is this conjunction now evaporating? Was it rooted in Western, Fordist managerialist, conceptions of the role of the 'welfare' state? (Harvey, 1989; Jessop, 1995). Is it tied to the existence of national governance systems and practices, now being eroded ('hollowed out') by transnational agreements? Should we then follow a so-called 'neo-liberal' pathway and leave the future to 'marketized processes', or to the energy of protest movements from civil society merged with capacities for communitarian self-regulation? Is 'planning' one idea anyway, or just a vague signifier which gets filled with all kinds of meanings and action potentials? Is there anything *general* we can say about the content of the planning idea and its relevance to twentyfirst-century urban and regional challenges? Is there any 'centre' to the planning field?[21]

A grounding for such an argument can be found in the pragmatist concept of a 'community of inquirers'. This emphasizes inquiry as a continually open, exploratory and evolving process, in which what is taken as

19 See the discussion in Holston (2008), and his decision to focus on the concept and practices of 'differentiated citizenship'. However, Holston knows little of the planning literature, or that on informal settlement upgrading.

20 See the European studies of 'megaprojects' (e.g. Salet and Gualini, 2007; see also Fainstein, 2010) and of spatial strategy-making practices (e.g. Albrechts et al., 2003; Healey, 2007; Healey et al., 1997; Salet et al., 2003).

21 John Friedmann noted, on reading a draft of this paper, that the planning theory 'literature proliferates but it has no centre' (email: 02.06.10). See also Friedmann (1998).

'truth for now' is neither fixed nor reflective of some transcendent 'truth' or 'reality' (Bernstein, 2010; Dewey, 1982). It is certainly not contained by the parameters of formal planning systems or by the work that those trained as professional planners do (Healey, 2010; Perera, 2009; Roy, 2009a). It evolves as people explore, experiment, debate and invent in relation to, and continually interacting with, the evolving, multi-systemic worlds-within-a world within which we are all, as humans in a wider world of non-human actors, situated and have our being.[22] From this ground, we can then take a look again at the current whirlpool of specific travelling planning ideas.

My argument rests on the relation of the idea of planning to the challenge of addressing the complexities of an open, interconnected, urbanized world.[23] Although some argue that the urge to plan is a general human characteristic, I prefer to see planning as specifically linked to certain material and cultural processes, that is to say the effort to shape urban and regional development pathways through some deliberate, collective governance efforts (see also Watson, 2009).[24] It is this challenge which underpinned the development of planning ideas and planning systems in urbanizing twentieth-century Europe. Today, such challenges are faced even more acutely in other continents, along with the increasing awareness of the dangers which our development activities have created for our local and global natural systems.[25]

These challenges, with their complexity of cross-cutting systemic forces and structures,[26] created and maintained by the efforts of active agency in all kinds of interactions across the world, cannot be faced just by light-touch self-regulation of markets or by communitarian retreat to the world of place-based, face-to-face community. To do so misses the crises and disjunctions resulting from interconnections in place and time, and across scales. Instead, meeting these challenges demands a combination of developing understanding of urban and regional dynamics (and especially the connection between 'parts' and 'wholes'), dreaming about possible pathways to better futures (Bertolini, 2009), and mobilizing the social energy to seek ways of moving away from nightmare scenarios towards more positive dreamscapes. Others have called this the mobilization of hope.[27] Looking at what happens when planning concepts and instruments 'travel' from place to place helps not only to identify the implicit 'pasts' that those of us in the 'planning community of inquirers' still carry with us, feeding our dreaming in negative and positive ways. It also helps in understanding what could be the 'contingent universals' of the present global era, how to go about learning from the contingent experiences of others, and how to recognize when it is more appropriate to invent home-grown 'wheels', that is, endogenously shaped responses to particular situations. Such transnational learning, if undertaken with a sensibility to contingent complexities, should help to make we planners less experientially innocent and to deepen our collective memories.

Working with Dynamic Contingency

To return to the Porto Alegre story, the practice of participatory budgeting was evidently invented by astute politicians and advisers, drawing on local experience (not just of general struggle against oppressive governments but of a prior history of neighbourhood organizing and activism). It was located in a particular moment – of local and national political history. It drew on political ideas, particularly those linked to a general idea of participatory democracy. It was then packaged up, projected into wider public discourse and started

22 I am not alone in arriving at this philosophical position. It is central to pragmatist philosophy (see Bernstein, 2010; see also Misak, 1999; Westbrook, 2005); see also Fainstein (2010) in the end who provides a similar grounding for her claim that the search for justice in urban governance has, in the present period, such a 'universal' standing. For the notion of non-human actors, see actor-network theory (e.g. Callon, 1986; Callon et al., 2009; Latour, 2005).

23 By 'urbanized', I do not mean just that more than 50% of the world population now lives in areas formally defined as urban. The term also refers to lifestyles and referents, which shape aspirations and cultures.

24 This orientation is very clear in Friedmann's lifetime work.

25 Many call this 'globalization'. But, as with all such general terms, it is important to look carefully at the meanings and practices it is being used to refer to, especially when used to analyse an empirical phenomenon.

26 *Pace* Latour (1993, 2005), I think we can talk about systems and forces, as long as we do not reify these into entities detached from agency.

27 See Forester (2009), Harvey (2000), Westbrook (2005).

'travelling' (Souza, 2001). It was in this packaging that it became deracinated. Initially, audiences shared many assumptions and understandings, so they knew the roots it came from. But this implicit understanding was not available in other parts of the world, except to careful readers of academic accounts, or those who did probing 'fieldwork' by organizing 'site visits' to Porto Alegre armed with some idea of what to look out for.[28] But if a planning idea has been shorn not just of its particular 'origin history', but any reference to some place of origin, then such site visits are not possible. Whitzman and Perkovic (2010) describe how the practice of undertaking Women's Safety Audits originated in feminist activism in Toronto, but, as it circulated around the world, partly through the advocacy of the UN, this origin and its radical edge were lost sight of. Roy (2010a, 2010b) writes the origin history of 'micro-credit', in order to confront the way a 'Washington consensus' appropriated the concept. She challenges it by re-emphasizing the more radical original concept developed in Bangladesh, which has the potential to help large numbers of very poor people.

This suggests that one angle for our critical attention when examining a 'travelling planning idea' is the extent to which it is attached to an 'origin narrative' which identifies the situation in which it came to life, and describes its particular history. It would also be helpful to accompany this with a story of 'how it got here', that is, to the place where it is being noticed and thought about. How did the idea get 'packaged up' and extracted from its 'origin site'? Through what channels has the idea of 'participatory budgeting' travelled to arrive in places such as the UK? What transformations and interpretations has it been subjected to on the way, and what forces are pushing it around?

Such narratives of the socio-political 'winds of fate' (see the poem at the start of this paper) should help those in other places to assess whether such an idea could have relevance to their place and their problems, and to look carefully at why such an idea is blowing their way. But there is another dimension of critical exploration, and that is to probe what a particular idea 'carries inside it'. Several of the cases in *Crossing Borders* (Healey and Upton, 2010) note how the critical edge of ideas such as 'compact cities', 'women's safety audits', 'micro-credit' institutions, etc. were dropped as they travelled, especially where, in their place of origin, their initial momentum derived from efforts to promote more attention to marginalized groups (such as women and their safety fears, the poor without money) or to neglected issues (such as spatially concentrated pollutions and the challenges of accessing the resources of the city for the very poor).

This raises the question of what kind of narratives to surround a travelling planning idea with? Are there ways to 'package' a planning idea so that an intended radical edge is not 'lost in passage' or 'in translation'? What is the legitimacy of attempting to reconfigure practices in some other place in the concepts and practices of a radical idea from somewhere else? Stories can be told and histories written in many different ways. Should we write in the tradition of political economy? Should we provide accounts of enlightened leaders and energetic promoters, of where they met and the mobilization efforts they engaged in? Should we observe the micro-dynamics of face-to-face encounters? Should we look more at cultures, institutions and social practices, through which meanings are produced and actions performed? It is here that our debates in planning theory prove helpful, in suggesting that there are different ways of writing accounts of the empirical world, through which different contingencies are given prominence. In the end, of course, we can never tell a complete narrative. Our knowledge is always to an extent partial and limited.[29] But the ambition is surely not the circulation of a 'complete and perfectly true' story, but to enrich the material available to those seeking to learn about experiences elsewhere which could help them work out whether and how to make use of them. My hypothesis here is that transnational learning works most productively through rich narratives – in-depth cases – rather than through 'best practice' summaries or attempts at typologies which systematize qualities of context and try to match them with qualities of experiences. This is an important reason why 'thick' case narratives are celebrated in the planning field.[30]

The focus on case narratives, in this case about how a planning idea emerges and how it then gets to circulate, takes us deeply into the exploration of agency. Some of the chapters in *Crossing Borders* tell us

28 This is a common practice these days, both within and between countries. See, for example, Stead et al. (2010). The UK Church Action for Poverty interest in participatory budgeting involved an early visit by community activists to Porto Alegre.

29 Once again, I draw on pragmatist philosophy (Bernstein, 2010; Westbrook, 2005).

30 See Flyvbjerg (2001), Watson (2009); for the value of in-depth case narratives in the policy and management fields, see Czarniawska (2010).

about the authors' own experiences – MIT graduate Dan Abramson doing a PhD in China, John Friedmann reflecting on his consultancy work in Latin America in the 1960s and in China in the 2000s. Stephen Ward presents accounts of key planners working on the international advisory circuits in the mid-twentieth century. In such personal narratives, we are taken beyond accounts which refer to organizations-as-actors, to encounter the people who actually do the acting. These people may be tied to a particular organization, with a strong internal 'culture' and loyalty which shapes how members behave. But 'actors' often have several identities and loyalties, which cross-cut how they look at issues and work out how to act in a situation. Nor are they autonomous individuals. They are tied into social relations and cultures of all kinds. They are 'doing work' of some kind in a governance context.

Attention to this micro-level helps to make visible how moments of opportunity are perceived, how meanings are promoted, how social energy is mobilized and how experiments are developed.[31] But they also reveal something of the way in which such micro-practices interrelate with and co-evolve with wider forces. Although in the planning academic field, we have big debates about the existence and nature of such wider forces, we cannot avoid realizing that our futures, like our pasts, are shaped not just by our own efforts but by processes which are not amenable to our 'control', even though such processes may be affected by how we respond to the conditions they create. This suggests that, however we position ourselves in social theory, narratives of travelling planning ideas need to give some attention not just to the micro-dynamics, but to the wider structuring forces which create moments of opportunity and limits on inventive possibilities.[32] This should help to enhance our sensibility to 'the winds that blow' planning ideas from place to place, and to the modes of thinking and acting which remain embedded within them as they travel.

In a related paper (Healey, 2013), I consider the research resources which are useful in tracing the pathways of planning ideas as they are packaged up, travel, move in and out of the broad sea or ether of ideas available to our 'planning community', and become translated and transposed into new situations (see also McCann, 2008; Tait and Jensen, 2007). But what about the 'planning idea' itself? Despite all our exhortations to pay attention to contingencies and localized interpretations in the literature we create and use, the term 'planning' gets deployed as if the word, or the idea, had a discrete persona which could act. We refer to 'planners' as if it is clear who they are wherever we encounter them in the world. In this way, we generalize and essentialize planning and planners into some kind of 'universal' phenomenon. When we do this, are we merely 'talking among ourselves' in a kind of jargon, assuming that our listeners and readers understand that the words are just proxies for many different manifestations of 'planning' and 'planners'? Or are we binding into these words certain ideas and certain practices which we think get 'carried' by the terms and labels, as the words travel from place to place? Certainly, in many countries which experienced European imperial colonization in the twentieth century, the legacy of an inappropriate 'masterplanning' remains the dominant understanding of what these words (planning/planners) mean. This leaves those who believe that the planning idea means something more than this, which is of value to 21st-century cities, with a struggle to articulate an alternative normative view of the values and practices which *could be* embedded in the idea. The next section looks at some of the attempts to address this.

A Content for the 'Planning Idea' in the Twenty-first Century

The sub-field of planning scholarship known as 'planning theory' is perhaps the place where such attempts might be located. Within this field, as many have commented, the focus of debate has swung about over the past half century between a focus on 'development' (urban and regional development, the design and layout of neighbourhoods and cities, the justice of development outcomes, the sustainability of social practices and building forms) and a focus on 'governance processes' (their efficiency and effectiveness, their inclusivity and justice, their ability to connect past and present to shape futures). Although for some, the planning idea remains linked to a general human capacity to think about and then act to bring future possibilities into being, and for others, it centres on imagining future urban forms, most in the end conclude that the heart of the

31 These concepts are well developed in the social movement literature.

32 In other words, we need some way of addressing the interaction of structuring dynamics and human agency.

planning idea lies in a combined, co-evolving relation between place development and governance processes, between 'substance' and 'process'. Some see the idea of planning attached to a particular institutional location (those performing assigned 'planning tasks' in a formal government system), or to a type of specialization, work done by those trained as planners, or to a particular planning instrument, such as a 'plan' or 'land-use regulations'.

There is nevertheless a well-developed tradition of giving a broader meaning to the idea of planning. This shines through Friedmann's work. Initially, he argued that planning should be understood a form of 'societal guidance' (Friedmann, 1973). Later, he became concerned that the practices of societal guidance had become too strongly linked to a statist model of social formations, which was being undermined by the crisis of industrial capitalism. Watson (2009: 2260) has recently echoed such a concern. In the 1980s, Friedmann sought an alternative development path more linked to 'self-reliance in development' and 'the recovery of political community' (Friedmann, 1987: 10). In much of his later work, he focuses his thinking around notions of 'the good city', and the qualities which those promoting better urban conditions should strive for. Within the idea of planning 'in general', he identifies a commitment to seeking

> dynamic balances between the part and the whole, the technical and the normative, the pragmatic and utopian, the near present and the distant future, exchange values and use values; it allows us to be visionary with an emphasis on values which include social justice, ecological sustainability, civic empowerment, community and human flourishing; ... [the idea promotes] ... continuous social learning. (Friedmann, 2011: 11)

In 2009, UN-Habitat devoted one of its influential 'global reports' to urban planning (UN-Habitat, 2009). The report struggles with discovering various meanings for the planning idea, and arrives at the following general meaning:

> Since the earliest days of human settlement, people have consciously and collectively intervened in the nature and form of urban areas to achieve particular social, political and environmental objectives. This activity has been known as planning. (p. 3)

This is then qualified: 'urban planning is a significant management tool for dealing with the unprecedented challenges facing twenty-first-century cities and attaining the goals of sustainable urbanisation' (p. 3), involving

> a self-conscious collective (societal) effort to imagine or re-imagine a town, city, urban region or wider territory and to translate the result into priorities for area investment, conservation measures, new and upgraded areas of settlement, strategic infrastructure investments and principles of land use regulation. (p. 19)

Bertolini (2009: 309) puts such a view more succinctly and inspiringly:

> Planning (involves) the task of shaping conditions for other beings to be empowered, other imaginations to be expressed, other endeavours to unfold. It is the task of making the co-existence in space of a diversity of human projects and interactions possible. To me, this is the dream of planning.

Davoudi and Pendlebury (2010: 638), in their attempt to define the 'epistemic core' of the 'planning discipline', focus on 'space as the discipline's substantive object of inquiry', on the particular integrative approach to knowledge development and use, and the close connection between knowledge and action. In my own recent work, I have attempted to express the core ideas associated with a progressive idea of planning. Centring on planning as a form of collective action, or governance – 'place governance with a planning orientation' – I emphasize five attributes:

- an orientation to the future and a belief that action now can shape future potentialities;
- an emphasis on liveability and sustainability for the many, not the few;

- an emphasis on interdependences and interconnectivities between one phenomenon and another, across time and space;
- an emphasis on expanding the knowledgeability of public action, expanding the 'intelligence' of a polity;
- a commitment to open, transparent government processes, to open processes of reasoning in and about the public realm. (Healey, 2010: 19)

All such statements are subject to vigorous debate by fellow planning theorists, and there are many more statements to choose from.[33] Yet it is these attempts at filling the 'planning idea' with some more specific content which serve to identify some attributes that continue to cluster around it. First, the idea of planning has a normative dimension. It urges those who take it up to use it in the search for a 'better' future, and to pay attention to struggles over what futures and whose futures get to count. It demands consideration of what 'better' could and should mean. Secondly, it has a methodological dimension. It encourages attention to the complex ways in which specific issues, groups, phenomena interconnect into larger 'wholes' (structures, systems, etc.). It demands an integrative and synthetic capacity. Thirdly, it embodies a commitment to encouraging political communities to make more transparent the issues at stake, and subject them to probing inquiry drawing on what knowledge is available. Fourthly, it has a substantive focus, which centres on how people live their lives in association with all kinds of others, in complex, dynamic, spatially differentiated places and the forms and processes of collective action (governance) which evolve to address the challenges and problems which arise.

It is this general orientation which is frequently 'shorn off', both as the planning idea travels and gets to land in particular places, and as it institutionalizes wherever it is taken up. Practices once inspired by the planning idea can lapse into regulatory procedures, the purposes of which have long become tacit. Or the idea becomes attached to a narrow political project through which one group seeks to dominate another.[34] Or its practice comes to centre on a narrow definition of whose knowledge gets to count, as in political contexts dominated by technical elites. Such actual and potential distortions place a particular responsibility on the planning academic and practice community. We planning experts and academics should not seek to own the planning idea. Nor should we imagine that only we can do planning work. The idea of 'place governance with a planning orientation' can develop in many situations, quite outside links with the formal planning 'community'.[35] Our role in relation to the planning idea in general, surely, is continually to give it attention, to enrich, critique and develop it. I do not mean by this to try to fix it to a particular definition. It is the narrative of critical debate with which the planning idea needs to be surrounded. Our contribution is to promote awareness that there are debates to be had about the meaning of the planning idea – normative, methodological, political and substantive, and that these need to be attached as far as possible to the idea as it travels. In this way, when the planning idea 'lands' in a particular place, or becomes the focus of discussion in some political community concerned to act about the challenges they face, richer resources are made available for people to learn from and with. Further, where people have invented a practice which looks very like the planning idea, they can, through recognizing it as such, expand their collective experience and learning possibilities.[36] The planning idea then becomes a focus of attention and a many-stranded package of debate and hypotheses rather than a specific recipe (or technology) for doing governance work.

If this is so, then, as ourselves a 'community of inquirers', we need to think carefully from time to time about how best to express the focus and the strands of debate in ways which enhance the capacity for situated learning from the experiences and ideas of others. If, as I suggest, the normative, methodological, political and

33 See, for example, Sandercock's idea of 'Cosmopolis' (Sandercock, 1998: Chapter 8). To take another example, Fainstein (2010) argues that a core idea for the planning project in the twenty-first century should be the pursuit of justice, translated into the more specific concepts of equity, diversity and democracy. See also Campbell (2006). Hillier has commented that such statements leave out the significance of non-human 'actants' (Hillier, 2011b).

34 See accounts in Yiftachel (2009) and Harrison et al. (2007); and also Sager (2012).

35 See Healey (2010), Perera (2009), Roy (2009a).

36 This of course is no guarantee that such a political community will give attention to the richness of the idea. My meaning here is that those of us in the planning community should not ourselves narrow it down, except in so far as we tell critical stories about how the 'idea' has been translated and institutionalized in particular instances.

substantive attributes of the planning idea evolved during the twentieth century to address the challenges of living in an open, interconnected and urbanized world, in Western contexts, then the twenty-first century seems set to face these challenges in an even more widespread and acutely difficult form. But while there is much of value in the twentieth-century experience, it is likely to be the innovations and learning about the massively urbanizing rest of the world which refresh and expand the planning idea in the twenty-first century (Roy, 2009b; Watson, 2009). Those of us in Western contexts need to learn how to learn from these experiences, just as much as those who are faced with working out how to respond to challenges in particular situations in other parts of the world. The example of participatory budgeting provides a fascinating example of such learning.

So, to answer the question raised in the *Crossing Borders* book, there *is* something general we, as a community of inquirers concerned about the challenge of co-existence in the shared spaces of complex urbanized societies, can claim about the idea of planning, so long as we do not think of the idea as the aggregate of practices which go under a label of planning, or as an absolute universal good technology. The idea instead carries with it a normative orientation and a tradition of debate, honed by a continual interaction between situated practice experiences and theoretical development. It is contingent both in the way the orientation and debates develop and in working out how the idea may inspire particular practices to develop. This implies that the general ideas we construct about planning as a project need to be continually renewed and refreshed, so that, as it travels, it carries an active capacity to critique and unsettle practices which claim to be planning but which neglect key attributes of the contingently universal planning idea as evolved through critical debate. It is this debate which provides a temporarily firm ground for confronting and enriching situated experiences, a kind of 'status quo in motion' (Innes and Booher, 2010: 114). But speaking from such a ground, those asserting the value and the attributes of the planning idea for the tenty-first century also need to address respectfully and sensitively the capacity for local invention and interpretation. Without this, the idea could either ossify in places where it has long been institutionalized, fail to grow at all in the places where it lands, or bloom too vigorously, crowding out the capacity for local interpretation and invention.

In Conclusion

I have suggested that most ideas and examples of practices which circulate in and around the planning field are likely to be shaped by their origins and by the channels through which they have travelled. If this is the case, then it is helpful to maintain an awareness of origins and travel trajectories when assessing the potential value and impact of a specific idea or practice when it lands in a particular place. Put another way, this implies attention to the contingencies of all 'travelling ideas', (including those of planning theorists!). Such a conclusion of course merely reinforces and expands what many in the planning field have been arguing for some time, namely that doing place governance work in a planning way involves capabilities in situated, pragmatic learning. It is situated, in the sense of grappling with the specific challenges thrown up in a particular place. It is pragmatic, in the sense of working out what would work practically in that particular context. And it involves learning, as those doing such work probe local specificities as well as searching out knowledge and experiences from elsewhere.[37] In the *Crossing Borders* book, Stead and colleagues (2010) describe two cases from Eastern Europe involving the transfer of ideas about urban region transport management. In one case, those involved were very skilled in the art of situated, pragmatic learning, and themselves set out to probe more deeply the ideas being promoted to them by a western European aid agency. In the other case, the advice of this same agency was just accepted as appropriate. Stead and colleagues suggest that the greater intelligence of the former led to better outcomes, assessed by criteria linked to those which I have suggested are embodied in the idea of planning itself. To complete the analysis of travelling planning ideas, we therefore need to tell narratives of what happens when ideas get to 'land' in particular places.[38]

In these reflections, I have argued firstly that the study of travelling concepts and practice in the planning field provides a valuable angle through which the situatedness and contingencies of specific institutional sites of planning activity can be better understood. It helps us to ask deeper questions about what we assume to

37 See Campbell (2006), Forester (1999), Friedmann (1973), Innes and Booher (2010), Perera (2009).

38 Planning history scholarship has good examples here; see also McCann (2008), Tait and Jensen (2007).

be 'natural' and 'universal' in human behaviour and the way we go about governance activity with respect to socio-economic development, spatial organization and land development. I have also argued that such study can be usefully undertaken through the construction of narratives of origin, of travelling and of 'landing' in particular sites. Such narratives involve honing our skills in identifying contingencies.

But this does not dispense with the need for some form of universal ground on which to stand when making judgements about the appropriateness of an idea or practice when it lands in a particular place. There are questions which those who are part of a 'planning community of inquiry' have a responsibility to consider. We need to be able to address analytical questions such as: 'will this idea work here?' We also need to be able to address normative questions, such as: 'should this idea be "imported" at the present moment in this situation? Who and what could it promote? Who and what might be harmed?' I have suggested that our traditions of debate about the principles and values associated with the planning idea itself can provide a ground for making such judgements. However, this planning idea is not a timeless universal which holds for all times and all places. It is not expressed in 'Tablets' which can be brought down from a mountain top to inform our messy world.[39] It is a contingently universal stream of ideas, evolving with our contingencies and innovative energies and always provisional. Our ideas about planning and its value for the contemporary world are kept alive by continual review, reassertion and reinterpretation, as we rework them in the flow of human activity. I suggest that it is this continual probing and questioning which shapes the universal idea of planning and the values it carries. Thus, the particular narratives of contingency interact with the evolving narrative of the planning idea, with all its richness of critical debate and normative striving. The friction between these two narratives helps to give the field its creative energy.

Recognizing this interaction could perhaps help analysts and practitioners avoid the traps of over-localizing and of over-generalizing. All narratives tell us about something local, whether it is the 'local' of the offices of the World Bank or of a village in India (Latour, 1993, 2005). Every 'wheel' and every 'seed' is in the end uniquely crafted in a specific situation. Yet we can say something general about wheels and seeds. Such generalities are useful because they hone our intelligence and suggest what to look for in a specific situation – about purposes, about surfaces, about soils and about climate, about winners and losers, and about the spatial reach and scales of time over which effects might be experienced. However, the idea of planning is not a discrete object or thing, like a wheel or a seed, with a recognizable form and materiality. It is an evolving, fluid, mental concept, constructed from the meanings given to it in all kinds of specific locales and the encounters and tensions between these. So we must not only treat it as a contingent universal. We must also be on guard against essentializing a general idea of planning into a thing-like object or an actor-like subject. One implication of these reflections is that we need more, not less, attention to planning theory within our field of scholarship and education. But we urgently need such theorizing to engage with the experiences of the unfolding urbanizing and urbanized worlds around us (Fainstein, 2005; Roy, 2009b; Watson, 2009), just as our forebears did. We may reject the modernization thesis, but not the engagement with forms of urban development and the search for ways of promoting the flourishing of the many not the few in relation to our wider environments. Studies of place development, governance dynamics and of how 'travelling' planning concepts and practices are invented, circulated and made use of in such worlds help us in developing such theorizing.

> Thus there developed in our dining room, in the lamplight that is so congenial to them, one of those talks in which the wisdom is not that of nations but of families, taking hold of some event … and slipping it under the magnifying glass of memory … and places it in perspective at different points in space and time what, to those who have not lived through the period in question, seems to be amalgamated on a single surface … The Muse who has gathered up everything that the more exalted Muses of philosophy and art have rejected, everything that is not founded upon truth, everything that is merely contingent, but that reveals other laws as well, is History. (Proust, 2006: 987)[40]

39 These metaphors derive from both the biblical story of Moses and from *Zen and the Art of Motorcycle Maintenance* (Pirsig, 1974).

40 Proust was writing in the early twentieth century, and was coming to reject the notion that somehow 'truth' could be found at the intellectual, social and political apex of the society in which he lived. We could perhaps expand Proust's idea of the Muse of History into a historical/sociological view of how social development proceeds, in an evolutionary, emergent way, from the past into the future.

References

Abers, R. (1998) Learning democratic practice: Distributing government resources through popular participation in Porto Alegre, Brazil. In: M. Douglass and J. Friedmann (eds) *Cities for Citizens*. London: Wiley, 39–66.

Abers, R. (2003) Reflections on what makes empowered participatory governance happen. In: A. Fung and E.O. Wright (eds) *Deepening Democracy: Institutional innovation and empowered participatory governance*. London: Verso, 200–207.

Abramson, D.B. (2010) A trans-Pacific education in reverse: Reflections on an American with a Chinese doctorate in urban planning and design. In: P. Healey and R. Upton (eds) *Crossing Borders: International exchange and planning practices*. London: Routledge, 289–312.

Albrechts, L., Healey, P. and Kunzmann, K. (2003) Strategic spatial planning and regional governance in Europe. *Journal of the American Planning Association* 69(2): 113–29.

Baiocchi, G. (2003) Participation, activism and politics: The Porto Alegre Experiment. In: A. Fung and E.O. Wright (eds) *Deepening Democracy: Institutional innovation and empowered participatory governance*. London: Verso, 45–76.

Bernstein, R.J. (2010) *The Pragmatic Turn*. Cambridge: Polity Press.

Bertolini, L. (2009) The dream of planning. *Planning Theory and Practice* 10(3): 309–13.

Callon, M. (1986) Some elements of a sociology of translation: Domestication of scallops and fisherman in St. Brieuc Bay. In: J. Law (ed.) *Power, Action and Belief*. London: Routledge and Kegan Paul, 196–233.

Callon, M., Lascoumes, P. and Barthe, Y. (2009) *Acting in an Uncertain World: An essay on technical democracy*. Cambridge, MA: MIT Press.

Campbell, H. (2006) Just planning: The art of situated ethical judgement. *Journal of Planning Education and Research* 26: 92–106.

Castells, M. (1977) *The Urban Question*. London: Edward Arnold.

Czarniawska, B. (2010) The uses of narratology in social and policy studies. *Critical Policy Studies* 4(1): 58–76.

Davoudi, S. and Pendlebury, J. (2010) The evolution of planning as an academic discipline. *Town Planning Review* 81(6): 613–45.

Dewey, J. (1982) The pattern of inquiry. In: H.S. Thayer (ed.) *Pragmatism: The classic writings*. Indianopolis: Hackett Publishing Co., 316–34.

Fainstein, S. (2005) Planning theory and the city. *Journal of Planning Education and Research* 25: 121–30.

Fainstein, S. (2010) *The Just City*. New York: Cornell University Press.

Flyvbjerg, B. (2001) *Making Social Science Matter: Why social inquiry fails and how it can succeed again*. Cambridge: Cambridge University Press.

Forester, J. (1999) *The Deliberative Practitioner: Encouraging participatory planning processes*. London: MIT Press.

Forester, J. (2009) *Dealing with Differences: Dramas of mediating public disputes*. Oxford: Oxford University Press.

Friedmann, J. (1973) *Re-tracking America: A theory of transactive planning*. New York: Anchor Press.

Friedmann, J. (1987) *Planning in the Public Domain*. Princeton: Princeton University Press.

Friedmann, J. (1998) Planning theory revisited. *European Planning Studies* 6(3): 245–53.

Friedmann, J. (2011) *Insurgencies: Essays in planning theory*. London: Routledge.

Friedmann, J. and Weaver, C. (1979) *Territory and Function*. London. Edward Arnold.

Hardy, D. (2008) Dubai – planning miracle or desert mirage? *Town and Country Planning* 70(3): 143–6.

Harrison, P., Todes, A. and Watson, V. (2007) *Planning and Transformation: Learning from the post-apartheid experience*. London: Routledge.

Harvey D (1989) From managerialism to entrepreneurialism: the formation of urban governance in late capitalism. *Geografisker Annaler* 71B: 3–17.

Harvey, D. (2000) *Spaces of Hope*. Edinburgh: Edinburgh University Press.

Healey, P. (2007) *Urban Complexity and Spatial Strategies: Towards a relational planning for our times*. London: Routledge.

Healey, P. (2010) *Making Better Places: The planning project in the twentyfirst century*. London: Palgrave Macmillan.
Healey, P. (2013) Circuits of knowledge and techniques: The transnational flow of planning ideas and practices. *International Journal of Urban and Regional Research* 37: 1510–26.
Healey, P. and Upton, R. (eds) (2010) *Crossing Borders: International exchange and planning practices*. London: Routledge.
Healey, P., Khakee, A., Motte, A. and Needham, B. (eds) (1997) *Making Strategic Spatial Plans: Innovation in Europe*. London: UCL Press.
Hillier, J. (2007) *Stretching Beyond the Horizon: A multiplanar theory of spatial planning and governance*. Aldershot: Ashgate.
Hillier, J. (2011a) Strategic projects: from sustainability to resilience. In: S. Oosterlynck, J. van den Broek, L. Albrechts and A. Verhetsel (eds) *Strategic Spatial Projects: Catalysts for change*, London: Routledge, 212–22.
Hillier, J. (2011b) Strategic navigation across multiple planes: towards a Deleuzean-inspired methodology for strategic spatial planning. *Town Planning Review* 82(5): 503–28.
Hillier, J. and Healey, P. (eds) (2008) *Foundations of the Planning Enterprise: critical readings in planning theory, Vol. 1*. Aldershot, Hampshire: Ashgate.
Holston, J. (2008) *Insurgent Citizenship: Disjunctions of democracy and modernity in Brazil*. Princeton: Princeton University Press.
Huxley, M. (2006) 'The soul's geographer': Spatial rationalities of liberal government and the emergence of town planning in the twentieth century in England and Australia. *Geography*. Milton Keynes, The Open University.
Innes, J.E. and Booher, D.E. (2010) *Planning with Complexity: An introduction to collaborative rationality for public policy*. London: Routledge.
Jessop, B. (1995) Towards a schumpeterian workfare regime in Britain? Reflections on regulation, governance and the welfare state. *Environment and Planning* A 27: 1613–26.
King, A.D. (1984) *The Bungalow*. London: Routledge and Kegan Paul.
Latour, B. (1993) *We have Never Been Modern*. London: Harvester Wheatsheaf.
Latour, B. (2005) *Reassembling the Social: An introduction to actor-network-theory*. Oxford: Oxford University Press.
Mannheim, K. (1940) *Man and Society in an Age of Reconstruction*. London: Lund Humphries.
McCann, E.J. (2008) Expertise, truth, and urban policy mobilities: Global circuits of knowledge in the development of Vancouver, Canada's 'four pillar' drug strategy. *Environment and Planning A* 40: 885–904.
Miraftab, F. (2009) Insurgent planning: Situating radical planning in the global south. *Planning Theory* 8(1): 32–50.
Misak, C. (1999) *Truth, Politics and Morality: Pragmatism and deliberation*. London: Routledge.
Nasr, J. and Volait, M. (eds) (2003) *Urbanism: imported or exported? Native aspirations and foreign plans*. Chichester, Sussex: Wiley-Academy.
Perera, N. (2009) People's spaces: Familiarization, subject formation and emergent spaces in Colombo. *Planning Theory* 8(1): 51–75.
Pirsig, R.M. (1974/1999) *Zen and the Art of Motorcycle Maintenance*. London: Vintage.
Proust, M. (2006) *Remembrance of Things Past* (translated by S.C.Scott Moncrieff and Stephen Hudson). Ware, Herts: Wordsworth Editions Ltd.
Roy, A. (2009a) Why India cannot plan its cities: Informality, insurgence and the idiom of urbanization. *Planning Theory* 8(1): 76–87.
Roy, A. (2009b) Strangely familiar: Planning and the worlds of insurgency and informality. *Planning Theory* 8(1): 7–11.
Roy, A. (2010a) Poverty truths: The politics of knowledge in the new global order of development. In: P. Healey and R. Upton (eds) *Crossing Borders: international exchange and planning practices*. London: Routledge, 27–45.
Roy, A. (2010b) *Poverty Capital: Microfinance and the making of development*. London: Routledge.

Sager, T. (2012) *Reviving Critical Planning Theory: Dealing with pressure, neo-liberalism, and responsibility in communicative planning*. London: Routledge.

Salet, W. and Gualini, E. (eds) (2007) *Framing Strategic Urban Projects: Learning from current experiences in european urban regions*. London: Routledge.

Salet, W., Thornley, A. and Kreukels, A. (eds) (2003) *Metropolitan Governance and Spatial Planning: comparative studies of European city-regions*. London: E & FN Spon.

Sandercock, L. (1998) *Towards Cosmopolis*. London: John Wiley.

Schon, D. (1983) *The Reflective Practitioner*. New York: Basic Books.

Souza, C. (2001) Participatory budgeting in Brazilian cities: limits and possibilities in building democratic institutions. *Environment and Urbanization* 13(1): 159–84.

Stead, D.M. de Jong and Reinholde, Y. (2010) West-East policy transfer in Europe: The case of urban transport policy. In: P. Healey and R. Upton (eds) *Crossing Borders: international exchange and planning practices*. London: Routledge, 173–90.

Tait, M. and Jensen, O. (2007) Travelling ideas, power and place: the cases of urban villages and business improvement districts. *International Planning Studies* 12(2): 107–28.

UN-Habitat (2009) Planning Sustainable Cities: global report on human settlements 2009. London: Earthscan.

Vidyarthi, S. (2010a) Re-imagining the neighbourhood unit for India. In: P. Healey and R. Upton (eds) *Crossing Borders: International exchange and planning practices*. London: Routledge, 73–93.

Vidyarthi, S. (2010b) Inappropriately appropriated or innovatively indigenized?: Neighbourhood unit concept in post-independence India. *Journal of Planning History* 9(4): 260–76.

Watson, V. (2008) Down to earth: Linking planning theory and practice to the 'Metropole' and beyond. *International Planning Studies* 13(3): 223–37.

Watson, V. (2009) Seeing from the South: Refocusing urban planning on the Globe's central urban issues. *Urban Studies* 46(11): 2259–75.

Westbrook, R.B. (2005) *Democratic Hope: Pragmatism and the politics of truth*. Cornell: Cornell University Press.

Whitzman, C. and Perkovic, J. (2010) Women's safety audits and walking school buses: the diffusion/de-fusion of two radical planning ideas. In: P. Healey and R. Upton (eds) *Crossing Borders: International exchange and planning practices*. London: Routledge, 219–36.

Yiftachel, O. (2009) Theoretical notes on 'gray cities': The coming of urban apartheid. *Planning Theory* 8(1): 88–100.

Chapter 23
From Municipal Colonialism to Collaborative Governance? An Experiment in Therapeutic and Transformative Planning

Leonie Sandercock

Preamble[1]

For five years, from 2006, I was immersed in an action research project in a deeply divided community in north central British Columbia (BC), the western-most province of Canada. In 1999 the small town of Burns Lake (see Figure 23.1) – population around 5,000 people, almost equally divided between Indigenous and non-Indigenous – became the site of a stand-off between the Village (i.e. municipality) and the Ts'il Kaz Koh First Nation, ostensibly over a taxation dispute but actually the culmination of an 80-year running battle over appropriated land. The Village had been laid out in 1915 on land traditionally occupied by the Carrier (Dakelh) peoples (see Figure 23.2), of whom the Ts'il Kaz Koh is a subtribe. The Ts'il Kaz Koh had been forced onto Reserve land in 1914 and when the town site was surveyed the following year, it gobbled up a third of the small (400 acres) Reserve allocation. When the taxation dispute erupted in the late 1990s it was in the larger context of a slowly changing political and legal landscape of Indigenous/state relations at the national level, bumping up against an extreme reluctance at local level to acknowledge advances in Indigenous rights and title and Indigenous desires for recognition and self-determination. The failure of the Village (controlled by the old guard 'pioneer stock') to acknowledge the legitimacy and rights of the Ts'il Kaz Koh resulted in the latter taking the Village to the Supreme Court of BC and, to their own great surprise (having never previously experienced justice in the white folks' system), winning. Yet the Village proceeded to ignore the Supreme Court ruling and, in February 2000, shut off water and sewer services to the Reserve, in the middle of winter.

Members of Ts'il Kaz Koh took to the streets in peaceful protest, and the stand-off became national news, partly because it was a microcosm of similar disputes that had erupted in other small towns across Canada in the previous decade, some of which had escalated into blockades, violence, and deaths (the Oka crisis of 1990 being the most in/famous). This was a very fraught time in Indigenous/non-Indigenous relations in Canada, in the wake of the 5 volumes of findings of the *Royal Commission on Aboriginal Peoples* in 1996 which had thoroughly documented the history of colonization of Indigenous peoples in Canada and condemned the more than a hundred years of national policy of assimilation that had systematically set out to destroy Indigenous cultures (Inuit, Metis, and First Nations)[2] and to dispossess them of their former lands and ways of life.

The dispute that erupted in Burns Lake was thus part of a much larger national story of deep historic wounds, unresolved conflicts and ongoing discrimination, a story that raises some very challenging questions for Canadians: is there a way forward? Is recognition and reconciliation possible? How can we manage a better, more respectful and just co-existence in the shared spaces of the nation, cities, and small towns? And is

1 While this chapter is my own reflection on a five-year project, I would like to acknowledge my research and film-making partner throughout most of the project, Dr Giovanni Attili, University of Rome La Sapienza. When I refer to 'we' in the chapter, it is to the two of us, unless otherwise specified.

2 There are three distinct Indigenous groups in Canada: Inuit, Metis, and First Nations, each of whom had different legal standing under the Indian Act of 1876. Since the revised Canadian Constitution of 1982, which acknowledged Native rights and title, official legal language uses the umbrella term Aboriginal to describe all Native peoples, but Indigenous people never use this term. 'Indian' is a term used by the Indian Act but dismissed by Native people (except for their own internal/self-referential humorous purposes). 'Native' is more colloquial, and typically how Canada's Indigenous people refer to themselves, although with regional variations. 'First Nations' is a term most predominant in western Canada (Gray 2011).

Figure 23.1 Screen shot from Finding the Way showing location of Burns Lake in north central BC
(Graphic Giovanni Attili)

there a role for planning in finding a way forward? As I began to explore the local contours of this conflict – including the challenges facing a local and regional economy that had been dependent on logging, an industry now devastated by the pine beetle infestation – it became clear that since Burns Lake was still a deeply divided community, there was really no way forward (because of the changing political and demographic landscape) unless partnerships for economic and social development could be set in motion. But partnerships depend on relationships, and relationships, especially involving Indigenous peoples, depend on trust. And there was no trust in Burns Lake because past wrongs had never been acknowledged, let alone addressed. So it was impossible to talk about the present, let alone the future. Things were stuck in the past.

And that raised for me a very interesting planning challenge. I had been invited to the town as a potential 'storyteller' – thanks to a previous documentary about racism and immigrant integration in Vancouver (Attili and Sandercock 2007) – by one of the local anti-racism activists. She was one of a small inter-cultural group who had been challenging the status quo represented by the Village. She introduced me to most of the local

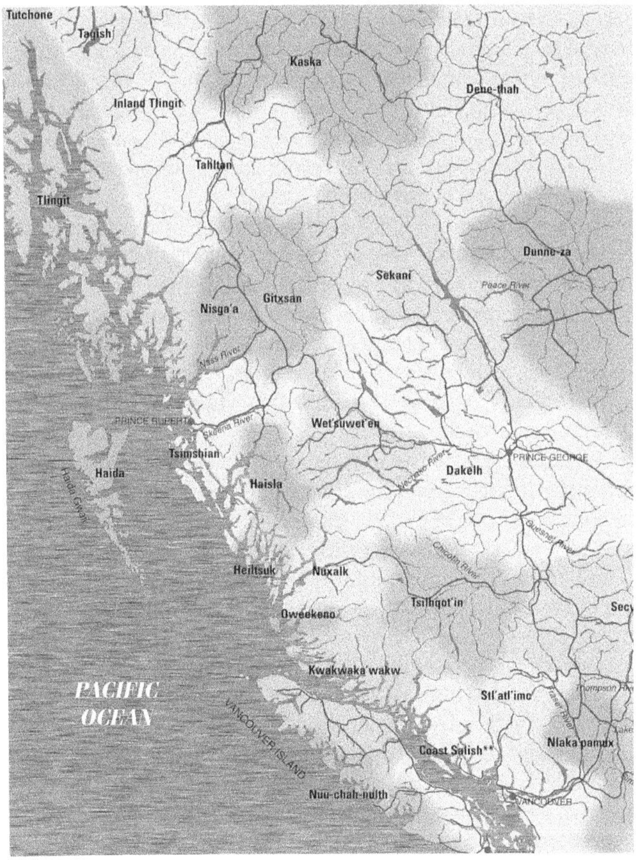

Figure 23.2 Map of traditional territories of First Nations in what is now known as British Columbia
Source: Government of BC, Ministry of Education

actors. And I began to wonder: could we make a planning intervention, through the making of a film, that might tell a new story, allow new voices to be heard, and begin to change the lenses through which people saw each other? Could the film become a catalyst for a set of community dialogues that would begin with the past, move into the present (asking 'is the past still present?'), and then ask what, if anything, people might like to change in order to move forward together? At the time when I began to ask this question, I was thinking of this planning intervention as, in part, a very necessary healing process, in line with some of the conclusions of the Royal Commission. Subsequently I have come to call this a 'therapeutic planning intervention' (Sandercock and Attili 2014), as I will explain shortly, but equally it could be thought of as an attempt at transformative planning along the lines theorized by Healey (2006).

* * *

As an occasion for both reflecting on the reach and building from the substance of Patsy Healey's exceptionally rich collection of writings, this chapter reflects on the extent to which ideas about collaborative and communicative planning and about governance transformation – ideas conceived in a European/western planning context – were helpful in this completely different cultural context. I will discuss why and how we decided to go about this particular kind of intervention, drawing on the collaborative and communicative planning literature and concerns around process design that have all been central features of Healey's work.

And I will reflect in hindsight on how Healey's conceptual framework for analysing transformative governance might help to understand the achievements and limitations of our therapeutic approach.

This chapter is addressed to that 'community of inquirers' (Healey 2012, 189) within the broad field of planning who, with Healey, see the planning project not just as 'managing our co-existence in the shared spaces of complex urban societies', but doing so for *transformative purposes*. For Healey the concept of transformative planning or governance necessarily involves shifts in social and institutional relations, along with or underpinned by shifts in cultural norms and expectations.

Proving that any planning intervention has achieved such a deep sense of transformation is challenging, if not impossible, in the short term. By their very nature, such shifts take time to become entrenched. There are too many 'hit and run' case studies in too many journals that claim too much in terms of 'transformation'. Short term impacts all too often unravel, but by that time the researcher has sped on to the next adventure in intervention. Healey's (2006) analytical schema of levels and dimensions of governance transformation – which distinguishes between 'episodes, processes, and cultures' – introduces some much needed rigour into claims for and writing about 'enduring change'. I use this frame to reflect as soberly as possible on my own case study in the penultimate section of this chapter.

My empirical focus, as already explained, is not a complex urbanized society but a small town remote from any major metropolitan area. Nevertheless I want to argue that the complexity of the conflicts in this particular place are every bit as profound and challenging for planners as one finds in a complex metropolitan region. In this reflection on a rather innovative planning intervention I am inevitably examining the relationship between theory and practice, (another central concern of Healey's work), as well as untangling why any local intervention needs much more than local or situated knowledge, more than practical and process skills. An emphasis on the latter is an Achilles heel of much writing in praise of communicative planning. Healey's work of the past fifteen years reminds us that a multi-scalar analysis of governance and legal regimes is not just crucial for but also a likely explanation of the limits of local interventions. So too, I will suggest in the final section, is an ability to think beyond western frames.

This chapter is situated within a normative orientation that seeks forms of planning intervention that can contribute to a larger societal project of healing and reconciliation, helping a deeply divided community 'to move toward a relationship of power sharing, mutual respect and joint problem solving' (*Report of the Royal Commission on Aboriginal Peoples*, vol. 1, part 1, 1996). I ask what kinds of healing happened as a result of our intervention: personal, intra-family, community, intercultural? What changes were there, if any, in social relations in this town? Why did we choose the making of a film as our mode of intervention? Is film an effective tool in therapeutic planning? Were there institutional and governance transformations as a result of this planning intervention, and if not, why not? I proceed from historical context to theory and methods and thence to reflection and analysis.

Historical Context

The displacement, dispossession, and disempowering of Native peoples began in the seventeenth century but was systematized in the latter part of the nineteenth century through the federal policy of assimilation (*Report of the Royal Commission on Aboriginal Peoples*, vol. 1, 1996). This policy was enforced through three 'technologies of power' (Jacobs 1996): the Indian Act, the Reserve system of Indian lands, and the Indian residential school system. The Indian Act of 1876 determined that 'Indians' were now wards of the state, deprived of their sovereignty, and dependent on the federal government for virtually every aspect of their lives, lands, finances and governance, from cradle to grave. During the second half of the nineteenth century, as westward expansion of settlers proceeded and displaced Native peoples from their former territories and livelihoods, the federal government made a series of treaties across the prairie provinces (seeking to extinguish Aboriginal claims to land), confining Natives to reserves, with promises of financial assistance.

But in British Columbia, these formerly sovereign peoples were simply forced onto reserves while the province refused to make treaties and refused to acknowledge or enter into any debate about Native title, in spite of ongoing Native resistance to colonial land appropriation (Tennant 1990; Galois 1992). By 1930 there were 1,536 small reserves scattered across the province, 'a reserve geography with no close North American equivalent' (Harris 2002, 263). Natives were now confined to a third of one per cent of the land base of the province (ibid.). This 'Reserve system of Indian lands' was in fact an apartheid land use policy which is still evident to this day throughout BC and Canada and is as visually striking as the geography of racial segregation used to be in the southern United States.

The third technology of power was the Indian residential school system, which forcibly removed Native children from their families, forbade them to speak their language or engage in any of their traditional cultural practices and, according to official documents, was intended to destroy the culture by 'killing the Indian in the Native child' (Molloy 1999). It has now been widely acknowledged that the cruelty with which children were treated, in addition to the direct attempt to destroy their cultural inheritance and family ties, has directly contributed to the ongoing dysfunction manifested in many Native communities to this day (such as high levels of addictions, violence and suicide rates) which has had long term, intergenerational impacts (*Report of the Royal Commission on Aboriginal Peoples*, vol. 1, 1996).

Our research took place in the traditional Carrier territory of north central BC. By the early twenty first century there were 203 First Nations recorded in BC under the (still extant) Indian Act, seventeen of them Carrier bands ('band' being the Indian Act terminology for formerly sovereign Nations and their subtribes). The Carrier peoples were forced onto reserves in the first two decades of the twentieth century. The Ts'il Kaz Koh Nation (more widely known as the Burns Lake Band) were allocated 400 acres of reserve land in 1914, but lost one third of it the following year when the town site for what is now the Village of Burns Lake was laid out. That expropriation marked the beginning of a century-long battle between the Band and the Village that culminated in the confrontation in 1999–2000 that ended up in the Supreme Court of BC. What had begun apparently as a municipal water and sewer issue was transformed into an indigenous/human rights issue when the Village shut off water and sewer services to the reserve over a taxation dispute that exposed the ongoing operations of colonialism at the municipal level, what Stanger Ross (2008) has called 'municipal colonialism' (the continuation of the larger ideological project of colonialism at municipal level, especially regarding land acquisition, with urban reserves as the object of desire). And that was the trigger for our involvement in this community.

This illegal exercise of power by the Village demonstrated their lack of recognition of Aboriginal rights and reflected a darker history of discrimination and segregation, a history in which Native people walked on one side of the street and non-Native people on the other; Native people were not served in local restaurants and stores, and were not employed in local businesses and industries. While efforts were made after the Supreme Court case to open a dialogue between the antagonists, these efforts soon fell apart through lack of trust and the ongoing failure to respect Native rights. For example, when a new high school was built in 2005, the Burns Lake Band asked for the return of the land on which the old high school had been built, which had been appropriated from their reserve. Instead, the School District starting tearing down the buildings, prompting the Band to occupy the site. Eventually the Band was given the land on a 99-year lease, and they wanted to turn it into a Gathering Place for all the clans. This new conflict rubbed the old wounds, and the ongoing planning process of trying to get development permits for the new premises through the Village's planning bureaucracy seemed to be a further illustration of the persistence of a colonial planning process and mentality.

But there were local forces for change, seen in a small group of anti-racism activists as well as a group of high school youth, Native and non-Native, who had collaborated on the writing and performing of an anti-racism song ('Leave it Behind'). The demographics were also shifting in favour of the Native population, which will soon be a numerical majority. And twenty years of federal legislation since the new Canadian Constitution of 1982, which had established Aboriginal rights and title, was creating a larger context of change, one that the Village appeared to be resisting. Our interest was, first, in understanding how such a deeply divided community had been created (through what relations of power?); whether there was a role for us as planners, engaging in a therapeutic planning intervention through the making of a film; and finally, whether such an intervention could help create more lasting, collaborative forms of governance. But before embarking on that story, here is a brief overview of our theoretical and methodological lenses.

Theories and Methods

This work is situated in a critical planning tradition which has focused on 'the dark side of planning' (Flyvbjerg 1996), that is, the operations of power through which planning comes to serve the interests of dominant cultures, classes, genders, and ethnicities (Sandercock 2003). I am particularly interested in the role of planning in settler societies in so far as it serves to render Indigenous populations invisible, to deny them rights, to exacerbate their geographic and economic marginalization, and to provide them with culturally inappropriate programs and services (Jojola 1998; Jackson 1998; Irlbacher Fox 2009; Porter 2010). A social justice orientation is central to this work, grounded in theories of what has variously been labelled insurgent (Sandercock 1998; Friedmann 2011), transformative (Lane and Hibbard 2005; Healey 2007) and empowerment (Reardon 2003) planning. There is a fundamental challenge to mainstream planning practice expressed in the concept of *decolonizing planning* (Jackson 1998; Cook 2008; Jojola 2008; Porter 2010; Erfan and Hemphill 2013; Sandercock et al. 2013). If, as these authors argue, planning has been deeply implicated in the colonization process, through its spatial technologies of power (such as mapping, surveying, land use and strategic planning) as well as its more recent processes of consultation or inclusion, which fail to acknowledge Aboriginal rights, title, or sovereignty, how might planning become a more ethical and socially just practice? How might it contribute to the process of healing and reconciliation identified as a national challenge in the findings of the Royal Commission on Aboriginal Peoples (1996)?

I agree with Jojola (2008) that this amounts to no less than a 'long-overdue paradigm shift' for mainstream planning. But this still begs the question of *how* to do such work, where to begin, and at what scale? And here I have been exploring two new directions: first an emergent approach that I have called 'therapeutic planning' (Sandercock 2003; Sandercock and Attili 2014), and second, experimenting with the use of more creative tools of engagement.

From Peter Marris's *Loss and Change* (1975) to Mel King's *Chains of Change* (1981), and more recently to the work of Howell Baum (1997), John Forester (1999, 2009), and Aftab Erfan (2013), there has been an undercurrent in planning that has been asking us to look at the role of emotions in planning practice and to be prepared to work with and through people's hopes, fears, memories, wounds, indelible experiences of loss and trauma. Drawing from the practice stories of Wendy Sarkissian and others, I began to outline a 'therapeutic planning process' in which a dialogic space is created for the unspeakable, for talk of fear and loathing as well as of hope and transformation (Sandercock 2003, 159–66). This involves the creation and design of a safe space in which conflicting parties can meet and speak without fear of being dismissed, attacked, or humiliated; a new space of recognition in which historic injustices are acknowledged, as a necessary prelude to addressing contemporary conflicts. This kind of planning work requires exceptional interpersonal, relationship-building and facilitation skills. It requires patience and the ability to establish trust. It involves difficult dialogue and negotiation across the gulf of cultural difference and historic injustice. When done well, it can open up new spaces of possibility, enable conversations that have previously been unthinkable, and lead to commitments to co-existence and collaboration hitherto deemed impossible (Forester 2009; Sarkissian and Benjamin-Mau 2009; Erfan 2013). And typically, this work requires its practitioners to be fluent in a range of ways of knowing and communicating: from storytelling to listening, to interpreting visual and body languages. In response to the challenges of this difficult work, a handful of practitioners and researchers have been exploring the role of the arts in this kind of community engagement and conflict resolution (LeBaron 2002; Sarkissian and Hurford 2010).

Attili and I have extended this methodological and epistemological exploration into the applications of multimedia, including film and video (Sandercock and Attili 2010, 2010a). We situate ourselves, like Healey, within a post-positivist framework that embraces story as a means of inquiry and a form of meaning making (Fischer and Gottweis 2012; Sandercock and Attili 2012). Our digital ethnographic approach emerges from a scholarly literature aimed at critically recognizing the controversial role anthropology historically played in portraying Native peoples (Banks and Morphy 1997; Sandercock and Attili 2014). More recently, visual anthropology has been informed by different reflective and transformative paths that intersect with different traditions of critical thinking and media practices, gradually shifting from being a colonial weapon to being a tool questioning each and every form of colonialism. This change has not been linear but it has progressively raised a new consciousness 'that engages with subjectivity, reflexivity and the notion of the visual as knowledge and a critical *voice*' (Pink 2003: 180). All these new levels of consciousness and reflexivity represent the ways in which visual anthropology is currently unlearning its own complicity with colonialism. This is where we self-reflectively locate our *digital ethnographic* approach (see Sandercock and Attili 2010), in addition to its more obvious affinity with communicative planning.

Our action research approach is intended as a contribution to transformative planning, helping to bring about a shift towards more equitable economic, social and political relations between Native and non-Native residents of Burns Lake. We were acutely aware that in order for such an *outcome* to happen, new more collaborative approaches to governance (i.e. *processes*) in the town needed to emerge (Healey 2007). We were 'hypothesizing' that such collaborative *processes* were not possible unless and until this healing work had been done, which in turn could contribute to changes in the local *culture*, the taken for granted ways of doing things (Healey 2006). We became involved in a local struggle for both reconciliation and the decolonizing of planning, and we brought with us these conceptual tools and some planning skills, along with technology and artistry (film as an expression of the power of story), and an ethical orientation to collaborative and empowering modes of planning. Thus began a three stage, five-year action research project using film as a therapeutic planning intervention.

What follows is a discussion of the role of film in a deeply divided community, asking to what extent it can open up a new space for dialogue about the past, present, and future. And, beyond dialogue, to what extent it can lead to action, to alternative imaginings that can re-shape the fragile co-existence of two peoples, Native and non-Native Canadians, towards reconciliation and partnership.

Planning to Film, Filming to Plan: Immersion in Burns Lake

I begin with the story of the research, filming, and editing of *Finding Our Way* (Attili and Sandercock 2010), explaining the 'who, what, and why' of our collaborative filmmaking, our partners and purposes. I then describe the action piece of the action research: how, with help from colleagues, we took the finished film back to the communities whose story it tells, organized screenings followed by dialogue circles, then evaluated that process. But first, a few words about the local power dynamics.

The more we talked to people, and the more we learned about the operations of power and privilege in Burns Lake, the more compelling we found the story to be, anticipating that it was a microcosm of Native/non-Native relations in non-metropolitan Canada. Our perception was that this twenty-first-century conflict was deeply rooted in the history of colonization. On the part of most non-Native residents, there was a lack of awareness of First Nations' lived experience of colonization, and particularly of the dysfunctional and intergenerational effects of the Indian residential schools. And on the part of First Nations, there was anger and frustration, since historic injustices had never been acknowledged, were ongoing, and were not being addressed. In other words, the past was very much still present, distorting what contemporary goodwill existed, and blocking a path forward. The more we talked with Native people, the more we heard about healing, about the kinds of healing they had undertaken both as individuals and as a community. What struck us was that there had been very little healing between Native and non-Native folks, even though a Community Healing Project had been initiated after the Supreme Court case. We wondered how planning issues (such as economic and social development) could be dealt with in the absence of some real work across the cultural divide.

Thus we conceived the film, during the research and interview stage, as a way of opening and encouraging a difficult dialogue, of changing the lens on the past, and necessarily 'unsettling' the settler society. Because we wanted to encourage dialogue, but starting from a fresh perspective (a new lens), we envisaged a collaborative film making process in which we would be creating the dialogic space for new stories to be told, by voices hitherto unheard.

We approached both the Ts'il Kaz Koh Nation (Burns Lake Band) and the Village seeking their collaboration. We explained our ethical protocol, which was to bring back rough cuts of the film to the community (Native and non-Native) at every step of the editing process to ask for their input; to offer every individual whom we interviewed the right to withdraw from the film if they didn't like how we used their words; and to bring the final cut back to the community and arrange a community-wide dialogue. The actual editing process was so collaborative that it took two and a half years, involving eight trips back to the community to share the footage with them, sometimes one-on-one (in the case of 'delicate' interviews such as that with the Mayor), other times in group screenings.

We asked our collaborators what they would like to see come out of such a process. The Mayor, Bernice Magee, guardedly agreed to cooperate after we made the case that the film could potentially contribute to a shift in understanding within the Village, between Native and non-Native residents, that might result in a less confrontational stance and more willingness to cooperate on joint projects for social and economic development. As she says in the film, 'working with First Nations is a challenge'. But she also says that 'it's a learning process', and that certainly turned out to be true even during the making of the film, as each 'side' in this conflict began to learn more about each other during the rough cut screenings. The Band was much more enthusiastic. They wanted their story to be told, and they trusted us to tell it. This trust, the Chief (Rob Charlie) explained to us much later, came about because 'you took the time to get to know us' (interview with Rob Charlie in the film's Epilogue). The Band was not sure what might change, if anything, as a result of the telling. But they recognized, as the Chief explains in Part 2 of the film, that they could not survive alone, and that they were willing to forgive, although not to forget, in order to move forward as a people.

The Action Part of the Action Research: Designing Community Dialogues

Initially we had conceived the screening and dialogues as the *action* part of our project, and the making, editing and rough-cut screenings of the film as the *research* part. On reflection, every stage of the film making was part of the 'action' in that it cumulatively brought diverse community members, hitherto antagonists, into new relationships with each other, simply by watching and listening to each other in the rough cuts. In other words, the therapeutic work that we had imagined beginning through the formal dialogue process that we were to design, actually began in the course of making the film, during which a lot of meaningful relationship building work across the cultural divide began to happen.

Four years after we entered this community, once we had the final cut of the film, we returned and formed an Advisory Committee (comprising the Mayor, the Chief of the Ts'il Kaz Koh Nation, the high school principal, the high school drama and dance teacher, who has a prominent role in the film through her work with youth, the Staff Sergeant of the Royal Canadian Mounted Police, two young Village councillors, and the anti-racism activist who had first invited me to Burns lake) to work with us to design the film screening and community dialogues. We envisaged two screenings and dialogue events, one for youth (with whom we had done specific participatory work during the filming stage) and one for the community at large. These community partners, after watching the film together, all agreed that this would not be an easy film for the community to watch, that it was not 'sugar-coated' and would likely provoke a range of reactions from surprise and anger to guilt and denial, empathy and hope. We collectively wondered how we might encourage the community to watch with an open mind and heart, and with some mindfulness about their own emotions as they watched. The Chief suggested that perhaps community leaders might introduce the film and convey this message, and (somewhat to our surprise) the Mayor readily agreed.

We then brainstormed about what the community dialogue process might look like, and collectively came up with the idea of breaking the audience into dialogue circles of eight to ten people, each of which would have two facilitators, a youth and an Elder, and a mix of Native and non-Native. Our Advisory Group helped us to recruit community members whom we then trained as facilitators. We also did a preview screening for the high school teachers, anticipating that the high school youth would be eager to discuss the film with them after the screening for the youth. The teachers were so engaged by the film (one said it 'lifted the veil of silence about what's been going on in this town') that they offered assistance as facilitators. The high school principal acknowledged that the school had been part of the problem, and he was eager to get the film into the classrooms.

The training sessions were in themselves profoundly moving events as this small sample of the local audience expressed their emotions and their support for using the film as an educational tool. The youth expressed anger that they had not been exposed to these stories in school. Some of the Native youth talked of growing up being ashamed of being Native but now, with the film's existence, they felt pride and wanted to share the film with their friends and family, immediately! Others spoke of growing up not wanting to hear their parents' stories of pain, especially the pain of the residential school experience, but now they understood why it was important to hear and discuss those stories and understand the intergenerational impacts. Each person spoke in these deeply felt ways and it was clear, during this preparatory stage, that further healing work was beginning.

Each of the two screenings began with food, and live music played by the local youths who had written the anti-racism song a few years earlier. As planned, the Chief and Mayor then introduced the film. On stage together, the Mayor said that the film told an important story and that the Village had made mistakes that needed to be acknowledged, while the Chief spoke of how, before the film process, he had given up hope in the town, but now his hope and vision of working together was revived. We then launched into the 90-minute screening, followed by the dialogue circles. The facilitated dialogues ran for 90 minutes, focusing on three questions: What was the most significant impact of the film on you? Is the past still present in this town? And if so, how might this community move forward together now, and what would your role be? One of the facilitators in each group took detailed notes, and the Mayor and the Chief, who moved about the room as witnesses during the dialogue circles, each made closing summaries of what they had heard and felt during this time. Both of them reflected on the spirit of hope for moving forward that they heard in the circles and saw on the flip charts in the form of visions for the future, including a vision for an intercultural Gathering Place for the town, and the desire of the youth to be involved in intercultural community activities.

Fifty high school youth (half of them Native) attended the youth screening and dialogue and 150 people (at least one third Native) attended the community-wide screening, half of whom remained for the dialogue, and beyond, to fill out our evaluation questionnaire.

Reflections on Healing

The primary question that framed our action research was whether some kind of therapeutic planning intervention could address the history of colonization in this deeply divided community and thereby open up a space of possibility for healing. The second, more methodological question, was whether film could serve as the catalyst to create this new space for dialogue and healing, by changing the lens through which people viewed each other. We have three ways of addressing these questions. The first (and least satisfactory) is through a grant-mandated survey questionnaire administered at the end of each of the two evenings of dialogues. The second is the extensive notes taken by facilitators for each dialogue circle, as well as our own observations as participants and witnesses. The third is more anecdotal and medium term, based on ongoing relationships in and with the community.

Doing an evaluation right on the heels of an event is common enough among planning practitioners, especially those working on government contracts. Not only did our provincial government funders require such an evaluation but they also dictated specific questions and even format, which made us very uncomfortable.[3]

Questions such as, on a scale of one to five, 'How well did this screening and dialogue help your community address racism?' or, 'How well did this event help your community identify pathways to working together across cultural differences?' can only elicit crude numerical replies. While we did receive very high 'scores' on these answers, we are reluctant to place too much weight on them, precisely because it is such an 'in the moment' response, and speaks at best to an immediate reaction and to *intentions* rather than subsequent actions. Qualitative answers (to questions we had added), from both youth and adults, to the question 'How will you act on what you have learned through the screening and dialogue?' contained many expressions of the desire to volunteer to work on community projects such as a Gathering Place, which was envisioned during the dialogues as an intercultural facility for the whole community. But these answers have the same limitations, being intentions rather than actions.

---

3 The province did not fund our research, but they did fund the community dialogues. The research was funded by the Social Sciences and Humanities Research Council of Canada, the Vancouver Foundation, and the Nechako Kitimaat Development Fund.

Questionnaires like these are also extremely limited in what they can reveal about the qualitative experience of an event. All of us in the dialogue project team[4] took away from the dialogue circles profound and visceral memories of levels of honesty and empathy, of the sharing of tears, of an energized and enthusiastic youth, of the opening of a space of hope. A high level of energy and commitment was released, especially among the youth, some of whom have since gone on to become local leaders, including one of the non-Native youth who, at age 21, was elected Mayor of the town in 2011. Some of the Native youth formed a Six Nations youth group after the community dialogues and started a Facebook page to inspire other youth to get involved in social change. Three days after the dialogues, the local newspaper ran an editorial, titled 'Inspiring' (*Lakes District News* 9 June 2010, p.4), which referred specifically to this commitment to social change evident among the youth who were there.

On reflection, too, the comments that the Mayor made immediately before and after the screening were a kind of local equivalent to the Prime Minister's apology in Parliament in 2008 for the Indian Residential Schools and the damage they had done to generations of Native children and families. While she did not *apologize* per se, she did admit some responsibility (on the part of the settler society) for 'mistakes' in the past. It may well be that 'apology' at a local level is an important complement to the apology at the national level and perhaps even in some ways more meaningful, since most of what affects the lives of people on a day-to-day basis happens at the local level. This could then be seen as a very important part of the ongoing healing process.

It would be reasonable to say that our community dialogues served as a voluntary public setting for opening up and cathartic sharing of some very personal stories about the past and the present, stories of both pain and shame. Some of the Native youth admitted to and talked through some of the personal and family struggles that they had previously kept private. Some older Native folks divulged for the first time, in public or private, a lifetime of pain as a result of their own or other family members' experiences of residential school, and this sharing had a profound effect on listeners. Some members of the settler society admitted shame at what had transpired both historically and in the very recent past, right under their noses. And the quality of listening and empathic engagement among participants was extraordinary. It seemed at the time, and was later confirmed to be, a turning point both for the tellers of these stories and for the folks hearing and receiving them, the beginning of new relationships in which older biases were now dropped and different and better relationships were now possible. This is what Bush and Folger (2004) call 'recognition', one of two pillars of conflict transformation. Recognition of others in their full humanity rather than as 'other' is significant in the context of our project, which called for a *re-setting of the social web of relationships*, something that is acknowledged to be central to the meaning of healing both within Indigenous communities and in conflict resolution *between* communities (LeBaron 2002; Waldram et al. 2008: Erfan 2013).

After the dialogues, the Mayor and the Chief as well as various community members took public responsibility for working on the Gathering Place as an intercultural project. To us, this affirmed that the film was effective as a catalyst for intercultural conversation, had created the space for that conversation, and was instrumental in shifting the historic conflict.

We have noted the energy and sense of possibility that emerged from the dialogues.

In August of 2010, three months after the dialogues, the Village of Burns Lake (prompted by the Mayor) adopted a motion of support in principle for one of the ideas that had emerged from the dialogues: the renovation of the old high school as a Gathering Place for the entire community. The Mayor personally drafted a motion that the Village Council write to the School District asking that they transfer ownership of the land back to the Ts'il Kaz Koh so that they could qualify for federal funds for capital improvements. And she urged the Chamber of Commerce and the Regional District Planning Office to write similar letters. This action by the Mayor would have been unimaginable four years earlier, when the interpersonal as well as intercultural conflict was so intense that she and the Chief of Ts'il Kaz Koh could literally not be in the same room together for more than five minutes.

4 At this point we partnered with Scott Graham (Social Planning and Research Council of BC) and Norma-Jean McLaren and Nathan Edelson (community development and planning consultants) in the design and facilitation of the dialogues and training of local facilitators.

During that same period immediately after the dialogues, Nathan Edelson, a Vancouver-based community economic development planner and colleague who had helped us conceive the dialogues, began working as a liaison between Ts'il Kaz Koh and the Village in moving the Gathering Place forward: getting architectural drawings as well as seeking capital funding. Bernice Magee retired as Mayor in 2011 and was replaced by Luke Strimbold, one of the non-Native youth who had volunteered as a facilitator of the community dialogues. While the former Mayor continues to serve as a staunch supporter on the committee that is pushing the Gathering Place project, the new young Mayor reflects the spirit of enthusiasm and intercultural partnership that came out of the youth dialogues.

From Healing to Transforming Governance?

In thinking through the extent to which our therapeutic planning intervention contributed to a governance transformation, I find Healey's analytical framework (developed in and for a completely different set of governance questions in the European Union) a helpful starting place. So much attempted 'change work' at local government level gets lost in ephemeral pilot projects or in innovations that disappear with a change of Council after an election. Excitement and new energy is quickly followed by disappointment and disillusion, apathy and despair at the political 'cycle'. Healey's analytical schema of levels and dimensions of governance – which distinguishes between *episodes, processes*, and *cultures* – asks how such 'episodes' of innovation might bring about enduring change (Healey 2006). Her answer is that innovative episodes must move 'from explicit formation episodes to arenas of investment and regulatory practice ... they have to become institutionalized in the routines of governance practices ... New concepts have to challenge and shift an array of already routinized governance processes, with their complex mixtures of conscious and taken-for-granted modes of practice (ibid. 305). New concepts have to jump boundaries and break through resistances, and this may involve a complexity of relations between different departments of local government, between administrators and politicians, between politicians and citizens, between the state and all kinds of lobby groups and power elites, each with their own relation to allocative power and regulatory power, as well as discursive power. Nor is shifting governance processes enough: for such processes are sustained and legitimated by cultural assumptions about appropriate agendas and practices. Shifts at this cultural level, 'promoted by long-term shifts in economic, socio-cultural and political relations, influence both those involved in particular innovative episodes and those involved in routine governance practices' (ibid.). The pressure for transformation may come from any of the levels, but meaningful transformation has to involve initiatives moving from the level of an *episode* to at least the level of governance *process*, and in some way to find resonance with *cultural* assumptions. In other words, enduring change requires institutionalization, as well as shifts in attitudes and behaviour.

If I use this frame to analyse what has been happening in Burns Lake, I would say that our therapeutic planning intervention was an episode that triggered new ways of seeing and new ideas about good governance, ideas that put an emphasis on partnership between indigenous and non-Indigenous residents in order to shape a different future. It also generated new social relationships across the cultural divide, and new young leaders, as well as transforming one old guard leader (the Mayor) and one senior administrator (the CAO of the Village) into supporters of change in the direction of more collaborative governance. As we've seen, within months, the Mayor was leading the Village Council in an entirely new direction, seeking support for the Gathering Place from higher levels of government and other power brokers (the Regional District, the School District that owned the land on which the old high school sat, the Chamber of Commerce). She was trying to mobilize support for a new way of doing things, and she formed a committee to move this forward, a committee that included Ts'il Kaz Koh. Within a year she and the committee had succeeded in having the land transferred back to Ts'il Kaz Koh, and conversations had begun with higher levels of government about capital funding. So resources were being mobilized, too.

The Village CAO, after some initial resistance to taking part in the dialogues in the belief that he must remain 'neutral', did participate, was deeply moved, and became an advocate for change within Council in terms of hiring policies and encouraging young Native people who were at the dialogues to get involved in Village politics and run for office. And the twenty-one-year-old Luke Strimbold, who became the new

Mayor, had been one of the youth whom we had trained as a facilitator for the dialogues. The son of a logging family, Luke had just finished college and had become an active and articulate change agent, representing new cultural values around reconciliation and recognition that were more in line with national political and legal changes in favour of Native rights and title.

We can at least say, then, that the level of governance processes in Burns Lake is 'in movement' since the dialogues, as exhibited by the changes in the 'old' leadership mindset as well as the generational change in actual leadership, symbolically ushering out the old guard pioneer stock in favour of a new generation with a very different set of attitudes and values regarding Indigenous/state relations and Indigenous/non-Indigenous social relations at local level. The new Mayor is eager to develop better working relationships with all of the six Nations living on Reserves in and around Burns Lake, and is welcome in all of them. Similar changes had been happening with the local RCMP (Royal Canadian Mounted Police) detachment, thanks to several progressive Staff-Sergeants, one of whom had served on our Dialogues Advisory Committee. And the high school teachers, enthusiastically supported by a new Principal (who had also helped us with the dialogues, including allowing us to host them in the high school gym) were eager to get our film into their classrooms and curricula, providing another avenue for longer-term *cultural* shifts.

To say that our therapeutic intervention was the cause of all this governance change would be claiming too much. We entered a political force field in 2006 in which there were already local forces for change and for healing. But they were embattled. Things were seemingly stuck. The attempt at healing in the immediate aftermath of the Supreme Court Case had quickly unravelled, even though there were new forces for change simmering in the shadows. The anti-racism song created by the high school youth and their accompanying video about racism in the town was evidence of this simmering, and was one of the reasons we were attracted to the town: for this paradox of the old guard in charge, and yet the pressures mounting for change.

Timing is always critical and we were gambling that the timing was right for our 'experiment' with film as a catalyst for healing. People in the town still talk about the film as a turning point in all sorts of ways, from the personal and familial to the political, (and in this case the personal was always already political, because of the history of colonization). We cannot be sure of the trajectory of this movement for change. Politics is a force field in constant motion. And progress is rarely linear. Economic development challenges for the region are huge. Local room for manoeuvre is limited, depending on access to resources from higher levels of government, as well as some very difficult choices about whether to support certain kinds of non-renewable resource development, including pipelines carrying oil from the Alberta tar sands through this territory. (Most Native leaders are opposed).

With all of those caveats, we can still say that there has been personal healing, intercultural healing, transformation in social relations and in governance processes, if not yet in the institutionalization of new ways of doing things. Our dialogic/communicative and collaborative processes seemed to have much to offer. So in one sense our gamble seems to have paid off, there has been transformation, *if we're evaluating it through the lens of western planning theory ...*

Beyond 'Western Eyes'

The shift from municipal colonialism to something approaching a more collaborative form of governance between Indigenous and non-Indigenous groups in the Village of Burns Lake constitutes progress, even 'transformation', in the conceptualization of western planning theory. But this conceptualization still 'frames' Indigenous peoples as stakeholders seeking inclusion in western political processes. While some of the 203 Indigenous Nations in BC and some of the more than 600 Nations across Canada do express this desire for inclusion, this is not the sum total of their aspirations. As Porter (2006, 389) has noted in the Australian context, 'conceptualizing Indigenous peoples as 'stakeholders' in planning processes fails to appreciate their unique status as original owners of country that was wrested from them by the modern, colonial state'. The emergent more collaborative forms of governance are one arena in which Ts'il Kaz Koh and other Nations in this traditional Carrier territory seek to participate. But of equal if not more importance and consequence are their struggles (through their Tribal Council) in the courts and at the treaty table, to achieve recognition and rights, specifically involving the return of lands that were never ceded.

These two paths can be understood as parallel and complementary strategies. The first is intended to improve relationships in the here and now, at the local level where people live. The second involves working on a longer time frame to achieve justice in relation to Native title and rights, and to shift the culture of planning from the current system, which 'allows' certain limited participation to Indigenous peoples, but always on terms dictated by the dominant culture, to something more like a genuine partnership in the managing of our co-existence in the shared living and working spaces of the nation. Progressive planners who seek to be allies of Indigenous peoples need to understand these distinctions between their short and the longer term aspirations in order to support them in these struggles in the most appropriate ways.

To return, in closing, to Healey's definition of the planning project as 'managing our co-existence in the shared spaces of complex urban societies', I would argue that this takes on a whole new meaning in the context of the current Indigenous rights and title claims in settler societies. While there is a robust environmental planning and resource management literature that has been asking what these claims mean for these domains of planning, the arena of urban planning has yet to come to grips with what co-existence might look like or what it might mean in urban settings. Indeed the question has only just begun to be asked (Howitt and Lunkapis 2010; Porter 2013), and can be framed in a variety of ways. What forms of planning governance might now be necessary and appropriate to fully and justly recognize 'the persistent footprint' of Indigeneity in settler cities? Not only do these cities exist on traditional and often unceded territory in which Indigenous peoples were shunted onto urban reserves over a hundred years ago, reserves which are now bursting at the seams and entirely inadequate in terms of realizing Indigenous aspirations for economic, social and cultural development. But there has also been a substantial 'rural to urban migration' of Indigenous peoples for at least half a century, as a result of which 84% of Maori people live in cities; 50% of First Nations, Inuit and Metis live in Canadian cities, and Australian cities harbour 43% of Aboriginal people. Surely then, the urban is the 'new frontier' of contestation for the recognition of Indigenous rights and title, the reclaiming of Indigenous spaces, and the reconciliation of Indigenous and non-Indigenous lifestyles and desires. Why has urban planning practice not noticed these 'seismic shifts in Indigenous rights recognition' (Porter 2013)? Surely, because we have been looking through western eyes.

What do Indigenous rights mean in an urban planning context? How might contemporary Indigenous title, governance, and aspirations to economic development and cultural revitalization be engaged in urban planning? What, actually, does co-existence imply? These are the profound challenges facing the next generation of planning theorists and urban planners in settler societies.

References

Attili, G. and L. Sandercock. 2010. *Finding Our Way*. 90 min documentary. (Vancouver: Moving Images).

Banks M. and H. Morphy (eds). 1997, *Rethinking Visual Anthropology*. New Haven. London: Yale University Press.

Baum, H. 1997. *The Organization of Hope*. Albany: State University of New York Press.

Bishop, A. 1994. *Becoming an Ally: Breaking the cycle of oppression*. Halifax: Fernwood.

Bush, R.A. and J.P. Folger. 2004. *Promise of Mediation: The transformative approach to conflict*. San Francisco: Jossey Bass.

Cook, J. 2008. 'Building on traditions of the past: The rise and resurgence of First Nations CCP', *Plan Canada*, 48(2): 13–17.

Dorries, H. 2012. 'Rejecting the "false choice": Foregrounding indigenous sovereignty in planning theory and practice'. Unpublished PhD dissertation, University of Toronto, Department of Geography.

Erfan, A. 2013. 'An experiment in therapeutic planning: Learning with the Gwa'sala-'Nakwaxda'xw First Nations'. Unpublished PhD dissertation, University of British Columbia, School of Community and Regional Planning.

Erfan, A. and J. Hemphill. 2013. 'Indigenizing and decolonizing: An alliance story', *Plan Canada* 53(2): 18–21.

Fischer, F. and H. Gottweis (eds). 2012. *The Argumentative Turn Revisited*. Durham, NC: Duke University Press.

Flyvbjerg, B. 1996. 'The dark side of planning: Rationality and Realrationalitat', in S. Mandelbaum, L. Mazza and R. Burchell (eds), *Explorations in Planning Theory.* New Brunswick, NJ: Center for Urban Policy Research Press, pp. 383–96.

Flyvbjerg, B. 2001. *Making Social Science Matter*. Cambridge: Cambridge University press.

Forester, J. 2009. *Dealing with Differences*. Oxford: Oxford University Press.

Friedmann, J. 2011. *Insurgencies: Essays in planning theory*. London: Routledge.

Fullilove, M. 2004. *Root Shock: How tearing up city neighborhoods hurts America, and what we can do about it*. New York: Random House.

Galois, R.M. 1992. 'The Indian Rights Association, native protest activity and the "Land Question" in British Columbia, 1903–1916', *Native Studies Review* 8(2a): 1–34.

Gray, L. 2011. *First Nations 101*. Vancouver: Adaawx Publishing.

Harris, C. 2002. *Making Native Space*. Vancouver: UBC Press.

Healey, P. 1997. *Collaborative Planning*. London: Macmillan (2nd ed. 2007).

Healey, P. 2006. 'Transforming governance: Challenges of institutional adaptation and a new politics of space', *European Planning Studies*, 14, 3: 299–321.

Healey, P. 2012. 'The universal and the contingent: Some reflections on the transnational flow of ideas and practices', *Planning Theory*, 11(2): 188–207.

Hibbard, M. and M. Lane. 2004. 'By the seat of your pants: Indigenous action and state response', *Planning Theory and Practice*, 5(1): 97–109.

Hibbard, M., M. Lane and K. Rasmussen. 2008. 'The split personality of planning: Indigenous peoples and planning for land and resource management', *Journal of Planning Literature* 23(1): 136–51.

Howitt, R. and G.J. Lunkapis. 2010. 'Co-existence: Planning and the challenge of Indigenous Rights', in J. Hillier and P. Healey (eds), *The Ashgate Research Companion to Planning Theory.* Farnham: Ashgate, pp. 109–33.

Innes, J. and D. Booher. 2010. *Planning with Complexity*. New York: Routledge.

Irlbacher-Fox, S. 2009. *Finding Dahshaa: Self-government, social suffering, and Aboriginal policy in Canada*. Vancouver: UBC Press.

Jackson, S. 1998. 'Geographies of co-existence: Native title, cultural difference and the decolonization of planning in north Australia'. Unpublished PhD dissertation. (Sydney, School of Earth Sciences, Macquarie University).

Jacobs, Jane M. 1996. *Edge of Empire: Postcolonialism and the City*. London: Routledge.

Jojola, T. 1998. 'Indigenous planning: Clans, intertribal confederations and the history of the All-Indian Pueblo Council', in L. Sandercock (ed.), *Making the Invisible Visible: A multicultural history of planning*. Berkeley: University of California Press, pp. 37–56.

Jojola, T. 2008. 'Indigenous planning: An emerging paradigm', *Canadian Journal of Urban Research*, 17(1): *Supplement.*

Walker, R., T. Jojola, and D. Natcher (eds). 2013. *Reclaiming Indigenous Planning*. Montreal: McGill Queens University Press.

Lane, M. and M. Hibbard. 2005. 'Doing it for themselves: Transformative planning by indigenous peoples', *Journal of Planning Education and Research*, 25(2): 172–84.

LeBaron, M. 2002. *Bridging Troubled Waters: Conflict resolution from the heart.* San Francisco: Jossey Bass.

Marris, P. 1974. *Loss and Change*. London: Routledge and Kegan Paul.

Milloy, J. 1999. *A National Crime: The Canadian Government and the Residential School System, 1879–1986*. Winnipeg: University of Manitoba Press.

Muckle, R. 2007. *The First Nations of British Columbia: An anthropological survey*. 2nd ed. Vancouver: UBC Press.

Patton, M. Quinn. 2010. *Developmental Evaluation*. Thousand Oaks, CA: Sage.

Pink, S. 2003, 'Interdisciplinary agendas in visual research: re-situating visual anthropology', *Visual Studies*, 18(2): 179–92.

Porter, L. 2006. 'Planning in (post)colonial settings: Challenges for theory & practice', *Planning Theory & Practice*, 7(4): 383–96.

Porter, L. 2010. *Unlearning the Colonial Cultures of Planning*. Aldershot: Ashgate.

Porter, L. 2013. 'Co-existence in cities: The challenge of Indigenous urban planning in the twenty-first century', in R. Walker, T. Jojola and D. Natcher (eds). *Reclaiming Indigenous Planning.* Montreal: McGill Queens University Press, pp. 283–309.

Regan, P. 2011. *Unsettling the Settler Within: Indian Residential Schools, truth telling, and reconciliation in Canada.* Vancouver: UBC Press.

Report of the Royal Commission on Aboriginal Peoples, 5 vols. 1996. Ottawa: Government of Canada.

Sandercock, L. 2003. *Cosmopolis 2: Mongrel cities of the 21st Century.* London: Continuum.

Sandercock, L. and G. Attili (eds). 2010. *Multimedia Explorations in Urban Policy and Planning: Beyond the Flatlands.* London: Springer.

Sandercock, L. and G. Attili. 2010a. 'Digital ethnography as planning praxis: An experiment with film as social research, community engagement and policy dialogue', *Planning Theory and Practice*, 11(1): 23–45.

Sandercock, L. and G. Attili. 2012. 'Unsettling a settler society: Film, phronesis and collaborative planning in small town Canada', in B. Flyvbjerg, T. Landman and S. Schram (eds). *Real Social Science*. Cambridge: Cambridge University Press, pp. 137–66.

Sandercock, L. and G. Attili. 2014. 'Changing the lens: Film as therapeutic planning practice', *Journal of Planning Education & Research*, 34: 19–29.

Sandercock, L., J. Cook, L. Sparrow and L. Grant. 2013. 'Indigenizing planning education, decolonizing planning practice: First steps', *Plan Canada*, 53(2): 48–51.

Sarkissian, W., N. Hofer, Y. Shore, S. Vadja, C. Wilkinson. 2009. *Kitchen Table Sustainability*. London: Earthscan.

Sarkissian, W, and W. Benjamin-Mau. 2009. *SpeakOut: The step by step guide to speakouts and community workshops*. London: Earthscan.

Sarkissian, W. and D. Hurford. 2010. *Creative Community Engagement*. London: Earthscan.

Stanger Ross, J. 2008. 'Municipal colonialism in Vancouver: City planning and the conflict over Indian reserves, 1928–1950s', *Canadian Historical Review* 89(4): 541–80.

Tennant, P. 1990. *Aboriginal Peoples and Politics: The Indian Land Question in British Columbia, 1849–1989*. Vancouver: UBC Press.

Tobey, M. 1981. 'Carrier', in J. Helm (ed.). *Handbook of North American Indians: Subarctic*, vol. 6. Washington, DC: Smithsonian Institution, pp. 413–32.

Waldram, J.B. et al. 2008. *Aboriginal Healing in Canada: Studies in therapeutic meaning and practice*. Ottawa: Aboriginal Healing Foundation.

Chapter 24

Collaborative Planning Crossing Borders: Practice in the Context of Egypt

Zeinab Noureddine Tag-Eldeen

Introduction

Planning demonstrates continual efforts to relate concepts of social science theory to planning practice in order to meet the continuous challenges posed by societal change (Healey 2006). This confirms the notion of planning as a social phenomenon (Connell 2011:270). Therefore, appreciation of the history and cultural values that characterise a society must be acknowledged by planners in the cross-cultural practices that inform planning knowledge. It is the moral and intellectual responsibilities therein that challenge those involved in such interactions (Healey and Upton, eds, 2010). This chapter presents the process of practising Collaborative Planning in an Egyptian context, such that the history and the cultural values of that context are recognised. It underlines the potential in a new context – in this instance, the case study of Zifta City – to provide different practical and theoretical interpretations of Collaborative Planning ideas. Collaborative Planning can withstand being 'carried' from one place to another due to its pliability, which arises from its emphasis on both content and process. According to Healey (2006), Collaborative Planning combines three elements: 1) interactions of people in places: which in Egypt is appreciated as supporting the requirement of the General Organization of Physical Planning (GOPP) to produce realistic and implementable plans that reflect the urban reality of Egyptian cities; 2) an institutionalist view of the dynamics of social change in a specific context: it is a significant element for building institutional capacity, which in the Zifta case helped to establish a link between formal and informal sectors, mobilise local resources and support program delivery; and 3) an 'interpretive' view of policy making and planning; all of which support Collaborative Planning's ability to be communicated in a new context.

The analytical framework of this work emphasises Collaborative Planning's core concepts as developed by Patsy Healey, and with special consideration to the philosophical ground that inspired Healey to elaborate her ideas. Revisiting the fundamental philosophical theories helps to understand the cultural and institutional assumptions that were embedded in the practice of a Collaborative Planning process in Zifta. The rules of communication within a Collaborative Planning process are analysed with reference to Habermas's (1987) communicative theory to understand patterns of communication in Zifta society and further, to make use of them to foster dialogue among the participants. Understanding the social system in Zifta city is achieved with reference to Giddens's (1984) structuration theory, which supports analysing society's formal and informal structures in the context of its rules and resources. The institutional structure of Zifta is examined with reference to Ostrom's (2005) institutionalist approach, and the role of planners is supported by the practical guidance of Crosby and Bryson (2005) on how to lead in a shared-power environment so as to facilitate collaboration.

A proactive approach was pursued, to set Collaborative Planning in dialogue with the people 'performing' social interaction in an Egyptian urban reality. Additional theoretical references are incorporated to explain the empirical findings and to broaden the interpretation of Collaborative Planning in a new context, as well to plan and act, and thus create change. Planning practice is used to identify the cultural values in Zifta associated with society's institutional structure and which guide its daily routines, and which (importantly) carry collaborative ideas. I have been influenced by literature addressing the significance of culture in organisational management (Flamholz and Randle 2011; Kotter and Heskett 1992; McDermott and O'Dell 2001; see also Sandercock 2003) for diverse ways of knowing culture in planning. The proactive approach is further emphasised by my dual role as an urban planning 'practitioner' leading the 'Zifta Demonstration Project' and as an 'action researcher' working on my doctoral study, with a special interest in the cross-cultural development of planning knowledge. While these roles overlapped and fed into each other, as a researcher guided by a specific analytical framework I have stepped back at various points in the project's implementation to reflect on events and analyse them in relation to the relevant theory. Thus, this alertness allowed for interplays between theory and the empirical material to be made explicit (see e.g. Alvesson and Kärreman 2007) throughout the progress of this work, with theory, in effect, thereby 'practised'. The role switches between research and practice enhanced my practitioner capacity for reflection-in-action (as expressed by Schön 1983). This helps to focus on knowledge production and community learning through interactive processes. It offers vivid and rich windows on events, experiences and relationships; and a mapping of ways of practising theory (Dyer and Wilkins 1991).

The planning process for Zifta has involved stakeholders from various groups of interests so as to obtain qualitative perspectives on, and deeper understanding of the factors that influence the city's urban conditions. As a result of the social interactive processes encouraged by the process, different types of knowledge have been transferred, and in a variety of directions. The interactive processes helped to transform and restructure power relationships between the different actors. This is demonstrated in the transformation of citizens' ideas into actions to be carried out collaboratively with the City Council. This confirms the transformative potential of Collaborative Planning as stressed by Healey (2006:265). Applying the practice of Collaborative Planning at multiple levels, (Zifta City Council and GOPP) gives the advantage of building institutional planning capacity as it addresses changes in both formal and informal relationships. Untangling the institutional structures of the society and incorporating them into a value-led process that favours the society's culture has helped to build consensus about a desired future for Zifta.

The Zifta project was given the GOPP's political support as it was considered a demonstration from which to learn[1]; this was a major factor in making the project possible. The interest of the GOPP in the Zifta Demonstration Project as a learning model is noticeable even after its completion. An evaluation of the project prepared by independent consultants for the GOPP appraised the community response in the planning process and the participation of different stakeholders. The evaluators describe the plan as 'fine-tuned in the light of feedback and experience. The guidelines provided are highly appreciated by the local authorities. It also fulfilled stakeholders' expectations and raised their awareness' (Bastawissi and Sherbini 2008:6).

Learning-oriented Planning Process

A learning-oriented planning process was devised that reflected the deliverables I needed to produce during the project for each of my assigned roles in the practice of Collaborative Planning in Zifta (See Figure 24.1). The design of this process developed gradually, according to the project's progress and the unfolding of the empirical findings. Three artificially demarcated phases were progressively explored. The 'getting in' phase was to familiarise myself and the planning team with the place. It included a project launch and the identification and mobilisation of stakeholders. The 'getting on' phase applied a mutual learning process among the stakeholders, to understand the context of Zifta, discover its assets and build consensus around its vision. In this socially interactive milieu, different categories of knowledge with technical, social and cultural characteristics were transferred in various directions, which contributed to the development of the Zifta Strategic Plan as well as to my overall research aim. 'Getting out' was the point where I terminated my inquiries, and as a practitioner I gradually transferred responsibility to local planners. This process incorporated learning-by-doing for the planners from GOPP participating in the planning team – led by myself in my practitioner role – the staff of the Zifta City Council and the citizens. The planning process at large was designed as learning-oriented in order to understand how things are done in Zifta and to produce a plan that responds to that reality.

1 Prior to the selection of Zifta as the site for a demonstration project, an assessment phase concluded that most plans were not implemented and development proceeded on the basis of unplanned decisions and resulted in a variety of informal developments. This was perceived as a major problem by the GOPP who, in response, tried to promote a new planning approach that involved the stakeholders and enhanced collaboration between the institutions concerned in order to develop plans that responded to the reality of Egyptian cities. The assessment phase created awareness among the governorates about the attempt underway to examine a new planning approach in Zifta. Later, it helped in obtaining the commitment of the El-Gharbieh Governor to support the Zifta Demonstration Project.

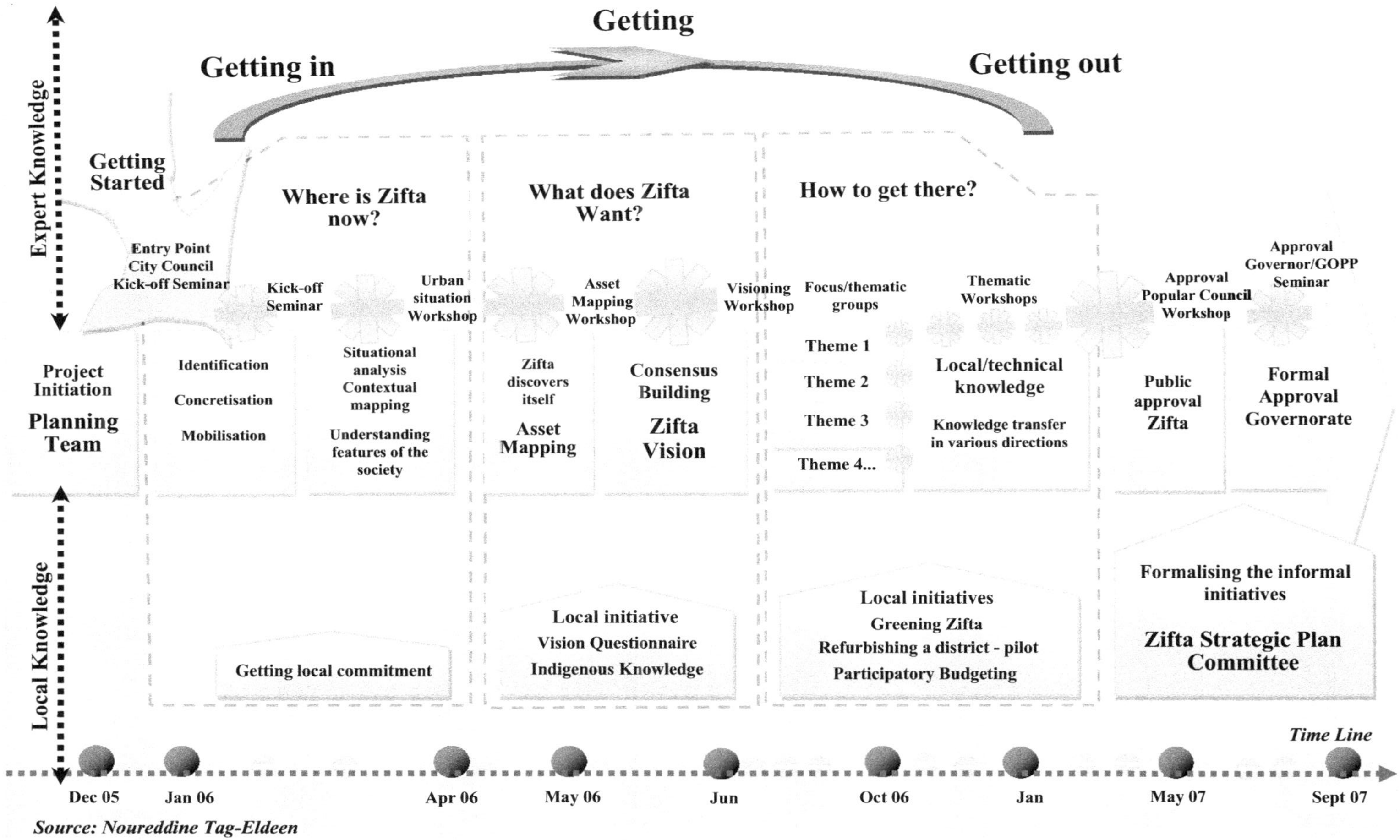

Figure 24.1 Learning-oriented planning process developed gradually, to respond to the dynamic of the project's progress

Source: Noureddine Tag-Eldeen 2012:67

Zifta Stakeholders' Landscape

Seeking inputs from citizens, politicians and various groups of interests has been significant in the Zifta Demonstration Project. To start the planning process it was particularly important to mobilise the support and interest of stakeholders from formal as well informal institutions. The GOPP political commitment encouraged the involvement of formal actors of the City Council, but the challenge was how to reach groups from outside the formal institutions. As an outsider, my entry into the city was arranged formally through the City Council, but it was reluctant to encourage citizen participation and substantially tried to manipulate the process, which made the first round of stakeholder identification very difficult. This necessitated undertaking a set of alternative courses of action to counter the obstruction posed by the established power structure. This included field visits conducted solely by the planning team and carrying out a number of informal conversations that enabled us to establish communication with people uninfluenced by City Council staff, and to identify project stakeholders with diverse interests, including, for example, youth, teachers, business people and investors, elderly people, NGOs, taxi drivers, vendors, school-age boys and girls, workshops owners, women's savings group members, community leaders and professionals. These actions increased our understanding of the socio-economic, physical and environmental conditions of the city, and the city's contextual settings. In line with Crosby and Bryson (2005), a relatively complete matrix of stakeholders' analysis was developed progressively, which outlined the stakeholders' interests and their potential roles in the project. The analytical matrix indicated the level of participation of stakeholders in relation to the needs of the planning process. The matrix also detailed the strategies required to gain their support and to reduce obstacles and include citizens' concerns.

Despite the GOPP's political commitments to the Zifta project, the general rules that define the political system in Egypt have played an essential role in the distribution of power across different government organisational levels and have contributed to struggles among different social classes. This was observed in the limited authority given to the City Council in Zifta and their inefficiency in delivering services to the citizens. This social struggle was also evident in the asymmetric relationships that characterise Zifta's social structure, arising from the tensions between the formal and informal rules and the unequal distribution of resources. The change towards an appropriate collaborative planning process necessitates political commitments that support its institutionalisation within the planning system in Egypt. Despite the will for changes in planning thinking expressed by planners and politicians at the GOPP, there is no policy in place that gives support to such changes.

Urban Situational Analysis

In addition to the quantitative survey carried out by specialised experts assigned by GOPP, several workshops were organised involving various stakeholders, to discuss urban issues covering the role of the major economic activities, socio-economic relationships in the city and their relation to the built environment and contemporary urban conditions. The aim was to obtain qualitative views about the factors influencing contemporary urban conditions in Zifta, factors that emerged from the city's historical development and geography, and that determined its relationships with its surroundings. Historically, Zifta was once an important centre for agriculture and agricultural industries in the El-Gharbieh Governorate. The development policy applied over the last five decades has been inadequate to cope with the challenges of rapid urbanisation. This has led to extensive encroachment on scarce and valuable agricultural land and increases in other types of economic activity that serve the building and construction sector. Many of these activities are undertaken within an informal economic system contributing greatly to the city's income at large. Relatedly, the social network provides an alternative welfare system consisting of faith-based organisations, individual donors, clubs, assemblies and associations and their human resources. Zifta has thousands of unemployed young graduates who are willing to play a role in the development of their city.

In addition, a range of opportunities needs to be considered when undertaking planning. These include: the availability of vacant urban spaces and agricultural areas within the city; the non-developed islands in the Nile with strong potential for development; the socio-economic integration and mutual benefits between Zifta and the surrounding villages; the interest of business people from Zifta and the surrounding villages in the city's development; and local awareness of the capacity of existing socio-economic and human capital, despite its fragmentation. A number of weaknesses and threats also needs to be addressed in the Zifta Strategic

Plan. These include: a poorly functioning basic infrastructure, the wastewater system (particularly in slum areas), the traffic system; rising poverty; a lack of skills among employees of the City Council; a lack of coordination among the various sectoral departments, which has decreased the efficiency of service provision; fragmented local capital sources and tension in the relations between formal and informal systems in the city; and a national economic (liberal-market) policy that does not protect small-scale farmers from the competitive power of large companies. The flow of financial capital out of Zifta to the bigger cities for the purchase of necessary goods was also described as a problem.

Knowledge Developed out of Interactive Social Processes

Knowledge is constructed through processes of social and institutional interaction; it is thus not neutral but is embedded in sets of social relations (Healey 2006). During the progress of practising Collaborative Planning in Zifta, distinctive kinds of knowledge were produced (see Figure 24.2), including: 1) socially oriented knowledge, constructed as a result of social agreement whereby the people share local values and history, 2) relevant knowledge obtained from the people's lived experiences; and 3) process and action-oriented knowledge produced in a milieu wherein experts' knowledge relevant to an issue has been combined with local knowledge. The planning process and the various types of actions and methods entailed the development of joint knowledge and concluded by way of transfers, as follows:

1. Knowledge transferred to Zifta via the application of the Collaborative Planning by the action researcher.
2. The Collaborative Planning theoretical foundations transferred for practical application and used as references from which to analyse and interpret certain practices.
3. Knowledge transferred from the stakeholders to the researcher in her attempts to learn about the context of Zifta with regard to its institutional structure and associated values.
4. Knowledge transferred between local people and experts, which contributed to the formulation of strategies for the development of Zifta.
5. Knowledge transferred between the local residents, the city council (informal/formal) to build institutional planning capacity.

Out of the socially interactive processes that the project promoted, participants learned about their assets and problems, and became actively engaged in generating local knowledge that reflected the reality of their city. Their knowledge provided the technical experts with qualitative and, sometimes, quantitative information about the city. However, the local knowledge was generated necessarily outside of the prescriptive frameworks such as are commonly provided by planners, which required that it be sorted to improve its legibility. Citizens' personal roles as played out in their city were integrated to produce knowledge and thus empowered them to take actions.

Knowledge Transferred Between Citizens Who Integrated Their Personal Roles

The participants brought their personal roles to the table and integrated them through discussions of Zifta's future development. The exchange of information between representatives from the formal and informal sectors played a role in validating the data obtained from both sides. The unemployed youth participated actively in the planning process, during which they established and formally registered an NGO for youth initiatives. They supported the planning team in organising the workshops, were responsible for documenting the results on a regular basis and disseminating them widely in the city, and also led several initiatives in collaboration with other groups' initiatives. The spirit of interaction in the planning process facilitated the consensus building process. The consensus issue was not about conflicts that required negotiated, consensual solutions; but rather, accepting formal/informal interactions and making use of the learning process in such a way as to formulate a strategic plan for the city's development.

This combination of expert and local knowledge suggested an emphasis on the integration of the socio-economic relations of Zifta with the surrounding villages in order to benefit from their interrelationships. The institutional structure, mutual relations between the individual and the society, and the role of family

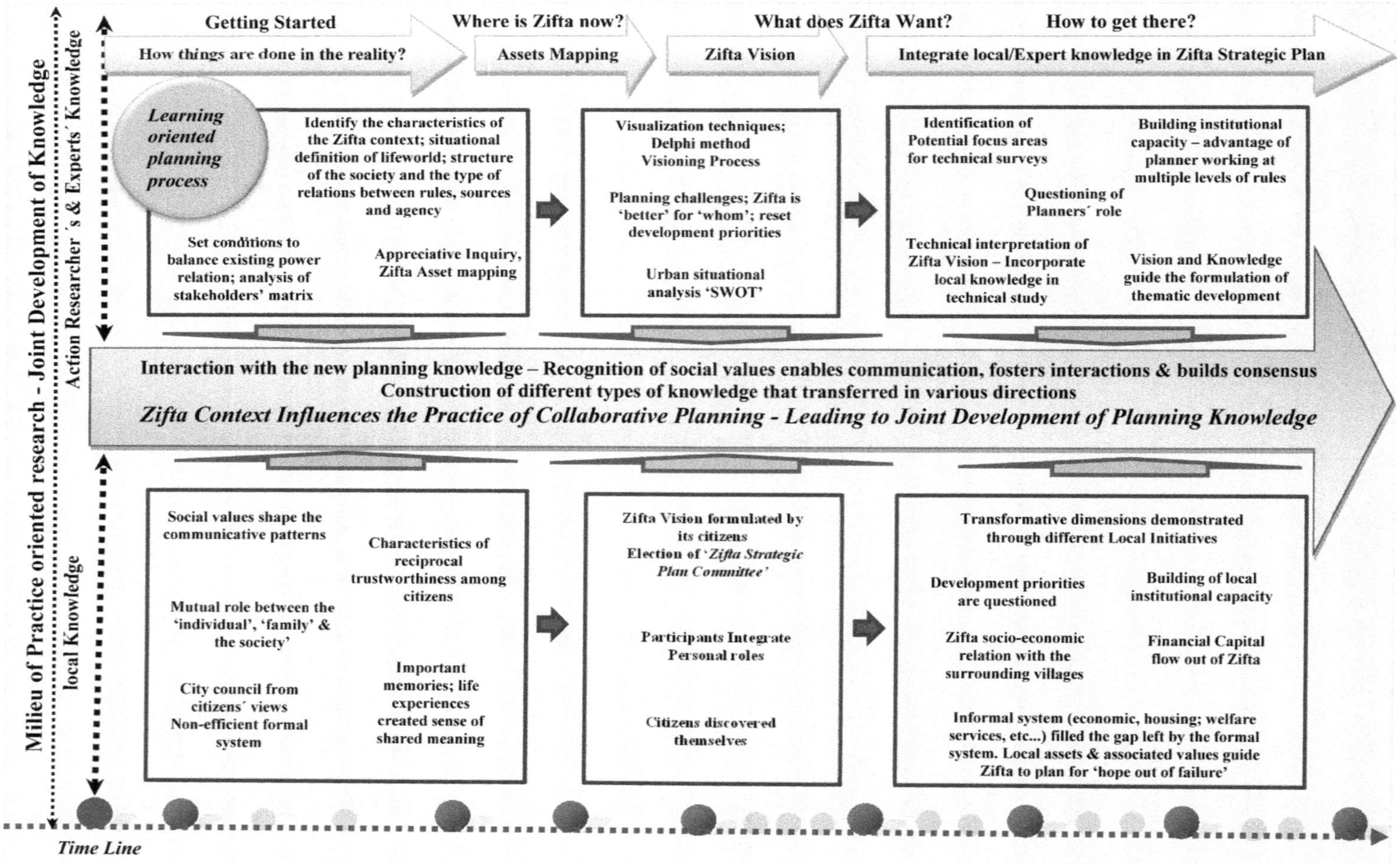

Figure 24.2 **Social interactions created an opportunity for a joint development of knowledge**
Source: Noureddine Tag-Eldeen 2012:147

as a core of society are all to be considered parts of a structural network that could be built upon in order to achieve small and medium-scale industrial processing and related economic activities. In this way, the hierarchal spatial relation of Zifta and its surrounding villages, and the socio-economic territorial logic of their interaction, might form a platform for building industrial-chains – of various industrial scales of operation but with complementary lines of industrial processing activity. This economic development concept will add value to the different crops produced locally: dates, flax, cotton, fruits and vegetables, for example, and will further contribute, through aggregating and managing the different lines of processing, to large scale industries. The industrial-chains concept – through the promotion of crop production, processing and services – can make use of and strengthen existing socio-economic mutual relationships between:

- The 'family', as the core of economic activity based on agricultural production and crop processing.
- The 'village', based on a number of family activities contributing to a larger scale of production. Small-size investments and services should be provided to improve production, such as vocational training and local markets.
- The 'city', comprising larger-sized economic units where different production and processing chains of families and villages are consolidated. At this level, different services are required, such as research centres, small, medium and large size investments, and regulations to facilitate marketing.

In such mechanisms, local initiatives along with local cultural values play a self-motivating role and become the dynamic behind local economic development. This can be achieved by building economic projects and various production structures on the basis of existing social networks.

Knowledge Interaction Created Actions That Built Up Citizens' Power

One example of a new interaction between citizens and the planning institution that emerged out of the Zifta project was a collective initiative to refurbish a locality (see Figure 24.3). This was undertaken by a number of stakeholders, involving local inhabitants in collaboration with the City Council, to demonstrate the constructive result of working together. This was a step towards further strengthening the relationship between the City Council and citizens. Local residents contributed with cash for buying building materials and equipment, hiring labour, or by contributing their own labour. The City Council provided the specialist equipment for paving the street. The interaction resulting from the Collaborative Planning process helped the transformation of stakeholder ideas into actions carried out collaboratively with the Council. The dialogue that Collaborative Planning processes facilitated between Zifta's inhabitants and institutions created a new power context, with long-term transformative potential and restructured institutional settings. This is summarised in Table 24.1.

Eventually, the knowledge gained about the city's capacity – knowledge drawn from the information given by the participants about the different potential skills in the city – led to the establishment of a network of local professionals and skilled labour from the city. A Zifta Strategic Plan Committee, elected during the visioning workshop, organised and maintained contact with the network. Thematic groups of stakeholders were organised to coordinate various types of expertise and interests. Each group was led by a planner from the planning team. The continuous incorporation of local and technical knowledge made it possible to transform the development ideas into a set of strategic goals and objectives that pointed to the central principles of how to achieve long term goals for Zifta, *Zifta Vision 2022*, which was approved by the Governorate's (El-Gharbieh) Governor in September 2007[2] (MHUUC 2007).

2 The Zifta Strategic Plan was documented by the planning team, including myself working as a practitioner, during the progress of the Zifta project. It was approved in September 2007. It consists of two volumes, Volume I is for the use of the City Council for the daily urban management of the city, while Volume II is more technical and consists of the detailed thematic technical studies. Volume I includes a summary of the thematic development, a list of the goals and objectives, and a summary of the action plan for the first five years. It includes physical maps of the land use and urban boundaries. It illustrates the implementation mechanism including participation, evaluation and updating of the plan. It includes the names and the role of the Zifta Strategic Committee members that were elected during the planning process. Volume II comprises a detailed technical study of the thematic development, goals and objectives of each development theme, in addition to physical maps and a detailed Action Plan for the first five years. It includes the implementation, participatory, evaluation and updating mechanisms of the plan (MHUUC 2007).

Citizens, planning team, city & popular councils celebrate the event in a public space in Zifta on 24 November 2006

Figure 24.3 Local initiative to refurbish the 'Zawyet el-Masry' quarter
Source: Noureddine Tag-Eldeen 2012:111

Table 24.1 Local initiatives taken during the different planning stages

Planning stage	Initiatives	Stakeholders who take initiatives	Transformative dimensions
Getting started – visioning process	Wider distribution of questionnaires on Zifta Vision	Youth group	Building consensus on Zifta Vision
Throughout the planning process	Organising workshops in weekends and evenings	Youth, teachers & investors groups	Widening the stakeholders network
Thematic development goals (environment)	Greening of Zifta 'Sadakah Jarieh' a charity campaign	Doctor's initiative based on shared values	Combination of moral and empirical arguments
Thematic development goals (local economy)	Gardening roofs, balconies and back yards	Families, youth, City Council, business community and elderly people	Strengthening the existing socio-economic role of the family
Visioning process and during the formulation of the thematic development goals (local economy)	Old stories told by elderly were relevant to vocational training for traditional handicrafts	Elderly group in collaboration with youth group	Creation of shared meaning to build consensus
Thematic development (local economy)	Technical study of water isolation before planting	Graduated engineers from youth group	Contribution to experts' knowledge
Thematic development (local economy)	Investigation of the number of vacant plots	Youth group, City council and investors	Provision of reliable data from local people
Thematic development (local economy)	Study the feasibility of industrial processing activity	Businessmen's group, and technical expert	Combination of local and expert knowledge
Contribute to the implementation of the plan	Participatory budgeting	Representatives of shop owners and the City Council	Support participation in decision making and transparency
Thematic development (improving the physical conditions of the city)	Refurbishing of a district	Citizens, City Council, businessmen community and youth group	Restructuring institutional settings – Formal/informal interaction
Thematic development (educational development)	Free Education program to support students in specific subjects	Teachers (female) in collaboration with families, youth and schools	Emphasise positive role of women in Egyptian society
Thematic development (social development)	Establishing of Youth NGO for youth initiatives	Youth in collaboration with the formal authority	Formal/informal interaction built trust in formal actions
Visioning, continued during the planning process and became legalised by the Governor to continue its role during implementation	Establishment of 'Zifta Strategic Committee' to represent the interests of Zifta's citizens during planning and implementation	Representative committee elected by the stakeholder groups	Building institutional capacity '*formalise the informal*'

Source: Noureddine Tag-Eldeen 2012:113.

Collaborative Planning Built Institutional Capacity

In the Zifta Demonstration Project, planning has addressed changes in formal/informal relationships, and taken advantage of working at multiple levels of rules at both the city and the GOPP levels (see Ostrom 2005:58–64 about the cost of the institutional changes and in relation to the different levels of rules). In Zifta two forms of institutions function in parallel: the formal and the informal. According to Healey (2006), a planner should become skilled in advising on appropriate process forms that 'fit' particular governance situations, in order to make changes in the way things happen. For Healey, the collaborative planning process is a unique construction in a specific situation, in which innovations of process design build capacities for change. The planning process in Zifta was controlled in such a way so as to provide arenas for formal and informal interactions. The participants from the City Council and the city reviewed their rules and how they might be applied, and mobilised their resources for each action situation. The outcomes of each operational situation were accumulated to form the local knowledge that was combined with technical information from the sectoral experts (see Figure 24.4).

The institutional capacity of planning was built through the formal/informal interaction arenas wherein new forms of rules and resources were produced. The interaction process created planning capacity at the Zifta level, which allowed more collaboration between formal and informal institutions. For Giddens, changes are created by what he calls 'human agency' (1984). In Zifta, at various levels of planning, the process enabled four categories of human agency to contribute to change: the citizens of Zifta; the society or community of Zifta, which is formed by larger numbers of participants who worked collaboratively for making changes; the GOPP, representing the decision makers at the central level who supported the process; and myself as an outsider, action researcher acting through the existing social system to facilitate that change. All interactions involved rules and resources of formal and informal types and enabled the different categories of human agency to contribute to the change as follows:

- Citizens participating in the planning process created expectations for change. In the course of their participation in the interaction arenas at a city level, stories of their personal roles and experiences were formulated and integrated into the knowledge produced.
- The community gained advantages when the expectations of citizens to create change were accumulated via the larger network of participants. Some of the expectations for change were achieved through realistic examples when local initiatives were created and implemented, involving formal and informal rules and resources.
- Myself as action-researcher, an external agent, led the planning process and built (together with the other categories of human agencies) a platform for change. My dual role equipped me with skills to mobilise local capacity towards changes. As action-researcher I enabled changes in rules and resources by supporting individuals through providing them with legitimate means to participate, act and contribute to change.
- The GOPP supported the Zifta Demonstration Project and the advantages of the dual role researcher/practitioner to support institutional changes at low cost.

The Structure of Zifta Society

Reflecting on Giddens' (1984) theory of structuration, society's structure constitutes a continuing pattern of actions structured along certain lines established by rules and resources. Zifta society is structured along types of rules that are formally stated by the City Council (e.g., through laws and regulations); and resources are allocated by this formal system (e.g., through services, production means, etc.). These rules are rarely implemented due to the lack of an implementation mechanism and the limitations on resources. Other informal rules, which stem from the norms and values of the society, are implemented and regulate the use of available resources. The combined formal and informal rules and resources create a social structure of asymmetrical social relationships with three interrelated dimensions.

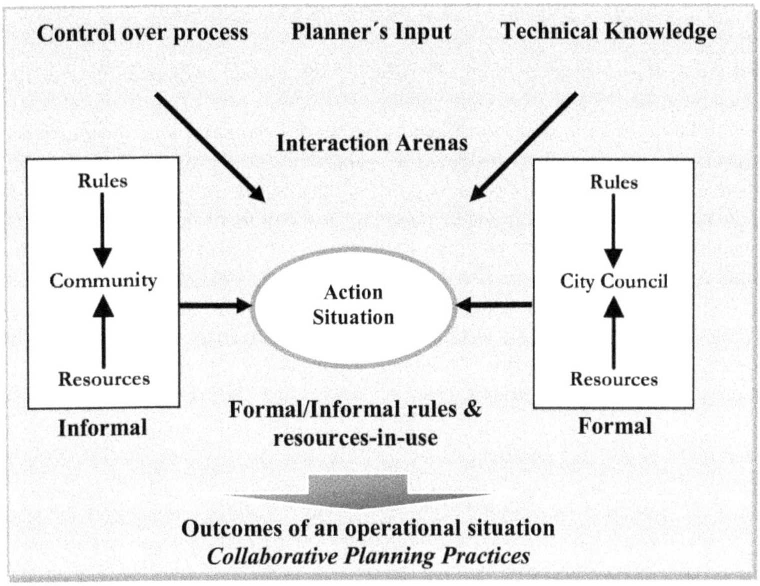

Figure 24.4 Formal/informal institutions interaction
Source: Noureddine Tag-Eldeen 2012:148)

1. The gap left by the government in terms of the provision of social welfare services.
2. The existence of a shadow and parallel welfare system, within the informal structure of the society, rooted in the profound ideological value of faith-based[3] and non-governmental organisations in Zifta, as well as other contributions from society's collective spirit of solidarity and mutual responsibility.
3. The inadequate official policies applied by the formal system has led to an increased poverty rate and a widened gap between rich and poor, thus strengthening the relationship between the people and the informal system.

The existing relational network in Zifta establishes a context in which human agency (on the behalf of citizens) is not strongly constrained by formal rules and is never without resources to make changes. The asymmetrical positioning of individuals, governed by the formal and informal rules, creates different interaction routines, which lead to a reproduction of social practices that strengthen inequalities but sustain mutual relations inspired by the society's collective spirit which does not allow the poor to lack all access to resources.

Cultural Values Associated with the Society Structure

The values associated with Zifta's society – namely those that carry collaborative characteristics – have strength and blend socio-cultural and religious influences. One can observe a common thread of values running through the social fabric of the city, which are reflected in the society's interrelationships, language and practices.

3 Faith-based organizations, namely those belonging to the Muslim Brotherhood Movement, have benefited from the gap in welfare provision of services left by the government. Their community-based activities later underpinned their political agenda. This was clearly understood by the Egyptians after the Revolution of 25 January, 2011, when they were elected to rule the country benefiting from their popularity at the grassroots level.

The Reciprocal and Mutual Role of the Individual, the Family and the Society

During the workshops, the roles of participants reflected the mutual responsibilities of the individual in relation to his/her own family and to his/her society. The strong role of agricultural activities in Zifta's economy allows all family members to contribute to the family economy. The independence of a growing child is not valued because it may be considered as a threat to the family livelihood. This underlines the interdependence and cohesion of social networks, in which the family (rather than the individual) is the core of society. The collective character of the society strengthens a culture of family relatedness (Zaharna 2009). The collectivist society is manifest through close, long-term commitment to a member's 'group', that being a family, extended family or extended relationships. Such norms are not always appreciated, since they may create pressures on the family members, especially when resources are limited. However, it was observed that the collective actions taken by groups of producers, vendors and business people to create and manage the Friday Market, held regularly in the centre of the city, emphasise the role of the existing social network and collective characteristics in the urban economy.

Communication Patterns Fostering Dialogue

Communication in Zifta includes social terms that emphasise closeness, and support the interests of a pro-family society. The 'we' consciousness is clear in expressing matters of common interest and problems. The mutual roles and responsibilities of the individual, family and society form the social relationships and shape the communication patterns in Zifta. Habermas' (1987) communicative theory is conceptually linked to the textually based speech of a society. Habermas suggests that speakers and listeners will refer to certain criteria, or a 'lifeworld', in order to judge and complement communicative actions. The lifeworld can be internally constructed in the process of communication, through which participants correlate their mutual understanding and interpretations, their standpoints and their critical views on the validation of their communication. For society in Zifta, the validation of communicative action is judged by its accordance with the values of that society, values that are reflected in daily practices and manifest in different forms of actions. Reference to these values was frequently made by participants in the Zifta Demonstration Project, when they discussed common concerns – e.g., the reasons behind their contributions to, or dependence on, a charitable organisation, the collective spirit based on solidarity value and influenced by society's shared history; or the culture of relatedness and the mutual responsibility among Zifta's inhabitants. These together can be considered as local references to measure truth, rightness, and sincerity or, more precisely, references to validate communicative actions and thus to potentially build consensus on common concerns.

Incorporating Values Helped to Build Consensus

The praxis of cultural values was materialised through three planning actions intended to build consensus on a realisable vision of the future for Zifta, termed the 'City Vision'. These actions were: memory of successful events in history; assets kept to the forefront of participants' minds; and imagination for hope, which helped the building of self-confidence in the city's capacity and created the inspiration for positive change. The practical actions of this stage included Asset Mapping, which aimed to strengthen the capacity of the people to discover their city. This covered social, economic, physical, environmental, and natural resources. Participants took part in several workshops and were encouraged to tell positive success stories thereby discovering the assets of the city in all its development sectors, so as to provide a platform from which to move towards formulating the City Vision. This process was undertaken through the Visioning Workshop, which took place over a two-day period (see Figure 24.5). The participants contributed during the process with relevant narratives – stories were told, historical events were highlighted and participants exchanged life experiences – which created shared meanings that helped in building consensus. This influenced their way of thinking and how they moved towards building a consensus on their desired vision. The Zifta Slogan as formulated by the stakeholders – *Zifta is a place for Freedom, Dignity and Leadership* – demonstrated the citizens' wishes, defined with reference to the vocabularies of the lifeworlds and in relation to the common values that had proved to be successful in their shared memory. In their view, the effective management of the

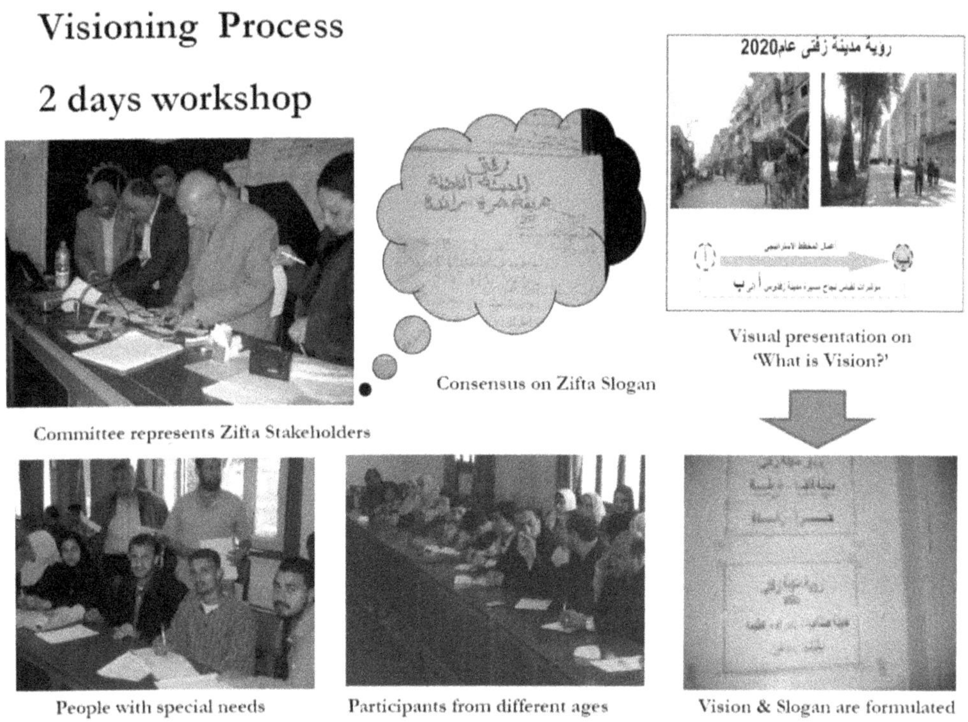

Figure 24.5 **Visioning Process**
Source: Noureddine Tag-Eldeen 2012:104

available resources could contribute to the city's development. This was further explained in the 'Zifta Vision' formulated by the participants as: '*Zifta is a Green Industrial City, and Great Through its Creative Citizens*'. The vision was technically interpreted to form the base of the thematic development in the planning process. The statement 'green industrial' indicates the desires of the stakeholders to combine the economic activities of crop production and the processing of agricultural products. This statement motivated people's ambitions for a green city by planning for urban agricultural activities that sustain the socio-economic relationships between Zifta and its surrounding villages. They desired to base the development drawn into the Zifta Strategic Plan on the capacity of the city's 'creative citizens'. This is to be done by mobilising the values of the society to empower the people and improve their performance, emphasising the role of citizens in decision-making, planning and implementation of the plan. The formulation of the Zifta Vision played an interesting role in further motivating the stakeholder contributions.

Conclusion

The action research project discussed in this chapter adopts an approach that makes use of the existing institutional setting in Zifta by recognising the specificities of that setting and incorporating its values when practising Collaborative Planning. Thus, the practising of that new planning knowledge – realised by way of setting it into a dialogue with the Egyptian people – demonstrates that advantage can be taken of its flexible characteristics. The Zifta Demonstration Project created arenas for people to uncover their epistemic standpoints and provide meanings to define what planning meant in relation to the development of their city. The ability of Collaborative Planning to be contextualised in a new setting refers to its emphasis on process, content and context; it is 'a contested practice in evolution' (Healey 2006:315) because it is 'not a simple recipe'(Healey 2006:320). This ability has contributed to the evolution of a new way of practising Collaborative Planning in Zifta, where 1) the process was designed as learning-oriented and communicating the new planning ideas to participants; 2) the content was directed towards drawing a plan based on the existing citizens' resources, the socio-economic assets of the city and linking the formal and informal activities to bind these assets; and 3) the context was understood by untangling the city's institutional structure and its associated values.

During the practice of Collaborative Planning, cultural values were integrated as an invisible strategic asset to mobilise people's participation. Important events were recalled from the common history of the people and their significance used to build shared meanings of common contemporary problems. Further, values, such as the mutual responsibility of individuals, reduced the tension between an individual's interest and the collective's interests. The interactive process created by the project led to the construction of a map of knowledge by gathering knowledge that was transferred in various directions. The interaction between different types of knowledge concluded with the production and approval of the Zifta Strategic Plan. The arenas constructed for social interaction and the integration of social values in the planning processes empowered people and motivated them to examine different ways to achieve change. This was demonstrated in the initiatives taken collectively during the planning process.

Additionally, the ability to adopt a dual role of both action-researcher and practitioner contributed to achieving the overall aim of the research. Given that practice focuses on solving specific problems and that research is more concerned with questioning planning practices, this project would have not been possible to achieve if I had worked solely as either a researcher or practitioner. The dual position provided me with a multi-level awareness, which was reflected in the practical and the theoretical strands of the research that I carried out.

Afterword

As I entered the final stage of writing my thesis, Egypt experienced the most significant turning point in its recent history. On 25 January 2011, a popular uprising driven by the youth movement demanded political freedom. Due to the limitations of time and the research timeframe, these events are not addressed in my work. However, my observations in Zifta were sufficient to indicate that change was unavoidable. Whilst the formal system was failing to provide notable development to society, it was also destroying the qualitative dimensions of life in Zifta. What was remarkable was the slogan developed by participants during the visioning process for the Zifta Strategic Plan: 'Freedom, Dignity, and Leadership'. As their slogan reveals, Zifta's citizens believe in their capacities. They aspire to freedom and human dignity, through which they wish to mobilise their institutional capacities to achieve a leadership role. As the youth of Zifta were the engine of the local initiatives of the Zifta Strategic Plan, so were the youth of Egypt the engine of the revolution.

References

Alvesson, M. and Kärreman, D. (2007) 'Constructing mystery, empirical matters in theory development' *Academy of Management Review*. 32(4):1265–81.

Connell, D. (2011) 'Schools of planning thoughts: Exploring differences through similarities' *International Planning Studies*. 15(4): 269–80.

Crosby, B.C. and Bryson, J.M. (2005) *Leadership for the Common Good, Tackling Public Problems in a Shared-Power World.* 2nd Edition. San Fransisco: Jossey-Bass.

Dyer, W.JR. and Wilkins, A.L. (1991) 'Better Stories, not better constructs, to generate better theory: A rejoinder to Eisenhardt'. *Academy of Management Review*. 16(3): 613–19. http://voynnetf.free.fr/mrq/02dyerwilkinsstories.pdf [accessed 10 September 2010].

Flamholz, E. and Randle, Y. (2011) *Corporate Culture: The ultimate strategic asset*. Stanford, CA: Stanford University Press.

Giddens, A. (1984) *The Constitution Of The Society*. Cambridge: Polity Press.

Habermas, J. (1987) *The Theory of Communicative Action*. Cambridge: Polity Press.
Healey, P. and Utpon, R. (eds) (2010) *Crossing Borders: International exchange and planning practices*. New York: Routledge.
Healey, P. (2006) *Collaborative Planning: Shaping places in fragmented societies*. New York: Palgrave Macmillan.
Kotter, J. and Heskett, J. (1992) *Corporate Culture and Performance*. New York: The Free Press.
McDermott, R. and O'Dell, C. (2001) 'Overcoming cultural barriers to sharing knowledge'. *Journal of Knowledge Management*. 5(1): 76–85.
MHUUC (2007) *Zifta Strategic Plan 2007–2022*. Cairo: The Ministry of Housing, Utilities and Urban Communities.
Noureddine Tag-Eldeen, Z. (2012) 'Cross-cultural knowledge development, the case of collaborative urban planning in Egypt'. PhD Thesis, The Royal Institute of Technologies, Stockholm.
Ostrom, E. (2005) *Understanding Institutional Diversity*. Princeton: Princeton University Press.
Sandercock, L. (2003) *Cosmopolis II, Mongrel Cities of the 21st Century*. London: Bloomsbury Academic.
Schön, D. (1983) *The Reflective Practitioner. How professionals think in action*. London: Ashgate.
Zaharna, S.R. (2009) 'an associative approach to intercultural communication competence in the Arab world'. In: Deardorff, A. (ed.) *The Sage Handbook of Intercultural Communication Competence*. Thousand Oaks, CA: Sage Publications. pp. 179–95.

Chapter 25

Transforming Land into Property: Varieties of Institutionalism and the Production of Urban Space

André Sorensen

In this chapter, I examine the significance of the sociological institutionalist (SI) ontology and epistemology underlying Patsy Healey's ground-breaking work since the 1990s, and highlight the distinctive and valuable insights this approach yields through a comparison with a historical institutionalist (HI) perspective. The premise is that, as Healey has shown so clearly, ideas are powerful agents of change, and planning theory is not merely a meta-conversation separate from everyday life, but is a part of planning culture that shapes the way we understand both routine and complex processes, and has powerful impacts on the kinds of questions we ask, and how we produce answers that help us to think and to act.

In section one I outline the central role of sociological institutionalist ideas in the major turning point in Healey's thought of the early 1990s which led to her 'Collaborative Planning' breakthrough, in section two I sketch out the similarities and differences between HI and SI, and in section three I outline the contours of an historical institutionalist conceptual framework, and apply this to land development planning processes, with a focus on the different set of questions and cases that this suggests compared to SI. A concluding section highlights the differences between these two approaches, and the distinctive contributions of each.

The suggestion is not that a historical institutionalist approach is better than a sociological institutionalist approach, nor that one should supplant the other, but that the two approaches lead to different questions and different insights, and may complement each other.

Planning, Agency, and the Institutional Turn

The sociological institutionalist approach developed by Patsy Healey and applied to place governance processes since the 1990s is one of her major contributions. Her particular conception of SI is original, and worth better understanding. The groundbreaking innovation comes in the early 1990s, with her distinctive fusion of Giddens and Habermas. Before this her framing was clearly institutionalist, but without the emphasis on the transformative potential of collaborative planning processes that is so compelling in her later work.

From her earliest work, Patsy Healey framed her research in terms of the possible roles of planning in shaping urban change, the roles of institutions in shaping planning action, and the possibility for governance processes to contribute positively to transformative change. For example, in 'Planning and Change' (1974), a major product of her PhD research at LSE, Healey is concerned with cases of institutional innovation represented by attempts to introduce more formal planning approaches in situations of rapid urbanization and weakly institutionalized planning systems in Venezuela and Columbia, and associated attempts to create a more formal planning system that might improve on the outcomes seen during the rapid urbanization period of the 1960s. But her fundamental questions are about the potential for planning to contribute to social and institutional change. She asks, 'What then is the relationship of planning with structural change and structural cohesion? Will planning be used to maintain, to change, or to create structures?' (Healey 1974: 150).

In this early work Healey defines institutions very broadly, equating institutions with 'social structures'. As she puts it 'social structures – and here institutions, systems, forms, patterns may often be used synonymously with structure – consist of a set of repeated events of interactive behaviour, the forces leading to this repetition defining or giving cohesion to the structure' (Healey 1974: 148). She concisely outlines a theory of institutional continuity and change, with institutions gaining significance as the cohesive social forces that generate continuity through repeated instances of interaction. Structural change is a product of the weakening of or loss of power of the former forces of cohesion. Healey then notes that such breakdown can result from both exogenous and endogenous factors. She suggests that while exogenous factors can obviously cause change, internal contradictions in any system can also lead to change. As she puts it, 'although structural change can be generated by pressures external to the structure, such change can also be generated from within' (Healey 1974: 149).

The major question for Healey, here as elsewhere, concerns the role of planning in shaping institutional change. She suggests that in stable situations planning can either be introduced to maintain the existing order, or it can be used subversively, to undermine the existing order. On the other hand, in unstable or changing situations, planning may be introduced to 'impose a particular structure or a particular direction' (Healey 1974: 221). Her conclusion is that for planning to be reasonably effective, some degree of social cohesion in society, and a degree of conformity of values between the society and planning concepts applied are both necessary (Healey 1974: 229). But she also suggests that planning can serve useful functions in shaping ideas, and propagating values and ideals different than those currently dominant.

Healey's development of her original sociological institutionalist conception of planning emerges out of her work on the actors and institutions of property development of the 1980s (Healey 1983; Healey, McNamara, Elson and Doak 1988; Healey and Barrett 1990; Healey and Nabarro 1990; Healey 1992c). This work focused on the roles and effectiveness of planning as a means of managing land and property development processes, with a focus on plan implementation in the context of existing economic and political drivers, the roles of different actors in land development processes, and varied local governance capacities.

In a seminal paper, Healey and Barrett (1990) develop a fully integrated approach to planning research drawing on ideas from sociological institutionalism, and Giddens' approach to structure and agency. After reviewing the existing work in the field, Healey and Barrett concluded that a major shortcoming was that most analysis had 'treated the conversion of economic and social processes into land use change and built form as essentially unproblematic' (Healey and Barrett 1990: 94). Instead, they argue that a focus on actors, their motivations and resources, the rules they follow, and the ideas and values that motivate them produces a quite different understanding of land development processes. They propose a sweeping research agenda to investigate the role of planning and property development in shaping the built environment, with an emphasis on '(a) the *resources* for development, as channelled via the financial system and the interrelation of supply and demand; (b) the politico-juridical *rules* which limit the construction of development opportunities; and c) the *ideas and values* people hold about what they should build, what they would like to occupy and what kind of environment they seek' (Healey and Barrett 1990: 94).

In this and other papers in the early 1990s, Healey sought a way out of the dilemmas of planning during the last third of the twentieth century, after the failures of comprehensive planning, new towns, and mass social housing estates to deliver better urban environments, and with the rise of neoliberal ideas critiquing the failure of the post-war Keynesian welfare state, and political economy critiques suggesting that planning served primarily the interests of capital accumulation, and the then emergent postmodern critique that increasingly diverse values meant that all positions were merely opinions and no preferred future vision could be selected. She suggested that while neoclassical economics reduced agency to a conception of individuals as consumers, as little more than mobile sets of preferences acting as individuals in both markets and political processes, political economy approaches too often presented structural forces as all-encompassing, leaving little room for either incremental change or individual agency.

The key to unlock this impass was a new approach to institutions, structure, and agency based on the ideas of Giddens (1984). This sociological institutionalist approach was valuable in challenging the deterministic relationship between structure and agency assumed by then-current scholarship, providing a more positive interpretation of the possibility of agency in the face of powerful institutional structures (Healey and Barrett 1990: 97). This is a fundamental turning point in Healey's work, in which she first sets out a fully developed institutional theory of planning action, frames institutions in terms of structure and agency in a Giddensian manner, and makes the key shifts towards the ground-breaking ideas presented in 'Collaborative Planning' (Healey 1997a).

The series of papers on the institutions and actors engaged in property development in the late 1980s and early 1990s included a creative fusion of the work of Giddens on structure and agency, and Forester (1989) and Habermas (1984) on the role of communication in planning and democratic dialogue. From this fertile mixture Healey arrives at the ontological and epistemological foundations for much of the rest of her work. From sociological institutionalism comes the insight that power is communicated and maintained through institutions, conceived of as taken-for-granted and received assumptions about the way the world works and who has power, as well as about what are appropriate goals and ideals. Giddens' (1984) work was also crucial because it provides a positive interpretation of the possibility of agency, situating the role of actors as participants in governance processes embedded within larger structuring forces, but with nonetheless, the potential to shape outcomes. Although actors' values and goals are always shaped and constrained by the institutions within which they work, they are also always creating and reproducing the structures within which they are working.

This fundamental and emancipatory insight from her synthesis of Giddens and Habermas is central to Healey's project. Structures communicate power and shape not only what we can do and the context and meaning of our actions, but are also reproduced by our own actions. We actively shape and give form to the structures of everyday life through our own activities, and so can work towards changing those structures. As Healey puts it: 'We make structural forces, as we are shaped by them. So we 'have power' and, … can work to make changes by changing the rules, changing the flow of resources, and, most significantly, by changing the way we think about things' (Healey 1997a: 49). *Ideas and values* therefore become crucial.

Healey's contribution is to show convincingly that governance processes around shared spaces are a particularly important venue for such transformative democratic practices. She has shown us in detail how planning practices might contribute to transformative agendas. As she puts it 'managing our co-existence in shared spaces … becomes an exercise not merely in consensus-building but in local *culture-building*, and in creating a public realm. It is an interactive and discursive effort, through which new understandings and institutional capacities are built' (1997: 48).

Much of the rest of Healey's work over the last two decades elaborates and provides evidence for this core set of insights, from work on building social capital and institutional capacity through collaborative planning approaches (Healey 1998; 1999), to social innovation and creativity in governance (Healey 2004; Gonzalez and Healey 2005; Healey 2006b), and the role of spatial planning and communicative practices in a revitalization of democratic practice (Healey 1992b; 2006a; 2013). This is a major achievement.

A Different Ontology

It is clear that the adoption of a sociological institutionalist ontology and epistemology was fundamental to Healey's major breakthrough in the early 1990s, and to much of her subsequent work. This has been an extremely productive avenue of research, and we are all richer in our understanding of the governance of shared places for it. But several 'new institutionalisms' were developing in the 1980s and 90s, of which the three most prominent were historical institutionalism and rational choice institutionalism, both in political science, and sociological institutionalism emerging from organizational analysis in sociology (Hall and Taylor 1996). The sociological institutionalism of Giddens (1984) and Dimaggio and Powell (1991) and others was most influential for Healey, and shapes her concerns in 'Collaborative Planning' with the potential and practice of actors reshaping their world in the face of the structuration of accepted ideas and ways of doing things, working in contexts saturated with power relations.

A historical institutionalist approach provides a different standpoint, and different insights into questions of structure and agency, continuity and change, and institutional design and innovation. The suggestion here is *not* that this is a better approach, but a different one, which provides different insights into the roles of planning and institutions in place governance processes. Like taking a photo of a familiar scene with infra-red film, the contours are the same, but different elements stand out. Crucially, historical institutionalism prompts us to examine different kinds of institutions, and ask different questions about them. In the next two sections I outline the different insights into the relationship between structuring forces and human agency in the production and reproduction of the built environment that an HI approach generates.

The key elements distinguishing institutionalisms are: 1. the definition of what counts as 'institutions'; 2. the explanation of the role of institutions in shaping preferences and behaviours; and 3. the analysis of processes and drivers of institutional continuity and change. The focus here is on sociological institutionalism

(SI) and historical institutionalism (HI). This section draws particularly on Hall and Taylor (1996), an influential paper which Healey has cited many times, which was the first to systematically define the major similarities and differences between the sociological, historical, and rational-choice 'new institutionalisms', and the work of Kathleen Thelen (Thelen and Steinmo 1992; Thelen 2003; Mahoney and Thelen 2010), and Paul Pierson (Pierson 1997; 2000; 2004).

The first, and fundamental, difference between SI and HI is that SI defines institutions much more broadly. For SI institutions include 'not just formal rules, procedures or norms, but the symbol systems, cognitive scripts, and moral templates that provide the 'frames of meaning' guiding human action' (Hall and Taylor 1996: 947). This broader definition effectively includes culture as an institution, so conceptions of institutional change must include the possibility of changes to broader social understandings and 'frames of meaning'. Healey adopts a similar definition, citing Giddens and DiMaggio and Powell:

> Institutions are the frameworks of norms, rules and practices which structure action in social contexts (Giddens, 1984; DiMaggio and Powell, 1991). They are expressed in formal rules and structures, but also in informal norms and practices, in the rhythms and routines of daily collective life. They structure the interactional processes through which preferences and interests are articulated and decisions made. They are a kind of 'soft infrastructure' of the governance of social life. (Healey 2006a: 64)

This very broad conception of institutions fits well with and is generative of Healey's project.

In contrast, HI tends to define institutions much more narrowly. According to Hall and Taylor, historical institutionalists define institutions as

> the formal or informal procedures, routines, norms and conventions embedded in the organizational structure of the polity or political economy. They can range from the rules of a constitutional order or the standard operating procedures of a bureaucracy to the conventions governing trade union behaviour or bank-firm relations. In general, historical institutionalists associate institutions with organizations and the rules or conventions promulgated by formal organization. (Hall and Taylor 1996: 938)

Similarly and more recently, Streek and Thelen (2005: 9) define institutions as 'collectively enforced expectations with respect to the behaviour of specific categories of actors or to the performance of certain activities'. Both these definitions see institutions as particular sets of rules or shared expectations within particular settings. This clearly excludes broader systems of shared understanding such as symbol systems, cognitive scripts and moral templates.

These different definitions each focus attention on a particular category of institutions for those studying planning and urban governance processes. For Healey, the broader SI definition led to a focus on multiple processes of governance and planning practice that involved ranges of stakeholders in processes of consultation, collaboration, negotiation, and joint working, in which ideas, visions, and practices of engagement and dialogue could be worked out. The narrower HI definition suggests a focus on particular sets of rules, regulatory practices, codes, and enforceable standards and requirements. This points to a different but no less important part of the machinery of urban governance, one that is particularly concerned with the regulation of urban property, the built environment, and infrastructure, as discussed below.

The second set of major differences between SI and HI involve the conception of relationships between institutions and individual behaviour. As Hall and Taylor put it, for SI 'institutions do not simply affect the strategic calculations of individuals, as rational choice institutionalists contend, but also their most basic preferences and very identity' so 'institutions influence behavior not simply by specifying what one should do but also by specifying what one can imagine oneself doing in a specific context' (Hall and Taylor 1996: 948). For SI, institutions shape individual action by shaping belief systems and cognitive templates, norms and expectations. As DiMaggio and Powell put it, institutional frameworks 'create the lenses through which actors see the world, and the very categories of structure, action, and thought' (DiMaggio and Powell 1991: 13), so 'Institutions do not just constrain options: they establish the very criteria by which people discover their preferences' (DiMaggio and Powell 1991: 11). One consequence is that much sociological institutionalist work emphasizes continuity, and the tendency of institutions to stasis and the reproduction of particular institutional

Table 25.1 Key Differences between SI and HI

	Sociological Institutionalism	Historical Institutionalism
Definition of institutions	Not just formal rules and procedures, but also ideas, cognitive scripts and moral templates that shape meaning and action in social settings	Focus on the formal or informal rules and procedures established and enforced in particular institutional and organizational settings
How institutions shape behaviour	Institutions shape actors' most basic preferences and very identity, so they influence behaviour not simply by setting rules but also by shaping belief systems and cognitive templates, norms and expectations	Institutions shape behaviour directly, by setting out rules and standard operating procedures, and also by structuring the actions of those seeking to change or prevent change to those rules
Analysis of factors structuring institutional continuity and change	As institutions include a diverse range of ideas, cognitive scripts and moral templates, processes of change are varied, but are usually seen to be slow and incremental, starting with changes of ideas incrementally influencing changes to cultures and values	Continuity is seen to be primarily a result of positive feedback effects that result in path dependence, such as when political actors who benefit from an institution mobilize to prevent change. Patterns of change are structured by characteristics of the institutions, e.g. whether significant discretion in enforcement allows incremental change without institutional reform; and the decision setting, e.g. whether there are clear veto points where change may be blocked
Which institutions are of interest?	Urban governance cultures, shared ideas and belief systems about place values, identity, and interests	Rules, codes, procedures, and regulations that directly order and shape behaviour
What major questions does each approach suggest?	How can processes of governance innovation be fostered that generate new shared ideas and possibilities for engagement in particular places and planning practices? How can place governance processes become not merely consensus-building exercises but also processes of engagement and public deliberation, leading to the development of shared ideas, strategic visions and local culture-building processes?	What kinds of urban institutions tend to become harder to change over time, and why? How do particular institutional frameworks shape behaviour and outcomes? When were particular institutions established, and how did the timing and sequencing of particular periods of institutional development shape institutional outcomes? Who is interested in particular institutions, who seeks to change them, and who seeks to prevent change?

forms even when these are not particularly efficient, through processes of 'institutional isomorphism' or the reproduction of institutional arrangements that are seen as legitimate (DiMaggio and Powell 1983; Mahoney and Thelen 2010: 5). Institutional arrangements persist in part because individuals cannot conceive of any other arrangement, or discount as unrealistic or unattainable those alternatives that they can imagine.

Here Healey, drawing on Giddens, makes a valuable and original contribution to SI. Her starting point is that indeed, individual identities, rules of behaviour and social relations are structured by relations of power and modes of thought generated in the past. Such 'structures operate in routinized and taken-for-granted ways, as we use them in the flow of daily life' (Healey 1997: 45). But Healey goes on to suggest that all such structures are also human creations, and that Giddens' idea that we are all 'active agents' means that we not are not only shaped by those structures, but we also actively constitute them and remake them through everyday life. As Healey puts it 'We live through culturally-bound structures of rules and resource flows, yet human agency, in our continually inventive ways, remakes them in each instance, and in remaking the systems, the structuring forces, we also change ourselves and our cultures' (Healey 1997: 47). And Healey's major contribution to SI was to go on and substantiate this important claim through empirically grounded research over the next decade and more. This research demonstrated the significance of processes of collaborative place governance, in which multiple actors work together to reshape not only local policy, but also change the local conversation and people's ideas and shared knowledges of place (Healey 1996; 1997b; 1998; 1999; 2006b).

Historical institutionalists understand the influence of institutions on behaviour differently. In their paper Hall and Taylor (1996) sought to position historical institutionalism as a possible bridge between sociological institutionalism and rational choice insitutionalism, suggesting that then-current HI approaches included both rational choice oriented 'calculus approaches' and broader 'cultural approaches' to the impacts of institutions on behaviour (Hall and Taylor 1996: 939). That particular project was criticized by Hay and Wincott (1998) among others, as mixing up incompatible social ontologies, and appears to have had little larger impact.

For my purposes here it is sufficient to note that the major difference between SI and HI conceptions of the influence of institutions on behaviour stems from the fact that the HI definition of institutions is much narrower, with the emphasis on particular sets of enforceable rules and particular organizations and settings. In this case, institutions affect behaviour by defining the rules of the game within particular settings, and also by setting out specific conditions necessary to change those rules. That is, if the institution in question is a particular public policy, legal framework, or organizational rule, then it shapes behaviour both by demanding conformance with the rules, but also by shaping the strategies of those interested in either changing the rules, or in preventing them from being changed (Mahoney and Thelen 2010).

This brings us directly to the different conceptions about how institutions change, a much more complex question. As noted, SI defines institutions very broadly, including norms, understandings and cognitive templates as institutions. This broad definition of institutions means that they are often considered to change relatively slowly. As Roland puts it: 'a prime example of a slow-moving institution is culture, including values, beliefs, and social norms, which tend to change gradually' (Roland 2004). Roland compares these to political or policy arrangements, which are often easier to revise, and are more subject both to exogenous shocks and endogenous incremental changes.

Even when change occurs, for example through the impact of exogenous factors, such as a new interpretive frame imported from outside which helps actors to shift their understanding and analysis, ideas about what constitutes appropriate and legitimate change can limit the range of options considered (DiMaggio and Powell 1983). As Hall and Taylor (1996: 949) put it: 'sociological institutionalists argue that organizations often adopt a new institutional practice, not because it advances the means-ends efficiency of the organization but because it enhances the social legitimacy of the organization or its participants'. For SI therefore, the emphasis tends to be on continuity, and the reproduction of institutions over time and into new settings. The major mechanisms of change are the gradual evolution of culture and ideas and also the need to adapt to new circumstances.

Again, Healey's work is an original and valuable contribution. Her work focuses primarily on transformative change, conceived within a sociological institutionalist ontology. In Healey's cases of transformative change, two major modes of change are seen as important. First, through collaborative governance activity to work out particular practices and policy around the governance of shared spaces. And second the suggestion that through engagement in such practices, new institutional capacity may be developed, new forms of social capital may be produced, new cognitive templates generated, and new cultures of place governance evolved.

It is on issues of continuity and change that HI is most different from SI. The key differences between HI and SI conceptions of institutional change arise from the definition of institutions, and the focus of HI on particular organizations, sets of rules, and ways of working, rather than on broader norms and conventions. HI asks why some institutions tend to change relatively little, while others change more readily, who is involved in change processes, who works to promote change and who is more likely to resist, and what are their strategies and motivations. HI research tends towards a power-political approach, with institutions seen to have profound distributional consequences, and therefore being contested, with those advantaged by the current institution more likely to seek to prevent change, and others – the losers of previous rounds of institutional change – with incentives to promote change (Hall and Taylor 1996; Lowndes 2009; Mahoney and Thelen 2010).

There are three main contributions of HI to understanding processes of institutional change: ideas of path dependence, the concept of critical junctures and developmental pathways, and more recent examinations of incremental and endogenous institutional change. HI is known primarily for the concept of path dependence, that some institutional arrangements once established have a tendency to become harder to change over time, creating a path dependent institution. But over the last decade or so a primary focus on HI has been on processes of incremental and endogenous institutional change that occur outside of critical junctures (Thelen 1999; 2003; Pierson 2004; Streeck and Thelen 2005; Mahoney and Thelen 2010).

First, the concept of path dependence. The basic idea of path dependence is that some institutions become increasingly difficult or costly to change over time. In such cases, even where multiple possibilities may have been present before a particular institution became established, once this occurs it becomes ever more difficult to go back to the earlier available possibilities. Pierson urges us to define path dependence relatively narrowly, as the concept has been popularized as meaning simply that 'history matters'. Pierson (2004: 21) defines path-dependent processes more narrowly as 'social processes that exhibit positive feedback and thus generate branching patterns of historical development'. Where positive feedback exists, each step down a particular pathway increases the likelihood of further steps along the same pathway, and increases the cost of reverting to some previously available option. An example of positive feedback is where a policy or institution provides a flow of resources or rewards to certain actors, those actors will then have an incentive to try to protect or enhance their benefits from the system, and thus create positive feedback effects in favour of institutional continuity. A famous example is pay-as-you-go pension systems, which use current contributions to pay for current pensions. Over time, as people continue to pay into the system, they have ever greater incentives to prevent major changes to weaken the system, as without a constant stream of new members and continued premiums, they will receive no pension when their time comes.

The greater the rewards provided by the institution, the more likely that a coalition in support of institutional continuity will emerge. Indeed, the very existence of a path dependent institution may be considered to be evidence that such positive feedback effects are providing a flow of resources sufficient not only to induce mobilization of a coalition in favour of continuity, but also that they have been large enough to ensure that those actors have been successful in defeating efforts to change the institution in the past. For example, a government in power may prefer the voting system that resulted in their election, and work to block voting reform which reduces the likelihood of their re-election. Among spatial planning institutions, greenbelts appear highly path dependent in this sense, as landowners near (but not inside) the greenbelt see significant amenity benefits, increased property values, and protection from further nearby development and congestion, and will often work hard to prevent any change to the greenbelt's protected status. Other factors being equal, the longer a greenbelt policy is in place the more likely that adjacent landowners will support it, if the protected status of the greenbelt land is seen to enhance their own property and/or amenity values.

The idea of path dependence is not that particular policies become frozen, but that where positive feedback exists, institutions follow branching patterns of institutional development in which particular contingent choice points create new branches, or developmental pathways. Positive feedback then makes it more likely that subsequent incremental changes will lead further along the same branch, and it becomes gradually more costly to change back to policy options that were equally possible before the choice point. Once a particular policy system has started down a particular track, the costs of reversal can become very high, so even though a reversal of policy is possible, it becomes less likely over time, and the range of future options will also be different than if another branch had been chosen.

Of course, not all institutions are path dependent, and just because some institution or technology has remained unchanged for an extended period does not necessarily mean that it is path dependent, if it is the only viable option. Those claiming path dependence must be able to show that other alternatives are available, and must be able to explain the positive feedback effects that generate continuity. In terms of political and policy processes this will normally mean identifying the actors who benefit from a particular institution, and who support continuity.

Second, the concept of critical junctures and developmental pathways is important. It is clear that in many cases institutions are the product of transformative moments of major change, usually described as 'critical junctures' (see Collier and Collier 1991; Capoccia and Kelemen 2007). In cases where institutions are path dependent, then such critical junctures of institution formation become very important as the moments when new branches or policy trajectories are determined. Katznelson describes critical junctures as moments when existing political and institutional structures fail to provide adequate solutions to pressing problems, and thus lose governance legitimacy and their ability to determine action and interpretation, creating opportunities for actors of all sorts to play greater roles in developing new institutions (Katznelson 2003: 277). Of course, even during such critical junctures agency does not operate under conditions of its own making, and even at such moments, the universe of conceivable alternatives is much smaller than the full range of hypothetical possibilities or alternatives (Katznelson 2003: 282). Research on path dependent processes seeks to understand these moments of institution-formation, the reasons for particular institutional choices, and which actors were involved in decision-making processes. The concept of 'developmental pathways' describes the periods of relative continuity in between critical junctures, during which path dependent institutions may continue to evolve, but within the broad parameters established during the last critical juncture. 'Process tracing' is an increasingly widely used approach to discovering the origins of institutions and the critical junctures during which they were established (Brady and Collier 2004; Collier 2011).

The third major contribution to understanding processes of institutional change is that proposed in recent work on incremental and endogenous institutional change which suggests that such processes may be even more common than the punctuated equilibrium model suggested by critical junctures, but in which established institutional structures still have a decisive impact on patterns and processes of change. This research focuses on the examination of endogenous change and the ways it can be structured by differential potentials for change among institutions, different patterns of compliance with and enforcement of institutional rules, and the nature and extent of veto opportunities for those seeking to prevent change, which may have profound impacts on the likelihood and nature of change, and which are hypothesized to shape the strategies of actors both for and against such change (see Thelen 2003; Hacker 2004; Streeck and Thelen 2005; Mahoney and Thelen 2010; Sorensen 2011b). For a review of this literature, and an application to planning history, see Sorensen (2015).

In both the earlier 'punctuated equilibrium' analysis of critical junctures and developmental pathways, and the more recent work on differentiated processes of incremental and endogenous change, HI has focused on processes of change, not simply on ideas of stability as suggested by the term 'path dependence'.

A Historical Institutionalist Approach to Property Development

The major suggestion in this chapter is that an HI approach generates a different perspective on institutions and institutional change dynamics than SI approaches. The sociological institutionalist concern with cognitive templates and systems of meaning leads intuitively to Healey's focus on the nature of urban governance cultures, processes of engagement and public deliberation, and the development of shared ideas and strategic visions. In contrast, the narrower definition of institutions of HI focuses attention on the sets of rules, codes, regulations, and financing mechanisms of particular urban organizations and systems, the fine-grain of regulation that supports and structures the complex processes of creating and maintaining shared urban spaces.

Cities are extraordinarily dense collections of such institutions, including land and property development rules, building standards, rules for the financing, use and maintenance of infrastructure networks, governance and policy-making systems, parking standards, tree protection regulations, public education systems (including tax systems to support them), animal bylaws, fire codes, property rights, mortgage systems, property insurance requirements, street and sidewalk standards, water use and disposal regulations, garbage and waste management systems, public health rules, and many other formally codified and collectively enforced sets of rules that have been developed as a part of efforts to collectively create, manage, and add value to urban space. This fine-grain of administrative systems, standard operating procedures, and regulatory frameworks both shapes the production of new urban space and has decisive impacts on the function of the urban space that already exists.

From a historical institutionalist perspective, the specific rules that structure urbanism and urbanization are created in contingent moments of institution building, and then sometimes become locked in for extended periods and become hard to change, even though they may not produce particularly desirable outcomes. For example, since the late nineteenth century in most states of the US it has been much easier to create new municipalities on the urban fringe than it has been for central cities to annex adjacent territory (Mandelker 1958; Teaford 1979). This has been a major factor behind inner-city decline, disparities between cities and suburbs, and seemingly insurmountable obstacles to regional planning. US institutions regulating municipal government incorporation have clearly had decisive impacts on patterns of urban governance and urbanization, contributing to urban sprawl (Fulton, Pendall, Nguyen and Harrison 2001; Carruthers 2003), deepening income disparities between central cities and their suburbs (Orfield 1997; Rusk 2003), and have contributed to making regional governance and planning extremely difficult (Norris 2001; Ward and Jonas 2004). Both the geographies of suburban municipal government, and the permissive rules allowing municipal incorporation in most states appear to be highly path dependent (though not unchanging), with positive feedback effects generated by the perceived benefits of municipal autonomy and defensive localism afforded by municipal fragmentation (Weir 1995; Cox 2002). There are also clearly multiple alternatives to this approach, as higher level governments in the UK, Canada, Australia and elsewhere have retained and continually exercised powers to amalgamate and reorganize local government units and geographies (Meligrana 2004).

Similarly, in the early twentieth century US, in most states progressive era reformers created special purpose local authorities with appointed or elected boards and taxation or borrowing powers to build major water and wastewater infrastructure, removing this function from the hands of elected municipal councils and mayors. In 1987 there were 29,532 Special Districts (mostly for water and sewer provision, but also for fire protection, parks and recreation, conservation, libraries, ports, and many other functions), and 14,721 School Districts, working alongside 19,200 municipalities and 3,041 counties (Norton 1993, 401). Fragmentation of responsibility for urban governance among a variety of bodies was a compromise solution to the conflicts created by large-scale debt-financed spending on infrastructure, democratically elected municipal governments, corrupt 'machine' politics, and the perceived need by urban reformers and elites for efficiency and professionalization of infrastructure management. This proliferation of governance units has had decisive impacts on patterns of urbanization, and is a prime example of a path-dependent institutional structure. A critical juncture in the early twentieth century led to a new approach to infrastructure management, the special purpose body, with its own revenue streams and debt instruments, governance structures, and incentives that has itself since proven to be highly path-dependent. Canada again provides an excellent comparator here, as even though municipal powers, revenue streams and urban forms are similar to the US, the functions of water supply and waste-water management remained core municipal responsibilities, and were used effectively in the post-war period to regulate land development in the suburbs (White 2003; Sorensen 2011a).

Where path dependence exists, the timing and sequencing of the creation of particular administrative, infrastructure-management, and other institutions appears to be of decisive importance, as such institutions tend to embody the political logics, power relations, and 'best-practice' ideas currently dominant in the particular arrangements and compromises that are adopted. It is clear, for example, that municipalities that established water supply systems in the early twentieth century had a very different menu of institutional and technical options than those newly establishing waterworks in the early twenty-first century, after decades of initiatives to promote the privatization of such infrastructure. Similarly, Rodgers argues that a major factor facilitating the successful creation of municipal streetcar systems in the UK and Germany was that local governments that had already established municipal gas distribution companies gained valuable institutional capacity and large profits and were much more likely to be successful in managing much more complex transit systems. He argues that in the US the timing of the municipal enterprise movement was later, gas and electrical systems were virtually all already privately owned, and that this contributed to the fact that in the US most attempts to establish municipal streetcar systems failed, and streetcar systems were almost all privately owned until they were dismantled in the early postwar period (Rodgers 1998). In each of these cases, it is not just the physical infrastructure and sunk costs that generate path-dependent trajectories, but also institutional structures, configurations of related institutions, and the timing and sequencing of institution genesis that are decisive.

Each of these cases supports the suggestion that municipal governance systems are sometimes powerfully path dependent, and that a historical institutionalist perspective may provide useful insights. The suggestion here is that HI provides a productive way forward for Healey's questions of the 1980s about the roles of planning in property development processes, that is quite different than that suggested by SI conceptions. While there has been enduring attention to theorizing the role of capital investment in property at the macro level (e.g. Harvey 1982), issues of mortgage financialization and urban property markets (Weber 2010; Rolnik 2013), and to the micro level of municipal actors, developers, and development industry in the production of property (Weiss 1987; Healey 1992a, 1992c; Fainstein 1994), surprisingly little attention has been paid to the market-shaping activities of the state as the definer, regulator, recorder, evaluator, and guarantor of urban property, and provider of shared infrastructure. To a greater degree than in perhaps any other market, the state is fundamental and indispensable to the existence and value of urban property, and to the viability of property markets. A key function of urban planning regulations, infrastructure management and financing arrangements, and governance institutions is to regulate the production of urban space.

States play a particularly central role in regulating the production of new urban property rights, whether this occurs through development of greenfields at the urban edge, or through redevelopment of already urban parcels into more intense uses. When existing properties are developed or redeveloped and divided into smaller pieces, new properties and new property rights are created. Such development creates unique new 'bundles' of property rights that are defined by development regulations and permitting processes. The contents of such bundles of property rights are varied, but include attributes such as whether and what size of building may be built, what uses a building may be put to, and whether it can be further subdivided and additional housing units may be built on it (if not then it has no development rights). The project of planning has often been described as one of developing controls over the private development of land in the public interest (Krueckeberg 1995; Jacobs and Paulsen 2009).

Significantly, over time the specific contents of the bundle of property rights of newly created properties have changed, as planning regulations have evolved, but in most cases the property rights of existing properties and districts remain untouched, or 'grandfathered'. For example, in most developed countries there was little regulation of land development in the nineteenth century, so new urban areas were developed incrementally, with a mixture of houses, shops along larger streets, businesses and small factories intermixed. In the early twenty-first century regulation of land development is everywhere much stricter, and bundles of property rights have become more restricted, with most new properties permitted only a very narrow range of possible building uses and sizes, and with much tighter building codes. To the very considerable extent that urban property values and meanings are influenced by surrounding property and public goods and patterns (Fischel 2001; Fennell 2009), property rights are also shaped by larger patterns of neighbourhood design and land use, and these have also changed radically over time. Property produced at different times and in different places therefore has different property rights, as systems regulating the production of property, and patterns of new development, have evolved. Crucially, the new urban space – whether on the urban fringe or as redevelopment – that is created during any particular period tends to bear an indelible imprint of the regulatory system that existed at the moment of creation of new property rights, creating specific and evolving landscapes of property, infrastructure, and regulation at different times and in different places.

An HI research agenda on the development of property in urban settings would focus on the evolution of the institutions that regulate the production of property, the planning system, infrastructure design, finance and management systems, development control processes, land registries and the evolving specification of the bundles of property rights that regulatory systems enable and protect, and those that they reject. Of particular interest would be critical junctures of institution formation of these systems, the extent of path dependence or openness to incremental change, and an examination of key actors and processes of institutional development and change.

An HI perspective leads us to see cities as spatial arrangements of people and capital investments in urban property and the varied interests that these give rise to, that are produced by and sustained with complex and costly infrastructures, and have developed in the context of specific governance structures, powers, and compromises. While these three types of institutions work together as a single system for the production and maintenance of urban space, they each have distinctive and characteristic mechanisms of path-dependence and patterns of critical junctures of new institution formation and incremental change. Urban infrastructure is not just sets of hardware, it is always a politically mediated and contested set of institutional practices that have specific geographies, histories, and relationships to property. The systems that produce and maintain urban space are highly varied between cities, and between countries, and with great variation in the social, economic and environmental qualities of the urban property created, and in the distribution of costs and benefits among actors. Systems that regulate the production and management of urban property have always been important, but in the current period, with a near doubling of global urban population expected by 2050 to 6.25 billion, and a doubling of total urban area projected during the next 20 years (Angel 2012), such systems are fundamental to the global future of humanity.

The suggestion here is that the institutions associated with the development of urban space are in some cases highly path dependent. If so then they may be neither particularly 'efficient' in the sense of producing good urban forms, nor responsive to the changing social and political priorities of the people and places they produce, as in the cases of municipal fragmentation and infrastructure governance institutions in the US discussed earlier. And if so, an historical institutionalist approach suggests that it is important to examine such enduring sets of institutions, and the power relations that they help to shape. If some urban institutions are path dependent, then the political dynamics present in the critical juncture that created them can have longer impacts, for as Lowndes puts it concisely: 'institutional rules embody power relations by privileging certain positions and certain courses of action over others – they express 'patterns of distributional advantage'' (Lowndes 2009: 95). Not all urban institutions are path dependent, but many are, and where they are they have profound impacts on the planning systems and approaches that become established and the landscapes of property that these produce, and the way these each evolve over time.

Seen this way, planning and land development control systems are the result of particular contingent moments of institution-building – or institutional fixes – specific compromises that were shaped by the political dynamics of the particular critical juncture of institution formation. The suggestion is not a simplistic equation of institutions with the interests of particular actors during critical junctures, but to recognize that when particular institutions endure, it is valuable to ask why, and who benefits from the particular patterns of distributional advantage they generate.

Finally, if the ways that new property is created were fundamentally similar between countries, then institutional configurations would be of much less interest. But such development control and planning systems are very different in different jurisdictions, even between countries that are similar in other respects, as might be expected in cases where branching patterns of institutional development are common. An HI approach would examine and compare both the critical junctures of institution formation and the patterns of incremental change in between major change points, and the characteristic patterns of property, infrastructure, and governance institutions that emerge in different places.

Conclusions

It seems likely that at some point, historians of social science will recognize Patsy Healey's body of work as a major contribution to sociological institutionalist thought, even though her work is seldom cited in this way today. The major contribution Healey's sociological institutionalist project has made to planning thought *is* widely celebrated, however. Here the suggestion has been that Healey's project of bringing institutionalism into planning theory can be extended by also framing questions from an historical institutionalist perspective. HI provides a somewhat different viewpoint, and different insights into questions of structure and agency, continuity and change, and institutional design and innovation. An HI approach would complement Healey's major contributions to SI and planning about the potential of collaborative governance processes in transforming local governance cultures and generating shared strategic visions, with a focus on the fine-grain of rules, codes, and regulations that continue to shape processes of urban change and property development, often long after the particular contingent moments of institutionalization in which they were first decided.

The focus here has been on institutions related to land and property development and capital investment in cities, picking up where Healey left off in the early 1990s, but with an HI lens. An HI approach might have provided a way forward for Healey's work on land and property development of the 1980s, but HI was still in its infancy, and lacked its own 'Healey' to develop its potential insights for cities and planning.

A major insight of this approach is the understanding that some urban institutions tend to be path dependent, and that if they do, then it becomes important to understand the contingent moments of institutional genesis, patterns of positive feedback, and the engagement of different actors in policy processes. This leads to a research agenda that emphasizes an investigation of the roles of institutions in regulating planning and property development practices that shape the built environment, with an emphasis on (a) the *institutions* that structure land and property development and management; (b) the kinds of *positive feedback effects* which tend to reinforce continuity and produce path dependent trajectories in some institutions; (c) the characteristic *patterns of incremental and endogenous change* of different institutions; (d) the *timing and sequencing* of moments of institution-building; and (e) the characteristic *landscapes of property* that are created by differing institutional structures, in different places and at different times.

An understanding of examples of path dependent institutions also suggests that planners should be extremely careful about institutional design, and avoid creating systems that become difficult to change, or perhaps in some cases, to deliberately design institutions to be difficult to change, as for example is true of most constitutions. Either way, it is important to better understand the significance of positive feedback effects on urban institutions, and the significance of institutional arrangements and continuities in the production of urban space.

References

Angel, S. (2012). *Planet of Cities*. Cambridge, MA: Lincoln Institute of Land Policy.

Brady, H.E. and D. Collier (2004). *Rethinking Social Inquiry : Diverse tools, shared standards*. Lanham, MD:, Rowman & Littlefield.

Capoccia, G. and R.D. Kelemen (2007). 'The study of critical junctures: Theory, narrative, and counterfactuals in historical institutionalism'. *World Politics* 59(3): 341–69.

Carruthers, J.I. (2003). 'Growth at the fringe: The influence of political fragmentation in United States metropolitan areas'. *Papers in Regional Science* 82: 475–99.

Collier, D. (2011). 'Understanding process-tracing'. *Political Science and Politics* 44(4): 823–30.

Collier, R.B. and D. Collier (1991). *Shaping the Political Arena*. Princeton, NJ: Princeton University Press.

Cox, K. R. (2002). *Political Geography: Territory, state, and society*. Oxford, UK; Malden, MA: Blackwell.

DiMaggio, P.J. and W. Powell (1983). 'The iron cage revisited: Institutional isomorphism and collective rationality in organizational fields'. *American Sociological Review* 48: 147–60.

DiMaggio, P.J. and W.W. Powell (1991). Introduction. In: W.W. Powell and P.J. DiMaggio (eds). *The New Institutionalism in Organizational Analysis*. Chicago and London: University of Chicago Press: 1–38.

Fainstein, S.S. (1994). *The City Builders: Property, politics, and planning in London and New York*. Oxford, UK; Cambridge, MA: Blackwell.

Fennell, L.A. (2009). *The Unbounded Home: Property values beyond property lines*. New Haven, CT: Yale University Press.

Fischel, W.A. (2001). *The Homevoter Hypothesis: How home values influence local government taxation, school finance, and land-use policies*. Cambridge, MA: Harvard University Press.

Fulton, W., R. Pendall, M. Nguyen and A. Harrison (2001). 'Who sprawls most? How growth patterns differ across the U.S'. *Survey Series*. Washington D.C.: Centre on Urban & Metropolitan Policy.

Giddens, A. (1984). *The Constitution of Society*. Cambridge: Polity Press.

Gonzalez, S. and P. Healey (2005). 'A sociological institutionalist approach to the study of innovation in governance capacity'. *Urban Studies* 42(11): 2055–69.

Hacker, J.S. (2004). 'Privatizing risk without privatizing the welfare state: The hidden politics of social policy retrenchment in the United States'. *American Political Science Review* 98(2): 243–60.

Hall, P.A. and R.C.R. Taylor (1996). 'Political science and the three new institutionalisms'. *Political Studies* 44(5): 936–57.

Harvey, D. (1982). *The Limits To Capital*. Chicago; Oxford: University of Chicago Press; B. Blackwell.

Hay, C. and D. Wincott (1998). 'Structure, agency and historical institutionalism'. *Political Studies* 46(5): 853–1027.

Healey, P. (1974). 'Planning and change'. *Progress in Planning* 2(3): 143–237.

Healey, P. (1983). *Local Plans in British Land Use Planning*. Oxford and New York: Pergamon Press.

Healey, P. (1992a). 'Development plans and markets'. *Planning Practice and Research* 7(2): 13–20.
Healey, P. (1992b). 'Planning through debate: The communicative turn in planning theory'. *Town Planning Review* 63(2): 143–62.
Healey, P. (1992c). *Rebuilding the City: Property-led urban regeneration*. London: E. & FN Spon.
Healey, P. (1996). 'The communicative turn in planning theory and its implications for spatial strategy formation'. *Environment and Planning B* 23: 217–34.
Healey, P. (1997a). *Collaborative Planning: Shaping places in fragmented societies*. Houndmills, Basingstoke: Macmillan Press.
Healey, P. (1997b). *Making Strategic Spatial Plans: Innovation in Europe*. London and Bristol, PA: UCL Press.
Healey, P. (1998). 'Building institutional capacity through collaborative approaches to urban planning'. *Environment and Planning A* 30(10): 1531–46.
Healey, P. (1999). 'Institutionalist analysis, communicative planning, and shaping places'. *Journal of Planning Education and Research* 19: 111–21.
Healey, P. (2004). 'Creativity and urban governance'. *Policy Studies* 25(2): 87–102.
Healey, P. (2006a). The new institutionalism and the transformative goals of planning. *Institutions and Planning*. Amsterdam; Boston: N. Verma: Elsevier: 61–87.
Healey, P. (2006b). 'Transforming governance: Challenges of institutional adaptation and a new politics of space'. *European Planning Studies* 14(3): 299–320.
Healey, P. (2013). 'Re-enchanting democracy as a mode of governance'. *Critical Policy Studies* 6(1): 19–39.
Healey, P. and S.M. Barrett (1990). 'Structure and agency in land and property development processes: Some ideas for research'. *Urban Studies* 27(1): 89–104.
Healey, P., P. McNamara, M. Elson and J. Doak (1988). *Land Use Planning And The Mediation Of Urban Change : The British planning system in practice*. Cambridge and New York, Cambridge University Press.
Healey, P. and R. Nabarro (1990). *Land and Property Development in a Changing Context*. Aldershot, UK and Brookfield, VT: Gower.
Jacobs, H.M. and K. Paulsen (2009). 'Property rights: The neglected theme of 20th-Century American planning'. *Journal of the American Planning Association* 75(2): 134–43.
Katznelson, I. (2003). Periodization and preferences: Reflections on purposive action in comparative historical social science. In: J. Mahoney and D. Rueschemeyer (eds). *Comparative Historical Analysis in the Social Sciences*. Cambridge: Cambridge University Press: 270–301.
Krueckeberg, D.A. (1995). 'The difficult character of property: To whom do things belong?' *Journal of the American Planning Association* 61(3). 301–9.
Lowndes, V. (2009). 'New institutionalism and urban politics'. In: J.S. Davies and D.L. Imbroscio (eds). *Theories of Urban Politics*.Los Angeles, CA: Sage: 91–105.
Mahoney, J. and K.A. Thelen (2010). *Explaining Institutional Change: Ambiguity, agency, and power*. Cambridge and New York: Cambridge University Press.
Mandelker, D.R. (1958). 'Municipal incorporation on the urban fringe: Procedures for determination and review'. *Louisiana Law Review* 18: 628–60.
Meligrana, J. (2004). *Redrawing Local Government Boundaries: An international study of politics, procedures and decisions*. Vancouver: UBC Press.
Norris, D.F. (2001). 'Prospects for regional governance under the new regionalism: Economic imperatives versus political impediments'. *Journal of Urban Affairs* 23(5): 557–71.
Norton, A. (1993). *International Handbook Of Local And Regional Government: A comparative analysis of advanced democracies*. Brookfield, VT: Edward Elgar.
Orfield, M. (1997). *Metropolitics: A regional agenda for community and stability*. Washington, D.C.; Cambridge MA: Brookings Institution Press; Lincoln Institute of Land Policy.
Pierson, P. (1997). Path dependency, increasing returns, and the study of politics. *Centre for European Studies – Harvard*. Not available as a pdf.
Pierson, P. (2000). 'Increasing returns, path dependence, and the study of politics'. *American Political Science Review* 94(2): 251–67.

Pierson, P. (2004). *Politics in Time: History, institutions, and social analysis*. Princeton, NJ: Princeton University Press.

Rodgers, D.T. (1998). *Atlantic Crossings: Social politics in a progressive age*. Cambridge, MA, and London: The Belknap Press of Harvard University Press.

Roland, G. (2004). 'Understanding institutional change: Fast-moving and slow-moving institutions'. *Studies in Comparative International Development* 38(4): 109–31.

Rolnik, R. (2013). 'Late neoliberalism: The financialization of homeownership and housing rights'. *International Journal of Urban and Regional Research* 37(3): 1058–66.

Rusk, D. (2003). *Cities Without Suburbs*. Washington, D.C.: Woodrow Wilson Center Press.

Sorensen, A. (2011a). 'Toronto megacity: Growth, planning institutions, sustainability'. In: A. Sorensen and J. Okata (eds). *Megacities: Urban Form, Governance, and Sustainability*. Tokyo: Springer Verlag: 245–72.

Sorensen, A. (2011b). 'Uneven processes of institutional change: Path dependence, scale, and the contested regulation of urban development in Japan'. *International Journal of Urban and Regional Research* 35(4): 712–34.

Sorensen, A. (2015). 'Taking path dependence seriously: A historical institutionalist research agenda in planning history'. *Planning Perspectives* 30(1): 17–38.

Streeck, W. and K. Thelen (2005). 'Introduction: Institutional change in advanced political economies'. In: W. Streeck and K. Thelen (eds). *Beyond Continuity: Institutional Change in Advanced Political Economies*. Oxford: Oxford University Press: 1–39.

Teaford, J. C. (1979). *City and Suburb: The political fragmentation of metropolitan America, 1850–1970*. Baltimore, MD: Johns Hopkins University Press.

Thelen, K. (1999). 'Historical institutionalism in comparative politics'. *Annual Review of Political Science* 2: 369–404.

Thelen, K. (2003). 'How institutions evolve: Insights from comparative historical analysis'. In: J. Mahoney and D. Rueschemeyer (eds). *Comparative Historical Analysis in the Social Sciences*. Cambridge: Cambridge University Press: 208–240.

Thelen, K.A. and S. Steinmo (1992). 'Historical institutionalism in comparative analysis'. In: S. Steinmo, K. A. Thelen and F. Longstreth (eds). *Structuring Politics: Historical institutionalism in comparative analysis*. Cambridge and New York: Cambridge University Press: 1–32.

Ward, K. and A.E.G. Jonas (2004). 'Competitive city-regionalism as a politics of space: A critical reinterpretation of the new regionalism'. *Environment and Planning A* 36: 2119–39.

Weber, R. (2010). 'Selling city futures: The financialization of urban redevelopment policy'. *Economic Geography* 86(3): 252–74.

Weir, M. (1995). 'Poverty, social rights, and the politics of place in the United States'. In: S. Leibfried and P. Pierson (eds). *European Social Policy: Between Fragmentation and Integration*. Washington, D.C.: Brookings Institution.

Weiss, M.A. (1987). *The Rise of the Community Builders: The American real estate industry and urban land planning*. New York:Columbia University Press.

White, R. (2003). *Urban Infrastructure and Urban Growth in the Toronto region, 1950's to the 1990's*. Toronto: Neptis Foundation.

PART V
Epilogue

Epilogue

People, Place and Shaping Futures: Struggling for Practically-relevant Understanding

Patsy Healey

The project of this book came as a surprise to me. I have always thought that my contribution as a planning scholar had to be judged by the critical evaluation of readers, but I had not imagined that anyone would attempt the interpretive task of making sense of the different things I have thought about, researched and written over the years. I have to thank Jean Hillier for being intrigued enough to initiate this project and determined enough to track down the various strands of work I have engaged in, and also for the many intense, and sometimes critical, discussions we have had over the years. There are terms I came to use in the 2000s which I would not have introduced into my work without these discussions. My thanks also to Jonathan Metzger for his grasp of my work and for our various, mostly e-discussions, and to all the authors of this volume, for their critical appreciation of what I have tried to do as a planning scholar.

Yet it is a curious experience to reflect on one's own corpus of academic work. Revisiting it, I discover echoes of concerns which still pre-occupy me, and wonder what has changed through decades of researching, teaching, university administration, various forms of practice consultancy and building institutions for the planning academic community. I am also reminded of the many people I have worked with – colleagues, research assistants, doctoral students, those involved in the practices I came to research, who have continually challenged and tested me as we worked together puzzling how to make sense of particular situations. These multiple engagements have meant that my interests roamed around a range of concerns rather than focusing on a narrow problem agenda. Yet they helped to enrich my knowledge of the world and its ways, inspiring me to revisit and re-articulate my spread of interest.

That interest, I came to realise later, was set out in my doctoral project. I took a winding way to this endeavour (Healey forthcoming). I set out to do a PhD as a way of finding out what the planning project could be about. I was already exposed to the contradiction between the idealistic view of 'planning' as a way of building a better society which I held in my student years and the experience of working in the planning department of a London Borough. I have always been interested in the transformative potential of planning ideas and practices, and in the 1960s I found myself in an institutionalised practice where, as so often happens, the participants could never really explain why it had the form that it did. This led me to puzzle, in a vague and unfocused way, about how political ideas became institutionalised. Through the accident of personal life history, I was able to explore this tension in the context of rapidly urbanising Latin American societies in the late 1960s. In those days in the UK, doctoral students were very much left to set out their own agendas in their own way. For me, the experience was a voyage of discovery. The 'planning literature' at the time was limited and scattered among conversations about urban and regional design, about regional economic development and, newly emerging, about technically-robust future-shaping policy processes. One aspect of my voyage involved delving into the London School of Economics library, which contained much of this material but also rich resources of social science literature on social, economic and political development. Since then, I have understood that the planning field needs to be continually refreshed through interacting with a range of related disciplinary discourses in order to develop understanding of how people relate to places and engage (or not) in future-shaping governance activity. But equally important in building my thinking was my fieldwork in Latin America. I approached this as a largely untutored geographer, mapping urban development activity, trying to grasp the processes by which land was being converted into urban uses. But I slowly came to see that there were not only different groups actively evolving different ways of 'building the city'. There were also different modes of practice connecting public sector activity, business and citizens. Planning institutions, emphasising fairness and technical justifications, struggled to get leverage in this situation. Because I had worked in a planning office in the UK, I knew that I had to get into this fine grain of how planning work was actually done. I always remember being in the well-regarded municipal planning office in Valencia, Venezuela, and noticing that planning applications sometimes got submitted and approved on the same day. Meanwhile, the technician who looked after the photocopier made sure that maps of where public development might next take place were passed to squatter groups, so that they could invade and claim compensation for any displacement before the 'state' arrived. A shocked westerner might see this as 'corruption', but I had sufficient anthropological understanding to recognise that these activities reflected alternative modes of governance, co-existing within the planning office itself. This layering of institutionalised practices reflected and helped to form the wider political and economic context in which urban development was being undertaken. My perception that it was important to keep these micro-practices in view was re-inforced. Micro-politics, to my mind, is as important as grand narrative politics in shaping how futures evolve.

Through this formative experience, an extract from which appears in Part IV (Healey, 1974), I became absorbed by how people – squatter groups, developers, infrastructure providers, urban managers, people who self-identified as planners – struggled to shape urban places, often in conflict with each other, and in political and economic situations not directly of their own choosing. I also, somewhat accidentally, evolved a way of writing about these processes which was not in the structured analytical format established in geography and political science at the time.[1] My thesis was more in the form of a historical-geographical narrative, a style I have used since in presenting much of the empirical research work I have done. These days, our field is dominated by such 'case narratives', but at that time research findings tended to be extracted and processed to fit analytical schema derived from positivist concepts of the relation between the world and accounts of it.

Doctoral work thus laid out an agenda of interest for me, but I had few intellectual resources with which to address it. I have never lost interest in the potentially transformative role of planning ideas and in the interaction between ideas and practices. My concern is not just with the relation between 'knowledge and action', as we so often assert in the planning field, but in *that* knowledge and *those* ideas which help to alter trajectories. In that relation, it is critical, it seems to me, to understand the complex socio-political processes through which ideas shape practices and vice versa. The literature I found most inspiring to me as a PhD student had echoes of the institutionalist approaches developed in the 1930s,[2] and my work ever since has been in the mode which some now call 'sociological institutionalism'. Such terms did not come back into the social sciences strongly until the late 1980s.

After the PhD, I set out on another voyage of intellectual discovery in the sociological literature – from phenomenology to organisational sociology, to get a better idea of institutions as social practices. This also helped me to develop a more 'interpretive' and less positivist approach to conceptualisation, research focus and analysis. I became interested in how governance processes could become so institutionalised that those involved hardly saw them as anything other than taken-for-granted practices. In this way, a planning idea initially aimed to be transformative could become tamed and distorted to quite different purposes. How far such practices were challenged then depended on the organisational cultures of work groups. With such ideas, I revisited my 'sites of experience' as a planner in London (see Part II, Healey and Underwood, 1978). This empirical work, along with the PhD and teaching challenges, became the basis for the 'mapping' of strands of planning ideas which helped me navigate through the increasing recognition among planning scholars that our field encompasses many different intellectual strands, which practising planners draw on to focus and justify what they do. This eventually became the 'map' of planning theories which appeared in Healey et al. (1982).

Because of my interest in the micro-processes of governance activity, it has sometimes been suggested that I have failed to attend to the 'substantive' aspects of urban development. This was clearly not the case in my PhD, and I have always argued that substance and process are intimately connected. But at various stages it is helpful to focus in depth on one aspect or another of the planning field of interest. One aspect, which to my mind is not given the attention it should in our field, is the dynamics of land and property development processes. I embarked on exploring these dynamics and their interaction with planning interventions on

1 My PhD supervisor and home department at LSE was based in geography; my external examiner was a political scientist.

2 My thanks to Frank Moulaert, who was more aware than I of the trajectory of these ideas, for reminding me of this heritage in the planning field.

completing the work on London Borough planners. Part III.1 provides some examples of this work, which preoccupied me during the 1980s. Once again, research inquiry into practice situations combined with reading especially the political economy literature[3] provided the foundation for what became an 'institutionalist' understanding of these processes.

By this time, I was beginning to develop a clearer understanding of how I found myself seeing and interpreting the planning project and the worlds in which, in various forms, such a project played out. My perception, shaped by reading in phenomenology and anthropology, is of a world of relations rather than atomised objects and individuals, in which meanings and ways of thinking are as significant as material flows. In this epistemology and ontology of a relational world, individual phenomena are tied into multiple webs of relations, each with a particular spatial and temporal reach. In some situations, the power dynamics of relations are dominated by some hegemonic force, but such force rarely completely overwhelms relations operating in quite different ways, which may then surface to challenge and undermine apparently stable systems. Thus, in a world in flux and evolution, periods of institutional stability are continually challenged by the tensions which any stabilisation suppresses. People struggle in particular situations to shape futures as well as to live them. Our concern, it seems to me, as a community of planning scholars, is to develop the critical understandings and moral implications of such struggles as they focus on places and their qualities, and to enrich the array of ideas in circulation about what shaping place futures might involve. With these perceptions, I embarked on my attempt to draw the various strands of my thinking together in what became, after a decade, *Collaborative Planning* (1997/2006).

This book, perhaps because of the title,[4] now seems to be understood as the promotion of a particular kind of governance process, inspired by a Habermasian understanding of communicative rationality. Yet my aim had been to set out the implications of a relational view of urban and environmental dynamics and the implications for governance processes. The works of geographers David Harvey and Doreen Massey, and sociologists Anthony Giddens and Claus Offe, were as important to me in building this foundation as were those of Habermas. If governance processes are to be understood in a relational way, and if structural forces and agency energy are seen as in continual interaction, then it is important to look at what goes on in these relations. Habermas helps us explore governance relations particularly as these play out in the public sphere and how the sphere of politics, the state and the law can be held to account in democratic ways through the cultivation of particular public sphere qualities.[5] This perspective challenges approaches to political life which deal predominantly in individuals and their specific interests rather than in institutionalised discourses and practices. But it also questions approaches which focus primarily on structural forces as drivers of transformative change.

The perspective I developed belonged within, though often in uneasy combination, the penetration of post-positivist and post-structural philosophy into the planning field. For me, it was inspiring to find planning scholars elsewhere, and particularly in the US, who were moving in similar directions, though with less of an interest in place-making than I had (see Part II, Healey and Gilroy, 1990, and Part III.2, Healey, 1992). But I was also strongly influenced by the UK's experience in the 1980s as the poster-child of 'neo-liberal' ideology. As with any political community, the UK has its own distinctive history with many tensions and contradictions. But my perception then and now is of an ongoing struggle to find a way for a changing and increasingly diverse civil society to get leverage on, and sometimes provide alternatives to, what is done by government in its name. In my country, the spatial planning field as an institutionalised practice provides a set of socio-political sites across which these struggles play out. The experimentation with partnerships and new governance forms encouraged by successive governments in recent decades provide another, as do initiatives arising directly from civil society with which I am now involved. In the planning field, we have used terms like 'participation', 'collaboration', 'deliberation', community 'engagement' and 'empowerment', and most

3 Doreen Massey, David Harvey, Allan Scott and Michael Ball did pioneering work in the 1970s in analysing land and property development processes from a political economy perspective.

4 Actually the creation of my publisher. My own working title had been 'planning inside politics'.

5 Habermas' concerns in this respect seem to me to be much better expressed in his *Between Facts and Norms* (1996) than in his earlier work.

recently 'co-production', to highlight the potential for transformative change within these arenas, and to chart the many disappointments as these fail to realise our normative hopes.

During the 1990s, I took the opportunity to research the micro-dynamics of several of these experiences (for example, Part III.2, Healey, 1997). Located by this time in Newcastle University, and in a region with a proud history but many challenges as traditional industries disappeared, I and colleagues were able to observe how political initiatives and policy programmes intended to have a transformative effect – on people's lives, on the local economy and on how governance was done – actually played out as citizens, business people and public servants were drawn into new kinds of relationships. These studies, where we were sometimes participant –observers and sometimes qualitative researchers, had a clear normative focus in our concern to bring into policy deliberation the voices and views of those living in an area and experiencing its problems. They also helped to develop a more precise analytical vocabulary for describing such experiences (see Part IV, Healey 2006). By the end of the 2000s, and drawing on a major study on strategic spatial planning I was able to undertake as a post-retirement researcher (Healey 2007), I found myself able to make two statements about what is involved in struggling for better place futures in ways which are more richly democratic than the practices our government systems in the UK and elsewhere had settled into by the early 2000s. The first (Part II, Healey, 2012) focuses on how to act in the ongoing flow of practice – whether that of activists, professionals or officials, in such a way that more progressive directions could be opened up rather than closed down. In my view, it is really important to explore the contribution of place governance practices to building a lively and informed public sphere, in the multiple polities which are drawn into such activity. The second (Healey 2010) is a statement of what the planning project now seems to me to be, illustrated with examples from different areas of place-making work. In both these contributions, I underline the importance of a core skill of planners who act strategically. This skill centres on being able to combine a strategic grasp of the contextual dynamics of particular challenges and situations with a real awareness of the particularity of moments of opportunity in which cracks can open in institutionalised discourses and practices. It demands a shrewd understanding of how to act to make a difference in these moments. And it is a skill not confined to those with formal planning training and experience.

In *Collaborative Planning*, I argued that the planning project has as its focus the challenge of 'managing co-existence in shared places', or as Leonie Sandercock (2002) has put it, living with strangers as neighbours. I still think that this challenge is as important as it was in the 1980s and in the rapidly-urbanising Latin American situations I studied forty years ago. Compared to then, we now have a substantial planning literature to draw on, rich in case examples and in conceptual debate. There is a tendency to think that our academic endeavours are continually moving our thinking forward into new terrains. I certainly thought this as a younger planning scholar. But these days, as I reflect on my own voyages, and read more history and biography, I often notice that there are few new ideas and perceptions around. What changes is the world around us and the intellectual climate in which we are positioned. Some planning ideas flourish because they are 'in their time'; others fail to thrive not having found their time. The same has applied to my own work. My kind of approach was deeply out-of-time in the 1970s, but slowly came to find its time in later decades. It is noticeable now, for example, that other social sciences, notably economics, are giving much more attention to institutional issues and 'place effects'. As planning scholars, we need to look outwards to other disciplines, partly for our own inspiration, but also to contribute to the discussions in other fields, as they come to notice planning activity and the significance of place qualities and spatial relations. We also need to maintain our own insights whilst recognising that they are not always appreciated or understood.

But there is a danger, now that there is such a richness of research and theoretical development in our field, that we read the world emerging around us too much through the weight of our own scholarship and lose the capacity to experience it directly and learn from this. It is all too easy to become absorbed in theoretical discourses, detached from engagement in the practical world around us, a danger so well described in Pirsig's *Zen and the Art of Motorcycle Maintenance* (1974/1999). Involved now in the practical work of local community development, I draw a lot on my intellectual store. But I use it not as a recipe for how to act. Instead, it provides me with an array of 'warning signs', which encourage me to keep my critical antennae honed to the wavelengths of what is going on around me. Perhaps our biggest challenge today is how to enrich the place-shaping struggles in which people across the world are engaged. It is clear to me that such struggles will remain a major challenge in the present century, as residents of the massive urban complexes recently created in many parts of the world look for ways to make their places more liveable and sustainable into the future. Planning scholars can help in these struggles, partly through direct involvement, but also through research inquiry which highlights the context and contingencies of such struggles, thus challenging the over-ready acceptance by policy-makers of fashionable general solutions circulating in globalised policy discourses.

The role of planning scholarship is surely to enhance the abilities of others to shape the emerging future in ways which release their own sensitive antennae and creative imagination, while also encouraging them to notice and take responsibility for their relations with others in their own, and other, times and places. So I hope planning scholarship will continue to explore the complex relations between 'knowledge and action' (Friedmann 1987) as these play out in place governance practices, and that such work will be undertaken in ways which promote creative experimentation and the kind of critical debate which encourages the search for better futures rather than drowning hope in the weight of negative criticism. Our field these days is rich in ideas, analyses and recorded experiences. But it is important that we never lose sight of the inspiration which fuelled planning scholarship in the previous century, including my own – the challenge of relating theory and practice, of combining analytical accounts with normative concerns, fuelled with the ambition of encouraging more progressive outcomes than would otherwise occur. It seems to me that scholarship inspired by such a dynamic is needed more than ever in our complex and confused worlds.

References

Friedmann, J. (1987). *Planning in the Public Domain.* Princeton: Princeton University Press.

Habermas, J. (1996). *Between Facts and Norms.* Oxford: Polity Press.

Healey, P (forthcoming) Finding my way: A life of intellectual exploration and institution-building, in B. Haselsberger, et al. (eds) *Ecologies of Planning Thought* (working title)

Healey, P. (1997/2006). *Collaborative Planning: Shaping places in fragmented societies.* London: Macmillan.

Healey, P. (2006). Transforming governance: Challenges of institutional adaptation and a new politics of space. *European Planning Studies* 14(3): 299–319.

Healey, P. (2007). *Urban Complexity and Spatial Strategies: Towards a relational planning for our times.* London: Routledge.

Healey, P. (2010). *Making Better Places: The planning project in the twenty-first century.* London: Palgrave Macmillan.

Healey, P., G. McDougall and M. Thomas (eds) (1982). *Planning Theory: Prospects for the 1980s.* Oxford: Pergamon.

Pirsig, R.M. (1974/1999). *Zen and the Art of Motorcycle Maintenance.* London: Vintage.

Sandercock, L. (2000). When strangers become neighbours: Managing cities of difference. *Planning Theory and Practice* 1(1): 13–30.

Patsy Healey – Publications

1969 A technique for assessing the effect of planning policies on residential land and population (with W.F. Stanley), *Quarterly Bulletin of the Research and Intelligence Unit of the Greater London Council* 8.

1971 Barrios are not Romantic, *Architectural Design*, XLI (October).

1973 *Ideologies in Planning*, (edited with W.G. Zwirner), Kingston Planning Workshop Paper No. 2. Kingston-upon-Thames: School of Planning, Kingston Polytechnic.

1974 Planning and change, *Progress in Planning*, 2(3): 143–237.

The problem of ideology, *The Planner*, 60(3): 602–4.

Towards comparative planning studies; some comments based on research into urban planning in Latin America. In S. Hamnett (ed.) *The Nature and Purpose of Comparative Planning*, Working Papers in Planning, Theory and Education No.2. Delft: Technische Hogeschool.

Approaches to a Research Problem: Theory and the study of planners (with V. Evans), Kingston Planning Workshop Paper No.3. Kingston-upon-Thames: School of Planning, Kingston Polytechnic.

1975 *Job Experiences of Recently Graduated Planners*, Kingston Planning Workshop Paper No. 6. Kingston-upon-Thames: School of Planning, Kingston Polytechnic.

Urban Planning in a Venezuelan City, *Town Planning Review*, 46(1): 63–82.

Planners' Debates on Policy Work in a London Borough, Oxford Working Paper No.22. Oxford: Department of Town Planning, Oxford Polytechnic.

1976 Student Careers Monitoring Project Preliminary Report: 1975 Graduates (with M. Clegg, G. Crispin and C. Pountney), *Education for Planning Association Newsletter* 6(1).

London Borough Planning Policies: Co-ordinated Strategy or Policy Patchwork? CES Structure Planning Conference Papers CES CP16. London: CES.

Planning Departments in the London Boroughs (with Underwood, J.), *RTPI News*, Autumn.

The Content of a Planning Theory Course (edited with M. Hebbert and S. Hopkins), Oxford Working Paper No.25. Oxford: Department of Town Planning, Oxford Polytechnic.

1977 *The Organization and Work of Planning Departments in the London Boroughs*, (edited with J. Underwood), CES Structure Planning Conference Papers CES CP18. London: CES.

Current issues in Planning Education, *Education for Planning Association Newsletter*, 7(2).

1978 How many planning graduates have jobs? *Planning*: 250.

Professional ideals and planning practice (with Underwood, J.), *Progress in Planning* 9(2).

Graduate planning jobs – was 1977 an isolated trough? *RTPI News*, December.

1979 Networking as a Normative Principle with particular reference to local government and land use planning, *Local Government Studies*, 5(1): 55–68.

The Concept of a Local Plan, Oxford Working Paper No.36. Oxford: Department of Town Planning, Oxford Polytechnic.

Central Government's role in Green Belt definition with particular reference to Local Plans. In M. Elson (ed.) *Perspectives on Green Belt Local Plans*, Oxford Working Paper No.37. Oxford: Department of Town Planning, Oxford Polytechnic.

On Implementation: Some thoughts on the Issues raised. In C. Minay (ed.) *Report of a Seminar on Implementation in Planning*, Oxford Working Paper No. 43. Oxford: Department of Town Planning, Oxford Polytechnic.

Local Needs in Areas of Restraint (with Elson, M. and Gault, I.), Oxford Working Paper No. 42. Oxford: Department of Town Planning, Oxford Polytechnic.

1980 *The Implementation of Selective Restraint Policy* (with S. Evans and S. Terry), Oxford Working Paper No. 45. Oxford: Department of Town Planning, Oxford Polytechnic.

Job hunt will be more difficult, *Planning*, 386.
The development of planning education in the UK and its relevance as a model for other countries, *Ekistics*, 47(285): 416–20.
1982 Development Plans and Development Interests (with M. Elson). In *Structure and Regional Planning Practice*, PTRC 1981 proceedings.
The Implementation of Development Plans (with J. Davis, M. Wood and M. Elson), Report for the Department of the Environment. Oxford: Department of Town Planning, Oxford Polytechnic.
The role of development plans in implementating policies (with M. Elson), *The Planner*, November/December.
Banbury; The Implementation of Strategic Planning Policy in a Small Town in the Outer South East (with M. Wood and M. Elson), Oxford Working Paper No.63. Oxford: Department of Town Planning, Oxford Polytechnic.
Planning Theory: Prospects for the 1980s, (edited with G. McDougall and M. Thomas). Oxford: Pergamon.
Understanding land use planning: The contribution of recent developments in political economy and policy studies. In P. Healey, G. McDougall and M. Thomas (eds), *Planning Theory Prospects for the 1980s*. Oxford: Pergamon, pp. 180–93.
1983 Local Plans in British Land Use Planning, *Urban and Regional Planning Series* 31. Oxford: Pergamon.
British Planning Practice and Education in the 1970s and 1980s (with H.W.E. Davies), Reports for the SSRC Review of Planning Education and Research 1982, Published as Oxford Working Paper No.70. Oxford: Department Town Planning, Oxford Polytechnic.
'Rational Method' as a mode of policy formation and implementation in land use policy, *Environment and Planning B*, 10: 19–39.
Speed, Economy and Effectiveness in Local Plan Preparation and Adoption (with C. Fudge, C. Lambert and J. Underwood), Occasional Paper No.11. Bristol: School of Advanced Urban Studies.
Local Planning in Practice (3): Camden, Manchester (with C. Fudge), SAUS Working Paper No. 35. Bristol: School of Advanced Urban Studies.
The Limitations of Development Control Data: A Critique of Brotherton (with P.F. McNamara), *Town Planning Review*, 55(1): 91–7.
The Dangers of Complacency (with D. Watson), *Times Higher Education Supplement*, 20 January 1984.
Bridging the Gap between North and South: Towards a Common Approach to Inter-Regional Planning: a comment, *Third World Planning Review*, 6(4).
Changing Employment Prospects for Planners, *The Planner*, (February).
1985 The Rise and Fall of Planning (with M. Grant). In M. Loughlin, D. Gelfand and K. Young (eds), *Half a Century of Municipal Decline 1935–1985*. Sydney: Allen and Unwin, pp. 169–86.
The professionalisation of planning in Britain: Its form and consequences, *Town Planning Review*, 56(4): 492.
Linee di ricerca emergenti per la pianificazione locale degli usi del suolo, *Rivista di Urbanistica*, 11(4): 1–28.
Local plans in rural areas (with E.A. Hill). In A. Gilg (ed.) *Countryside Planning Yearbook 1985*. Norwich: Geo Books Limited.
Land Policy: Problems and Alternatives, (edited with S.M. Barrett). Aldershot: Gower.
The Political Economy of Land: Urban Development in an Oil Economy (with A. Gilbert). Aldershot: Gower.
Emerging directions for research in local land use planning, *Environment and Planning B: Plannning and Design*, 13: 103–20.
The Implementation of Planning Policy and the Role of Development Plans (with P.F. McNamara, A.J. Doak and M.J. Elson), Research Report for the Department of the Environment, 2 Vols.

1986 *Research on Local Land Use Planning*, (edited with C. Lambert), Oxford Working Paper No.91. Oxford: Department of Town Planning, Oxford Polytechnic.

Esperienza in Gran Bretanga, *Nuove Prospettive per la Pianificazione Urbanistica*, Quaderno 1 pp. 17–29.

The role of Development Plans in the British planning system, *Urban Law and Policy*, 8: 1–32.

Planning policies, policy implementation and development plans, *The Planner*, pp. 9–11.

The diffusion of the new town idea in the developing world. In I. Masser and R. Williams (eds) *Learning from other Countries*. Aldershot: Gower, pp. 121–33.

A ciascuno la sua informazione, *Agora*, III(9/12): 19–22.

1987 The future of Local Planning and Development Control, *Planning Outlook*, 30(1): 30–40.

1988 *Land Use Planning and The Mediation of Urban Change* (with P. McNamara, M. Elson and J. Doak), Cambridge: Cambridge University Press.

Imagining the city: Projects or policies, *Urban Design Quarterly*, 26: 7–13.

Gender Issues in Planning Education (with J. Hillier and S. Davoudi), Education for Planning Association Discussion Paper. Bristol: EPA.

The British planning system and managing the urban environment, *Town Planning Review*, 59(4): 397–417.

Planning tools and redistributive ends: Introduction, *Planning Outlook*, 31(2): 73–4.

Whatever happened to methodology in land use planning (in Chinese), *City Planning Review*, (China) 6: 23–6.

1989 Planning deregulated, *Local Government News*, March pp. 11–13.

The planning field welcomes women *The Times*, February.

Planning for Choice and Opportunity (with Working Party), Report of the RTPI Women and Planning Working Party. London: RTPI.

Directions for change in the British planning system *Town Planning Review*, 60(2): 125–49

Response to comments *Town Planning Review*, 60(3): 330–32.

Procesi di policy nella pianificazione urbanistica. In de A. Cugis (ed.) *Politica-politiche territoriali, Azione di governo ed uso pubblico della conoscenza.* Milano: Franco Angeli.

1990 Structure and agency in land and property development processes (with S.M. Barrett), *Urban Studies*, 27(1): 89–104; Also published in M.L. Greenhut and G. Norman (eds) (1994) *The Economics of Location*. London: Edward Elgar.

Policy processes in planning, *Policy and Politics*, 18(1): 91–103.

Places, people and polices, *Town and Country Planning*, 59(1): 9–10.

Democracy in the planning system, *The Planner*, 76(19): 14 15.

Places, people and policies: Plan making in the 1990s, *Local Government Policy Making*, 17(2): 29–39.

Towards a people-sensitive planning (with R. Gilroy), *Planning Practice and Research*, 52(2): 21–9.

Understanding land and property development processes: Some key issues. In P. Healey and R,. Nabarro (eds) *Land and Property Development Processes in a Changing Context.* Aldershot: Gower, pp. 3–14.

Using Planning Consultants: the Experience of the Tyne and Wear Development Corporation (with S. Davoudi), Urban Regeneration and the Development Process, Project Paper No. 2.

1991 The content of planning education programmes, *Environment and Planning B: Planning and Design*, 18(2): 97–110.

Urban regeneration and the development industry, *Regional Studies*, 25(2): 97–110.

Dilemmas of Planning Practice: Planners' Reflections on Ethics, Legitimacy and the Validation of Knowledge, (edited with H. Thomas). Aldershot: Avebury Technical.

From Shipyard to Housing Estate: Transformation at Hebburn, Urban Regeneration and the Development Process, Project Paper No. 3. Newcastle upon Tyne: Department of Town and Country Planning, University of Newcastle upon Tyne.

Die stadt und die umwelt: Problem und Perspektiven der Stadtpolitik, In W. Maharans, C. Ax and G. Buck (eds) *Stadt um Umwelt: Aspekte einer europaischen stadtpolitik.* Basel: Birkhawer Verlag.

The Education of Planners, Policy statement by the RTPI, including new principles for accrediting planning programmes. London: RTPI.

The Treatment of Planning Gain in the 'new' Local Plans, Working Paper No. 14. Newcastle upon Tyne: Department of Town and Country Planning, University of Newcastle upon Tyne.

Researching planning practice, *Town and Planning Review*, 62(4): 457–68

Models of the development process: A review, *Journal of Property Research*, 8: 219–38.

1992 An institutional model of the development process, *Journal of Property Research*, 9 (1): 33–44

A planner's day: Knowledge and action in communicative practice, *Journal of the American Planning Association*, 58 (1): 9–20 (included in J. Poon and P. Nijkamp (eds) (2006) *Social Planning*. Cheltenham: Edward Elgar).

Planning gain and the 'new' Local Plans (with M. Purdue and F. Ennis), *Town and Country Planning*, 61(2): 39–43.

Rationales for planning gain (with M. Purdue and F. Ennis), *Policy Studies*, 13(2): 18–30.

Planning gain and the grant of planning permission: Is the US test of the 'Rationale Nexus' the appropriate solution? (with M. Purdue and F. Ennis), *Journal of Environmental and Planning Law*, November pp. 1012–24.

Development plans and markets, *Planning Practice and Research*, 7(2): 13–20.

Rebuilding the City: Property-led Urban Regeneration, (edited with S. Davoudi, M. O'Toole, S. Tavsanoglu and D. Usher). London: E. & F.N. Spon.

The reorganization of state and market in planning, *Urban Studies*, 29(3/4): 411–34.

The education of planners: the new RTPI Policy Statement, *The Planner*, 78(8): 5–7.

Local planning implementation in the West Midlands, England in the 1980s. In M. Zoppi (ed.), *Il nodo Urbanistica.* Florence: Fondazione Callisto Pontello.

The European Expansion Strategies of British Property Consultancy Firms: A Survey (with P. Bland, I. Holland and R. Williams). Newcastle upon Tyne: Department of Town and Country Planning, Working Paper No 17, University of Newcastle upon Tyne.

Planning through debate, *Town Planning Review*, 63(2): 143–63 (included in S. Fainstein and S. Campbell (1996) *Readings in Planning Theory.* Oxford: Blackwell).

1993 Planning through debate. In F. Fischer and F. Forester (eds), *The Argumentative Turn in Planning and Policy Analysis.* Durham, NC: Duke University Press, pp. 233–53. (a revised version of the 1992 paper).

The communicative work of Development Plans, *Environment and Planning B: Planning and Design* 20(1): 83–194, (included in A. Khakee and K. Eckerberg (eds) (1993) *Process and Policy Evaluation in Structure Planning.* Umea: University of Umea, also in S. Mandelbaum and L. Mazza (eds) (1997) *Explorations in Planning Theory.* New Brunswick: State University of New Jersey, Rutgers).

Routes and settlement patterns (with A. Madanipour). In B. Farmer and H. Louw (eds), *Companion to Contemporary Architectural Thought.* London: Routledge, pp. 90–94.

The Longer Term Effects of the Tyne and Wear Metro (with S. Davoudi, A. Gillard, B. Pullen, S. Raybould, F. Robinson, D. Silcock, D. Usher and C. Wymer). Crowthorne, Berks: Transport Research Laboratory Report CR 357.

European planning systems: Diversity and convergence (with R.H. Williams), *Urban Studies*, 30 (4/5): 201–20.

Gains from Planning?: Dealing with the Impacts of Development (with M. Purdue and F. Ennis). York: Joseph Rowntree Foundation.

The Development Industry in Tyne and Wear (with S. Davoudi), Working Paper No 21. Newcastle-upon-Tyne: Department of Town and Country Planning, University of Newcastle-upon-Tyne.

Frameworks for Negotiating Development: Towards a Systematic Approach to Development Obligations (with F. Ennis and M. Purdue), Research Report. Newcastle-upon-Tyne: Department of Town and Country Planning, University of Newcastle-upon-Tyne.

Planners, plans and sustainable development (with T. Shaw), *Regional Studies*, 27(8): 769–76.

The treatment of 'Environment' in the British Planning System (with T. Shaw), Working Paper No 31. Newcastle-upon-Tyne: Department of Town and Country Planning, University of Newcastle-upon-Tyne.

1994 Environmental sustainability and the British planning system (with T. Shaw) In H. Voogd (ed.), *Issues in Environmental Planning*, pp. 31–57. London: Pion.

La ragioni per un premio al miglior progetto di marketting urbano. In G. Ave and F. Corsico (eds) *Marketing Urbano in Europa.* Torino: Torino incontra.

Urban policy and property development: The institutional relations of real-estate development in an old industrial region, *Environment and Planning A*, 26: 177–98.

Developing in a fragile property market: The development industry in Tyne and Wear, *Northern Economic Review*, 21: 13–22.

Development Plans: New approaches to making frameworks for land use regulation, *European Planning Studies*, 2(1): 39–57.

Trends in Development Plan-Making in Europe, Working paper no. 42. Newcastle-upon-Tyne: Department of Town and Country Planning, University of Newcastle-upon-Tyne.

Strategic Spatial Planning as a Process of Argumentation, Working Paper. Perth: School of Art, Construction and Planning, Curtin University.

The changing meanings of 'environment' in the British planning system (with T. Shaw), *Transactions of the Institute of British Geographers*, 19(4): 425–38.

J. Brian McLoughlin: An appreciation, *Town Planning Review*, 65(4): 341–7.

Alla ricerca della democrazia: nuovi modo di usare strumenti vecchi: forma e contenuti dei piani di sviluppo. In F. Archibugi and P. Bisognio (eds) *Per una teoria della pianificazione*, Prometheus16/17. Milano: Franco Angeli, pp. 90–108.

1995 City Challenge: Sustainable process or temporary gesture (with S. Davoudi), *Environment and Planning C: Government and Society*, 13(1): 79–95 (included in: R. Hambleton and H. Thomas (eds) *Urban Policy Evaluation.* London,: Paul Chapman, pp. 158–74).

Managing Cities, (edited with S. Cameron, S. Davoudi, S. Graham and A. Madani Pour). London: John Wiley.

The institutional challenge for sustainable urban regeneration, *Cities*, 12 (4): 221–30.

Le renouvellement des politiques urbaines spatialisees en Europe. In A. Motte (ed.) *Schema Directeur et projet d'agglomeration: l'experimentation de nouvelles politiques urbaines spatialisees 1981–1993.* Paris: Les editions Juris Service.

Community Mobilisation in Swan Valley: Claims, discourses and rituals in local planning (with J. Hillier), Working Paper No 49. Newcastle-upon-Tyne: Department of Town and Country Planning, University of Newcastle-upon-Tyne.

Integrating the concept of social diversity into public policy. In J. Beall, L. McFarlane and R. Ebbs (eds) *Women in the City: Housing, Services and the Urban Environment.* Paris: OECD, pp. 50–56.

Education for planning: A continuous process, *Proceedings of the ISOCARP conference, Prague 1994*, pp. 50–56.

The argumentative turn in planning theory and its implication for spatial strategy formation. In T. Pakarinen (ed.), *Are Local Strategies Possible – scrutinising sustainability.* Tampere: Tampere University of Technology.

Negotiating Development (with M. Purdue and F. Ennis). London: E & FN Spon.

Changing the way we do things here: The Cruddas Park Initiative (with J. Wood, R. Gilroy and S. Speak). Newcastle-upon-Tyne: Department of Town and Country Planning, University of Newcastle-upon-Tyne.

Greening the Red Rose County: Working towards an integrated sub-regional strategy (with A. Hull and S. Davoudi), Working paper No 58. Newcastle-upon-Tyne: Department of Town and Country Planning, University of Newcastle-upon-Tyne.

The British System. In L.J. Co and Y.H. Kim (eds) *Market Economies and Land Systems: A comparative study of Korea and dominant global models.* Honolulu: East-West Center, pp. 517–69.

1996 The communicative turn in planning theory and its implications for spatial strategy-making, *Environment and Planning B: Planning and design*, 23(2): 217–34 (included in S. Campbell and S. Fainstein (eds) (2003) *Readings in Planning Theory*, 2nd edition).

Consensus-building across difficult divisions: New approaches to collaborative strategy-making, *Planning Practice and Research*, 11(2): 209–18.

Negotiating development: Planning gain and mitigating impacts (with M. Purdue and F. Ennis), *Journal of Property Research*, 13: 143–60.

The communicative work of Development Plans. In S. Mandelbaum, L. Mazza and R. Birchell (eds) *Explorations in Planning Theory.* New Brunswick, NJ: Center for Urban Policy Research, Rutgers University, pp. 263–88.

Communicative micropolitics: A story of claims and discourses (with J. Hillier), *International Planning Studies*, 1(2): 165–84.

Strategie Cooperative per le regione urbane, *Urbanistica*, 106: 96–101.

Environmental concerns and economic imperatives in strategic plan-making (with S. Davoudi and A. Hull), *Town Planning Review*, 76(4): 421–36.

1997 Operationalising environmental considerations in the British Planning System (with T. Shaw). In D. Borri (ed.) *Evaluating Theory and Practice and urban-rural interplay in planning.* Elsevier: Amsterdam, pp. 167–92.

The State of the Borough: Main Report (with T. Norwood), Report to Sedgefield Borough Council.

The State of the Borough: Social Life (with R. Gilroy and T. Norwood) Report to Sedgefield Borough Council.

City fathers, mandarins and neighbours: Crossing old divides in new partnerships. In O. Kalltorp, I. Elander, O. Ericson and M. Franzen (eds), *Cities in Transformation: Transformation in cities.* Aldershot: Avebury, pp. 266–88.

Collaborative Planning: Shaping places in fragmented societies. London: Macmillan.

Making Strategic Spatial Plans: Innovation in Europe (with A. Khakee, B. Needham and A. Motte). London: UCL Press.

Rhetoric and reality in structure planning in Lancashire (with S. Davoudi and A. Hull). In *Making Strategic Spatial Plans: Innovation in Europe* (with A. Khakee, B. Needham and A. Motte), London: UCL Press, pp. 153–72.

Negotiating development and the new circular 1/97: a positive view of planning obligations, *Town and Country Planning*, 66(4):100.

Building sustainable futures in small and medium-sized cities in Europe. In V. Mega and R. Petrella (eds) *Utopias and Realities of Sustainable Development.* Dublin: European Foundation of Living and Working, pp. 79–88.

A strategic approach to sustainable urban regeneration, *Journal of Property Development*, 1(3): 105–10.

Social exclusion, institutional capacity and the neighbourhood. In H. Vestergaard et al. (eds) *Housing in Europe.* Horsholm: Danish Building Research Institute, pp. 88–110.

Collaborative approaches to urban planning and their contribution to institutional capacity-building, *International Journal of Urban Sciences*, (Korea) 1(2): 167–183.

Civic rights, *Parliamentary Review*, October pp. 32–33.

Medium-Sized Cities in Europe (with S. Cameron and S. Davoudi), Dublin: European Foundation of Living and Working, pp. 256.

Comment on Britton Harris' 'The theory of planning and its profession', *Environment and Planning B: Planning and Design*, 24(4): 490–92.

Introduction: Planning theory, political economy and the interpretive turn: the debate continues, *Planning Theory*, 17: 10–12.

Situating communicative practices: Moving beyond political economy, *Planning Theory*, 17: 65–82.

1998 Collaborative planning in a stakeholder society, *Town Planning Review*, 69(1): 537–57 (included in M. Wegener, K. Button and P. Nijkamp (eds) (2006) *Planning History and Methodology*. Cheltenham: Edward Elgar).

Building institutional capacity through collaborative approaches to urban planning, *Environment and Planning A*, 30: 1531–46.

Movement and mobility in the post-Fordist city (with A. Gillespie and K. Robins). In D. Banister (ed.) *Transport Policy and the Environment.* London: E and FN Spon, pp. 243–66.

Regulating property development, *Journal of Property Research*, 15(3): 211–27.

The place of 'Europe' in contemporary spatial strategy-making, *European Urban and Regional Studies*, 4: 139–53.

Planning practices and planning theories: An intertwined personal agenda, *Environment and Planning B: Planning and Design*, 25: 118–23.

Institutionalist theory, social exclusion and governance. In A. Madanipour, G. Cars and J. Allen (eds) *Social Exclusion in European Cities.* London: Jessica Kingsley, pp. 53–74.

British land use planning and the planning system. In L.J. Kho and Y.H. Kim (eds) *Ten Paradigms of Market Economies and Land Systems.* Seoul: KRIHS, pp. 137–78.

1999 Sites, jobs and portfolios: Economic development discourses in the planning system, *Urban Studies*, 56(1): 27–42.

European developments in contemporary spatial strategy-making (with A. Khakee, A. Motte and B. Needham), *European Planning Studies*, 7(3): 339–54.

Relational concepts of place and space: Issues for planning theory and practice (with S. Graham), *European Planning Studies*, 7(5): 339–54.

Territorial integration and 'plan-led' planning (with G.Vigar), *Planning Practice and Research*, 14(2): 153–69.

Institutional capacity-building, urban planning and urban regeneration projects (with C. de Magalhães and A. Madanipour). In M. Sotorauta (ed.) *Urban Futures: A loss of shadows in the flowing of spaces?* Special issue of *Futura*, Helsinki: Finnish Society of Futures Studies.

The future governance of cities. In F. Gaffikin and M. Morrissey (eds) *City Visions*, London: Pluto Press, pp. 183–92.

La negoziazione tra comuni nell'ambito di un piano strategico metropolitano. In R. Camagni (ed.) *La pianificazione sostenibile delle aree periurbane.* Bologna: Il Molino, pp. 101–46.

A model for urban regeneration. In I. Latham and M. Swenarton (eds) *Brindleyplace: A model for urban regeneration.* London: Rightangle Publishing, pp. 104–9.

Deconstructing communicative planning theory: A reply to Tewdwr-Jones and Allmendinger, *Environment and Planning A*, 31: 1129–35.

Institutionalist analysis, communicative planning and shaping places, *Journal of Planning Education and Research*, 19(2): 111–21, (included in H. Priemus, K. Button and P. Nijkamp (eds) (2006) *Classics in Planning: Land Use Planning.* London: Edward Elgar).

2000 Connected cities, *Town and Country Planningm* 69(2): 55–7.

Planning, Governance and Spatial Strategy in Britain (with G. Vigar, A. Hull and S. Davoudi). London: Macmillan.

Transport Planning and Metropolitan Governance (with G. Vigar, M. Steele, J.D. Nelson and A. Wenban-Smith). London: Landor.

European Spatial Planning and Urban-Rural Relationships: the UK Dimension, (ESPRIN Team). London: DETR 2000.

Planning in relational space and time: Responding to new urban realities. In G. Bridge and S. Watson (eds) *A Companion to the City.* London: Blackwell, pp. 505–16.

Planning theory and urban and regional dynamics: A comment on Yiftachel and Huxley, *International Journal of Urban and Regional Research*, 24(4): 917–21.

New partnerships in planning and implementing future-oriented developments in European Metropolitan Regions. In *Informationen zur Raumentwicklung*, special issue on *European Metropolitan Regions*11/12: 745–50.

Editorial, *Planning Theory and Practice*, 1(1): 7–10.

2001 New approaches to the content and process of spatial development frameworks. In OECD (ed.) *Towards a New Role for Spatial Planning*. Paris: OECD, pp. 143–63.

Towards a more place-focused planning system in Britain. In A. Madanipour, A. Hull and P. Healey (eds) *The Governance of Place*. Aldershot: Ashgate, pp. 265–87.

Editorial, *Planning Theory and Practice*, 2(1): 7–10.

Editorial, *Planning Theory and Practice*, 2(3): 255–8.

Planning theory: The interaction with institutional contexts, Entry in the *Urban Studies and Planning* Section of the International Encyclopedia of the Social and Behavioural Sciences (url:www.iebs.com).

2002 Place, identity and governance: Transforming discourses and practice, In J. Hillier and E. Rooksby (eds) *Habitus: A sense of place*. Aldershot: Ashgate, pp. 173–202, (2nd Edition, 2005, pp. 189–218).

Urban Governance, institutional capacity and social milieux, (edited with G. Cars, A. Madanipour and C. de Magalhães). Aldershot: Ashgate.

Att utforska planeringspraktiken – samspelet mellan teorie och praktik, *Plan*, 56(1–2): 8–11.

Spatial planning as a mediator for regional governance – Conceptions of place in the formation of regional governance capacity, In D. Furst and J. Kneiling (eds) *Regional Governance: New Modes of Self-Government in the European Community*, Studies in Spatial Development, No 2 ARL. Hannover: Akademie fur Raumforschung und Landesplanung, pp. 13–25.

Developing environmentally respectful policy programmes: Five key principles (with G. Vigar), in *Journal of Environmental Planning and Management*, 45(4): 517–32.

Editorial, *Planning Theory and Practice*, 3(2): 127–30.

On creating the city as a public realm resource, *Urban Studies*, 74(10): 1777–92.

Shaping city centre futures: Conservation, regeneration and institutional capacity (with A. Madanipour, C. de Magalhães and J. Pendlebury), CREUE Occasional Paper. Newcastle-upon-Tyne: SAPL, University of Newcastle-upon-Tyne.

Urban-rural relationships, spatial strategies and territorial development, *Built Environment*, 28(4): 331–9.

2003 Strategic spatial planning and regional governance in Europe (with L. Albrechts and K. Kunzmann), *Journal of the American Planning Association*, 69(2): 113–29.

Editorial, *Planning Theory and Practice*, 4(1).

Editorial, *Planning Theory and Practice*, 4(3).

Citta e istituzioni: Piani collaborative in societa frammentate (trans D. Borri). Bari: Edizioni Dedalo.

Place, identity and local politics: Analysing partnership initiatives (with C. De Magalhães, A. Madanipour and J. Pendlebury), In M. Hajer and H. Wagenaar (eds) *Deliberative Policy Analysis: Understanding Governance in the Network Society.* Cambridge: University Press, pp. 60–87.

'My Voice: My Place': Tracking transformations in urban governance (with J. Coaffee), *Urban Studies*, 40(10): 1979–99.

Collaborative planning in perspective, *Planning Theory*, 2(2): 101–23.

Rhetoric and reality in spatial planning in the Seoul Metropolitan Area (with Soon-Tak Suh), In Kyu-Bang Lee (ed.) *Global City Region: Integrated planning and sustainable policies of Korea.* Seoul: KRIHS, pp. 411–32.

2004 The treatment of space and place in the new strategic spatial planning in Europe, *International Journal of Urban and Regional Studies*, 28(1): 45–67.

Editorial, *Planning Theory and Practice*, 5(1).

Towards a 'social democratic' policy agenda for cities, In C. Johnstone and M. Whitehead (eds) *New Horizons in British Urban Policy.* Aldershot: Ashgate, pp. 159–72.

Creativity and urban governance, *Policy Studies*, 25(2): 87–102, (reprinted in *DISP* 158, 3/2004, pp. 11–20.

L'istituzionalizzione dalla capacita degli attore colletivi, *Urbanistica*, 123: 34–7.

Editorial, *Planning Theory and Practice*, 5.3.

2005 In search of the city in spatial strategies: Past legacies, future imaginings (with G.Vigar and S. Graham), *Urban Studies*, 42(8): 1391–410.

Editorial, *Planning Theory and Practice*, 6(1).

On the project of 'institutional transformation' in the planning field: Commentary on the contributions, *Planning Theory*, 4(3): 301–10.

A sociological institutionalist approach to the study of innovation in governance capacity (with S. Gonzalez), *Urban Studies*, 42(11): 2055–70.

Network complexity and the imaginative power of strategic spatial planning, In L. Albrechts and S. Mandelbaum (eds) *The Network Society: A new context for planning?* London: Routledge, pp. 146–60.

Governance capacity, policy networks and territorial specificities, In L. Albrechts and S. Mandelbaum (eds) *The Network Society: A new context for planning?* London: Routledge, pp. 307–11.

2006 *Collaborative Planning: Shaping places in fragmented societies*, 2nd Edition. London: Palgrave Macmillan.

Editorial, *Planning Theory and Practice*, 7(1).

Relational complexity and the imaginative power of strategic spatial planning, *European Planning Studies*, 14(4): 525–46.

Transforming governance: Challenges of institutional adaptation and a new politics of space, *European Planning Studies*, 14(3): 299–319.

Territory, integration and spatial planning, In M. Tewdwr-Jones and P. Allmendinger (eds), *Territory, Identity and Spatial Planning: Spatial governance in a fragmented nation.* London: Routledge, pp. 64–79.

Place development in a relational world, *Town and Country Planning*, 75(11): 318–21.

2007 *Urban Complexity and Spatial Strategies: Towards a relational planning for our times.* London: Routledge.

Re-thinking key dimensions of strategic spatial planning: Sustainability and complexity, In G. De Roo and G. Porte (eds) *Fuzzy Planning: The role of actors in a fuzzy governance environment.* Aldershot: Ashgate, pp. 21–41.

The new institutionalism and the transformative goals of planning, In N. Verma (ed.) *Planning and Institutions.* Amsterdam: Elsevier, pp. 61–87.

The 'collaborative planning' project in an institutional and relational perspective: A note, *Critical Policy Analysis*, 1(1): 123–30.

European integration and the planning academy: Reflections on the AESOP project, *Town Planning Review*, 78(5): iii–vii.

2008 *Critical Essays in Planning Theory, Vol 1: Foundations of the planning enterprise*, (edited with J. Hillier). Ashgate: Aldershot.

Critical Essays in Planning Theory, Vol 2: Political economy, diversity and pragmatism, (edited with J. Hillier). Ashgate: Aldershot.

Critical Essays in Planning Theory Vol 3: Contemporary movements in planning theory, (edited with J. Hillier). Ashgate: Aldershot.

Knowledge flows, spatial strategy-making and the roles of academics, *Environment and Planning C: Government and Policy*, 26: 861–81.

Interface, *Planning Theory & Practice*, 9(3).

Editorial, *Planning Theory & Practice*, 9(4).

Making Choices that Matter: The practical art of situated strategic judgement in spatial strategy-making, In J. van den Broek, F. Moulaert and S. Oosterlynck (eds) *Empowering the Planning Fields: Ethics, creativity and action.* Leuven: Acco, pp. 23–41.

2009 The pragmatic tradition in planning theory, *Journal of Planning Education and Research*, 28(2): 277–92.

City regions and place development, *Regional Studies*, 43(6): 831–43, (included in M. Neuman and A. Hull (eds) *The Futures of City Regions* 2011. London: Routledge, pp. 43–55).

In search of the 'strategic' in strategic spatial planning, *Planning Theory & Practice*, 10(4): 439–57.

2010 Making better places: The planning project in the 21st century. London: Palgrave Macmillan.

Crossing Borders: International exchange and planning practices, (edited with R. Upton). London: Routledge.

Re-thinking the relations between planning, state and market in unstable times, In P. Pinho and V. Oliveira (eds) *Planning in Times of Uncertainty.* Porto: FEUP edicoes, pp. 11–22.

2011 Foreword, In John Friedmann's *Insurgencies: Essays in planning theory*. London: Routledge, pp. xi–xiv.

Forgotten classics special: John Friedmann's *Re-tracking America*: A theory of Transactive planning 1973, *Regional Insights*, 1(2).

L'élaboration des concepts de planification 'communicative' et 'collaborative', In G. Bettoni (ed.) *Gouverner les territoires: antagonismes et partenariats entre acteurs publiques.* Paris: IGPDE, pp. 61–93.

2012 The universal and the contingent: Some reflections on the transnational flow of planning ideas and practices, *Planning Theory*, 11(2): 188–207.

The idea of 'communicative' planning: Practices, concepts and rhetorics, In B. Sanyal, L. Vale and C. Rosan (eds) *Planning Ideas that Matter.* Cambridge, MA: MIT Press, pp. 333–57.

Re-enchanting democracy as a mode of governance, *Critical Policy Studies*, 6(1): 19–39.

Performing place governance collaboratively: Planning as a communicative practice. In F. Fischer and H. Gottweis (eds) *The Argumentative Turn Re-Visited.* Durham NC: Duke University Press, pp. 58–82.

2013 Comment on Albrechts and Balducci 'Practising strategic planning', *disP*, 194(49:3): 48–50.

Circuits of knowledge and techniques: The transnational flow of planning ideas and practices, *International Journal of Urban and Regional Research*, 37(5): 1510–26.

Quelques questions clé pour la fabrique des strategies metropolitaines. In C. Lefevre, N. Rosenau and T. Vitale (eds) *Les Defis de la gouvernance metropolitaine*. Paris: L'oeil d'or, pp. 129–45.

2014 *The Routledge Handbook of Planning Research Methods* (edited with E. Silva, N. Harris and P. van den Broek). London: Routledge.

2015 Planning theory: The interaction with institutional contexts, entry in the *Urban Studies and Planning Section*, International Encyclopedia of the Social and Behavioural Sciences, www.iebs.com (revised).

Interpretive understandings in the planning field. In M. Bevir and R. Rhodes (eds) *The Routledge Handbook of Interpretive Political Studies*. Routledge: London

Civic capacity, place governance and progressive localism. In S. Davoudi and A. Madanipour (eds) *Reconsidering Localism*. Routledge: London.

Index

References to illustrations are in **bold.**

For Product Safety Concerns and Information please contact our EU representative GPSR@taylorandfrancis.com
Taylor & Francis Verlag GmbH, Kaufingerstraße 24, 80331 München, Germany

www.ingramcontent.com/pod-product-compliance
Ingram Content Group UK Ltd.
Pitfield, Milton Keynes, MK11 3LW, UK
UKHW050753180926
475701UK00007B/1012
* 9 7 8 0 3 6 7 7 8 7 3 6 3 *